Introduction to Functional Nanomaterials

This book provides a comprehensive review of nanomaterials, including essential foundational examples of nanosensors, smart nanomaterials, nanopolymers, and nanotubes. Chapters cover their synthesis and characteristics, production methods, and applications, with specific sections exploring nanoelectronics and electro-optic nanotechnology, nanostructures, and nanodevices. This book is a valuable resource for interdisciplinary researchers who want to learn more about the synthesis of nanomaterials and how they are used in different types of energy storage devices, including supercapacitors, batteries, fuel cells solar cells in addition to electrical, chemical, and biomedical engineering.

Key Features:

- Comprehensive overview of how nanomaterials can be utilised in a variety of interdisciplinary applications.
- Explores the fundamental theories, alongside their electrochemical mechanisms and computation.
- Discusses recent developments in electrode designing based on nanomaterials, separators, and the fabrication of advanced devices and their performances.

M. Anusuya, M.Sc., M.Phil., B.Ed., Ph.D. is a specialist in material science, thin-film technology, nanoscience, and crystallography. She works as a registrar of Indra Ganesan Group of Institutions, Trichy, Tamil Nadu. As an administrator and teacher, with more than 25 years' experience, for her perpetual excellence in academics, she has been recognised with many awards. She has received over 45 awards in academic and social activity. She has published more than 30 research papers in national and international journals, 16 chapters in edited books, been granted 15 patents, presented 50 papers in the conferences, and organised more than 200 webinars, both nationally and internationally.

Fabian I. Ezema is a professor at the University of Nigeria, Nsukka. He earned a PhD in physics and astronomy from the University of Nigeria, Nsukka. His research focused on several areas of materials science, from synthesis and characterisations of particles and thin-film materials through chemical routes with emphasis on energy applications. For the last 15 years, he has been working on energy conversion and storage (cathodes, anodes, supercapacitors, and solar cells, among others), including novel methods of synthesis, characterisation, and evaluation of the electrochemical and optical properties. He has published about 180 papers in various international journals and given over 50 talks at various conferences. His h-index is 21 with over 1,500 citations, and he has served as reviewer for several high-impact journals and as an editorial board member.

Introduction to Functional Nanomaterials

Edited by
M. Anusuya
Fabian I. Ezema

CRC Press
Taylor & Francis Group
Boca Raton London New York

CRC Press is an imprint of the
Taylor & Francis Group, an **informa** business

Front cover image: SkillUp/Shutterstock

First edition published 2025
by CRC Press
2385 NW Executive Center Drive, Suite 320, Boca Raton FL 33431

and by CRC Press
4 Park Square, Milton Park, Abingdon, Oxon, OX14 4RN

CRC Press is an imprint of Taylor & Francis Group, LLC

Library of Congress Cataloging-in-Publication Data
Names: Anusuya, M, editor. | Ezema, Fabian I, editor.
Title: Introduction to functional nanomaterials / edited by M. Anusuya and
Fabian I. Ezema.
Description: First edition. | Boca Raton, FL : CRC Press, [2025] | Includes bibliographical references and index. |
Identifiers: LCCN 2024013318 | ISBN 9781032800103 (hbk) | ISBN
9781032801025 (pbk) | ISBN 9781003495437 (ebk)
Subjects: LCSH: Nanostructured materials.
Classification: LCC TA418.9.N35 I597 2025 | DDC 620.1/15--dc23/eng/20240405
LC record available at https://lccn.loc.gov/2024013318

ISBN: 978-1-032-80010-3 (hbk)
ISBN: 978-1-032-80102-5 (pbk)
ISBN: 978-1-003-49543-7 (ebk)

DOI: 10.1201/9781003495437

Typeset in Times
by SPi Technologies India Pvt Ltd (Straive)

Contents

Contributors

Srinivasa Acharya
Department of Electrical and Electronics Engineering
Aditya Institute of Technology and Management
Tekkali, India

S. Afnan
PG and Research Department of Physics, Cauvery
College for Women (Autonomous) (affiliated
to Bharathidasan University)
Trichy, India

E. Agalya
Department of Food Service Management and
Dietetics
Cauvery College for Women (Autonomous)
Tiruchirappalli, India

Richa Agarwal
Department of Applied Sciences
KIET Group of Institutions
Ghaziabad, Delhi-NCR
Ghaziabad, India

E. Ajith Jubilson
School of Computer Science and Engineering
VIT-AP University, Amaravati
Guntur, India

A. Anu Kuttan
Department of Aeronautical Engineering
Noorul Islam Centre for Higher Education
Kanyakumari, India

M. Anusuya
Department of Physics
Indra Ganesan College of Engineering
Trichy, India

Saravanakumar Arthanari
Department of Pharmaceutics
Vellalar College of Pharmacy
Maruthi Nagar, Thindal, India

M. Ayisha Zeenath
PG and Research Department of Physics
Cauvery College for Women (Autonomous)
Trichy, India

R. Babu Ashok
Department of Electrical and Electronics
Engineering
Karaikal Polytechnic College (PIPMATE-Govt. of
Puducherry)
Karaikal, India

Sonia H. Bajaj
Department of Computer Science and Engineering
G H Raisoni Institute of Engineering and
Technology
Nagpur, India

Paranthagan Balasubramanian
Department of Electrical and Electronics
Engineering
Saranathan College of Engineering
Tiruchirappalli, India

Ravi Prakash Balasundaram
Department of Processing and Food
Engineering
Agricultural Engineering College and Research
Institute, Tamil Nadu Agricultural University
Kumulur, Tiruchirappalli, India

Harish Kumar Banga
Fashion & Life Style Accessory Design
National Institute of Fashion Technology (NIFT)
Kangra, India

Saravana Bavan
Department of Mechanical Engineering
Dayananda Sagar University
Bangalore, India

G. Bharath Reddy
Department of Mechanical Engineering
CVR College of Engineering
Hyderabad, India

M. K. Chaanthini
Department of Mechanical Engineering
E.G.S. Pillay Engineering College
Nagapattinam, India

S. Chandra
Department of Chemistry
PSG College of Arts and Science
Coimbatore, India

Devchand Chaudhari
Department of Computer Science and Engineering
Government College of Engineering
Maharashtra, India

Santhosh Kumar Chinnaiyan
Faculty of Pharmacy
Karpagam Academy of Higher Education
Eachanari, Coimbatore, India

Apparao Damarasingu
Department of Mechanical Engineering
Aditya Institute of Technology and Management
Tekkali, India

Ramesh Desikan
Department of Renewable Energy Engineering,
    Agricultural Engineering College and Research
    Institute
Tamil Nadu Agricultural University
Coimbatore, India

M. Devaiah
Department of Mechanical Engineering
Geethanjali College of Engineering and
    Technology
Telangana, India

R. M. Devarajaiah
Department of Mechatronics Engineering
Acharya Institute of Technology
Bengaluru, India

P. K. Dhal
Department of Electrical and Electronics
    Engineering
Vel Tech Rangarajan Dr. Sagunthala R&D
    Institute of Science and Technology
Chennai, India

Dharmalingam Ganesan
Department of Mechanical Engineering
Vel Tech Rangarajan Dr. Sagunthala R&D
    Institute of Science and Technology
Avadi, Chennai, India

C. Divya
Centre for Information Technology and
    Engineering
Manonmaniam Sundaranar University
Tamil Nadu, India

J. Eindhumathy
Department of Electronics and Communication
    Engineering
Saranathan College of Engineering
Tamil Nadu, India

D. Elayaraja
Department of Mechanical Engineering
Indra Ganesan College of Engineering
Trichy, India

J. Femila Roseline
Department of Electronics and Communication
    Engineering
Saveetha School of Engineering
Chennai, India

Koli Gajanan Chandrashekhar
Department of Mechanical Engineering
Sanjeevan Engineering & Technology Institute
Panhala, Waghave, India

Amos Gamaleal David
Department of Mechanical engineering
Panimalar Engineering College
Chennai, India

R. Gayathri
PG and Research Department of Physics, Cauvery
    College for Women (Autonomous) (Affiliated
    to Bharathidasan University)
Trichy, India

Amuthaselvi Gopal
Department of Processing and Food Engineering,
    Agricultural Engineering College and Research
    Institute
Tamil Nadu Agricultural University
Coimbatore, India

K. M. Govindaraju
Department of Chemistry
PSG College of Arts and Science
Coimbatore, India

Praveen Gunasekaran
Department of Processing and Food Engineering
Agricultural Engineering College and Research
    Institute, Tamil Nadu Agricultural University
Kumulur, Tiruchirappalli, India

R. Halima
Department of Biotechnology
Sir M Visvesvaraya Institute of Technology
Hunasamaranahalli, Bangalore, India

Vishwanath Hokrani
Department of Tool and Dye making
Government Tool Room and Training Centre
Bagalkote, Kudalasangama, India

Deepa Jaganathan
Department of Food Technology
Hindustan College of Engineering and
    Technology
Coimbatore, India

S. K. Jameer Basha
School of Computer Science and Engineering
VIT-AP University, Amaravati
Guntur, India

R. Jeeva
Department of Computer Science and Engineering
Thamirabharani Engineering College
Tirunelveli, India

R. Jeyabharath
Department of Electrical and Electronics
    Engineering
K.S.R. College of Engineering
Tamil Nadu, India

S. Jeyabharathi
Department of Microbiology
Cauvery College for Women (Autonomous)
Trichy, India

A. Joseph Arockiam
Department of Mechanical Engineering
Arasu Engineering College
Kumbakonam, India

Gitanjali Jothiprakash
Centre for Post-Harvest Technology
Agricultural Engineering College and Research
    Institute, Tamil Nadu Agricultural University
Coimbatore, India

A. Kadirvel
Department of Mechanical Engineering
R.M.K. Engineering College
Tamil Nadu, India

E. Kavitha
Department of Physics
Dr. MGR Educational and Research Institute
Chennai, India

G. G. Kavitha Shree
Centre for Post-Harvest Technology
Agricultural Engineering College and Research
    Institute, Tamil Nadu Agricultural University
Coimbatore, India

Sarvani Jowhar Khanam
Department of Solar Cell and Photonics
    Laboratory, School of Chemistry
University of Hyderabad
Hyderabad, India

V. Koushick
Department of ECE
Vel Tech Rangarajan Dr. Sagunthala R&D
    Institute of Science and Technology
Chennai, India

P. Krishnan
St. Joseph's College of Engineering
Chennai, India

Rohit Kumar
Cyber security
Coer University
Vardhmanpuram, India

Sharan Kumar
Department of Mechanical Engineering
Sharnbasva University
Kalaburagi, India

P. Lakshmi Prabha
Department of Chemistry
Shrimati Indira Gandhi College
Trichy, India

P. Lalitha
Department of Physics, Indra Ganesan College of
    Engineering
Tiruchirappalli, India

S. Leena Nesamani
Department of Computer Science and Engineering
Vel Tech Rangarajan Dr. Sagunthala R&D
    Institute of Science and Technology
Avadi, Chennai, India

Barla Madhavi
Department of Mechanical Engineering
IFHE, Faculty of Science and Technology
Hyderabad, India

Venkateswaran Madhu
Department of Electrical and Electronics
    Engineering
Lendi Institute of Engineering and Technology
Vizianagaram, India

K. Mahesh Dutt
Department of Mechanical Engineering
Dayanandasagar Academy of Technology &
    Management
Bangalore, India

K. Malathi
Department of Computer Science and Engineering
Saveetha Institute of Medical and Technical
    Sciences
Chennai, India

J. Manikandan
Department of Chemistry
PSG College of Arts and Science
Coimbatore, India

P. Manikandan
Civil Engineering
Federal TVT Institute
Ethiopia

M. Monisha
Electronics and Communication Engineering
Vels Institute of Science Technology and
    Advanced Studies VISTAS
Pallavaram, Chennai, India

Venkatasami Murugesan
Department of Processing and Food Engineering
Agricultural Engineering College and Research
    Institute, Tamil Nadu Agricultural University
    Coimbatore, India

Ramya Nandakumar
Department of Computer Science and
    Engineering
Saranathan College of Engineering
Trichy, India

N. M. Nanditha
Department of Electronics and Communication
    Engineering
Sathyabama Institute of Science and Technology
Chennai, India

K. Nikitha
Department of Food Service Management and
    Dietetics
Cauvery College for Women (Autonomous)
Tiruchirappalli, India

V. R. Niveditha
Department of Computer Science and
    Engineering
Sathyabama Institute of Science and Technology
Chennai, India

R. G. Padmanabhan
Department of Mechanical Engineering
Arasu Engineering College
Kumbakonam, India

M. Padmarasan
Department of Electrical and Electronics
    Engineering
Panimalar Engineering College
Chennai, India

Moumita Pal
Department of Computer Science and
    Engineering
Stanley College of Engineering and Technology
Hyderabad, India

Ram Prakash Ponraj
Department of Electrical and Electronics
    Engineering
Saranathan College of Engineering
Tiruchirappalli, India

M. PrasannaBlessy
Dr. Mahalingam College of Engineering & Tech.
Pollachi, India

A. M. Prasanna Lakshmi
Department of MCA
Velagapudi Ramakrishna Siddhartha Engineering
    College
Kanuru, Vijayawada, India

P. V. Premalatha
Department of Civil Engineering
M.I.E.T. Engineering College
Tiruchirappalli, India

R. Premalatha
Electrical and Electronics Engineering,
    S.A. Engineering College
Poonamallee, Veeraraghavapuram, India

J. Priyadharshini
PG and Research Department of Physics
Cauvery College for Women(Autonomous)
    (affiliated to Bharathidasan University)
Trichy, India

A. Purna Chandra Rao
Department of Electrical and Electronics
    Engineering
QIS College of Engineering and Technology,
    Vengamukkapalem, Ongole
Andhra Pradesh, India

J. Raffiea Baseri
Department of Chemistry
PSG College of Arts and Science
Coimbatore, India

Satheesh Ragunathan
Department of Electrical and Electronics
    Engineering
Saranathan College of Engineering
Tiruchirappalli, India

K. Rajan
Department of Mechanical Engineering
Dr. MGR Educational and Research Institute
Chennai, India

Selvamani Rajendran
Department of Mathematics
Karunya Institute of Technology
Coimbatore, India

K. Rajesh
AMET University,
Chennai, India

Arulmari Ramaraj
Department of Processing and Food
    Engineering
Agricultural Engineering College and
    Research Institute, Tamil Nadu Agricultural
    University
Kumulur, Tiruchirappalli, India

G. Ramya
Department of Electrical and Electronics
    Engineering Faculty of Engineering &
    Technology
SRM Institute of Science & Technology
Chennai, India

B. Rathinambal
Department of Microbiology
Cauvery College for Women (Autonomous)
Trichy, India

Vijay Ravindran
Department of Electrical and Electronics
    Engineering
Saranathan College of Engineering
Tiruchirappalli, India

M. Renugadevi
Department of Physics
St. Joseph's College (Autonomous),
Tiruchirappalli, India

S. Roseline
Department of Mechanical Engineering
MAM College of Engineering and Technology
Trichy, India

J. Sadhik Basha
Department of Process Engineering
National University of Science & Technology
    (IMCO)
Sohar, Oman

Madona B. Sahaai
Department of Electronics and Communication
    Engineering
Vels Institute of Science, Technology & Advanced
    Studies VISTAS
Pallavaram, Chennai, India

V. Saravanan
Department of Physics Indra Ganesan College of
    Engineering
Trichy, India

G. A. Shabeen Taj
Department of Computer Science and
  Engineering
Government Engineering College
Karnataka, India

C. Sharanya
Electronics and Communication Engineering
Sathyabama Institute of Science and Technology
Chennai, India

M. Sharanya
Electrical and Electronics Engineering
Malla Reddy College of Engineering and
  Technology
Maisammaguda, Secunderabad, India

Mahesh R. Shukla
Department of Mechanical Engineering
MKSSS Cummins College of Engineering for
  Women
Nagpur, India

Titus Sigamani
Department of Electrical and Electronics
  Engineering
K. Ramakrishnan College of Engineering
Trichy, India

Balkeshwar Singh
Mechanical Engineering
Program of Manufacturing Engineering, Adama
  Science and Technology University
Adama, Ethiopia

A. Sinthiya
Department of Physics
St. Joseph's College (Autonomous),
Tiruchirappalli, India

M. Sivadharani
Department of Microbiology
Cauvery College for Women (Autonomous)
Trichy, India

Yeshwant M. Sonkhaskar
Department of Mechanical Engineering
Shri Ramdeobaba College of Engineering and
  Management
Nagpur, India

Mahaveer Sree Jayan Madasamy
State Key Laboratory of Mechanics and Control
  of Aerospace Structures
Nanjing University of Aeronautics and
  Astronautics
Nanjing, People's Republic of China

Nidamanuri Sreenivasa Babu
University of Technology and Applied Sciences-
  Shinas
College of Engineering and Technology
Engineering Department
Al-Aqur, Shinas, Sultanate of Oman

D. R. Srinivasan
Department of Mechanical Engineering
JNTUA College of Engineering
Anantapur, India

Sriramajayam Srinivasan
Department of Agricultural Engineering
Agricultural College and Research Institute, Tamil
  Nadu Agricultural University
Killikulam, Vallanad, India

Mukuloth Srinivasnaik
School of Engineering
Jawaharlal Nehru University JNU
New Delhi, India

E. Srividhya
Department of Computer Science and Engineering
Sathyabama Institute of Science and Technology
Chennai, India

Karthikeyan Subburamu
Centre for Post-Harvest Technology, Agricultural
  Engineering College and Research Institute
Tamil Nadu Agricultural University
Coimbatore, India

Mohanraj Subramanian
Department of Pharmacology
Vellalar College of Pharmacy
Thindal, Erode, India

C. Sudha
Computer Science and Engineering
GITAM School of Technology, GITAM deemed
  to be University
Hyderabad, India

S. Sudhahar
Department of Physics
Alagappa University
Karaikudi, India

A. Suresh
Department of Sciences and Humanities
Brilliant Group of Technical Institutions
Hyderabad, India

P. Suresh
Department of Electrical and Electronics
    Engineering
St. Joseph College of Engineering
Chennai, India

S. Sudhahar
Alagappa University
Tamilnadu, India

M. Swapna
Department of Computer Science and
    Engineering
Stanley College of Engineering Technology for
    Women
Hyderabad, India

Aenikapati Swetha Priya
Department of Electronics and Communication
    Engineering
Amrita School of Engineering
Amrita Vishwa Vidhyapeetham
Bangalore, India

Hariharan Thangavel
Department of Food Science and Nutrition,
    Community Science College and Research
    Institute
Tamil Nadu Agricultural University
Madurai, India

B. Thanuja
Department of Food Service Management and
    Dietetics
Cauvery College for Women (Autonomous)
Tiruchirappalli, India

R. Thenmozhi
Department of Microbiology
Indra Ganesan College of Arts and Science
Trichy, India

P. Thirumurugan
Department of ECE
Annai Veilankannis College of Engineering
Chennai, India

R. Umamageswari
Department of Electrical and Electronics
    Engineering
Annai Mira College of Engineering and
    Technology
Arapakkam, India

P. Veena
Department of Electrical and Electronics
    Engineering
K.S.R. College of Engineering
Namakkal, India

E. Veeramanipriya
Vivekanandha College of Engineering for Women
    (Autonomous)
Thiruchengode, India

R. Venkatesh
Department of Physics
PSNA College of Engineering and Technology
Dindugal, India

C. Vennila
Department of Electronics and Communication
    Engineering
Saranathan College of Engineering
Trichy, India

P. Vignesh
Centre for Additive Manufacturing
Chennai Institute of Technology
Chennai, India

K. Vinoth Bresnav
Department of Electrical and Electronics
    Engineering
Excel Engineering College
Namakkal, India

Lifeng Wang
State Key Laboratory of Mechanics and Control
    of Aerospace Structures
Nanjing University of Aeronautics and
    Astronautics
Nanjing, People's Republic of China

1

Transparent Contact Innovations for Solar Cells

D. R. Srinivasan
JNTUA College of Engineering, Anantapur, India

A. Kadirvel
R.M.K. Engineering College, Tamil Nadu, India

R. Umamageswari
Annai Mira College of Engineering and Technology, Arapakkam, India

Mukuloth Srinivasnaik
Jawaharlal Nehru University JNU, New Delhi, India

E. Kavitha
Dr. MGR Educational and Research Institute, Chennai, India

1.1 Introduction

1.1.1 Background

Solar energy is a renewable and sustainable power source harnessed from the sun's radiation. Photovoltaic cells, commonly known as solar panels, convert sunlight into electricity, while solar thermal systems utilise sunlight for heating applications [1]. With minimal environmental impact and infinite potential, solar energy plays a pivotal role in addressing climate change and fostering energy independence. Constant technological advancements continue to enhance the efficiency and affordability of solar solutions, making them increasingly vital components in the global energy landscape.

Innovations in solar energy include breakthroughs in transparent contact technology, a pivotal development enhancing the efficiency and aesthetics of solar panels. These innovative transparent contacts allow for improved light transmission, optimising the absorption of sunlight by photovoltaic cells. By minimising energy losses and maximising solar exposure, these advancements contribute to the overall effectiveness of solar energy systems [2]. This transparent contact innovation holds the promise of revolutionising solar panel design, making them not only more efficient but also adaptable to various surfaces, such as windows and building facades. As solar technology continues to evolve, these transparent contacts represent a significant stride towards seamlessly integrating renewable energy solutions into our everyday surroundings.

1.2 Fundamentals of Solar Cells

1.2.1 Photovoltaic Effect

The photovoltaic effect, at the core of solar cell technology, operates on the principle of transforming light energy into electrical energy through a sequential process. It begins with the absorption of photons – individual units of light – by a semiconductor material [3]. As photons with sufficient energy strike the material, they elevate electrons to higher energy states, creating electron-hole pairs within the semiconductor. Guided by the internal electric field, either intrinsic or induced by an external bias, these separated charges move in opposite directions, resulting in the generation of an electric current. This current becomes accessible for practical use by connecting an external circuit to the semiconductor. In solar cells, often constructed with layers of doped silicon, the efficiency of this photovoltaic conversion process is paramount, influencing the overall effectiveness of solar energy technologies in converting sunlight into usable electric power.

1.2.2 Components of Solar Cells

The major components of a solar cell include the following:

Semiconductor Material: Typically made of silicon, the semiconductor material absorbs photons from sunlight to initiate the photovoltaic process.

Doping Materials: N-type (negative) and P-type (positive) doping introduce impurities to create regions with excess electrons or electron 'holes,' essential for generating an electric current.

Metal Contacts: Placed on the semiconductor surface, metal contacts collect the generated electric current and provide a pathway for electrons to flow into an external circuit.

Antireflective Coating: This coating minimises light reflection, enhancing the absorption of sunlight by the semiconductor material.

Busbars and Grids: Conductive pathways on the cell's surface collect and direct the generated electricity towards the electrical contacts for efficient current collection and distribution.

Encapsulation: Solar cells are encapsulated to protect them from environmental factors, such as moisture and mechanical stress, while allowing sunlight to reach the semiconductor.

Junction Box: Housing electrical connections, the junction box protects the solar cell from environmental elements and facilitates the connection of multiple cells into a solar panel.

1.3 Importance of Transparent Contacts

Transparent contacts play a pivotal role in enhancing the efficiency and functionality of solar cells, particularly in optimising their optical and electrical properties [4]. Typically made of conductive materials, these components allow light to pass through, addressing a critical challenge in solar cell design – maximising light absorption without compromising electrical conductivity.

Maximising Light Transmission: Transparent contacts are designed to minimise light reflection and absorption, ensuring a higher percentage of incident sunlight reaches the semiconductor material beneath. This is crucial for optimising the photovoltaic process, where the absorption of photons initiates the generation of electric current.

Reducing Energy Losses: By facilitating efficient light transmission, transparent contacts contribute to the reduction of energy losses within the solar cell. This is particularly important for improving the overall efficiency of solar energy conversion, making the technology more effective in harnessing sunlight for electricity generation.

Enhancing Aesthetics and Integration: Transparent contacts enable the creation of solar panels with a sleek and visually appealing design. This is especially valuable for applications where aesthetics matter, such as integrating solar technology into windows, building facades, or other transparent surfaces. The ability to maintain transparency while serving an electrical function opens up new possibilities for the seamless integration of solar cells into various environments.

Adaptable Design: The use of transparent contacts allows for more flexible and adaptable solar cell designs. Solar cells can be incorporated into unconventional surfaces, such as glass or transparent plastics, expanding the range of applications for solar technology beyond traditional rooftop installations.

Improved Performance in Tandem Cells: Transparent contacts are crucial in the development of tandem solar cells, where multiple layers of solar cell materials with different absorption properties are stacked. They enable the transmission of light through the upper layers, ensuring efficient absorption by the subsequent layers and maximising the overall energy conversion efficiency.

1.4 Existing Transparent Contact Technologies

1.4.1 Indium Tin Oxide (ITO)

ITO serves as a key component in solar cell technology, functioning as a transparent contact with a unique combination of optical transparency and electrical conductivity. Comprising indium oxide and tin oxide, ITO allows sunlight to penetrate while efficiently collecting and transporting electric charge within the solar cell [5]. Its thin films, deposited through techniques like sputtering or chemical vapour deposition, offer flexibility for applications in lightweight and flexible solar panels. Renowned for stability and resistance to environmental factors, including moisture and oxidation, ITO ensures the durability and longevity of transparent contacts in solar cell systems. Beyond solar cells, ITO finds widespread use in electronic devices like flat-panel displays, touchscreens, and light emitting diodes (LEDs), showcasing its versatility in various optoelectronic applications [6]. While effective, the relatively higher cost of indium has prompted ongoing research for alternative, cost-effective, and environmentally sustainable transparent conductive materials in the pursuit of advancing solar energy technology.

1.4.2 Conductive Polymers

Conductive polymers, often termed 'synthetic metals,' represent a remarkable class of materials that seamlessly integrate the flexibility and adaptability of polymers with the electrical conductivity typically associated with metals [7]. Characterised by conjugated pi-electron systems, these polymers enable the free movement of charge carriers, endowing them with the ability to conduct electricity. Noteworthy for their flexibility, conductive polymers can be shaped into various forms, including films, fibres, and coatings, offering versatility in design and application. Synthesised through methods such as oxidative polymerisation, they allow for precise control over properties like conductivity by adjusting factors such as chain length and doping levels. Widely applicable, conductive polymers have found their way into diverse fields, contributing to flexible displays, organic solar cells, sensors, and even applications in biomedicine, such as tissue engineering and drug delivery systems [8]. The process of doping enhances their conductivity further, offering a tuneable material for specific needs. As research advances, conductive polymers continue to hold promise for innovative solutions in emerging technologies.

1.4.3 Graphene-Based Contacts

Graphene-based contacts represent a cutting-edge development in materials science, particularly within the realm of electronic devices and energy technologies. Graphene, a single layer of carbon atoms arranged in a hexagonal lattice, possesses extraordinary electrical, thermal, and mechanical properties [9]. When employed as contacts in electronic devices, such as solar cells or transistors, graphene's high electrical conductivity allows for efficient charge transport. The two-dimensional nature of graphene facilitates excellent transparency, making it an ideal material for transparent conductive films. Additionally, graphene exhibits exceptional mechanical strength and flexibility, contributing to the durability and adaptability of devices [10]. The unique combination of these properties positions graphene-based contacts as promising elements for enhancing the performance and versatility of various electronic and energy applications. Ongoing research endeavours continue to explore innovative methods for integrating graphene into electronic components, marking its role as a forefront material in the pursuit of advanced technologies.

1.5 Recent Innovations in Transparent Contacts

1.5.1 Nanomaterials and Nanostructures

The incorporation of nanomaterials and nanostructures in the realms of solar energy and electronic devices carries transformative implications, ushering in unprecedented prospects for innovation. Operating at the nanoscale, materials unveil unique properties capable of significantly elevating the efficiency and functionality of technologies. Nanostructured surfaces, boasting increased surface area, facilitate enhanced light absorption in solar cells, thereby contributing to heightened energy conversion efficiency [11]. Materials such as nanowires and carbon nanotubes, celebrated for their exceptional electrical conductivity, pave the way for efficient charge transport within electronic devices, resulting in elevated performance. Quantum dots, with their tuneable optoelectronic properties, provide meticulous control over light absorption, influencing both the solar energy absorption spectrum and the colour purity in displays. Moreover, the mechanical flexibility inherent in nanostructured materials fosters the creation of flexible and lightweight electronic devices, unlocking pathways for unconventional applications. Nanomaterials assume a pivotal role in propelling advancements in energy storage technologies, highly sensitive sensors, and the development of nanocomposites to augment the mechanical and thermal properties of materials. Amidst the immense potential benefits, the responsible consideration of safety, environmental impact, and ethical concerns becomes paramount as nanotechnology continues to shape the landscape of solar energy and electronic devices.

1.5.2 Perovskite Solar Cells

Perovskite solar cells have emerged as a revolutionary technology in the realm of photovoltaics, ushering in a new era for efficient and cost-effective solar energy conversion. Constructed using perovskite-structured compounds, typically hybrid organic-inorganic lead or tin halide materials, these cells offer notable advantages in terms of fabrication simplicity, low-cost materials, and the potential for high efficiency. The use of perovskite materials allows for processing through straightforward solution-based methods, rendering them scalable and adaptable for large-scale production.

A standout feature of perovskite solar cells lies in their exceptional light absorption properties. Perovskite materials boast a high absorption coefficient, enabling them to efficiently capture sunlight across a broad spectrum. This characteristic, coupled with their tuneable bandgap, facilitates the customisation of perovskite solar cells to match specific absorption wavelengths, thereby optimising energy conversion [12].

Despite rapid progress and significant breakthroughs in efficiency, challenges persist, particularly concerning stability issues related to moisture and temperature. Ongoing research is dedicated to addressing these challenges to ensure the commercial viability and long-term durability of perovskite solar cells.

The versatility of perovskite materials extends beyond traditional silicon-based solar cells, as they can be seamlessly incorporated into flexible and lightweight substrates. This flexibility opens up innovative possibilities for integrating perovskite solar cells into various applications, including portable electronic devices, building-integrated photovoltaics, and even clothing.

While perovskite solar cells hold immense promise, continuous research endeavours aim to further enhance their stability, scalability, and environmental impact. The potential for low-cost, high-efficiency solar energy solutions position perovskite solar cells as a focal point in research and development, with the ultimate goal of contributing to a sustainable and renewable energy future.

1.5.3 Quantum Dot Technology

Quantum dots, nanoscale semiconductor particles endowed with distinctive electronic and optical properties, have emerged as transformative components in the realms of solar energy and electronic devices. These minute structures showcase quantum confinement effects, allowing precise control over their electronic characteristics through size modulation [13]. In the solar energy domain, quantum dots exhibit promises in augmenting photovoltaic performance, thanks to their size-dependent bandgap that enables the tailoring of the absorption spectrum for efficient light harvesting across a diverse range of wavelengths.

Incorporating quantum dots into solar cells for the creation of multijunction devices represents a noteworthy strategy, enhancing the capture of different segments of the solar spectrum more effectively than conventional materials. This approach carries the potential to significantly elevate the efficiency of solar energy conversion. Furthermore, quantum dots can be engineered to facilitate the generation and separation of electron–hole pairs, a pivotal process in the utilisation of solar energy.

Within electronic devices, quantum dots contribute to advancements in displays, LEDs, and transistors. Their capacity to emit specific colours of light, dictated by their size, positions them ideally for generating vibrant and energy-efficient displays [14]. Quantum dot LEDs have gained traction as alternatives to traditional light sources due to their efficiency and colour purity. Additionally, quantum dots find application in transistors, optimising charge transport and influencing the performance of electronic circuits.

Despite their remarkable properties, challenges such as stability and toxicity necessitate addressing for widespread adoption. Ongoing research is directed towards developing stable and environmentally friendly quantum dots to ensure their safe integration into various applications.

1.6 Challenges and Future Directions

1.6.1 Challenges

Within the realm of solar energy innovations, various technologies, including perovskite solar cells, transparent contacts, and quantum dots, face common challenges that require concerted efforts for sustainable progress. Issues of stability and durability take precedence, particularly when confronted with environmental factors like moisture and temperature variations [15]. The presence of certain materials, such as lead in perovskite compositions, introduces environmental and health concerns, necessitating the development of eco-friendly alternatives. The universal challenge of scaling up production to meet the demands of large-scale energy applications requires consistent quality, reproducibility, and cost-effectiveness. Bridging the gap between laboratory-scale efficiency and commercial viability remains a shared hurdle, demanding breakthroughs in reliability and widespread adoption. The integration of innovative technologies into existing infrastructure poses compatibility challenges, mandating seamless adaptation to current grid systems and architectural designs.

1.6.2 Future Directions

The trajectory of solar energy technologies relies on collaborative endeavours to address shared challenges and explore innovative directions. Material innovation takes centre stage, with ongoing research focused on discovering new compositions that enhance stability, efficiency, and environmental sustainability across various technologies. Tandem solar cells, amalgamating different materials for enhanced performance, provide a promising avenue for achieving higher efficiencies and warrant further exploration. Advances in manufacturing techniques, such as roll-to-roll processing and printing methods, hold the key to scalability and cost reduction across diverse technologies. Critical for extending the lifespan of solar technologies are durability enhancements, encapsulation methods, and protective coatings. Regulatory frameworks and industry standards will play a pivotal role in fostering confidence, regulatory compliance, and accelerated commercialisation. Essential for building a supportive environment, addressing concerns, and propelling solar innovations towards a sustainable and widely embraced energy future are cross-disciplinary collaboration, educational outreach, and public awareness campaigns.

1.7 Conclusion

A detailed discussion in the chapter about solar energy innovations encompasses a comprehensive exploration of the challenges faced by various technologies, along with promising future directions for the field.

First, the challenges in solar energy innovations, spanning perovskite solar cells, transparent contacts, and quantum dots, reveal common hurdles demanding collaborative solutions. The critical issues of stability and durability are illuminated, emphasising the universal concern posed by environmental factors like moisture and temperature variations. The toxicity of materials, such as lead in perovskite compositions, underlines the imperative for eco-friendly alternatives. Scaling up production emerges as a universal challenge requiring consistency, reproducibility, and cost-effectiveness. The chapter underscores the shared hurdle of bridging the efficiency gap between the laboratory scale and commercial viability, stressing the need for breakthroughs in reliability and widespread adoption. Integrating innovative technologies into existing infrastructure surfaces as a challenge demanding seamless adaptation to current grid systems and architectural designs.

Moving forward, the discussion delves into the future directions of solar energy technologies. The paramount focus on material innovation is explored, with an emphasis on ongoing research geared towards discovering compositions that enhance stability, efficiency, and environmental sustainability across various technologies. Tandem solar cells, blending different materials for enhanced performance, are positioned as a promising avenue for achieving higher efficiencies and warrant further exploration. The pivotal role of advances in manufacturing techniques, such as roll-to-roll processing and printing methods, is highlighted as key to scalability and cost reduction across diverse technologies. Durability enhancements, encapsulation methods, and protective coatings are deemed critical for extending the lifespan of solar technologies. Regulatory frameworks and industry standards are identified as pivotal in fostering confidence, regulatory compliance, and accelerated commercialisation. The chapter underscores the importance of cross-disciplinary collaboration, educational outreach, and public awareness campaigns for building a supportive environment, addressing concerns, and propelling solar innovations towards a sustainable and widely embraced energy future.

In essence, the detailed discussion provides a nuanced understanding of the challenges and promising future avenues within the dynamic landscape of solar energy innovations. It encourages a holistic perspective, emphasising collaboration, innovation, and adaptability to drive the transformative potential of solar technologies.

REFERENCES

[1] Khan, J. and Arsalan, M.H., 2016. Solar power technologies for sustainable electricity generation–A review. *Renewable and Sustainable Energy Reviews*, *55*, pp. 414–425.

[2] Ezema, F.I., Anusuya, M., and Nwanya, A.C. (Eds.), 2023. *Materials for Sustainable Energy Storage at the Nanoscale* (1st ed.). CRC Press. https://doi.org/10.1201/9781003355755

[3] Ramalingam, K. and Indulkar, C., 2017. Solar energy and photovoltaic technology. *Distributed Generation Systems*, Butterworth-Heinemann, pp. 69–147.

[4] Husain, A.A., Hasan, W.Z.W., Shafie, S., Hamidon, M.N. and Pandey, S.S., 2018. A review of transparent solar photovoltaic technologies. *Renewable and Sustainable Energy Reviews*, *94*, pp. 779–791.

[5] Arockiam, A.J., Subramanian, K., Padmanabhan, R.G., Selvaraj, R., Bagal, D.K. and Rajesh, S., 2022. A review on PLA with different fillers used as a filament in 3D printing. *Materials Today: Proceedings*, *50*, pp. 2057–2064.

[6] Ginley, D.S. and Perkins, J.D., 2010. Transparent conductors. In David S. Ginley (ed.) *Handbook of Transparent Conductors*. 1–25. Boston, MA: Springer US.

[7] Maziz, A., Özgür, E., Bergaud, C. and Uzun, L., 2021. Progress in conducting polymers for biointerfacing and biorecognition applications. *Sensors and Actuators Reports*, *3*, p. 100035.

[8] Chandrashekhar, K.G., Laxmaiah, G., P. RamKumar and Ramesh, B., 2023. Load-bearing characteristics of a hybrid Si_3N_4-epoxy composite. *Biomass Conversion and Biorefinery*, pp. 1–9.

[9] Gazzi, A., Fusco, L., Orecchioni, M., Ferrari, S., Franzoni, G., Yan, J.S., Rieckher, M., Peng, G., Lucherelli, M.A., Vacchi, I.A. and Chau, N.D.Q., 2020. Graphene, other carbon nanomaterials and the immune system: Toward nanoimmunity-by-design. *Journal of Physics: Materials*, *3*(3), p. 034009.

[10] Wang, X. and Shi, G., 2015. Flexible graphene devices related to energy conversion and storage. *Energy & Environmental Science*, *8*(3), pp. 790–823.

[11] Arockiam, A.J., Rajesh, S., Karthikeyan, S., Thiagamani, S.M.K., Padmanabhan, R.G., Hashem, M., Fouad, H. and Ansari, A., 2023. Mechanical and thermal characterization of additive manufactured fish scale powder reinforced PLA biocomposites. *Materials Research Express*, *10*(7), p. 075504.

[12] Koh, T.M., Wang, H., Ng, Y.F., Bruno, A., Mhaisalkar, S. and Mathews, N., 2022. Halide perovskite solar cells for building integrated photovoltaics: Transforming building façades into power generators. *Advanced Materials*, *34*(25), p. 2104661.

[13] Xu, Q., Cai, W., Li, W., Sreeprasad, T.S., He, Z., Ong, W.J. and Li, N., 2018. Two-dimensional quantum dots: Fundamentals, photoluminescence mechanism and their energy and environmental applications. *Materials Today Energy*, *10*, pp. 222–240.

[14] Joseph Arockiam, A., Rajesh, S. and Karthikeyan, S., Development of fish scale particle reinforced PLA filaments for 3D printing applications. *Journal of Applied Polymer Science, 141*, p. e55132.

[15] Shaikh, P.H., Nor, N.B.M., Sahito, A.A., Nallagownden, P., Elamvazuthi, I. and Shaikh, M.S., 2017. Building energy for sustainable development in Malaysia: A review. *Renewable and Sustainable Energy Reviews, 75*, pp. 1392–1403.

2

Investigation of Crystalline Electron Conductors

V. Koushick
Vel Tech Rangarajan Dr. Sagunthala R&D Institute of Science and Technology, Chennai, India

J. Eindhumathy
Saranathan College of Engineering, Tamilnadu, India

C. Divya
Manonmaniam Sundaranar University, Tamilnadu, India

C. Vennila
Saranathan College of Engineering, Tamilnadu, India

2.1 Introduction

A crystalline material is characterised by an organised crystal shape. A crystal is a solid composed of atoms, ions, or molecules arranged in a three-dimensional structure. Each crystal structure within a specific crystal system is defined by a component of the cell. The crystal's smallest recited component is a component of mobile. When thinking about crystals, it's far easier to ignore the actual atoms, ions, or molecules and instead focus on the geometry of periodic arrays. The crystal is then signified as a lattice, which is a three-dimensional collection of factors with similar spacing (lattice points). Two-dimensional lattices can be found in wrapping paper throughout the festive season. Lattices on your bathroom floor are another example; Figure 2.1 represents the lattice structure and point. A crystal system defines each crystal lattice. There are seven crystal systems in three dimensions: Hexagonal, cubic, monoclinic, triclinic, orthorhombic, tetragonal, and rhombohedral. The Bravais lattices are a hard and fast arrangement of systems [1, 2]. Crystalline solids, regularly known as crystals, incorporate exceptional internal systems that bring about separate flat surfaces, or faces.

The faces connect at the substance's distinctive angles. When bombarded by x-rays, every arrangement produces a distinct pattern that can potentially be used for determining the material. The feature angles are independent of the dimension of the crystal and represent the regular repetitive pattern of the module particles, fragments, or ions in region.

When a crystalline crystal is split (see Figure 2.2), repellent forces source it to disruption along fixed planes, resulting in fresh surfaces that connect at the same ratios as the initial crystal. The angles that intersect the surfaces meet of a covalent stable, including a cut diamond, are not arbitrary; rather, they are strongminded by the position of the particles of carbon in the crystalline. The electrical crystal deforms, causing a single plane to glide across each other [3]. The layers divide due to the repulsive interactions among ions with similar charges. Figure 2.2 displays the slicing of an ionic composite crystal along an ion plane. Crystals have generally sharp, established temperature boundaries because all of the constituent particles, ions, and atoms are comparably separated from each other and have the same kind of peers; that is, the consistency of the lattice of a crystal offers local circumstances that are identical. As an outcome, the solid-molecular connections are constant, and a comparable amount of power is required to break each interface simultaneously [4].

DOI: 10.1201/9781003495437-2

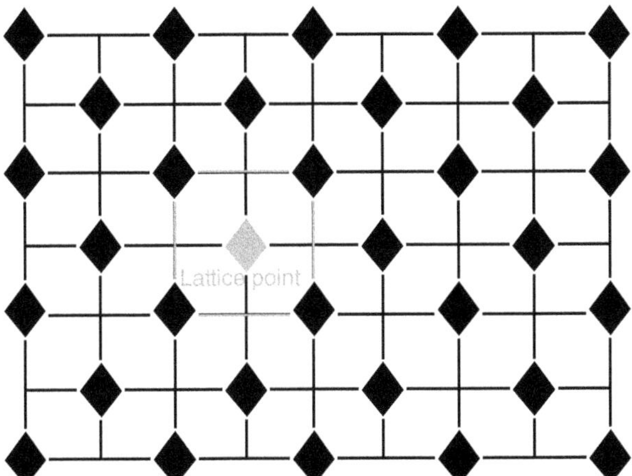

FIGURE 2.1   Structure of lattice crystal and lattice point.

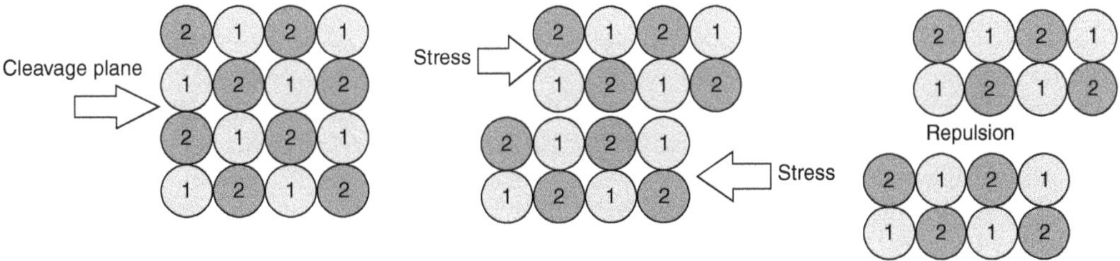

FIGURE 2.2   Cleaving an ionic compound crystal along an ion plane.

Nebulous objects have two distinguishing characteristics. When sliced or shattered, they crop wreckages with uneven, usually bowed exteriors, and when exposed to x-rays, they exhibit ill-defined designs as their mechanisms are not placed in an even array. A glass is a nebulous, transparent substance. If the fluid state is abruptly pushed back, practically any substance can solidify in a nebulous state. Some substances, on the other hand, are basically nebulous because their mechanisms are not fit composed well enough to form a constant crystalline lattice or because they include scums that disturb the lattice. While the biochemical configuration and essential physical components of a quartz crystal and quartz glass are identical (equally they are SiO_2 and made up of linked SiO_4 tetrahedra), the atomic configurations in space differ. Although the silicon and oxygen molecules in crystalline quartz are highly structured, the atoms in quartz glass are organised almost arbitrarily [5]. Quartz glass is formed when melted SiO_2 is rapidly chilled (4 K/min), whereas bulky, flawless lechatelierite minerals sold in inorganic plants have been cooled over millions of years.

Aluminium, on the other hand, crystallises significantly faster. Only until the fluid is chilled at the exceptional rate of 4×10^{13} K/s does nebulous aluminium emerge, preventing the atoms from organising themselves into a regular pattern. Atoms are organised in an even way in a construction composed of linked tetrahedra. The local environment, which includes the numbers and distances of neighbours, fluctuates across a nebulous solid. A certain amount of thermal energy is required to overcome each of these connections. As a result, unlike crystalline materials, which have a well-defined melting point, nebulous solids soften gradually across a wide temperature range. Long-term storage of a nebulous solid at temperatures around its melting point allows the molecules, atoms, or ions to progressively reorganise into a more highly ordered crystalline form [6]. Crystals have distinct melting points, whereas nebulous substances do not have it. Figure 2.3 depicts the crystal structure of quartz.

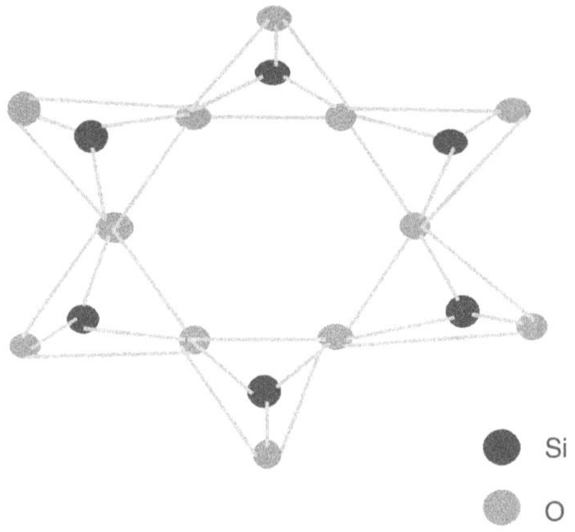

FIGURE 2.3 Crystal structure of quartz.

2.2 Crystals Fundamental

Since a crystalline solid is made up of echoing patterns of its component elements in 3D (a quartz lattice), we may illustrate the complete quartz by demonstrating the construction of the tiny alike elements that, when composed in a slanted position, make the quartz. This fundamental recapping unit is referred to as a unit cell. A single stamp, for illustration, is the part cell of a sheet of indistinguishable postage stamps, while a sole block is the part cell of a pile of bricks. This section goes through the atomic configurations in numerous unit cells [7].

Figure 2.4(a–c) depicts three 2D lattices representing the unit cell's potential configurations. The relative locations or orientations of the unit cells inside the lattice vary, but they are all feasible alternatives

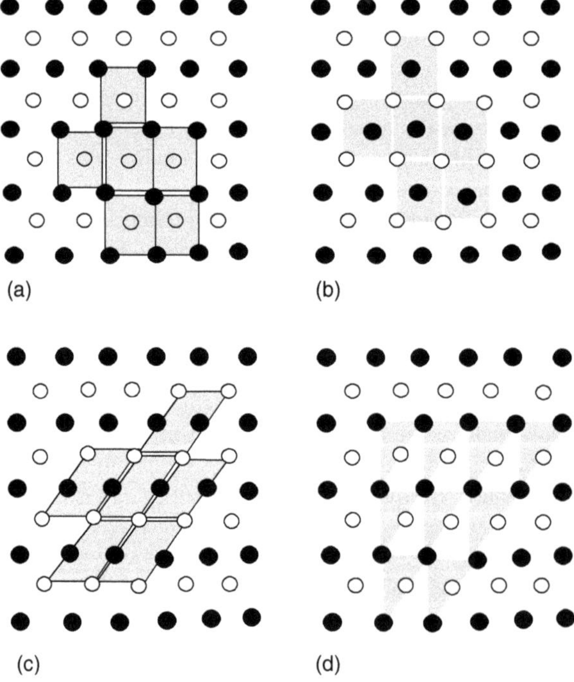

FIGURE 2.4 2D unit cells.

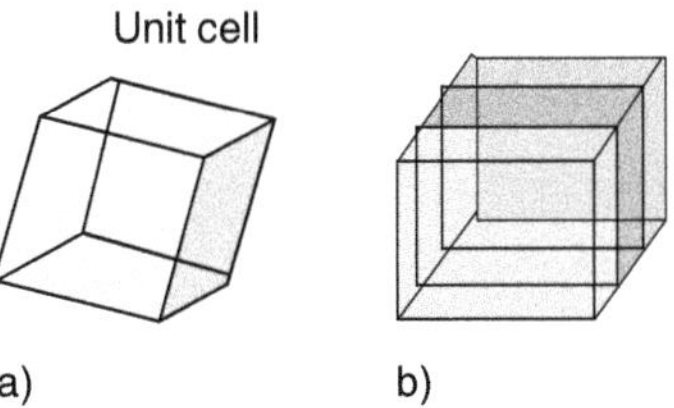

FIGURE 2.5 3D unit cells.

since recapping them in whatever way fills the overall design of dots. Figure 2.4(d) shows that, because repeating the triangle in space takes up just half the space in the pattern, it is an invalid unit cell. Unit cells are easily visible in two dimensions. In many circumstances, more than one unit cell can be used to depict a specific architecture, as seen in the Escher image in the section icebreaker and Figure 2.4 for a 2D crystal lattice. Typically, the least part cell that wholly defines the directive is selected. The solitary condition for a binding part cell is that it must generate an even framework when repeated in space [8].

As a result, the unit cell in Figure 2.4(d) is not a feasible alternative since it does not crop the necessary lattice (triangular holes) when repeated in space. The notion of part cells is stretched to a 3D lattice, as shown schematically in Figure 2.5. Figure 2.5(a) shows a 3D part cell, and Figure 2.5(b) shows the resultant 3D even lattice.

2.3 Unit Cells

There are 7 basically diverse sorts of unit cells, which vary based on how long and at what angles the edges are in respect to one another (Figure 2.6). Each of the six sides of a unit cell has a parallelogram-shaped side. The concepts we describe apply to non-cubic unit cell substances; however, we emphasise on cubic part cells, which have all flanks spanning identically and all angles are 90°. The universal faces of the 7 rudimentary unit cells are depicted in Figure 2.6 [9]. The spans of the part cell ends are designated

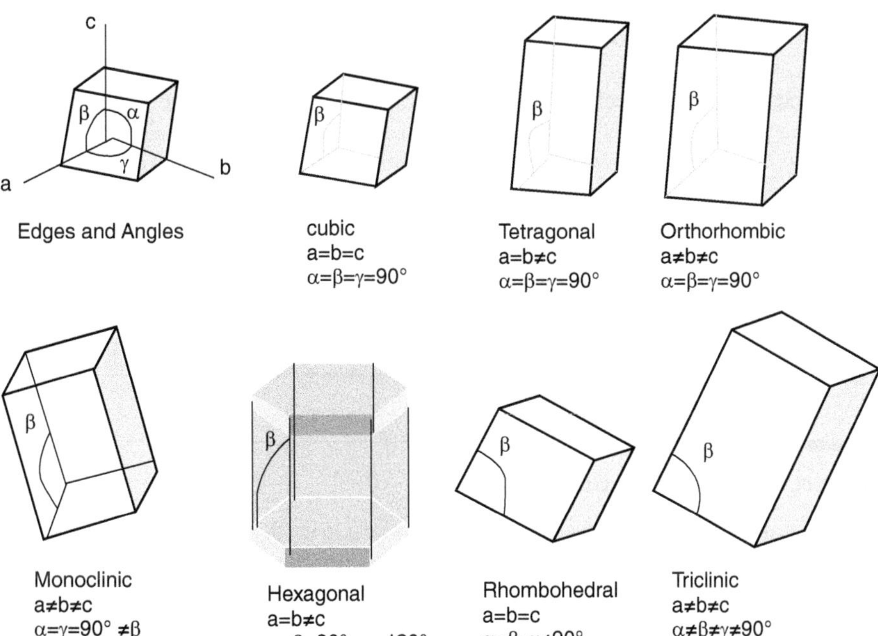

FIGURE 2.6 Structure of basic unit cells.

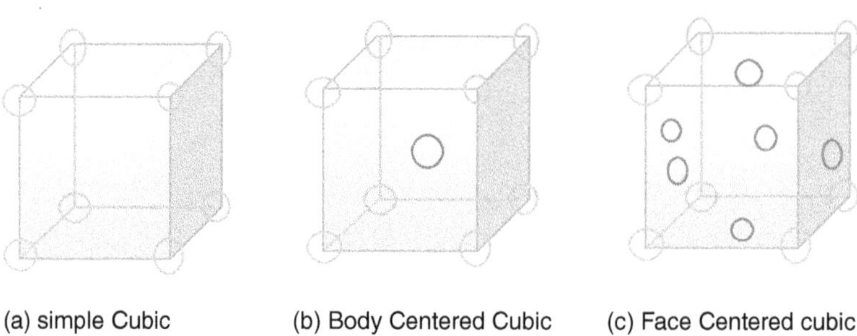

(a) simple Cubic   (b) Body Centered Cubic   (c) Face Centered cubic

FIGURE 2.7 Structure of cubic unit cells.

as a, b, and c, and the angles are as follows: b and c, a and c, and a and b. Cubic: $a = b = c$, $\alpha = \beta = \gamma = 90°$. Tetragonal: $a = b$, $\alpha = \beta = \gamma = 90°$. $A = \beta = \gamma = 90°$. Orthorhombic. $A = \gamma = 90°$; monoclinic. $A = b$, $\alpha = \beta = 90°$, $\gamma = 120°$. Rhombohedral: $\alpha = \beta = \gamma$, $a = b = c$. All three are distinct.

If the cubic unit cell includes 8 component particles, molecules, or atoms positioned at the cube's angles, it is called modest cubic (Figure 2.7(a)). The unit cell is body-centred cubic (BCC) (Figure 2.7(b)) if it also contains an alike component in its core. The unit cell is face-centred cubic (FCC) if there are components in the centre of each face in addition to those at the cube's corners (Figure 2.7(c)). Figure 2.7 depicts the three categories of cubic unit cells. There are three representations for the three classes of cubic part cells: Simple cubic (a), BCC (b), and FCC (c): A ball-and-stick model, a space-filling cutaway perfect that displays the share of a piece particle that deceits in the part cell, and a collective of some part cells. A basic cube is made up of 8 quarters of spheres. The BCC is made up of 8 quarters of spheres, each having a sphere at the core. FCC is composed of 8 quarter and 6 half spheres [10].

A solid, as seen in Figure 2.7, is composed of a huge quantity of part cells organised in 3D. A majority substantial, such as density, must thus be linked to its part cell. Since density is the form of substance per part volume, we may compute the mass of the bulk substantial from the mass of a sole part cell. To do so, we must first determine the volume of the part cell, the molar form of its constituents, and the number of constituents per unit cell. As demonstrated in Figure 2.7, when counting particles or molecules in a unit cell, those on a corner, face, or edge donate to more than one part cell. For example, one atom communal to two nearby unit cells on a part cell's face is tallied as 1212 atoms per part cell. Similarly, one atom on the boundary of a part cell is shared by 4 adjoining unit cells, each of which underwrites 14 atoms. An atom in the corner of a part cell is shared by all 8 part cells, adding 1818 atoms to each [11].

The statement that particles lying on a unit cell's edge or corner count as 1414 or 1818 atoms per unit cell is true for all part cells excluding the hexagonal one, which has three unit cells sharing each perpendicular edge and 6 sharing each angle (Figure 2.7), resulting in 13 and 16 atoms per part cell for atoms in these places. Atoms that are completely limited within a part cell, such as the particle at the core of a BCC part cell, solely belong to that unit cell. Excluding the hexagonal part cells, particles on the faces underwrite 12 particles to each part cell, particles on the boundaries underwrite 14 particles to each part cell, and particles on the corners give 18 particles to each part cell [12].

2.4 Structure of Crystalline Bonds

Precipitation of solids is caused by various chemical interactions that often result in predictable and determinable lattice patterns that make up the solid. Chemical bonds contain structured patterns and three-dimensional structures that contribute to the overall crystal structure. This is what we mean when we talk about chemical bond types for substances in the context of crystal formation. A naturally occurring substance or a man-made material is a crystal solid with a crystalline structure that may be generated by four distinct types of linkages. Each of the several types of bonds that occur at the atomic and molecular levels is distinguished by different features according to the nature of their bonds. Among the several forms of bonding are

- Ionic – Held by electrostatic forces amid positively and negatively charged particles.
- Metallic – The bonding of metallic atoms with identical electronegativity that constitute a metallic substance.
- Covalent – When two atoms bond by sharing electron pairs.
- Molecular – Intermolecular forces hold molecular atoms or molecules together.

The sort of bond that makes up a material has a direct impact on its overall qualities. The melting and boiling points of various types of crystalline solids are tabulated in Table 2.1.

2.4.1 Ionic Crystals

Negatively charged anions and positively charged cations alternate in the ionic crystal structure (see Figure 2.8). Ions can be moreover monatomic or polyatomic. Ionic crystals arise when Group 1 or 2 metals interact with Group 16 or 17 non-metals or non-metallic polyatomic particles [13]. Ionic crystals are

TABLE 2.1

Crystalline Solids: Melting and Boiling Points

Type of Crystalline Solid	Examples (Formulas)	Melting Point (°C)	Normal Boiling Point (°C)
Ionic	NaCl	801	1413
	CaF_2	1418	1533
Metallic	Hg	−39	630
	Na	371	883
	Au	1064	2856
	W	3410	5660
Covalent network	B	2076	3927
	C (diamond)	3500	3930
	SiO_2	1600	2230
Molecular	H_2	−259	−253
	I_2	114	184
	NH_3	−78	−33
	H_2O	0	100

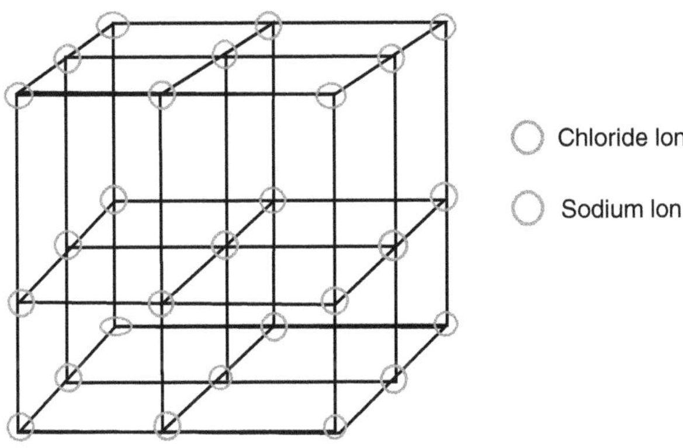

FIGURE 2.8 NaCl crystal.

tough and hard, having high molecular values. When solid, ionic substances do not carry electricity, but when melted or liquid, they do.

2.4.2 Metallic Crystals

In metal cations, metallic crystals are enclosed by a 'sea' of moveable valency electrons (see Figure 2.9). These electrons, also known as delocalised electrons, can move freely throughout the crystal and are not bound to any single element. As a result, metals carry electricity well, as can be observed in Figure 2.9. The melting points of metallic crystals vary extensively, as seen in Table 2.1.

2.4.3 Covalent Network Crystals

At the crystal's lattice points, each atom in a covalent network crystal is covalently fused to its adjacent neighbour particles (see Figure 2.10). The atoms in the three-dimensional covalently connected network are extraordinarily dense. Network solids include quartz, diamond, transition metal, metalloid oxides, and various metalloids. Network solids are secure and stiff, with hot temperatures. They never conduct electricity because they are made up of atoms rather than ions. Rhombus is a dense grid made up of carbon particles that are covalently connected to one another in three dimensions. In a tetrahedral shape, each carbon atom forms a single covalent connection [14].

2.4.4 Molecular Crystals

Weak intermolecular interactions at the crystal's lattice points hold a typical molecular crystal together (see Figure 2.11). Intermolecular interactions in nonpolar crystals can be dispersion forces, but dipole–dipole forces can exist in polar crystals. Hydrogen bonding holds molecular crystals together, for example, ice. When a hydrogen gas is chilled and hardened, the lattice points transform into single particles somewhat molecules. In all cases, the intermolecular forces that hold the atoms composed are distant

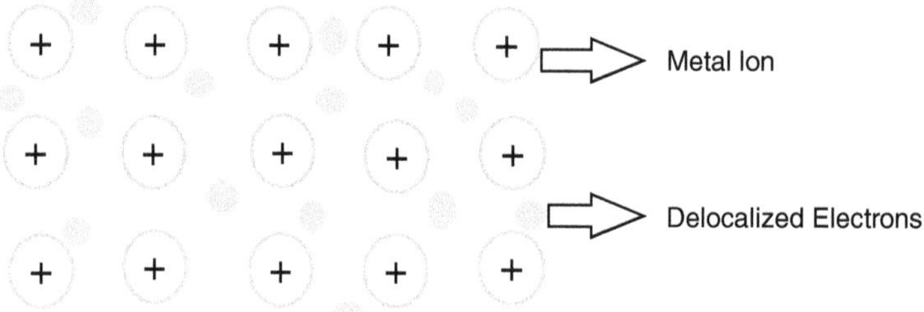

FIGURE 2.9 Metallic crystal lattice with free electrons.

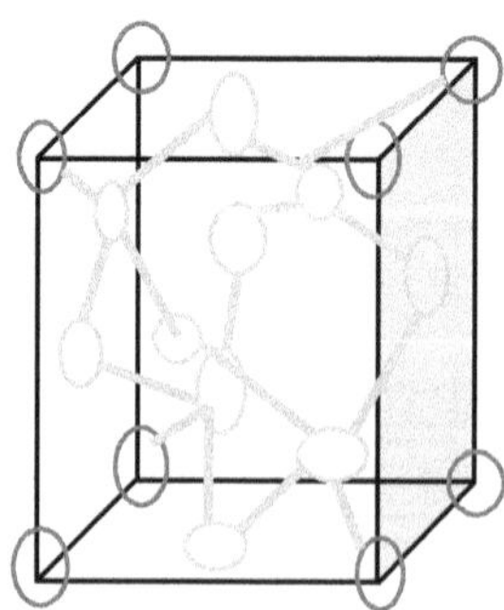

FIGURE 2.10 Covalent crystal network.

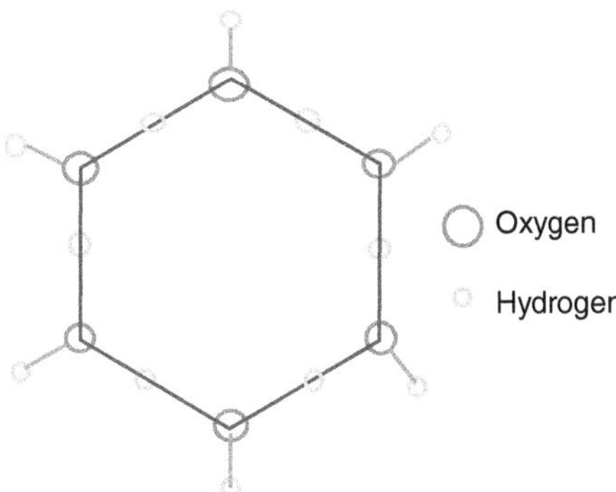

FIGURE 2.11 Ice crystal structure network.

TABLE 2.2

Properties of the Major Classes of Solids

Ionic Solids	Molecular Solids	Covalent Solids	Metallic Solids
Bad conductors of temperature and electricity	Bad conductors of temperature and electricity	Bad conductors of temperature and electricity[a]	Good conductors of temperature and electricity
Relatively high melting point	Low melting point	High melting point	Melting points depend strongly on electronic configuration
Hard but brittle; shatter under stress	Soft	Very hard and brittle	Easily deformed under stress; ductile and malleable
Relatively dense	Small density	Small density	Usually high density
Dull surface	Dull surface	Dull surface	Shiny

[a] There is a lot of exclusion. Diamond, for example, has the maximum thermal conductivity of any recognised material, while graphite has actual high electrical conductivity privileged than the carbon planes.

less than ionic or covalent bonds [15]. As an outcome, molecular crystal tender and hot temperatures are much inferior. Molecular crystals are deprived electrical rods since they lack particles or free electrons. Table 2.2 summarises the universal structures of the 4 main clusters of solids.

2.5 Conductivity of Metals

Conduction electrons are abundant in metals. Aluminium particle contains 3 valence electrons in an outside shell that is only partly filled. In metallic aluminium, the 3 valence electrons per particle develop transfer electrons. Regardless of temperature or pollutants, the number of conduction electrons stays endless. Metals conduct electricity at all temperatures; however, most metals' conductivity is maximum at small temperatures. Monovalent particles, like lithium or gold, provide one valence electron, whereas divalent atoms, like magnesium or calcium, contribute two. As previously set up, the amount of transfer electrons alone does not constitute conductivity; electron mobility also plays a role. Silver is a better conductor than aluminium, which has three conduction electrons per atom, since silver's higher mobility compensates for its fewer electrons [16].

Metal atoms, such as aluminium and sodium, move all of their valence electrons to the conduction band. The resulting ions are minute, inhabiting only approximately 10–15% of the volume of the crystal. The remaining region is open for conduction electrons to wander. A basic model, which frequently accurately represents the behaviour of conduction electrons, portrays them as interrelating neither with particles nor with one another. The electrons are represented as free atoms moving freely over the crystal.

Arnold Johannes Wilhelm Sommerfeld, a German physicist, was the first to suggest this idea. It works effectively for simple metals, such as aluminium, magnesium, calcium, zinc, and lead, whose conduction electrons are provided from sp-shells. They are dubbed simple because Sommerfeld's basic theory accurately describes them.

The periodic table has 3 rows of transition metals: Scandium through nickel in the first row, yttrium through palladium in the second row, and lanthanum plus hafnium through platinum in the third row. As the atomic number grows, electrons in these rows fill d-states in the atom's outer shell [17]. Transition metal atoms in crystal form are metals with fascinating features. The ion centre is more strongly attached to d-electrons than to sp-electrons. Although sp-valence electrons develop transmission electrons and flow easily across the crystal, d-electrons incline to break close to the ion. Ions in close proximity can form covalent bonds with d-electrons. Most of the time, these d-states are just half filled. Electrons in these d-states can conduct as well as those in the sp-states, but the Sommerfeld model of free particles does not well reflect electron mobility in the d-states. Instead, electrons proceed from particle to particle via the common covalent bonds of the d-electrons. Because these alloys contain transmission electrons from sp-states and others from d-states, some electrons drift easily according to the Sommerfeld model while others move through the bonds. Each electron alternates amid these two styles of conduction, resulting in extraordinarily complex electron movement [18].

When an electric charge is applied, metal electrons accelerate and return to the electrical current. Electrons occasionally scattered from defects in the crystal, and the quantity of sprinkling impacts flexibility. Electrons do not scatter from particles in the crystal that are in the predicted location in the crystal frame. Rather than scattering from the host ions, the electrons shift to accept them. However, if one ion is absent, displaced, or of a dissimilar species, the electron will scatter from this flaw. Particles oscillate about their frame location, increasing temperature increasing the magnitude of the oscillation. The tremor may displace the particle from its crystal position, creating a flaw from which an electron will scatter. Metal resistivity increases at high temperatures due to an upsurge in ion atmospheres in the crystal and the associated surge in bit [19].

2.5.1 Conducting Properties of Semiconductors

Semiconductors have conducting characteristics in between insulators and metals. Semiconductors are insulators in certain circumstances and metals in others. Semiconductors, like insulators, have no conduction electrons in a pristine crystal with no current variations. Conduction electrons are supplied by impurities or thermal fluctuation of electrons from atomic shells. The nature of the traps is what distinguishes insulators from semiconductors. A trick is a defect's local electron energy state. While insulator traps forcefully bind conduction electrons, semiconductor traps only faintly bind the electrons. Thermal fluctuations can return a trapped transmission electron to the transmission band in a semiconductor. The bulk of additional electrons originate in the transmission band rather than tricks at ambient temperature. The major distinction between semiconductors and insulators is the inability of traps to retain electrons. At normal temperature, a semiconductor possesses an adequate amount of transference electrons to give excellent electrical conductivity. Because electron mobility is extremely significant in many semiconductors, a uniformly distributed quantity of electrons that conduct can often be sufficient to attain high conductivity [20].

Silicon contains four valence electrons, while phosphorus takes five. Four of a phosphorus atom's five valence electrons enter covalent bonds when it alternates for an atom in a silicon crystal frame. The fifth is an extra, which sits in a shallow trap near the phosphorous spot. Thermal variations, on the other hand, quickly stimulate it to the conduction band. At ambient temperature, silicon has about one conduction electron for every phosphorus contamination. The conductivity of silicon may be controlled by varying the quantity of impurities. Arsenic and antimony are two more substitutional elements that operate as electron donors to silicon's conduction band. When a significant amount of conduction electrons is introduced into a semiconductor by impurities, the electrical characteristics change to metallic. The grave attention of scums N_c varies depending on the kind of contamination. For contamination absorptions less than the critical number N_c, the transmission electrons get involved in solutions at extremely low temperatures, and the semiconductor forms an insulator. When the impurity attention exceeds N_c, the conduction electrons are not confined in tricks at little temperatures, and the semiconductor shows metallic transmission. $N_c = 2 \times 10^{18}$ scums per cubic centimetre for phosphorus impurities in silicon. Although this appears to be a big quantity, it signifies around one phosphorus atom for every 100,000 silicon particles. On a percentage basis, a little number of phosphorus particles will alternate silicon from an insulator to a metallic electrode. Other semiconductors have alike characteristics. In gallium arsenide, the grave attention of impurities for metallic transmission is 100 epochs inferior than in silicon [21].

Gallium atoms, like phosphorus atoms, can be employed as silicon's substitutional impurities. Covalent bonds need three electrons from each atom. Because a tetrahedral structure requires four electrons, one electron is missing from a complete set of covalent bonds for each gallium atom. A hole is the name given to the missing electron. Holes can move across the crystal in the same way as ion vacancies do, except that there is an electron position in this case. An electron from a neighbouring covalent bond can cross over and seal the unoccupied electron state, causing the hole to transfer to the surrounding bond. The hole has a positive charge because it lacks an electron. In reaction to an external voltage, the mobility of holes is roughly equivalent to that of conduction electrons. A semiconductor may have a high mass of scums that generate holes, and their mobility results in a high electrical conductivity. A p-type semiconductor has more holes than conduction electrons. An n-type semiconductor has more conduction electrons than holes. The symbols p and n refer to the charge of the atoms, which is positive for holes and negative for electrons [22].

Thermal fluctuations can excite an electron from a covalent bond and transform it to a conduction electron. As a result of the missing electron, a hole forms in the bond. As a result, thermal fluctuations generate electron–hole pairs. Normally, the electron and hole divide and drift away in space. Gregory Hugh Wannier, a Swiss-American physicist, was the first to propose that electrons and holes may weakly link together. This bonded state, known as a Wannier exciton, exists; the hole is positively charged, the electron is negatively charged, and opposites attract. In tests using electromagnetic radiation, the exciton is easily detected. It has a short lifetime, ranging from a nanosecond to a microsecond contingent on the semiconductor. The electron's proclivity to re-join a covalent bond state, which removes both the hole and the conduction electron, accounts for its short lifetime. Because the two particles are spatially close, this recombination of electron and hole is easily performed from the exciton state. When an electron and a hole emerge from the exciton state due to current fluctuation, they move apart from one other. Recombination is thus less likely and occurs only when the nomadic particles come near to one another again. Recombination can also happen at faulty locations [23]. The first particle is linked to the defect and then the second particle. The electron and hole are once again in near proximity, allowing the electron to reoccupy the covalent link.

Electron scattering limits electron mobility in semiconductors, just as it does in metals. For crystals with few defects, mobility is restricted at low temperatures by defect scattering and at moderate and high temperatures by ion vibrations. The resistivity of semiconductors with few flaws is high because the quantity of conduction electrons is minimal. Impurities are added to semiconductors to enhance the quantity of conduction electrons. Unfortunately, this increases scattering from contaminants, reducing mobility [24].

Photoconductors are semiconductors with low impurities. Photoconductivity happens when a substance's electrical conductivity is augmented by revealing it to light. Light is a kind of electromagnetic radiation with a limited frequency range. Light quanta are engrossed by the semiconductor, generating electron–hole pairs that allow electrical conduction. Additional penetrating energy produces more electron-hole pairs, increasing conductivity. Because each semiconductor engrosses light at a distinct frequency range, dissimilar semiconductors are used as photoconductors at unlike frequencies. Zinc oxide (ZnO) is a fascinating material in terms of conductivity. It possesses ionic and covalent connections and crystallises as wurtzite. Isolators are solitary crystals of great purity. Zinc oxide is the most piezoelectric of all materials and is often used as a transducer in electrical devices. (Piezoelectricity refers to a crystal's ability to become polarised when subjected to pressure.) When aluminium impurities are present in the crystal, zinc oxide becomes an excellent semiconductor. Ohm's law is obeyed by polycrystalline ceramics made of semiconducting zinc oxide. The electrical current in zinc oxide ceramics with modest concentrations of other oxides, such as chromium and barium oxides, is the most nonlinear of any identified material. When the exponent n exceeds 100, the current I becomes proportionate to a power of the voltage V_n

in certain voltage ranges. Varistor, which is an acronym for variable and resistor, is the name given to this material. Zinc oxide varistors are frequently used as voltage surge prevention circuit components [25]. There is a little current until a critical voltage of about 330 volts is touched, at which point the current begins to rise abruptly and nonlinearly. Zinc oxide was also used as a white pigment in paint, which is an innovative use. However, it has been substituted with the whiter titanium dioxide (TiO_2).

2.6 Magnetism

Electrons are continually spinning, and because they are charged, their rotation generates a slight magnetic instant. Magnetic instants are tiny magnets that have opposite north and south poles. The instant is travelling southward from northward. Because electron spins in nonmagnetic materials are randomly arranged, electron moments cancel. When two electrons' moments are oriented in conflicting directions, their properties tend to abandon. A magnet is formed when a huge quantity of electrons align their distinct instants in an identical way. Magnetism occurs in just a trivial proportion of crystals. The forces that bring into line electron turns are minor. The explanation of magnetism is separated into three components. They are as follows:

1. Atoms with valency electrons in the f or d shells make up the majority of magnets. Angular momentum is represented by the atomic shell notation, where s stands for zero units, p for one, d for two, and f for three. In comparison to those in f-shells, electrons in d-shells are more tightly bonded to the ion.

2. In each electron orbital, there can be 2 electrons, with one having an upward turn and the other having a downward spin. The d-shell, when completely filled, has a total of five orbital states and can accommodate ten electrons. On the other hand, the f-shell has 7 states and can hold 14 electrons. According to the empirical principle, which suggests that electrons should occupy states with the highest spin and magnetic moment, electrons are added individually to the d-states. If the first electron in a d-state has a spin, the following four electrons will also start to spin. The sixth electron is required to have a downward rotation because the d-shell can only accommodate a maximum of five electrons with upward rotation. Likewise, the f-shell allows up to 7 electrons with identical rotation before accepting electrons with conflicting rotation. The sequence in which electrons occupy atomic shells is governed by Hund's rubrics, with the initial priority being to maximise the overall spin. Atoms with partly occupied d- and f-shells often exhibit a net magnetic moment and a non-zero total spin. Magnetic ions form the basis of magnetic crystals. The first rule of Hund is closely connected to a phenomenon known as electron exchange. As mentioned earlier, the Pauli exclusion principle dictates that it is not possible for 2 electrons with a similar turn orientation to inhabit similar interplanetary location simultaneously. Since electrons carry a charge, they repel each other. When 2 electrons originate near to each other, a substantial quantity of repulsive energy is generated. Electrons tend to avoid such close contact because physical processes favour the lowest energy state. The Pauli principle states that when electrons have parallel spins, they tend to stay away from each other. This configuration helps to keep the electrons apart and minimises repulsive energy, which is why electrons in the similar shell favour having parallel spins. The perception of electron conversation forms the base of magnetism and provides an explanation for the presence of strong magnetic moments in ions like iron. Divalent iron (Fe_{2+}), for example, possesses six positions within the d-electron configuration that maximise both electron spin and magnetic moment.

3. The collective magnetic properties observed in a crystal arise from the collaboration of individual ions, each having a fixed magnetic moment, aligning their moments. In ferromagnetic crystals, all the magnetic moments of the constituent particles are associated in the same direction, resulting in the crystal itself having a magnetic moment that equals the sum of those individual moments. In order for the individual ions to align their magnetic moments cooperatively, a magnetic force must act upon them. This strength is also generated by electron conversation.

When the d-orbitals of neighbouring particles slightly intersect, covalent bonds are formed. This allows the d-electrons from different ions to be shared with their neighbours through covalent bonding. As a result of this electron exchange, the spins of the neighbouring ions often become aligned. When all adjacent pairs are aligned, all the ions in the crystal become aligned as well. The force within an individual ion's atomic shell is significantly stronger than the exchange force between neighbouring ions. Despite its limitations, this force is strong enough to induce ferromagnetism.

2.6.1 Ferromagnetic Materials

Iron is a commonly found ferromagnetic material. However, not all iron bars exhibit magnetism; whether a bar is magnetic or not depends on the arrangement of regions within it. In a crystal, a domain refers to a section where all the particles are aligned ferromagnetically in the similar plane. A bar can consist of different regions, each having its own magnetic alignment. In cases where these regions have opposite magnetic directions, the bar as a whole does not possess a net magnetic moment. In other words, the magnetic moments of the domains cancel out each other [26]. However, when an iron bar is exposed to a strong magnetic field, it becomes magnetic. The bar then forms a single domain as a result of the field, with all the magnetic moments aligning along the external field. Rather than the domain boundaries rotating, it is the moments within the domains that move. When one domain expands, the connected section produces, whereas the others contract. After being detached from the magnetic field, the iron bar retains its magnetisation for a significant period of time. Most iron bars are composed of multiple tiny grains of single crystals arranged randomly, making them polycrystalline. Each grain can consist of a single domain, an area can have multiple domains, or a large grain can comprise numerous domains. The magnetic arrangement of ferromagnetic materials undergoes a transition at a specific temperature known as the Curie temperature (Tc). This temperature marks the point at which the magnetic properties change. For instance, iron, cobalt, and nickel, which are three popular ferromagnetic materials, have Curie temperatures of 631 K, 1043 K and 1394 K individually. At temperatures below Tc, the crystal becomes magnetic as the magnetic fields of the ions align. However, at temperatures above Tc, the crystal loses its ferromagnetic properties as the atomic instants are no longer coordinated. Although the atomic moments exhibit some temporary ordering above Tc, it is not sustained in the long term. Instead, there is a localised ordering known as short-range order. Neighbouring magnetic moments often align in the same direction, following the alignment of a single moment. While this alignment persists over numerous frame locations, it is not upheld over very extensive reserves. The phenomenon of alignment occurring across extensive distances is known as long-range order. At temperatures slightly below Tc, the instants exhibit strong short-range order but relatively weak long-range order. This suggests that the bar does not possess a strong magnetic character. The inclination towards long-term order becomes more pronounced as the temperature decreases. As the temperature reaches Tc, which marks the start of long-term order, the magnetic properties of an iron bar diminish. Heating the iron bar above Tc leads to the loss of its magnetic properties. When the bar is cooled below Tc, the individual grains within it become magnetic. However, due to the random orientation of their attractive moments, the bar as a perfect doesn't exhibit magnetism. Demagnetisation may be achieved by heating the bar and allowing it to cool. On the other hand, the bar can be magnetised again by subjecting it to a strong magnetic field, such as by immersing it in one. Ferromagnetism is observed in many metals and insulators. Chromium bromide ($CrBr_3$) is an example of an insulator that exhibits ferromagnetic properties. This is because chromium, being trivalent, requires one electron to complete its external shell, resulting in the formation of ferromagnetically aligned moments in each trivalent chromium atom. Other examples are europium oxide (EuO) with a Tc of 77 K and gadolinium chloride ($GdCl_3$) with a Tc of 2.2 K, both of which display ferromagnetic behaviour.

2.6.2 Antiferromagnetic Materials

In numerous crystals, magnetic ions are arranged in patterns that do not exhibit ferromagnetic properties. Instead, they form an antiferromagnetic ordering, where moments flowing in one way are stable in the moments flowing in the conflicting way, resulting in a material with no overall magnetism. In such cases, the conversation collaboration between the particles has a conflicting sign, promoting different spin configurations. The effect of the exchange mechanism between the ions can be beneficial or detrimental, depending on factors such as the distance of the covalent bond and the bonding angles. The temperature at which antiferromagnetic materials undergo a transition is referred to as the Neel temperature (TN). Underneath this transition region, the particles in these materials exhibit antiferromagnetic ordering, but there is no long-term antiparallel direction present. Instances of antiferromagnetic crystals include iron oxide (FeO) with TN = 198 K, manganese oxide (MnO) with TN = 116 K, and manganese sulphide (MnS) with TN = 160 K. In the case of manganese oxide, the presence of two electrons in oxygen atoms and the divalent nature of manganese atoms result in its behaviour as an insulator. The magnetic attraction between manganese ions persists throughout. As the temperature drops under the Neel temperature, the atomic part cell undergoes a doubling in extent to accommodate 2 atoms of each ion type. The expanded cell size is essential to ensure that there is one moment on each side, as the magnetic impulses of the manganese atoms below TN are no longer equal and oriented in opposite directions.

2.6.3 Ferrimagnetic Materials

An alternative form of magnetic ordering is known as ferrimagnetism. While there are contrasting examples of ferrimagnetic materials, they do not nullify each other. A notable instance of a ferrimagnetic mineral is magnetite (Fe_3O_4). In magnetite, the secondary iron particle is divalent, whereas the primary iron particle is trivalent. The opposing characteristics of the two trivalent electrons align with each other, effectively nullifying their impact and resulting in a net magnetic moment for the divalent iron particle. The famous lodestone, which was one of the first materials known to possess magnetic properties, is composed of magnetite. Another category of ferrimagnets, such as $Y_3Fe_5O_{12}$, have a garnet structure. In these materials, iron particles are the primary magnetic entities. The net magnetic moment of an individual iron particle within its own unit cell consists of three points on one side and a pair on the other. Additionally, the presence of a rare-earth particle, such as gadolinium (Gd), replacing yttrium (Y), contributes to ferrimagnetic behaviour.

Ferrites are chemical compounds with the formula MFe_2O_4, where M represents a divalent particle, for example magnesium, manganese, cadmium, zinc, or nickel. The iron particle in the oxygen particle gains two electrons and ferrite is trivalent (ferric). Interestingly, M can also refer to magnetite (Fe_3O_4), which consists of divalent iron. The mineral known as spinel is represented by the chemical formula $MgAl_2O_4$, and its name is derived from its crystal shape. Ferrites are electrical insulators that possess magnetic ordering. Their insulating properties make them suitable for use as magnetic cores. Metallic ferromagnetic materials subjected to blinking magnetic fields experience significant thermal fatalities due to the presence of eddy currents. Ferrite magnets effectively minimise heat losses due to their high resistance, resulting in reduced energy dissipation. Moreover, they are capable of absorbing electromagnetic radiation with long wavelengths. Unlike insulators, which emit such radiation, metals tend to scatter it, distinguishing ferrite magnets as unique in this regard. The charge within small ferrite crystallites exhibits a spinning motion at the frequency of light in response to electromagnetic radiation. The frequency of light absorption depends on the particular crystal grain structure. In a polycrystalline material, the diverse variety of grain extents and forms allows for the collection of an extensive spectrum of radar frequencies. Ferrite crystals play crucial role in snooper paint, which is used to hide stealth aircraft from radar detection. However, regular flights can still detect the presence of an aircraft by receiving radar signals. Stealth aircraft, on the other hand, cannot be detected using this method as they reflect radar signals, rendering them invisible to radar-based detection techniques.

2.7 Kondo Effect

When magnetic ions are present as impurities within nonmagnetic crystals, they exhibit intriguing properties. These tiny magnets are irregularly dispersed throughout the crystal and tend to retain their attractive qualities. In the case of a metal host crystal, the presence of magnetic impurities introduces an interesting additional electrical resistance. The magnetic impurities scatter the conduction electrons, causing them to deviate from their usual path. If the spins of the conduction electron and the impurity align, they can switch during the scattering process. At temperatures below zero, the phenomenon of spin-flip scattering becomes particularly significant and even intensifies as the temperature decreases. Jun Kondo, a notable theoretical physicist from Japan, is recognised for his explanation of how magnetic impurities influence resistivity, leading to the naming of the Kondo effect in his honour. The Kondo temperature, a unique temperature point, is determined by the interplay between the impurity and the metallic substrate. At and below the Kondo temperature, resistivity shows a notable increase as the temperature decreases. An instance of a Kondo system is the presence of iron impurity in copper, where the Kondo threshold is observed at 24 K. The rise in resistivity with temperature in such systems is attributed to electron scattering caused by ion oscillations.

Magnetic impurities in copper or gold can be found in almost every transition metal atom. Each of these systems has its own distinct Kondo temperature, which can range from as high as 1000 K to a fraction of a Kelvin. Unlike other components of resistivity, the spin-flip component exhibits a unique behaviour where it becomes significant at low temperatures and decreases at higher temperatures.

2.8 Conclusion

Crystalline oxides exhibit charge transfer through ionic displacements only in the presence of particle vacancies. The behaviour of oxides in terms of electron-anion (n-type), hole-cation (p-type), amphoteric (n- and p-type), or ionic (cation–anion) conduction depends on the specific type of particle vacancies present. These positions in the oxide crystal lattice can either be charged or neutral. When the conditions are reducing (low pressure), oxygen-deficient nonstoichiometric oxides exhibit negative anion vacancies. On the other hand, under oxidising conditions (high pressure), oxygen-excess oxides that are not stoichiometric can have positive cation vacancies. Both stoichiometric and nonstoichiometric oxides can also have neutral ion vacancies due to thermal motion of ions or by doping them with cations that have charges different from the host crystal cations.

In crystalline oxides, the movement of neutral voids caused by heat enables the transportation of ions without any charge or mass transfer. The presence of mass exchange is essential for charge transfer in crystalline oxides, making electrical conduction a reliable gauge of mass transmission situations. When the accumulation of heat-induced, impurity-induced, and redox voids reaches a certain threshold, crystalline oxides undergo a state change resembling a pseudo-fluid. Following this alteration, anions and cations can migrate independently while the quartz lattice of the oxides remains unchanged. The Tammann temperature, which signifies the onset of chemical contact amid oxides and their surroundings, coincides with the temperature at which electrical conduction emerges in dielectric oxides at higher temperatures. This temperature also aligns with the transition of oxides into a pseudo-liquid state.

REFERENCES

[1] J. M. Williams, J. R. Ferraro, R. J. Thorn, K. D. Carlson, U. Geiser, H. H. Wang, A. M. Kini and M. H. Whangbo, *Organic Superconductors (Including Fullerenes)*, Prentice Hall, Englewood Cliffs, New Jersey, 1992.

[2] E. B. Yagubskii, I. F. Shchegolev, V. N. Laukhin, P. A. Kononovich, M. V. Karatsovnik, A. V. Zvarykina and L. I. Buravov, *Pis'maZh. Eksp. Teor. Fiz.*, 1984, *39*, 12.

[3] J. M. Williams, H. H. Wang, M. A. Beno, T. J. Emge, L. M. Sowa, P. T. Copps, F. Behroozi, L. N. Hall, K. D. Carlson and G. W. Crabtree, *Inorg. Chem.*, 1984, *23*, 3839.

[4] H. H. Wang, M. A. Beno, U. Geiser, M. A. Firestone, K. S. Webb, L. Nun, G. W. Crabtree, K. D. Carlson, J. M. Williams, L. J. Azevedo, J. F. Kwak and J. E. Schirber, *Inorg. Chem.*, 1985, *24*, 2465.

[5] T. J. Emge, H. H. Wang, P. C. W. Leung, P. R. Rust, J. D. Cook, P. L. Jackson, K. D. Carlson, J. M. Williams, M.-H. Whangbo, E. L. Venturini, J. E. Schirber, L. J. Azevedo and J. R. Ferraro, *J. Am. Chem. Soc.*, 1986, *108*, 695.

[6] T. J. Emge, H. H. Wang, M. K. Bowman, C. M. Pipan, K. D. Carlson, M. A. Beno, L. N. Hall, B. A. Anderson, J. M. Williams and H. H. Whangbo, *J. Am. Chem. Soc.*, 1987, *109*, 2016.

[7] N. Yoneyama, A. Miyazaki, T. Enoki and G. Saito, *Synth. Met.*, 1997, *86*, 2029.

[8] J. M. Williams, A. M. Kini, H. H. Wang, K. D. Carlson, U. Geiser, L. K. Montgomery, G. J. Pyrka, D. M. Watkins, J. M. Kommers, S. J. Boryschuk, A. V. Strieby Crouch, W. K. Kwok, J. E. Schirber, D. L. Overmyer, D. Jung and M.-H. Whangbo, *Inorg. Chem.*, 1990, *29*, 3272.

[9] A. M. Kini, U. Geiser, H. H. Wang, K. D. Carlson, J. M. Williams, W. K. Kwok, K. G. Vandervoort, J. E. Thompson, D. L. Stupka, D. Jung and M.-H. Whangbo, *Inorg. Chem.*, 1990, *29*, 2555.

[10] U. Geiser, A. J. Schultz, H. H. Wang, D. M. Watkins, D. L. Stupka, J. M. Williams, J. E. Schirber, D. L. Overmyer, D. Jung, J. J. Novoa and M. H. Whangbo, *Physica C*, 1991, *174*, 475.

[11] U. Geiser, J. A. Schlueter, H. H. Wang, A. M. Kini, J. M. Williams, P. P. Sche, H. I. Zakowicz, M. L. VanZile, J. D. Dudek, P. G. Nixon, R. W. Winter, G. L. Gard, J. Ren and M.-H. Whangbo, *J. Am. Chem. Soc.*, 1996, *118*, 9996.

[12] H. H. Wang, M. L. VanZile, J. A. Schlueter, U. Geiser, A. M. Kini, P. P. Sche, H. J. Koo, M. H. Whangbo, P. G. Nixon, R. W. Winter and G. L. Gard, *J. Phys. Chem. B*, 1999, *103*, 5493.

[13] M. Kurmoo, A. W. Graham, P. Day, S. J. Coles, M. B. Hursthouse, J. L. Caulfield, J. Singleton, F. L. Pratt, W. Hayes, L. Ducasse and P. Guionneau, *J. Am. Chem. Soc.*, 1995, *117*, 12209.

[14] T. K. Hansen, J. Becher, T. Jorgensen, K. S. Varma, R. Khedekar and M. P. Cava, *Org. Synth.*, 1995, *73*, 270.

[15] K. S. Varma, A. Bury, N. J. Harris and A. E. Underhill, *Synthesis*, 1987, 837.

[16] R. J. Willenbring, J. Mohtasham, R. Winter and G. L. Gard, *Can. J. Chem.*, 1989, *67*, 2037.

[17] G. W. Gokel, D. J. Cram, C. L. Liotta, H. P. Harris and F. L. Cook, in *Organic Synthesis*, ed. C. Johnson, John Wiley & Sons, New York, 1977, Vol. *57*, p. 30.

[18] T. J. Emge, H. H. Wang, M. A. Beno, J. M. Williams, M. H. Whangbo and M. Evain, *J. Am. Chem. Soc.*, 1986, *108*, 8215.

[19] D. A. Stephens, A. E. Rehan, S. J. Compton, R. A. Barkhau and J. M. Williams, *Inorg. Synth.*, 1986, *24*, 135.

[20] P. Guionneau, C. J. Kepert, G. Bravic, D. Chasseau, M. R. Truter, M. Kurmoo and P. Day, *Synth. Met.*, 1997, *86*, 1973–1974.

[21] H. H. Wang, J. R. Ferraro, J. M. Williams, U. Geiser and J. A. Schlueter, *J. Chem. Soc., Chem. Commun.*, 1994, 1893.

[22] J. E. Eldridge, C. C. Homes, J. M. Williams, A. M. Kini and H. H. Wang, *Spectrochim. Acta*, 1995, *51A*, 947.

[23] J. Dong, J. L. Musfeldt, J. A. Schlueter, J. M. Williams, P. G. Nixon, R. W. Winter and G. L. Gard, *Phys. Rev. B*, 1999, *60*, 4342.

[24] I. Olejniczak, B. Jones, Z. Zhu, J. Dong, J. L. Musfeldt, J. A. Schlueter, E. Morales, U. Geiser, P. G. Nixon, R. W. Winter and G. L. Gard, *Chem. Mater.*, 1999, *11*, 3160.

[25] H. J. Koo, M. H. Whangbo, J. Dong, I. Olejniczak, J. L. Musfeldt, J. A. Schlueter and U. Geiser, *Solid State Commun.*, 1999, *112*, 403.

[26] H. H. Whangbo and R. Hoffmann, *J. Am. Chem. Soc.*, 1978, *100*, 6093.

3

Advancements in EMI Shielding Strategies and Innovations with a Focus on Energy Storage Solutions

A. Purna Chandra Rao
QIS College of Engineering and Technology, Vengamukkapalem, Ongole, Andhra Pradesh, India

P. K. Dhal
Vel Tech Rangarajan Dr. Sagunthala R&D Institute of Science and Technology, Chennai, India

M. Padmarasan
Panimalar Engineering College, Chennai, India

Mukuloth Srinivasnaik
Jawaharlal Nehru University JNU, New Delhi, India

K. Rajan
Dr. MGR Educational and Research Institute, Chennai, India

3.1 Introduction

Electromagnetic interference (EMI) is a ubiquitous challenge in the realm of electronic devices, posing threats to their performance and reliability. Defined as the undesired disturbance caused by electromagnetic signals, EMI can result in malfunctions, signal degradation, and even system failures. Its pervasive impact spans a wide array of electronic applications, ranging from consumer electronics to critical components in industrial settings [1]. The importance of mitigating EMI cannot be overstated, particularly as the sophistication and prevalence of electronic devices continue to escalate. Uncontrolled electromagnetic interference can lead to compromised functionality and cross-talk between devices, undermining the seamless operation of complex systems. As electronic technologies advance, so does the need for innovative EMI shielding strategies to ensure the integrity of these systems [2]. This chapter seeks to explore the latest advancements in EMI shielding strategies, with a particular emphasis on their application in energy storage solutions. The connection between EMI shielding and energy storage is pivotal, given the integral role that both play in contemporary electronic systems. As energy storage technologies become increasingly prevalent, addressing the unique challenges of EMI shielding in this context is paramount for ensuring the efficiency and reliability of these systems [3].

The chapter unfolds with an examination of the fundamentals of EMI shielding, providing a comprehensive understanding of traditional methods and their limitations. It then navigates through the evolving landscape of EMI shielding, spotlighting emerging technologies and the impetus for innovation in modern electronic devices. Advanced materials and novel techniques take centre stage, showcasing their potential to revolutionise EMI shielding strategies. Real-world applications, case studies, and a discussion on future trends and challenges provide a well-rounded exploration of the dynamic field of EMI shielding [4].

By delving into these aspects, this chapter aims to serve as a valuable resource for researchers, engineers, and professionals navigating the intricate landscape of EMI shielding, particularly in the context of advancing energy storage solutions.

DOI: 10.1201/9781003495437-3

3.2 Overview of Electromagnetic Interference

EMI is a pervasive phenomenon that arises from the generation of electromagnetic waves by electronic devices, leading to disturbances and potential malfunctions in nearby electronic systems. These disturbances can compromise the performance of electronic devices, impacting signal integrity and causing unwanted cross-talk. EMI encompasses a broad spectrum of frequencies, ranging from radio frequencies to microwave frequencies, posing challenges across various applications [3].

3.2.1 Fundamental Principles of EMI Shielding

The fundamental principles of EMI shielding revolve around the prevention or reduction of electromagnetic radiation from electronic devices. EMI shielding aims to create barriers that absorb, reflect, or redirect electromagnetic waves, preventing their interference with neighbouring devices. The primary objective is to establish a controlled environment where electronic systems can operate without being adversely affected by external electromagnetic fields.

3.2.2 Traditional EMI Shielding Methods and Their Limitations

a) Metal Enclosures and Faraday Cages
    Metal enclosures, commonly known as Faraday cages, have been a traditional method for EMI shielding. These enclosures work by creating a conductive barrier that prevents the penetration of electromagnetic waves. However, they may be impractical in certain applications due to weight, size, and cost constraints [2].
b) Shielded Cables and Connectors
    Shielded cables and connectors are employed to minimise electromagnetic radiation from signal-carrying conductors. While effective in reducing EMI, their application is limited, and they may not address higher-frequency interference adequately [5].
c) Ferrite Chokes and Beads
    Ferrite chokes and beads are passive components that absorb high-frequency noise in electronic circuits. While useful for mitigating specific frequency ranges, their effectiveness diminishes with lower frequencies [6].
d) Conductive Coatings and Paints
    Applying conductive coatings or paints to electronic components or enclosures helps create a shield against EMI. However, their efficacy may be compromised by wear and tear, and they may not be suitable for all environments [7].

3.2.3 Limitations and Challenges

Traditional EMI shielding methods have inherent limitations. They may not adequately address the increasing complexity of electronic devices, especially those operating at higher frequencies. Additionally, the weight, size, and cost associated with traditional shielding materials may become prohibitive in certain applications. As electronic technologies advance, there is a growing need for more sophisticated and adaptable EMI shielding solutions to meet the evolving demands of modern electronics [1].

3.3 Advanced Materials for EMI Shielding

a) Conductive Polymers
    Conductive polymers have emerged as promising materials for EMI shielding due to their unique combination of electrical conductivity, flexibility, and lightweight properties. These polymers, doped with conductive additives, offer tuneable conductivity levels and can be applied as coatings

or integrated into composite materials. Polyaniline, polypyrrole, and polythiophene are examples of conductive polymers demonstrating effectiveness in absorbing and attenuating electromagnetic waves across a wide frequency range [8].

b) Graphene-Based Materials

Graphene, a single layer of carbon atoms arranged in a hexagonal lattice, and its derivatives have gained significant attention for EMI shielding applications. Due to their exceptional electrical conductivity, mechanical strength, and high surface area, graphene-based materials can effectively absorb and reflect electromagnetic waves. Graphene sheets, graphene oxide, and reduced graphene oxide can be integrated into composites, coatings, or even standalone films to enhance EMI shielding performance [9].

c) Nanocomposites

Nanocomposites, composed of nanoscale particles dispersed in a matrix, offer a multifaceted approach to EMI shielding. Incorporating conductive nanoparticles, such as metal nanoparticles or carbon nanotubes, into polymer matrices enhances electrical conductivity and EMI shielding effectiveness. The synergistic combination of materials at the nanoscale provides improved mechanical strength, flexibility, and thermal stability, making nanocomposites highly suitable for diverse applications [10].

3.4 EMI Shielding in Energy Storage Solutions

3.4.1 Overview of Energy Storage Technologies

Energy storage technologies play a pivotal role in modernising power systems, facilitating the integration of renewable energy sources, and enhancing grid reliability. Various technologies, including batteries, supercapacitors, and flywheel systems, contribute to storing and releasing energy efficiently. As these technologies become integral to diverse applications, ensuring their reliable operation becomes paramount.

3.4.2 Importance of EMI Shielding in Energy Storage Systems

EMI shielding holds heightened significance in the realm of energy storage systems due to the vulnerability of electronic components within these systems to electromagnetic interference. Unmitigated EMI can lead to operational disruptions, compromised battery performance, and safety concerns. As energy storage solutions become increasingly embedded in critical infrastructure and applications, safeguarding these systems from EMI-induced issues becomes imperative for ensuring reliability and longevity [11].

3.4.3 Challenges Specific to EMI Shielding in Energy Storage

Several challenges unique to EMI shielding in energy storage systems must be addressed. The high energy density and rapid switching frequencies inherent in certain storage technologies, such as lithium-ion batteries, pose challenges in designing effective EMI shielding solutions. Additionally, the need for lightweight and compact energy storage systems requires innovative shielding materials and strategies that do not compromise the overall efficiency and performance of the storage system [12].

3.5 Innovations and Case Studies

3.5.1 Recent Breakthroughs in EMI Shielding Innovations

In the dynamic landscape of EMI shielding, recent breakthroughs have introduced innovative materials and techniques, pushing the boundaries of traditional approaches. One notable innovation involves the utilisation of metamaterials – engineered structures with unique electromagnetic properties. Metamaterials exhibit unprecedented capabilities in manipulating and controlling electromagnetic waves, offering new possibilities for highly efficient EMI shielding.

1. Graphene-Based EMI Shielding in Consumer Electronics
   A multi-layered approach was employed where graphene sheets were strategically integrated into the structural components of electronic devices, forming a conductive network. The graphene layers were applied as coatings on critical components, creating a flexible and conductive barrier. The integration of graphene layers showcased a significant reduction in electromagnetic interference across a wide frequency spectrum. Measurements revealed an increase in structural integrity alongside enhanced EMI shielding effectiveness. The flexible and lightweight nature of graphene made it an ideal candidate for seamless integration into various consumer electronic applications, showcasing its adaptability and potential for wide-scale implementation [13].

2. Aerospace Application of Conductive Polymers for Satellite EMI Shielding
   Conductive polymers, specifically tailored for space environments, were applied as coatings to the outer shell of satellites. The conductive polymer coating formed a protective layer, creating an effective shield against electromagnetic interference. The polymers were engineered to maintain their conductivity in the harsh conditions of space.

   The integration of conductive polymers demonstrated a substantial reduction in electromagnetic interference during satellite communication. Measurements revealed enhanced signal integrity and communication reliability. The lightweight and space-adaptable nature of the conductive polymers made them a viable solution for aerospace applications where weight and performance are critical considerations [14].

3.6 Conclusions

In the realm of electromagnetic interference (EMI) shielding, recent breakthroughs, including the use of metamaterials, conductive polymers, and graphene-based materials, mark a significant shift in the approach to mitigating electromagnetic disturbances. These advancements offer more efficient and adaptable solutions for various applications, transcending traditional methods.

Metamaterials provide unprecedented control over electromagnetic waves, while conductive polymers and graphene-based materials exhibit exceptional effectiveness in diverse applications. These innovations signify the dynamic evolution of EMI shielding strategies.

Celebrating current advancements comes with a call for sustained research and development. With technology rapidly advancing, collaboration among researchers, engineers, and industry stakeholders is essential to explore novel materials and methodologies, ensuring EMI shielding keeps pace with emerging technologies.

The future of EMI shielding in energy storage solutions holds immense promise. As energy storage technologies become integral to daily life, tailored solutions addressing challenges specific to these systems are crucial. The integration of advanced materials and design approaches will shape the trajectory of EMI shielding in this evolving domain.

In conclusion, these advancements in EMI shielding unveil a landscape of possibilities, emphasising innovation as the key to unlocking new potentials. With a call for sustained exploration, EMI shielding is poised to contribute significantly to the seamless integration of electronic devices and energy storage solutions in our rapidly evolving technological landscape.

REFERENCES

[1] Clayton R. Paul, *"Introduction to Electromagnetic Compatibility,"* Wiley, 2006.

[2] Henry W. Ott, *"Electromagnetic Compatibility Engineering,"* Wiley, 2009.

[3] K. H. Ang and T. S. Lim, "EMI Issues in Power Electronics," *IEEE Transactions on Industrial Electronics,* 2006.

[4] R. Zhang, Y. He, and S. S. Venkatesh, "Graphene-Based Materials for Electromagnetic Interference Shielding," *2D Materials*, 2017.

[5] J. C. Balda, "Shielding: An Essential Component of EMC," *IEEE Transactions on Electromagnetic Compatibility*, 2005.

[6] W. M. McLyman, *"Ferrites for Inductors and Transformers: Theory and Applications,"* Wiley, 1998.

[7] Y. Zhang, et al., "Conductive Paints and Coatings: A Review," *Materials & Design*, 2019.

[8] D. J. Li, *"Conductive Polymer Composites: Principles and Applications,"* Wiley, 2020.

[9] Z. Wu, et al., "Graphene-Based Transparent Conductive Films for Flexible Electronics," *Advanced Materials*, 2014.

[10] A. B. Kunwar, et al., "Electromagnetic Interference Shielding Behavior of Polymer Matrix Nanocomposites: A Comprehensive Review," *Journal of Materials Science*, 2021.

[11] D. C. Sinclair, et al., "Electromagnetic Interference Shielding Effectiveness of Carbon Fiber Reinforced Polymer Composites," *Journal of Applied Physics*, 2007.

[12] M. N. Islam, et al., "Advanced Materials and Processing for Lithium-Ion Batteries," *Advanced Materials*, 2013.

[13] X. Li, et al., "Graphene on Metal Mesh as Flexible Transparent Conductive Electrodes for Highly Efficient EMI Shielding," *Small*, 2018.

[14] S. Yoon, et al., "Aerospace Applications of Conductive Polymers: A Review," *Progress in Polymer Science*, 2014.

4

Performance Evaluation and Statistical Analysis of Green Materials for CO_2 Reduction Using Fly Ash Mixture

P. Manikandan
Federal TVT Institute, Ethiopia, Africa

E. Veeramanipriya
Vivekanandha College of Engineering for Women (Autonomous), Thiruchengode, India

4.1 Introduction

In modern times, building infrastructure requires concrete, and the global concrete sector has an enormous impact on national economies and development. Concrete is the world's most consumed man-made material, coming in second only to water in terms of consumption. Due to its exceptional performance and versatility as well as the accessibility of raw materials locally, concrete is highly utilised in a variety of construction applications, which might be linked to its high consumption. When compared to other building materials like steel and glass, concrete is a more environmentally responsible choice. However, a larger amount of its constituents is manufactured and used for a variety of construction and restoration purposes; the use of concrete as a building material in large numbers exceeds its sustainability benefit.

Due to the extensive use of concrete, the concrete industry is one of the largest users of natural resources and a major contributor to worldwide anthropogenic carbon dioxide emissions. The primary source of carbon dioxide emissions from concrete is the manufacturing of Portland cement, which serves as the primary binder, as well as material transportation. Roughly 80% of carbon dioxide is released into the atmosphere for each mass of Portland cement generated during the production process.

Concrete production results in high carbon dioxide emissions as well as excessive material use, which strains these deposits and deforms the surrounding ecosystem. The total quantity of raw materials needed to make one tonne of Portland cement is about doubled throughout the production process alone. Furthermore, the largest user of freshwater is the concrete industry.

The large amount of garbage produced during the building and dismantling of concrete structures also greatly increases environmental damage. Growing urbanisation and population growth, particularly in emerging nations, are forecast in the upcoming years, which will lead to increased demand for concrete and a rise in emissions and raw material consumption. Many sectors and nations have launched various programs to cut the greenhouse gas emissions as a result of the growing global awareness of sustainability. In order to maintain the sustainability and preservation of our world, the concrete industry has recently implemented various initiatives.

The primary amounts of waste created by the commercial, industrial, and residential sectors are rising tremendously. This problem is getting worse because of the rising population and style of living. Around 1.3 billion tonnes of waste is produced worldwide each year, and by 2025, that number is predicted to rise to 4.3 billion tonnes [1]. The world is producing three times more electronic garbage (e-waste) than any other type of waste, and this growth is occurring at an extremely rapid pace [2, 3]. India has produced more than 8 lakh metric tons of e-waste annually during the last 20 years. In 2010, there were around

DOI: 10.1201/9781003495437-4

400 million tonnes of e-waste generated on average [4]. The Central Pollution Control Board (CPCB) calculated that 1.47 lakh tonnes, or 0.573 MT of e-waste is produced daily in India [5]. The handling of e-waste is a significant issue in developing nations, especially India.

Because of its adaptability, the utilisation of e-trash is always increasing. The environmental system is threatened by the large amount of e-garbage that is often disposed of in landfills or dumped into the ocean. These are the modes of transportation used by plastic in daily life, which over time has a contaminating effect. Moreover, e-waste produces a significant number of harmful emissions, including carbon [6]. For this reason, the use of e-wastes as an alternative is investigated in a number of industries, most notably the building sector.

The primary potential application of incorporating and utilising e-wastes is as a replacement for aggregate in concrete. In 2015, the aggregate consumption amounted to around 48.3 billion metric tonnes [7]. A limited number of studies have examined the performance of concrete with substituted material and the replacement of both coarse and fine aggregates with e-waste [8]. The building industry uses a variety of wastes to enhance the mechanical strength and durability of regular materials that have particular properties. Fly ash waste products, silica fume, and sludge are used to improve the properties of polymer concrete [9].

Laboratory studies were carried out by Magureanu et al. to assess the structural characteristics of low-grade aggregates treated with cement. The study found that combination of low-grade aggregates such as gravel and laterite in right amounts with Portland slag cement and sand can be utilised as a road basis or subbase [10].

Subramanian combined a three-pronged strategy to waste management: Plastic reduction, recycling, and reuse. He examined how much plastic is consumed annually based on the developing trends of the USA and India. The potential for a thorough examination of the ecological, economic, and technical elements of recycling is also discussed [11].

Four recycled concrete aggregates, whose compressive strengths ranged from 15 to 75 MPa, were crushed to obtain the aggregates for Nataatmadja's investigation [12]. The performance of the recycled concrete aggregate is equivalent to that of the new base coarse aggregates, according to the test results. Additionally, it is said that the 10% Fines test is a good evaluation tool for determining the parent material's compressive strength.

Shiou created the standards for the use of recycled concrete aggregate in pavement base course after evaluating the material's technical and physical characteristics. The findings indicate that when recycled concrete aggregate is used as the foundation material in flexible pavements, the abrasion loss should be less than 48%. Furthermore, recycled concrete material should not be subjected to sodium sulphate test [13].

Katz's investigation of the characteristics of recycled aggregates made from crushed concrete revealed that these aggregates' qualities were remarkably similar, even when they were crushed at different ages. Additionally, the size distribution of the recycled aggregates and their bulk specific gravity, bulk density, cement content, and crushing value are shown [14]. Using only recycled aggregates in place of all natural aggregates can result in a 16% reduction in concrete strength as well as lower performances in durability-related properties, according to Arlindo et al.'s analysis of the impact of recycled concrete aggregates on durability-related concrete properties. This is primarily because recycled aggregates have a higher porosity than natural aggregates [15].

Urban population growth would significantly demand the infrastructure currently in place for housing, transportation, recreation, health care, education, and other services. Given the current indicators of rapidly increasing urbanisation and population growth [9], it is anticipated that concrete production and consumption will rise significantly in order to meet the demands and requirements of infrastructure. It is estimated that by 2050, there will be a demand for more than 18 billion tonnes of concrete annually [16]. Therefore, the concrete industry must contribute significantly to ensuring that the planet is preserved by incorporating a number of sustainable efforts into all aspects related to the production and use of concrete. The performance assessment and statistical analysis of fly ash-mixed recycled polymer e-waste concrete are presented in this work. It provides an overview of the most recent developments made by the concrete sector to reduce carbon dioxide emissions.

4.2 Methodology

4.2.1 Materials

Typically, Portland cement of grade 43 is used for this kind of project. The IS 4031:1988 code is used to examine the various physical qualities of cement. The casting and curing processes require the use of clean, potable water with a pH of 7.5–8.0. ACC, Ltd. of Coimbatore, Tamil Nadu, procures the specimens from the Mettur Thermal Power Plant (MTTP) using fly ash, which has a specific gravity of 2.37. With a specific gravity of 2.76 and a grading of zone II according to IS 383-1970, well-graded river sand is progressively employed as fine aggregate. In addition, locally accessible, crushed and angular coarse material of 2.80 specific gravity is used. According to IS 383-197, tests are conducted on coarse aggregates' physical characteristics. Scrap material from electrical and electronic devices is recycled from their outer frames, and polymer e-waste is collected and crushed with a crushing machine to a size of between 10 and 12.5 mm. Crushed polymer e-wastes are cleaned with potable water and then exposed to the sun to dry.

4.2.2 Mix Proportions

Using design progress and IS10262:19882, the mix proportion for concrete in the M25 grade is estimated. Predicted mix percentage is 1:1.56:2.56, while the water to cement ratio is 0.45. Coarse aggregate made from leftover polymer e-waste is used to make concrete. To replace coarse aggregate with polymer e-waste, the following amounts are calculated based on the volume and concrete with a water–cement ratio of 0.45. Concrete contains 0%, 5%, 10%, 15%, 18%, and 20% of e-wastes, in that order.

4.2.3 Casting and Curing

The prepared polymer e-waste mix's slump ratio is assessed. All of the specimens are cast and cured using regular groundwater. Curing and casting of the specimens (cubes, cylinders, prisms, beams, T beams, and RC beams) take place at room temperature (about 27°C ± 2°C). Until the prescribed curing time is reached, the cast concrete specimens are remoulded and submerged in chemicals and clean water.

4.2.4 Experimental Test

4.2.4.1 Durability Tests on Concrete

The ability of cement concrete to withstand deterioration processes such as weathering, chemical attack, abrasion, and other processes is sometimes referred to as its durability. Durable concrete in the test will hold its original quality, shape, and suitability being exposed to the environment. The durability testing procedures outlined in ASTM C 666 have been used to evaluate the resilience of concrete specimens to aggressive situations like acid attack.

4.2.4.2 Chloride Attack Test

The primary factor affecting concrete's durability is chloride attack, since 40% of initial corrosion in reinforced concrete structures is caused by chloride attacks. The main objective of the chloride attack is strength loss via corrosion. Cement can have up to 0.05% of its total content of chloride, according to the Bureau of Indian Standards. The concrete samples that were cast with polymer e-waste in this investigation were kept at room temperature until they were demoulded, at which point they were placed in clean water to undergo their first curing. The polymer e-waste concrete samples were soaked in a 5% sulphuric acid (H_2SO_4) and hydrochloric acid (HCl) solution after 28 days. The durability factor is calculated after the chemical cure times of 28 and 60 days.

4.2.4.3 Sulphate Attack Test

Sulphate attack causes the hydration of cement and solution to react chemically, increasing the amount of cement paste in the polymer concrete. When the firm concrete is exposed to external sulphate salt, the calcium aluminate hydrate (C–A–H) can react. Sulphate attack may cause the concrete to gradually disintegrate, increasing the volume by up to 227%. The casted e-waste concrete samples in this investigation are soaked in a 10% concentrated sodium sulphate solution following a typical 28-day curing period.

4.2.4.4 Sulphuric Acid Test

Throughout a maximum of 60 days across the curing process, the cast concrete samples are submerged in a 3% concentrated solution of sulphuric acid. After 60 days of curing, the acid content (pH level 3) is checked regularly and kept constant. Because sulphate ions are present, the sulphuric attack is especially corrosive and heavily dependent on the high alkalinity of cement. The weights of the e-waste specimens are recorded after 60 days of acid curing and 28 days of normal curing.

4.3 Results and Discussion

4.3.1 Compressive Strength

A compression testing instrument with a capacity of 2000kN has been utilised to measure compressive strengths. Every test value is calculated as the mean of three samples. Table 4.1 lists the final compressive load and compressive strength of the conventional and e-waste concrete cubes for the 14- and 28-day curing periods. The pre-testing configurations are displayed in Figure 4.1(a) and (b).

E-waste concrete cubes are found to gradually improve in strength at 10% weight replacement with coarse aggregate in concrete during compressive strength testing, with e-waste employed as a coarse aggregate in the cubes ranging from 0% to 25%. Concrete's compressive strength steadily declined once over 10% of electronic waste is added. In comparison to normal concrete, 10% of e-waste in concrete resulted in an average percentage of compressive strength that is higher, as illustrated in Figure 4.2.

As e-waste is substituted into concrete, the density of the material gradually decreases. In comparison to coarse aggregate, e-waste has a lower specific gravity. When substituting e-waste, the volume of e-waste must be greater than the volume of coarse aggregate, and some concrete mix must be left over after casting the specimens. This implies that the quantity of cement, aggregates, and e-wastes used in the casting of e-waste concrete specimens cannot be squandered. Because of this, concrete specimens made of e-waste experience a synchronous weight loss when the garbage is replaced. Table 4.2 shows the density of the polymer e-waste concrete cubes for various weights.

The e-waste concrete was used in the compressive strength test at 0–20% replacement ratio. The highest compressive strength was achieved with an e-waste replacement range of 10%. When the proportion

TABLE 4.1

Compressive Strength Results for Concrete Cubes

	Load(kN)			Compressive Strength (N/mm²)		
% of e-Waste	14 Days	28 Days	60 Days	14 Days	28 Days	60 Days
0	612	679	685	27.20	30.17	30.44
5	639	683	692	28.40	30.35	30.76
10	683	781	788	30.36	35.02	35.02
15	661	778	782	29.38	34.71	34.76
18	655	699	706	29.11	31.06	31.38
20	585	660	665	26.00	29.33	29.56

　　　　　　　　　　　　　　　　　　　　Introduction to Functional Nanomaterials

(a)　　　　　　　　　　　　　　　　　　　　(b)

FIGURE 4.1　(a) Compressive strength testing of concrete specimen, (b) concrete cube specimen after testing.

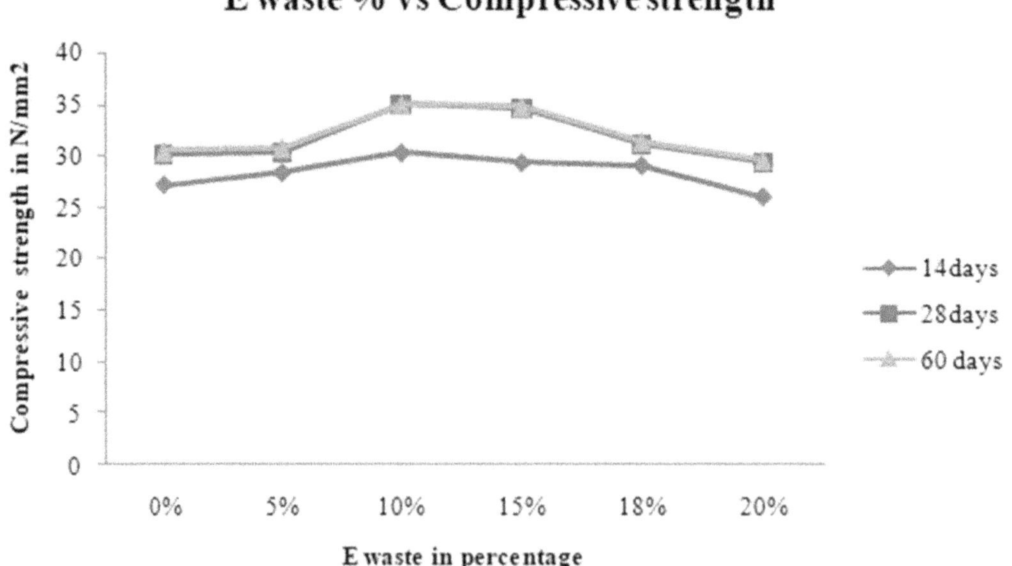

FIGURE 4.2　Compressive strength for cubes cured with water.

TABLE 4.2

Density of e-Waste Concrete Cubes

e-Waste (%)	Wet Weight (kg)	Dry Weight (kg)	Loss of Weight (%)
0	8.72	8.70	0.22
5	8.65	8.38	3.12
10	8.37	8.28	1.07
15	8.27	8.26	0.12
18	7.97	7.97	0
20	7.80	7.78	0.25

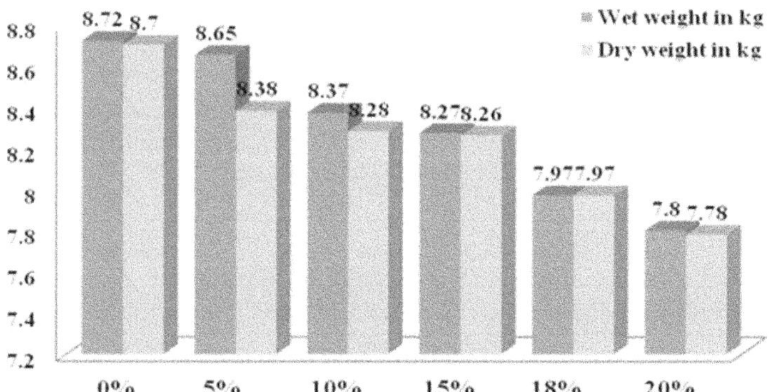

FIGURE 4.3 Density comparison for cubes.

of e-waste replaced by concrete surpassed 10%, the concrete's compressive strength quickly decreased. Waste concrete has a lower density because e-waste has a lower density than conventional concrete. In this instance, 10% more e-waste was added to the concrete. Because e-waste has a higher crushing strength than other materials, concrete should have a high compressive strength when adding it to the mix.

Concrete becomes more porous and becomes low density when electronic wastes are added in excess of 10%. Excess e-waste interferes with the concrete's hydration process and makes it less workable. Figure 4.3 shows the density comparison of cubes for different mixing proportions. The limited workability of e-waste concrete at maximal replacement causes the concrete to become porous on the inside, which also reduces the concrete's strength. Adding fly ash to e-waste concrete should make it more workable since it contains silica, which can react with calcium hydroxide to form C–S–H gel. It is found to improve the pore structure and raise the density of the e-waste concrete matrix [17].

4.3.2 Durability Tests

Chloride attack's effects on the polymer e-waste concrete mix and the controlled concrete mix with a 20% e-waste addition are noted. Figure 4.4 plots the results, which are shown in Table 4.3. After 60 days of curing, the average weight decrease for e-waste concrete is noted. The findings indicate that chloride has no effect on the e-waste in concrete.

The results of density variations for e-waste concrete cubes on 28th and 60th days are shown in Figure 4.5 and Table 4.4. In order to conduct chloride attack tests, samples are cured in H_2SO_4 and Na_2SO_4. According to the testing, e-waste concrete loses 3.6% of its compressive strength and 2.5% of its weight when compared to controlled concrete. It is established that adding e-trash waste to concrete during chemical curing could lead to a performance that is exceptional. When there is significant amount of e-waste in concrete, the outer layer of the concrete breaks down when acid strikes it. Additionally, when e-waste is added to concrete in amounts above 15%, pores appear on the surface of the concrete; these pores produce hollow spaces that lead to cracking.

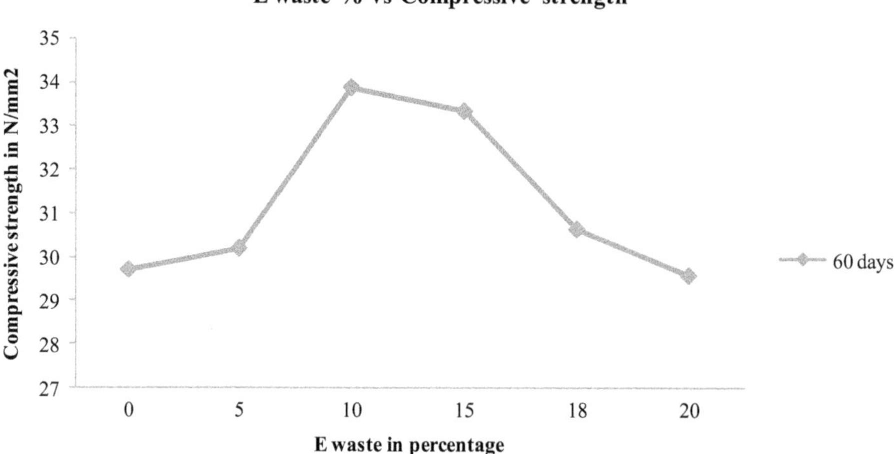

FIGURE 4.4   Compressive strength of cubes under 60 days of chloride curing.

TABLE 4.3

Compressive Strength of Cubes under 60 Days of Chloride Curing

e-Waste (5)	Load (kN)	Compressive Strength (N/mm²)	Loss of Strength (%)
0	668	29.69	2.4
5	679	30.18	1.8
10	762	33.87	3.2
15	750	33.33	4.1
18	689	30.62	2.4
20	665	29.56	0

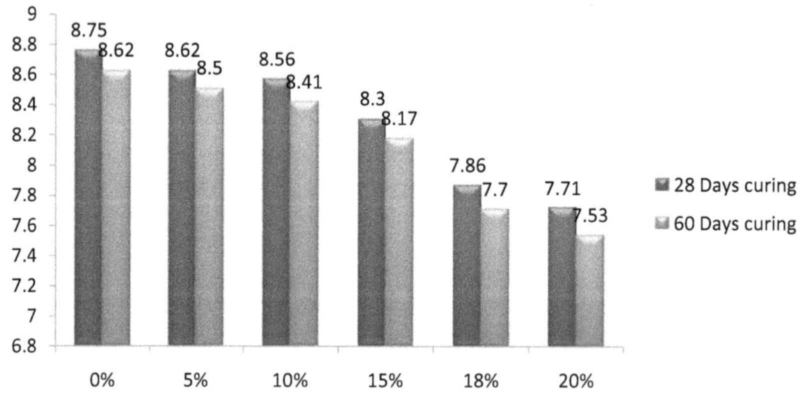

FIGURE 4.5   Density comparison for cubes under chloride curing.

The results of the investigation into the sulphate attack on controlled concrete and e-waste concrete are summarised, respectively, in Tables 4.5 and 4.6. The same results are viewed, respectively, in Figures 4.6 and 4.7. After 60 days of curing, the maximum weight loss of e-waste concrete is 15% replacement of course aggregate. Comparing the concrete with e-waste to controlled concrete, the results demonstrated the superior resistance of the former to H_2SO_4 solution.

TABLE 4.4

Density of Concrete under Chloride Curing

e-Waste (%)	Wt. at 28 Days Curing after Dry (kg)	Wt. at 60 Days Curing after Dry (kg)	Loss of Weight (%)
0	8.75	8.62	1.48
5	8.62	8.50	1.39
10	8.56	8.41	1.75
15	8.30	8.17	1.56
18	7.86	7.70	2.03
20	7.71	7.53	2.33

TABLE 4.5

Compressive Strength for Cubes (Na_2SO_4 for 60 Days of Curing)

e-Waste (%)	Load (kN)	Compressive Strength (N/mm²)	Loss of Strength (%)
0	665	29.56	2.8
5	686	30.49	0.8
10	775	34.44	1.6
15	749	33.29	4.2
18	702	31.20	0.5
20	661	29.38	0.6

TABLE 4.6

Density of Concrete Cured under Na_2SO_4 Solution

e-Waste (%)	28 Days Curing after Dry (kg)	60 Days Curing after Dry (kg)	Loss of Weight (%)
0	8.68	8.53	1.72
5	8.63	8.44	2.20
10	8.58	8.36	2.56
15	8.34	8.17	2.03
18	7.86	7.72	1.78
20	7.74	7.53	2.71

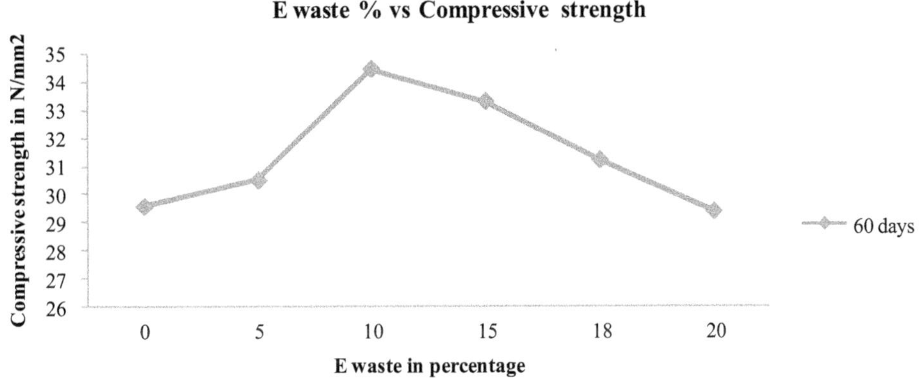

FIGURE 4.6 Compressive strength for cubes under Na_2SO_4 for 60 days of curing.

In different percentages, the test specimens are cast with and without e-waste. The e-waste concrete cubes are cured in clean, drinkable water for the first 28 days. The surfaces of the concrete cubes are cleaned and weighed after 28 days of water cure. Next, for a duration of 60 days, the e-waste concrete specimens are submerged in a 3% concentrated solution of sulphuric acid (H_2SO_4). Throughout the 60-day curing period, the solution's concentration is monitored and maintained on a regular basis. Weighing the e-waste specimens after 60 days of chemical curing allows us to calculate the weight loss of the e-waste concrete. The test results are presented in Figures 4.8–4.10 and recorded in Tables 4.7 and 4.8.

The results of compressive strength for e-waste concrete cubes at 14 and 28 days are shown in Figure 4.11.

Table 4.9 shows the saturated water absorption test results in controlled concrete and e-waste concrete. From saturation water absorption (SWA) test results, 20% replacement of e-waste in concrete produces highest observations of water. The increase in e-waste is directionally proportional to water observation of concrete.

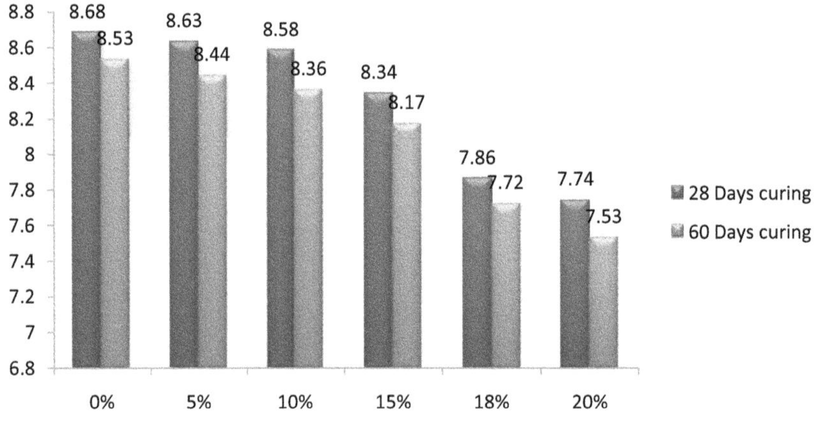

FIGURE 4.7 Density comparison for e-waste concrete cubes cured under Na_2SO_4.

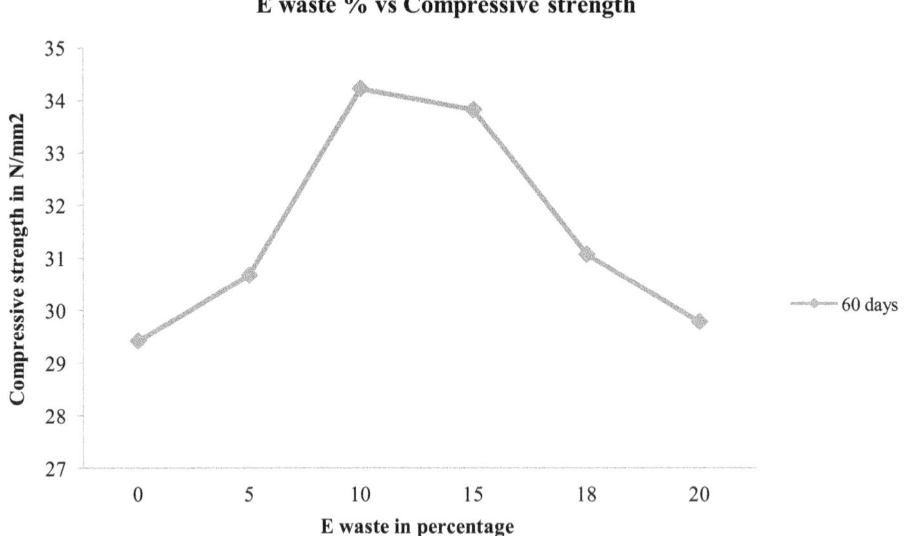

FIGURE 4.8 Compressive strength of cubes cured under H_2SO_4 solution for 60 days.

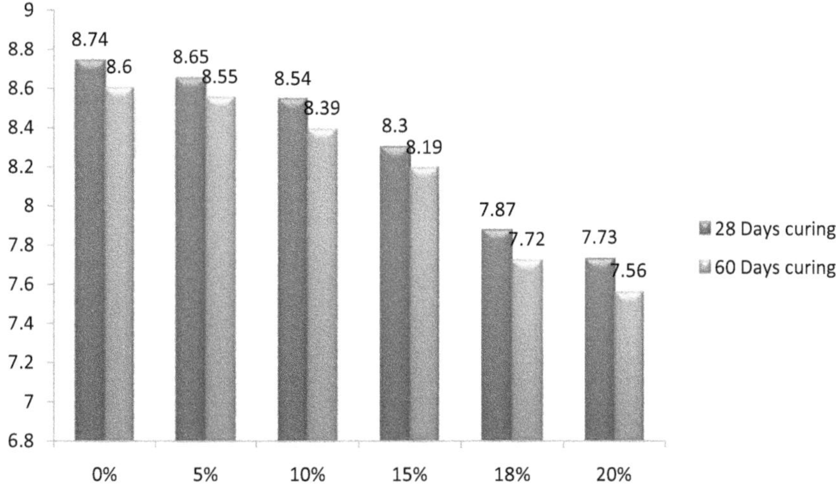

FIGURE 4.9　Self-weight comparison of cubes cured under H_2SO_4 solution.

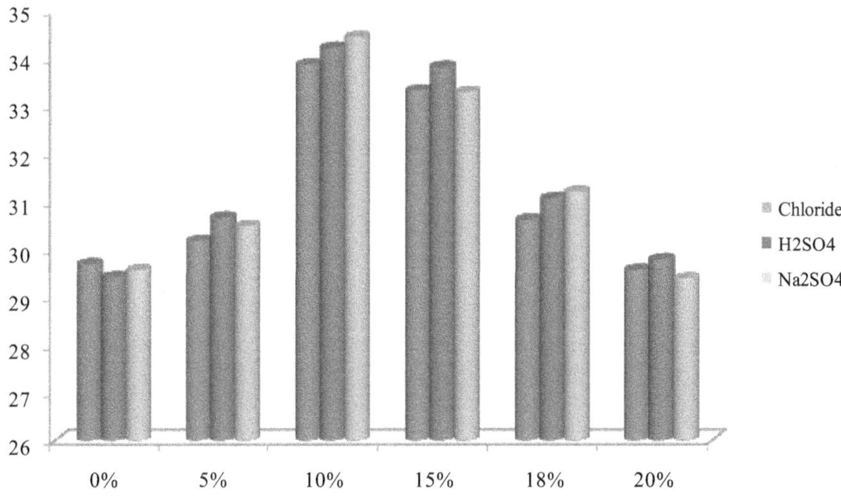

FIGURE 4.10　Compressive strength comparison chart.

TABLE 4.7

Compressive Strength of Cubes Cured under H_2SO_4 Solution for 60 Days

e-Waste (%)	Load (kN)	Compressive Strength (N/mm²)	Loss of Strength (%)
0	662	29.42	3.3
5	690	30.67	2.9
10	770	34.22	2.2
15	761	33.82	2.7
18	699	31.07	0.9
20	660	29.33	0.7

TABLE 4.8

Density of Concrete Cured under H_2SO_4

e-Waste (%)	28 Days Curing after Dry (kg)	60 Days Curing after Dry (kg)	Loss of Weight (%)
0	8.74	8.60	1.60
5	8.65	8.55	1.15
10	8.54	8.39	1.75
15	8.30	8.19	1.32
18	7.87	7.72	1.90
20	7.73	7.56	2.20

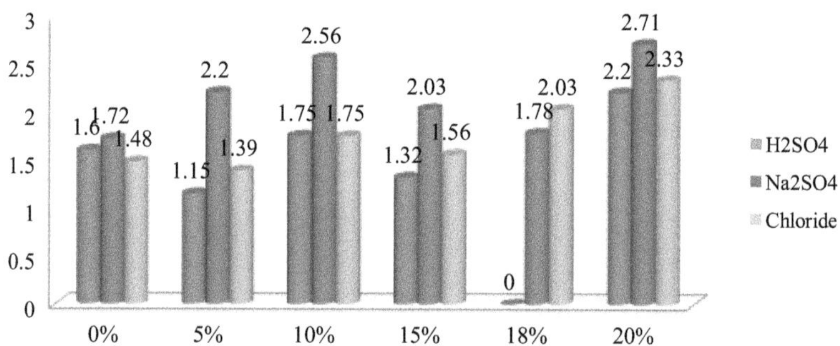

FIGURE 4.11   Density comparison for cubes under various curing condition.

TABLE 4.9

Results of SWA Test on e-Waste Concrete

e-Waste (%)	SWA (%)
0	1.24
5	1.98
10	2.46
15	2.82
18	3.23
20	3.67

4.3.3 CO_2 Emission Analysis

Table 4.10 provides a summary of the carbon footprint assessment for each concrete mix, represented in kg CO_2-e/m^3. This analysis covers the total cement, aggregates, and other non-significant emissions such water uses and explosive agents.

There are four stages in which carbon dioxide emissions might occur: Manufacture, building, use, and disposal. The transportation of recycled polymer e-waste concrete and its carbonisation, which can lower carbon emissions, are the two primary influencing variables in the first two phases of recycled polymer e-waste concrete. The study examined the carbon dioxide emissions throughout the production process of recycled polymer e-waste concrete. The key to the total CO_2 emissions during the production phase is the mix fraction of recycled concrete and the raw material transportation. This model's calculation boundary is to determine the CO_2 emissions per unit of recycled polymer e-waste concrete product produced during the process, beginning with the raw materials manufacture and concluding with the finished product.

TABLE 4.10

CO$_2$ Emissions from Recycled Polymer e-Waste Concrete

Name	Emission Factor (kg/t)	Carbon Emissions from the Materials (kg)	Transport Energy Consumption (L)	Carbon Emission Factor (kg/L)
Cement	855	246.58	0.4	2.03
Sand	2.95	1.25	1.2	2.04
Polymer e-waste mixing	1.85	1.80	1.3	2.05
Water	0.92	0.26	0.0	0.0
Fly ash mixing	0.01	0.00	0.6	2.04

TABLE 4.11

Error Analysis of Compressive Strength of Cubes under 60 Days of Chloride Curing

e-Waste (%)	Compressive Strength (N/mm²)	Predicted Value	Error	PE (%)
0	29.69	30.8443	−1.1543	0.7909862
5	30.18	31.0049	−0.8249	
10	33.87	31.16551	2.70449	
15	33.33	31.32611	2.00389	
18	30.62	31.42247	−0.80247	
20	29.56	31.48671	−1.92671	

TABLE 4.12

ANOVA for Chloride Curing

	Sum of Squares	df	Mean Square	F	Sig.
Regression	.005	1	.005	.071	.803
Residual	.308	4	.077		
Total	.313	5			

4.3.4 Statistical Analysis

The IBM SPSS 23 statistical application is used to carry out the statistical analysis which is given in Table 4.11. Using linear regression analysis and root mean square error (RMSE), the anticipated and experimental values of the compressive strength of cubes after chloride curing are compared. The chi-square results are displayed in ANOVA Table 4.12. It is discovered that the RMSE and chi square values are, respectively, 0.308 and 0.077. Throughout the experiment, the absolute difference between the predicted and experimental values is evaluated to determine the relative percentage of error (PE). When the relative percentage error is less than 10%, the outcome is good. The best outcome of polymer e-waste concrete is displayed in Table 4.11 with a computed PE of 0.7909862%.

The observed and predicted values for the compressive strength of cubes after 60 days of chloride curing are displayed in Figure 4.12. Using RMSE and chi square, the predicted and experimental values of compressive strength of cubes cured with Na$_2$SO$_4$ are compared using ANOVA (Table 4.13). It is discovered that the RMSE and chi square values are, respectively, 0.368 and 0.092. The best outcome of polymer e-waste concrete is displayed in Table 4.14 with a computed PE of 0.851122%. The observed and projected values for the compressive strength of cubes after 60 days of Na$_2$SO$_4$ curing are displayed in Figure 4.13.

The best result of polymer e-waste concrete is displayed in Table 4.15 with a computed PE of 0.89409%. After 60 days of H$_2$SO$_4$ curing, Figure 4.14 displays the experimental versus expected value for the

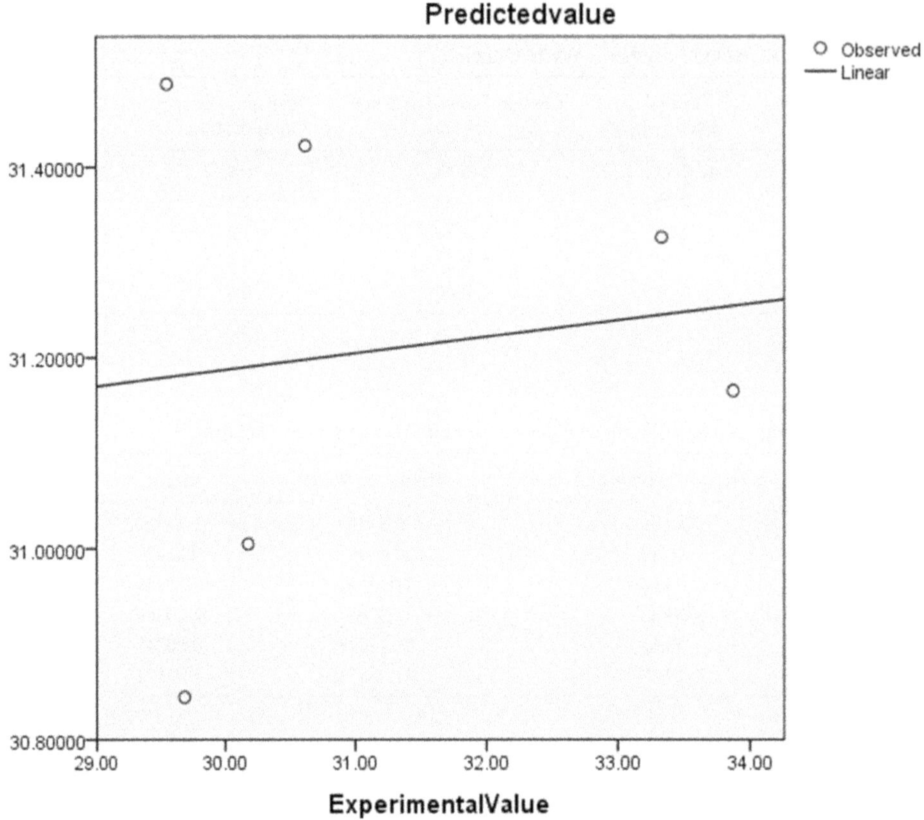

FIGURE 4.12   Experimental versus predicted value for compressive strength of cubes under 60 days of chloride curing.

TABLE 4.13

ANOVA for Na_2SO_4 Curing

	Sum of Squares	df	Mean Square	F	Sig.
Regression	.007	1	.007	.072	.802
Residual	.368	4	.092		
Total	.374	5			

TABLE 4.14

Error Analysis of Compressive Strength for Cubes under 60 Days of Na_2SO_4 Curing

e-Waste (%)	Compressive Strength (N/mm²)	Predicted Value	Error	PE (%)
0	29.56	30.9953	−1.4353	0.851122
5	30.49	31.1709	−0.6809	
10	34.44	31.34651	3.09349	
15	33.29	31.52211	1.76789	
18	31.2	31.62747	−0.42747	
20	29.38	31.69771	−2.31771	

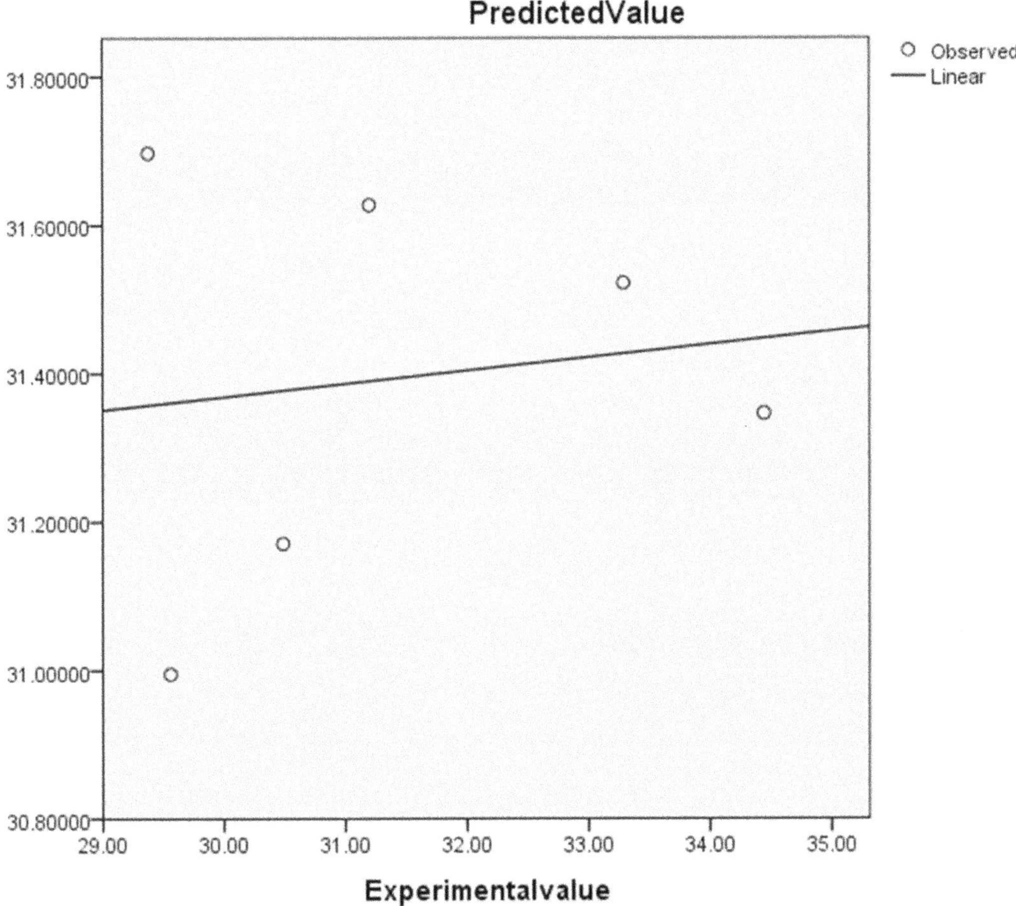

FIGURE 4.13 Experimental versus predicted value of compressive strength for cubes under 60 days of Na_2SO_4 curing.

compressive strength of cubes. Using RMSE and chi square, the predicted and experimental values of the compressive strength of cubes cured in H_2SO_4 are compared using ANOVA (Table 4.16). It is discovered that the RMSE and chi square values are, respectively, 0.468 and 0.117.

4.4 Conclusion

In the next decades, concrete will remain a highly preferred building material; for this reason, it is critical to both meet current and future demands for concrete and doing it in an environmentally responsible manner. The primary source of carbon dioxide emissions from the concrete industry is Portland cement. The industry will therefore significantly reduce its carbon dioxide emissions if Portland cement, which is used to manufacture concrete mixtures, is replaced, either completely or in part. The high cost and carbon emissions connected with the transportation of raw materials and concrete components from one place to another can be effectively mitigated by using locally available resources as components in concrete.

The current study highlights the utilisation of fly ash and regenerated polymer components from e-waste as coarse aggregate in concrete. It explored how to incorporate electronic wastes in better grade concrete and showed careful use of chemical combinations and minerals can result in polymer e-waste concrete with the required high strength. It can therefore meet the condition durability requirements. According to the results of the chloride attack tests, e-scrap concrete lost 3.6% of its compressive strength and 2.5% of its weight when compared to controlled concrete. The maximum weight loss of e-waste concrete after 60 days of curing, as determined by the sulphate attack test, is 15% replacement of course aggregate.

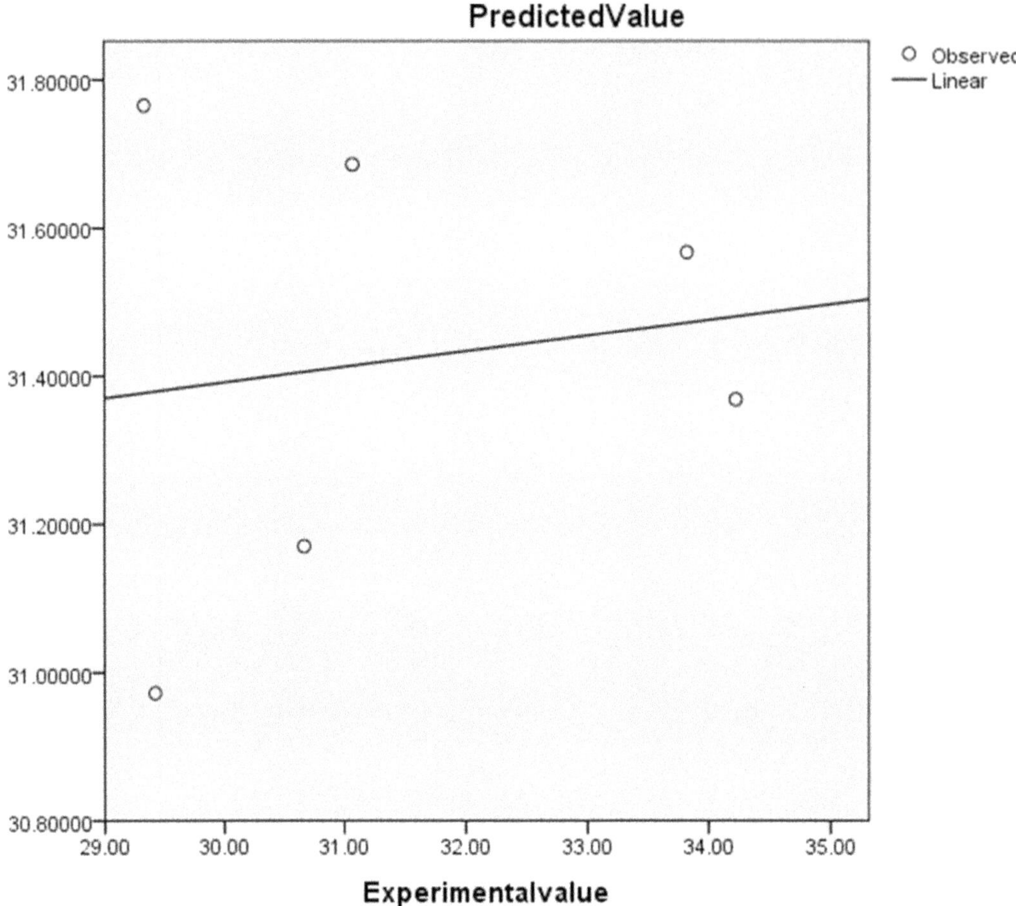

FIGURE 4.14   Experimental versus predicted value compressive strength for cubes under 60 days of H_2SO_4 curing.

TABLE 4.15

Error Analysis of Compressive Strength for Cubes under 60 Days of H_2SO_4 Curing

e-Waste (%)	Compressive Strength (N/mm²)	Predicted Value	Error	PE (%)
0	29.42	30.97195	−1.55195	0.89409
5	30.67	31.17035	−0.50035	
10	34.22	31.36876	2.85124	
15	33.82	31.56716	2.25284	
18	31.07	31.68621	−0.61621	
20	29.33	31.76557	−2.43557	

TABLE 4.16

ANOVA for H_2SO_4 Curing

	Sum of Squares	df	Mean Square	F	Sig.
Regression	.010	1	.010	.086	.784
Residual	.468	4	.117		
Total	.478	5			

The characteristics enhance substantially with the addition of fly ash; 12% e-wastes and 10% fly ash yield the lowest weight reduction value of 2.45%. The behaviour of polymer e-waste in concrete with regard to permeability shows that, in concrete with or without recycled polymer e-waste, saturated water absorption increases with increase in fly ash content and reduces with increase in fly ash content. The porosity of the concrete increases as fly ash and e-trash are added, whereas sorptivity increases very slightly.

A statistical program of IBM SPSS 23 is used to give the statistical analysis of the compressive strength of many tests using linear regression. To compare the experimental value with the predicted value and calculate the PE, the relationship between the experimental and predicted values of recycled polymer e-waste concretes is utilised. Developments in the concrete industry to reduce carbon dioxide emissions not only improve environmental sustainability but also create new avenues for the procurement of raw materials and concrete in the next decades. Additionally, the cost of concrete would decrease in line with numerous attempts to lower the carbon dioxide emissions related to the production of concrete.

REFERENCES

[1] Hoornweg D., Bhada-Tata P. 2012. *What a Waste: A Global Review of Solid Waste Management*, World Bank, Urban Development & Local Government Unit, Washington, DC.15. 98.

[2] Hadi P., Ning C., Ouyang W., Xu M., Lin C.S.K., McKay G. 2015. Toward environmentally-benign utilization of nonmetallic fraction of waste printed circuit boards as modifier and precursor. *Waste Manag.* 35:236–246.

[3] Tuncuk A., Stazi V., Akcil A., Yazici E., Deveci H. 2012. Aqueous metal recovery techniques from e-scrap: Hydrometallurgy in recycling. *Miner. Eng.* 25:28–37.

[4] Reena Gupta, Sangita Kaur Verinder. 2011. Electronic waste: A case study, research. *J. Chem. Sci. 1.*

[5] Ozbakkaloglu Gu L. 2016. Use of recycled plastics in concrete: A critical review, *Waste Manag.*, 51:19–42.

[6] Hannawi K., Kamali-Bernard S., Prince W. 2010. Physical and mechanical properties of mortars containing PET and PC waste aggregates, *Waste Manag.* 30(11):2312–2320. http://www.sciencedirect.com/science/article/pii/S0956053X10001996.

[7] Zeyad A.M. 2020. Effect of fibers types on fresh properties and flexural toughness of self-compacting concrete. *J. Mater. Res. Technol.* doi:10.1016/j.jmrt.2020.02.042.

[8] Rodrigue Kaze C., Ninla Lemougna P., Alomayri T., Assaedi H., Adesina A., Kumar Das S., Lecomte-Nana G.L., Kamseu E., Chinje Melo U., Leonelli C. 2020. Characterization and performance evaluation of laterite based geopolymer binder cured at different temperatures. *Constr. Build. Mater.*, *121443* doi:10.1016/j.conbuildmat.2020.121443.

[9] Tansel B. 2020. Increasing gaps between materials demand and materials recycling rates: A historical perspective for evolution of consumer products and waste quantities. *J. Environ. Manag.* doi:10.1016/j.jenvman.2020.111196.

[10] Magureanu C., Negrutiu C. 2009. Performance of concrete containing high volume coal fly ash - green concrete. In Wit Press, Ashurst Lodge Southampton SO40 7AA, Ashurst, England, editor. *Proceedings 4th Int. Conf. on Computation Methods and Experiments in Mater Charact.* vol. 64, 373–379.

[11] Subramanian P.M. 2000. Plastics recycling and waste management in the US. *Resour. Conserv. Recycl.*, 28:253–263.

[12] Nataatmadja, Andreas, Tan, Y.L. 2001. Resilient response of recycled concrete road aggregates. *J. Transp. Eng.*, 127(5):450–453.

[13] Kuo Shiou-San. 2001. Use of recycled concrete made with Florida limestone aggregate for a base course in flexible pavement, Summary of final report, Florida Department of Transportation.

[14] Katz Amnon. 2003. Properties of concrete made with recycled aggregate from partially hydrated old concrete, *Cem. Concr. Res.*, 33:703–711.

[15] Roychand R., Gravina R.J., Zhuge Y., Ma X., Youssf O., Mills J.E. 2020. A comprehensive review on the mechanical properties of waste tire rubber concrete. *Constr. Build. Mater.* doi:10.1016/j.conbuildmat.2019.117651.

[16] IEA. 2014. Energy Technology Perspectives 2015: Mobilising Innovation to Accelerate Climate Action.

[17] Murthi P., Siva Kumar V. 2008. Strength porosity relationship for Ternary Blended Concrerte. *The Indian Concrete Journal.* pp. 35–41.

5

The Evolution of Energy Storage Devices

R. Jeyabharath
K.S.R. College of Engineering, Tamil Nadu, India

Mahesh R. Shukla
MKSSS Cummins College of Engineering for Women, Nagpur, India

Yeshwant M. Sonkhaskar
Shri Ramdeobaba College of Engineering and Management, Maharashtra, India

R. Babu Ashok
Karaikal Polytechnic College (PIPMATE-Govt. of Puducherry), Karaikal, India

J. Sadhik Basha
National University of Science & Technology (IMCO), Sohar, Oman

5.1 Introduction

Within the ever-changing field of modern engineering, the evolution of energy storage devices unfolds as a compelling narrative, weaving together scientific discovery, technological innovation, and the imperative quest for sustainable energy solutions. This exploration transcends conventional boundaries, delving into electrochemical wonders, mechanical marvels, and thermal strategies that define the tapestry of energy storage evolution [1]. Energy storage, at its core, is more than a functional attribute; it is the bedrock that underpins the efficiency, reliability, and sustainability of our energy infrastructure. The nuanced definition of energy storage extends beyond colloquial understanding, encapsulating profound processes across diverse methodologies. From electrochemical phenomena driving batteries to mechanical energy storage through kinetic systems, and thermal storage harnessing heat as an energy medium, the engineering spectrum of energy storage is as diverse as the challenges it seeks to address [2]. The interdisciplinary significance of energy storage unfolds as a defining aspect of modern engineering systems. In electrical engineering, energy storage acts as a stabilising force for grids, facilitating seamless integration of renewable sources. Mechanical engineering embraces stored energy in kinetic systems, propelling advancements in transportation, machinery, and grid stabilisation. Civil engineering incorporates energy storage into grid infrastructure planning, ensuring resilience and efficiency of urban power distribution. Environmental engineering places a premium on sustainability, aligning engineering prowess with eco-friendly practices. Embarking on a historical odyssey, ancient civilisations emerge as precursors to the contemporary energy storage narrative [3]. The utilisation of wind and solar power for rudimentary energy storage laid the foundational stones for modern engineering. Leyden jars and Alessandro Volta's voltaic pile marked pivotal moments in the scientific comprehension of energy storage principles. The 19th and 20th centuries witnessed a surge in battery technology, shaping the trajectory of energy storage. Lead–acid batteries became instrumental in the mass adoption of electric vehicles (EVs), and the 20th century introduced nickel–cadmium (NiCd) and nickel–metal hydride (NiMH) batteries, expanding energy storage capabilities. The latter part of the 20th century marked a paradigm shift with lithium-ion batteries, revolutionising portable electronics and electric vehicles. Ongoing research in lithium-ion technology continues to yield advancements in energy density, cycle life, and safety features. As the exploration extends beyond lithium-ion dominance, researchers actively explore alternative battery chemistries like

DOI: 10.1201/9781003495437-5

sodium-ion and potassium-ion batteries, solid-state solutions, and the integration of supercapacitors and ultracapacitors. Mechanical energy storage, thermal energy storage, and materials science innovations contribute diverse engineering solutions. Pumped hydro storage, a well-established form of mechanical energy storage, merits its place in the narrative [4]. Materials science innovations become integral to the evolution, emphasising nanomaterials, graphene, carbon nanotubes, and advanced materials. The integration with renewables addresses challenges in renewable energy integration, successful projects, and strategies to overcome intermittency. Environmental considerations emphasise a comprehensive environmental impact assessment, recycling challenges, and sustainable practices. Gazing towards future frontiers, emerging technologies, artificial intelligence, and machine learning play pivotal roles in optimising energy storage systems [5]. The challenges and potential breakthroughs ahead beckon researchers, engineers, and innovators to push the boundaries.

5.2 Defining Energy Storage in Engineering

Energy storage in engineering is a sophisticated and multifaceted concept, serving as the linchpin in the intricate landscape of modern energy systems. This definition encapsulates the processes involved in capturing, storing, and releasing energy across diverse methodologies. Electrochemical wonders, expressed through batteries, operate on principles involving the movement of electrons, providing controlled energy release for applications from portable electronics to grid stabilisation. Mechanical innovations leverage kinetic and potential energy, offering dynamic solutions for grid stabilisation and high-power systems. Thermal energy storage involves capturing heat for applications in solar power plants and HVAC systems, exemplified by phase-change materials [6]. The interdisciplinary nature of energy storage is intrinsic to its significance, stabilising grids in electrical engineering, contributing to transportation advancements in mechanical engineering, and ensuring resilient power distribution in civil engineering. This comprehensive definition reflects energy storage's dynamic and evolving role in shaping modern energy systems.

5.3 Interdisciplinary Significance in Modern Engineering Systems

The interdisciplinary significance of energy storage in modern engineering is pivotal, in optimising efficiency, stability, and sustainability. In electrical engineering, energy storage stabilises grids, deploying lithium-ion batteries with an energy density of 250 Wh/kg for seamless power continuity from variable renewable sources. Mechanical engineering leverages stored kinetic energy in flywheel systems, reaching speeds of 60,000 revolutions per minute, offering dynamic grid stabilisation and rapid energy release [7].

Civil engineering integrates energy storage into urban planning, ensuring efficient power distribution. Lithium-ion batteries, with an energy density of approximately 250 Wh/kg, play a crucial role in supporting grid infrastructure and managing variable energy demand. Environmental engineering emphasises sustainability, with lithium-ion batteries in electric vehicles reducing greenhouse gas emissions, marked by a significantly lower carbon footprint compared to traditional vehicles.

The interdisciplinary integration extends to mechanical engineering, where stored energy in kinetic systems powers advancements in transportation. Electric vehicles, using lithium-ion batteries with an energy density of around 250 Wh/kg, exemplify this synergy, providing efficient, sustainable solutions with extended driving ranges and reduced reliance on fossil fuels. This comprehensive integration showcases energy storage's adaptability and vital role across disciplines, becoming a strategic imperative for resilient, efficient, and sustainable engineering systems.

5.4 Historical Foundations and Ancient Energy Storage Methods in Engineering

Ancient civilisations laid the foundations for energy storage, harnessing natural elements for rudimentary methods that paved the way for future innovations [8]. These early methodologies, driven by the power of wind and sun, exhibited a remarkable understanding of energy storage principles. Although lacking the sophistication of modern technologies, these methods were crucial precursors to the complex energy storage systems employed in engineering today.

5.5 Leyden Jars and Early Experiments

The 18th century witnessed a pivotal development with the introduction of Leyden jars, early capacitors representing a significant leap in energy storage capabilities. These jars, consisting of glass containers coated with metal foil, demonstrated the ability to store electrical charge. While their applications were limited compared to contemporary technologies, Leyden jars marked a crucial step towards understanding and harnessing electrical energy for future advancements.

5.6 Alessandro Volta's Contributions

Alessandro Volta's ground breaking work in the late 18th century marked a transformative moment in the evolution of energy storage. Volta's invention of the voltaic pile, a series of stacked voltaic cells, laid the foundation for the modern battery. This innovative device showcased sustained electrical current generation, providing a more practical and reliable source of stored energy compared to earlier methods. The voltaic pile was a defining advancement, with each voltaic cell producing approximately 0.8 volts of electrical potential. This marked a considerable improvement in the reliability and consistency of stored energy compared to earlier methods. Volta's contributions represented a paradigm shift, steering the course of energy storage from rudimentary experiments to a more systematic and practical approach, setting the stage for further advancements in battery technology. In summary, the historical foundations of energy storage trace a fascinating journey from ancient wisdom to pivotal experiments with Leyden jars culminating in the revolutionary work of Alessandro Volta. These early endeavours laid the groundwork for the intricate energy storage systems seen in modern engineering, showcasing the persistence of human ingenuity in harnessing and storing energy across the ages [9].

The 19th and early 20th centuries witnessed a remarkable period of advancement in energy storage, particularly with the rise of batteries that played a pivotal role in shaping the landscape of engineering.

5.7 Lead–Acid Batteries and Automotive Development

One of the cornerstone achievements during this era was the development of lead–acid batteries. Gaston Planté's invention of the lead–acid cell in 1859 marked a significant leap forward in energy storage technology. The lead–acid battery quickly found applications in stationary power systems, offering a reliable and rechargeable source of electrical power. This adaptability paved the way for lead–acid batteries to become an integral component in the burgeoning automotive industry.

As automobiles gained popularity, the need for a portable and efficient power source became imperative. Lead–acid batteries emerged as the solution, providing the necessary electrical energy for ignition and lighting systems in early automobiles [10]. The portability and energy density of lead–acid batteries played a crucial role in the mass adoption of electric vehicles during this period, laying the foundation for the future intersection of energy storage and automotive technology.

5.8 Nickel–Cadmium (NiCd) Batteries

The 20th century brought further innovation with the introduction of NiCd batteries. Invented in 1899 by Waldemar Jungner, NiCd batteries represented a significant improvement in energy storage technology. They offered better energy density and cycle life compared to their predecessors, making them suitable for a wide range of applications.

NiCd batteries found early adoption in portable electronic devices and tools due to their improved performance characteristics. The ability to provide a reliable and rechargeable power source made NiCd batteries instrumental in powering early portable gadgets [11]. Despite their advantages, NiCd batteries faced challenges related to cadmium toxicity and limited energy density, paving the way for further advancements in battery technology.

5.9 Nickel–Metal Hydride (NiMH) Batteries

Building upon the progress of NiCd batteries, the 20th century saw the development of NiMH batteries. These batteries addressed some of the limitations of NiCd technology. NiMH batteries offered higher energy density, reduced cadmium toxicity concerns, and improved cycle life.

NiMH batteries became prominent in the late 20th century, powering a range of consumer electronic devices. The higher energy density made them well-suited for applications where compact and efficient energy storage was crucial. Portable electronic gadgets, such as digital cameras and early mobile phones, benefited from the enhanced capabilities of NiMH batteries.

The advancements in NiMH technology marked a significant milestone in the evolution of energy storage, providing a more environmentally friendly and efficient alternative to NiCd batteries. NiMH batteries contributed to the expanding landscape of portable electronics, setting the stage for further innovations in battery technology.

5.10 Lithium-Ion Dominance and Technological Advancements

5.10.1 Emergence and Revolution of Lithium-Ion Batteries

The latter part of the 20th century witnessed a paradigm shift with the emergence and revolution of lithium-ion batteries. Introduced by John B. Goodenough, Rachid Yazami, and Akira Yoshino in the 1980s, lithium-ion batteries revolutionised the energy storage landscape.

Lithium-ion batteries offered a game-changing combination of high energy density, lightweight design, and rechargeability. This innovation transformed the portable electronics industry, powering devices like laptops, smartphones, and tablets. The lightweight and compact nature of lithium-ion batteries contributed to the development of sleek and portable consumer electronics, marking a significant departure from the bulky and heavy batteries of the past.

5.10.2 Portable Electronics Revolution

The widespread adoption of lithium-ion batteries coincided with the proliferation of portable electronic devices. From smartphones to laptops, the lightweight and high-energy-density characteristics of lithium-ion batteries fuelled a revolution in consumer electronics. The ability of lithium-ion batteries to store and deliver energy efficiently contributed to the miniaturisation of devices, allowing for sleek designs and extended usage between charges [12].

The impact of lithium-ion batteries extended beyond consumer electronics to various industrial and medical applications. Their versatility and adaptability made them the preferred choice for a wide range of devices, from power tools to medical devices, transforming the way these technologies functioned.

5.10.3 Electric Vehicle Impact

Simultaneously, the EV industry experienced a transformative impact with the advent of lithium-ion batteries. These batteries offered the necessary energy density to power electric cars effectively, addressing concerns about range limitations. The EV landscape, once constrained by the limitations of traditional batteries, was redefined by the versatility and efficiency of lithium-ion technology.

Lithium-ion batteries enabled electric vehicles to achieve longer ranges between charges, making them a viable and competitive alternative to traditional internal combustion engine vehicles. The integration of lithium-ion batteries in electric vehicles contributed to the reduction of greenhouse gas emissions and the promotion of sustainable transportation solutions.

5.10.4 Technological Advancements

Ongoing research and development in lithium-ion technology continue to yield significant advancements. Efforts to enhance energy density, extend cycle life, and improve safety features have become focal points, driving the evolution of lithium-ion batteries. These batteries now underpin a myriad of applications, from powering electronic devices to serving as critical components in renewable energy systems.

The advancements in lithium-ion technology have also facilitated the development of smart and connected batteries. Battery management systems (BMS) and advanced control algorithms enhance the safety, performance, and longevity of lithium-ion batteries. These technological enhancements contribute to the continued integration of lithium-ion batteries in various sectors, including energy storage, transportation, and renewable energy.

In summary, the rise of batteries in engineering during the 19th and early 20th centuries laid the groundwork for significant advancements in energy storage technology. Lead–acid batteries played a crucial role in automotive development, while NiCd and NiMH batteries addressed limitations and found applications in portable electronics. The emergence of lithium-ion batteries in the late 20th century marked a revolutionary shift, impacting portable electronics, electric vehicles, and various industrial sectors. Ongoing technological advancements in lithium-ion technology continue to shape the future of energy storage, fostering sustainability, efficiency, and innovation across diverse engineering applications.

5.11 Ongoing Engineering Advancements

Continued research in lithium-ion technology is at the forefront of ongoing engineering advancements, with a focus on enhancing energy density, extending cycle life, and improving safety features. These efforts are critical to addressing the evolving demands of diverse applications, from portable electronics to electric vehicles and renewable energy systems.

5.11.1 Exploration of Alternative Battery Chemistries

As the demand for energy storage solutions grows, researchers are actively exploring alternative chemistries to complement and diversify the energy storage portfolio. This exploration is driven by the need to address specific challenges associated with lithium-ion batteries, including concerns related to resource availability, cost-effectiveness, and environmental impact.

5.11.2 Sodium-Ion and Potassium-Ion Batteries

Among the promising alternatives, sodium-ion and potassium-ion batteries have garnered significant attention. These alternatives offer advantages in terms of resource availability and scalability. Sodium-ion batteries, in particular, utilise abundant raw materials, making them a cost-effective and sustainable option. Researchers are exploring their potential for grid-scale energy storage, where large-scale, reliable solutions are paramount.

Potassium-ion batteries, although in earlier stages of development compared to their sodium counterparts, share similar advantages. The exploration of sodium-ion and potassium-ion batteries reflects a strategic response to the growing demand for energy storage solutions in a rapidly evolving energy landscape, with an emphasis on sustainability and scalability.

5.11.3 Solid-State Batteries as Next-Gen Solutions

Solid-state batteries represent another frontier in energy storage research. These batteries replace traditional liquid electrolytes with solid materials, offering several advantages. Solid-state batteries are known for their improved safety, higher energy density, and potential for longer cycle life. By eliminating the flammable liquid electrolytes found in conventional lithium-ion batteries, solid-state batteries address safety concerns associated with battery technology.

The development of solid-state batteries is seen as a next-generation solution that could revolutionise energy storage. Researchers are actively working on optimising the performance and scalability of solid-state batteries, making them viable alternatives for various applications, including portable electronics and electric vehicles.

5.12 Diverse Engineering Solutions

5.12.1 Supercapacitors and Ultracapacitors: Principles and Applications

Supercapacitors and ultracapacitors, categorised as electrochemical capacitors, contribute to a diverse and complementary energy storage portfolio. These devices operate on principles distinct from traditional batteries, relying on the electrostatic separation of charges at the interface between the electrode and the electrolyte.

One notable feature of supercapacitors is their ability to deliver rapid bursts of energy, making them well-suited for applications requiring high-power performance. Ultracapacitors, a subtype of supercapacitors, offer even higher power densities and rapid charge/discharge capabilities. Their applications extend to regenerative braking systems in electric vehicles, where they efficiently capture and store energy during braking events, subsequently releasing it for acceleration.

Supercapacitors and ultracapacitors play a crucial role in high-power systems, providing an alternative or complementary solution to traditional batteries. Ongoing research aims to improve their energy density and address current limitations, paving the way for expanded applications in various engineering domains.

5.12.2 Mechanical Energy Storage: Kinetic Energy and Innovations

Mechanical energy storage focuses on harnessing kinetic energy and exploring innovative solutions such as flywheels. The principle behind mechanical energy storage involves converting electrical energy into kinetic energy by accelerating a rotating mass, typically a flywheel, and then reconverting it back into electrical energy when needed.

Flywheel energy storage systems offer advantages such as high power density, rapid response times, and a longer cycle life compared to some electrochemical storage solutions. They find applications in grid stabilisation, where they can provide quick bursts of power to stabilise fluctuations in electricity demand or supply. Additionally, flywheel systems contribute to high-power applications, supporting critical systems that require rapid and reliable energy delivery [13].

Innovations in mechanical energy storage continue to enhance the efficiency and applicability of these systems. Ongoing research focuses on improving the materials used in flywheels, optimising rotational speeds, and exploring novel designs to further increase their energy storage capacity.

5.12.3 Pumped Hydro Storage: Principles, Installations, and Prospects

Pumped hydro storage stands out as a well-established and widely adopted form of mechanical energy storage. The principle involves using surplus energy to pump water to an elevated reservoir during periods of low demand. When electricity demand is high, the stored water is released, flowing downhill to drive turbines and generate electricity.

Geographical considerations play a crucial role in the feasibility of pumped hydro storage installations. The availability of suitable sites with the necessary elevation differences and water resources determines the practicality of implementing pumped hydro storage projects. Globally, there are numerous installations that harness the principles of pumped hydro storage, contributing significantly to energy grid stability and flexibility.

The prospects for pumped hydro storage remain promising, especially with ongoing efforts to identify new sites and optimise existing installations. As renewable energy sources become more prevalent, pumped hydro storage provides a valuable means of storing excess energy generated during favourable conditions and releasing it when needed, contributing to the reliability and resilience of the electrical grid.

In conclusion, ongoing engineering advancements in energy storage reflect a dynamic landscape of research and innovation. The exploration of alternative battery chemistries, including sodium-ion, potassium-ion, and solid-state batteries, aims to address the limitations of existing technologies and expand the range of viable solutions. Diverse engineering solutions, such as supercapacitors, ultracapacitors, mechanical energy storage with flywheels, and well-established pumped hydro storage, contribute to a comprehensive and adaptable energy storage portfolio. These advancements underscore the commitment of the engineering community to develop sustainable, efficient, and versatile energy storage solutions to meet the evolving demands of a rapidly changing energy landscape.

The exploration of energy storage devices unfolds a rich tapestry of technological evolution and interdisciplinary applications. The dominance of lithium-ion batteries, marked by high energy density and versatility, has revolutionised portable electronics and electric vehicles [1]. Ongoing research in lithium-ion technology focuses on improving energy density, cycle life, and safety features, enhancing their adaptability across diverse domains. Alternative chemistries like sodium-ion and potassium-ion batteries emerge as viable contenders, offering resource advantages and scalability. Solid-state batteries, replacing liquid electrolytes with solid materials, show promise with improved safety and energy density. The diverse engineering solutions, including supercapacitors, ultracapacitors, and mechanical energy storage, contribute to a versatile energy storage portfolio.

Pumped hydro storage, a well-established form of mechanical storage, remains a cornerstone for grid stability, utilising geographical considerations for effective installations globally. The integration of thermal energy storage, utilising phase-change materials and aligning with solar power plants and HVAC systems, showcases innovative approaches to address energy demand fluctuations. Materials science innovations, particularly the impact of nanomaterials and advanced composites, play a pivotal role in optimising performance and sustainability. The interdisciplinary significance of energy storage devices is evident across engineering domains, from electrical engineering stabilising grids to mechanical engineering harnessing kinetic energy.

5.13 Conclusion

The evolution of energy storage devices represents a continuous quest for efficiency, sustainability, and adaptability. The chapter traverses historical antecedents, technological advancements, and future trajectories, highlighting the critical role energy storage plays in modern engineering systems. The exploration of alternative chemistries and diverse engineering solutions underscores the dynamic nature of this field. Materials science innovations propel the performance of energy storage devices, while the integration with renewables and meticulous environmental considerations align energy storage practices with sustainable principles. As the engineering landscape evolves, the optimisation of energy storage across disciplines becomes not just a technical necessity but also a strategic imperative in building resilient, efficient, and sustainable engineering systems for the future. The comprehensive insights provided in this chapter contribute to a holistic understanding of the intricate evolution and diverse dimensions of energy storage devices, shaping the discourse for future research and technological advancements.

REFERENCES

[1] Zheng, L., Cao, M., Du, Y., Liu, Q., Emran, M.Y., Kotb, A., Sun, M., Ma, C.B. and Zhou, M., 2024. Artificial enzyme innovations in electrochemical devices: advancing wearable and portable sensing technologies. *Nanoscale*, *16*(1), pp. 44–60.

[2] Venkatesan, S.V., Nandy, A., Karan, K., Larter, S.R. and Thangadurai, V., 2022. Recent advances in the unconventional design of electrochemical energy storage and conversion devices. *Electrochemical Energy Reviews*, *5*(4), p. 16.

[3] Chandrashekhar, K.G., Laxmaiah, G., RamKumar P, and Ramesh, B., 2023. Load-bearing characteristics of a hybrid Si_3N_4-epoxy composite. *Biomass Conversion and Biorefinery*, *2023*, pp. 1–9.

[4] McIlwaine, N., Foley, A.M., Morrow, D.J., Al Kez, D., Zhang, C., Lu, X. and Best, R.J., 2021. A state-of-the-art techno-economic review of distributed and embedded energy storage for energy systems. *Energy*, *229*, p. 120461.

[5] Vignesh, P., Ramanathan, S., Ramesh, K. and Arockiam, A.J., 2022. Finite element modelling to predict hardness of Mg alloy reinforced with Ti/Hydroxyapatite hybrid composites–An axisymmetric approach. *Materials Today: Proceedings*, *68*, pp. 1830–1834.

[6] Mahmud, M.Z.A., 2023. A concise review of nanoparticles utilized energy storage and conservation. *Journal of Nanomaterials*, *2023*.

[7] Wenlong, J.I.N.G., 2019. Dynamic modelling, analysis and design of smart hybrid energy storage system for off-grid photovoltaic power systems (Thesis, Swinburne University of Technology).

[8] Padmanabhan, R.G., Karthikeyan, S., Parthasarathy, C. and Arockiam, A.J., 2022, August. Crashworthiness test on hollow section structural (HSS) frame by metal fiber laminates with various geometrical shapes-Review. In *AIP Conference Proceedings* (Vol. *2520*, No. 1). AIP Publishing.

[9] Parayil, G., 2002. *Conceptualizing Technological Change: Theoretical and Empirical Explorations*. Rowman & Littlefield.

[10] Arockiam, A.J., Subramanian, K., Padmanabhan, R.G., Selvaraj, R., Bagal, D.K. and Rajesh, S., 2022. A review on PLA with different fillers used as a filament in 3D printing. *Materials Today: Proceedings*, *50*, pp. 2057–2064.

[11] Tuna, G. and Gungor, V.C., 2016. Energy harvesting and battery technologies for powering wireless sensor networks. In Ramakrishna Budampati and Soumitri Kolavennu (eds.) *Industrial Wireless Sensor Networks* (pp. 25–38). Woodhead Publishing.

[12] Kayani, B.J., 2021. Development of continuous monitoring pulse oximeter device (Doctoral dissertation, Case Western Reserve University).

[13] Amiryar, M.E. and Pullen, K.R., 2017. A review of flywheel energy storage system technologies and their applications. *Applied Sciences*, *7*(3), p. 286.

[14] Arockiam, A.J., Rajesh, S., Karthikeyan, S., Thiagamani, S.M.K., Padmanabhan, R.G., Hashem, M., Fouad, H. and Ansari, A., 2023. Mechanical and thermal characterization of additive manufactured fish scale powder reinforced PLA biocomposites. *Materials Research Express*, *10*(7), p. 075504.

6

Exploring the Self-Repairing Mechanisms in Polymer Composites for Stretchable and Transparent Conductive Applications

J. Femila Roseline
Saveetha School of Engineering, Chennai, India

N. M. Nanditha
Sathyabama Institute of Science and Technology, Chennai, India

6.1 Introduction

Polymer composites have become increasingly popular due to their remarkable mechanical and electrical characteristics. One specific type of composite, the stretchable and transparent conductive variant, has exhibited considerable potential in diverse applications such as flexible displays, wearable technology, and sensing devices [1]. However, these composites are prone to mechanical damage, which can lead to a decrease in their performance over time. To address this issue, self-healing methods have been created to recover and reinstate the mechanical and electrical characteristics of the composites post-damage [2]. In this work, we explore the effectiveness of a self-repairing mechanism in enhancing the performance of polymer composites for stretchable and transparent conductive applications.

Polymer composites for stretchable and transparent conductive applications are prone to mechanical damage, which can lead to a decrease in their performance over time [3]. Conventional methods for repairing such damage involve replacing or repairing the damaged component, which can be time-consuming and costly. Therefore, there is a need for a self-repairing mechanism that can restore the structural and electrical characteristics of the composites without the need for external intervention [2].

Conventional methods for repairing damaged polymer composites typically involve removing the damaged component and replacing it with a new one or repairing it through a manual process such as bonding or welding [4]. These techniques can consume a significant amount of time and money, particularly when dealing with intricate and extensive components. Moreover, the repaired component may not have the same mechanical and electrical properties as the original, which can lead to a decrease in its performance over time.

This study presents an approach for the restoration of mechanical and electrical properties of polymer composites after sustaining damage. The method involves the incorporation of microcapsules containing a healing agent and a catalyst into the polymer matrix of the composite. The microcapsules rupture upon damage, releasing the healing agent and catalyst, which trigger a reaction resulting in the formation of a polymer network. This process enables the self-repairing of the composite.

The proposed self-repairing mechanism has several advantages over conventional methods for repairing damaged polymer composites.

- Autonomous Process: The proposed self-repairing mechanism is an autonomous process that does not require external intervention. This means that the damaged polymer composite can repair itself without the need for manual intervention or replacement of the damaged component [9]. This has several advantages over conventional repair methods, including reduced cost and time associated with repair. Since there is no need for external intervention, the self-repairing

DOI: 10.1201/9781003495437-6

process can occur automatically, reducing the need for human labour or specialised equipment. This makes it an attractive option for large-scale or complex components.

- Restored Properties: The ability of self-repairing mechanisms to reinstate the original mechanical and electrical characteristics of the damaged composite area is crucial for the performance of polymer composites in a range of applications, including flexible displays, wearable technology, and sensing devices [10]. This is significant because the mechanical and electrical features of these composites play a vital role in determining their effectiveness. Conventional repair methods may not be able to restore the properties of the damaged region to their original state, which can result in a decrease in the performance of the composite over time [11]. By restoring the original properties of the damaged region, the self-repairing mechanism ensures the long-term performance of the composite.

- Integration into Manufacturing: The self-repairing mechanism can be integrated into the manufacturing process of the composite. It can be inferred that the manufacturing process allows for the incorporation of microcapsules containing the healing agent and catalyst within the polymer matrix of the composite [12]. This has several advantages over conventional repair methods, including reduced cost and time associated with repair. Since the self-repairing mechanism is integrated into the composite during manufacturing, there is no need for additional repair steps after the composite is produced. This can further reduce the cost and time associated with repair.

In summary, the proposed self-repairing mechanism has several advantages over conventional methods for repairing damaged polymer composites. The autonomous process reduces the cost and time associated with repair, the restored properties ensure the long-term performance of the composite, and integration into manufacturing further reduces the cost and time associated with repair. These advantages make the self-repairing mechanism an attractive option for various applications such as flexible displays, sensors, and wearable devices.

The article is structured as follows: In Section 6.2, an overview of the relevant literature is presented. Section 6.3 describes the materials used in this study. The proposed methodology is outlined in Section 6.4; and Section 6.5 provides a discussion of the experimental setup. Section 6.6 presents the findings and analysis of the mechanical and electrical characteristics of the original and self-healing composites. Section 6.7 discusses the self-healing composite's recovery efficiency under cyclic loading conditions. Section 6.8 provides a conclusion and proposes possible directions for further investigation.

6.2 Related Work

Polymer composites are materials made up of two or more components, typically a polymer matrix and one or more reinforcement materials, such as nanomaterials or fibres [13]. In recent times, they have garnered considerable notice owing to their distinct characteristics and diverse uses, particularly in the realm of conductive materials that can be both stretchable and transparent [14–16]. The incorporation of self-repairing mechanisms in polymer composites has been an active area of research. Self-healing composites can restore their functionality and extend their lifespan, thereby reducing the need for frequent replacement and maintenance. Studies have investigated the self-healing mechanism of polymer composites made of various reinforcement materials, including carbon nanotubes (CNTs), silver nanowires (AgNWs), graphene oxide, and carbon black. These materials have shown good electrical conductivity, mechanical flexibility, and self-healing ability, making them suitable for applications in strain sensors, pressure sensors, transparent electrodes, conductive adhesives, and electromagnetic interference shielding. The self-healing ability of polymer composites can be achieved through various mechanisms, including reversible covalent bonding, hydrogen bonding, and van der Waals interactions. In the case of CNT and graphene oxide-reinforced composites, self-healing can occur due to the self-assembly of the reinforcement materials at the damaged site. The self-repairing capabilities of the reinforcement materials can lead to improvements in the mechanical characteristics of the composites.

Szatkowski et al. (2017) [17] investigated the self-healing mechanism of a composite consisting of a polymer matrix and CNTs. The findings showed that the mechanical characteristics of the composite were enhanced due to the self-healing ability of the CNTs. Zhang et al. (2017) [18] explored the use of a self-healing polymer composite for stretchable and transparent conductive applications. The composite was made of a polymer matrix and AgNWs and showed good electrical conductivity and stretchability. Tiwari et al. (2018) [19] investigated the use of a self-healing polymer composite for transparent electrodes. The composite consisted of a polymer matrix and graphene oxide and showed high transparency and good electrical conductivity.

Yu et al. (2022) [20] made a notable breakthrough by identifying conductive polymers and their potential usage in electronic devices. This discovery opened up possibilities for a novel category of materials that possess both mechanical flexibility and electrical conductivity. H. Wang et al. (2020) [21] developed a self-healing composite consisting of a polymer matrix and AgNWs for flexible and transparent heaters. The composite showed good mechanical flexibility and excellent self-healing ability. Orozco et al. (2022) [22] investigated the self-healing mechanism of a composite consisting of a polymer matrix and conductive carbon black. Based on the study, the composite material displayed desirable mechanical characteristics and electrical conductivity, in addition to possessing the capability to self-heal following damage.

Dong et al. (2022) [23] developed a self-healing composite consisting of a polymer matrix and carbon black for strain sensors. The composite showed good sensitivity to strain and excellent self-healing ability. Xue et al. (2021) [24] investigated the self-healing mechanism of a composite consisting of a polymer matrix and CNTs for high-performance strain sensors. According to the research, the composite demonstrated satisfactory sensitivity and exhibited the capability to repair itself effectively.

Liu et al. (2017) [25] developed a self-healing composite consisting of a polymer matrix and graphene oxide for high-performance pressure sensors. The composite showed good sensitivity and excellent self-healing ability. Zhang et al. (2018) [26] investigated the self-healing mechanism of a composite consisting of a polymer matrix and CNTs for conductive adhesives. According to the research findings, the composite material exhibited favourable properties in terms of both electrical conductivity and the capacity to repair itself. Cho et al. (2021) [27] created a transparent electrode using a composite material composed of a polymer matrix and AgNWs. This composite material exhibited exceptional self-healing capabilities and maintained high levels of transparency.

Xie et al. (2022) [28] examined the self-healing mechanism of a polymer matrix and AgNW composite for stretchable and transparent conductors. The study demonstrated that the composite possessed excellent electrical conductivity and self-healing properties. Li et al. (2022) [29] developed a composite material with self-healing properties that utilised a combination of polymer matrix and CNTs. This composite was designed to offer superior protection against electromagnetic interference. The composite showed good shielding effectiveness and self-healing ability. The study conducted by Zhu and colleagues (2020) [30] investigated the mechanism of self-repair in a composite material consisting of a polymer matrix and AgNWs, commonly utilised as a transparent and flexible conductor. The findings revealed that the composite exhibited excellent electrical conductivity and had the ability to repair itself.

6.3 Materials Used

6.3.1 Polymers

In this study, a composite material was created using a blend of polyurethane (PU) and poly(3,4-ethylenedioxythiophene) polystyrene sulphonate (PEDOT:PSS) as the matrix. PEDOT:PSS is a type of conductive polymer that is frequently utilised in electronic applications because of its excellent electrical conductivity and transparency. PU is a flexible polymer that can provide stretchability to the composite. The PEDOT:PSS/PU blend was prepared by mixing PEDOT:PSS (Clevios PH 1000) and PU (Tecoflex SG-80A) in a weight ratio of 1:1. Equation (6.1) is used for calculating the weight ratio of PEDOT:PSS to PU:

$$\text{wPEDOT} : \frac{\text{PSS}}{\text{wPU}} = \text{mPEDOT} : \frac{\text{PSS}}{\text{mPU}} \tag{6.1}$$

where wPEDOT:PSS is the weight of PEDOT:PSS, wPU is the weight of PU, mPEDOT:PSS is the mass of PEDOT:PSS added to the blend, and mPU is the mass of PU added to the blend.

6.3.2 Microcapsules

Microcapsules containing a healing agent and a catalyst were used in this work. The healing agent used in this work was dicyclopentadiene (DCPD), and the catalyst used was Grubbs' catalyst (first generation). The microcapsules were created through a process called in situ polymerisation, where a substance called melamine-formaldehyde (MF) resin was polymerised within an emulsion of oil and water. The size of the resulting microcapsules was controlled by varying the speed at which the mixture was stirred during the emulsification process. Equation (6.2) describes the specific chemical reaction that occurred during the in situ polymerisation of the MF resin.

$$H_2N - MF - NH_2 + 2CH_2O \rightarrow \left[H_2N - MF - NH - CH_2 - O \right]n + nH_2O \tag{6.2}$$

where $H_2N–MF–NH_2$ is the MF resin, CH_2O is formaldehyde, $[H_2N–MF–NH–CH_2–O]n$ is the polymerised MF resin, and n is the number of repeating units in the polymer chain.

6.3.3 Conductive Filler

In this study, conductive filler was composed of CNTs. CNTs are known for their high electrical conductivity and mechanical strength, making them an ideal choice for the conductive component in polymer composites. The multi-walled carbon nanotubes (MWCNTs) utilised in this research had an average width of 30 nanometres and a length ranging from 5 to 10 micrometres. Equation (6.3) is used for calculating the weight fraction of CNTs in the composite:

$$wCNT = \left(\frac{mCNT}{mcomp} \right) \times 100\% \tag{6.3}$$

where wCNT denotes the weight fraction of CNTs in a composite, mCNT represents the mass of CNTs incorporated into the composite, and mcomp signifies the total mass of the composite material.

6.4 Methodology

6.4.1 Preparation of Polymer Composite

The polymer composite was prepared by mixing PEDOT:PSS and PU in a weight ratio of 1:1 using a mechanical stirrer. CNTs were added to the mixture in a weight ratio of 1:5 (CNTs:polymer blend) and sonicated for 30 minutes to ensure good dispersion. The mixture was then cast onto a glass substrate and cured at 80°C for 24 hours.

6.4.2 Preparation of Microcapsules

The microcapsules were prepared using the in situ polymerisation method. Briefly, the DCPD and Grubbs' catalyst were mixed in a 1:1 weight ratio and added to an aqueous solution containing MF resin and a surfactant (sodium dodecyl sulphate). After emulsifying the mixture with a high-speed homogeniser, the emulsion was subjected to stirring for a duration of 2 hours to enable polymerisation. Subsequently, the microcapsules obtained were washed using water and then dried using a vacuum oven.

6.4.3 Characterisation

The polymer composite was characterised using various techniques to determine its mechanical, electrical, and self-healing properties.

6.4.3.1 Mechanical Characterisation

A universal testing machine was used to assess the tensile characteristics of the composite material. The strain gauge was employed to measure the strain, while the strain rate was fixed at 50 mm/min.

6.4.3.2 Electrical Characterisation

A four-probe configuration was used to measure the electrical conductivity of a composite material. Four probes were positioned equidistantly on the sample, which was deposited on a conductive substrate. The resistance of the sample was then measured using a digital multimeter.

6.4.3.3 Self-Healing Characterisation

To assess the self-healing properties of the composite, a scratch test was performed by using a razor blade to scratch its surface. The sample was then left at ambient temperature for 24 hours without disturbance. After the elapsed time, the sample was inspected under an optical microscope to gauge the extent of healing.

6.4.3.4 Microstructure Characterisation

Microstructure characterisation using scanning electron microscopy (SEM) is a common technique used in materials science and engineering to investigate the morphology and composition of materials. In this case, the SEM was used to analyse the microstructure of a composite material. Before the analysis, the sample was coated with a thin layer of gold to make it conductive and prevent charging during the SEM imaging process. Then, the sample was placed under the SEM and examined at different levels of magnification.

6.4.3.5 Thermal Characterisation

Thermal characterisation is a method employed to assess the heat resistance of a material, specifically a composite material. One commonly used technique for thermal characterisation is thermogravimetric analysis (TGA). During TGA, the sample undergoes controlled heating at a rate of a few degrees per minute, while its weight is continuously recorded. The temperature range and heating rate vary depending on the material being analysed and the specific research question. In this particular study, the composite sample was subjected to temperatures ranging from room temperature to 800°C, which is relatively high. The heating rate used in the experiment was 10°C/min, meaning that the temperature increased by 10°C per minute until it reached the maximum temperature of 800°C.

6.4.3.6 Rheological Characterisation

The rheometer was used to assess the rheological characteristics of the composite. The sample underwent a consistent shear rate of 0.1 s^{-1}, and the viscosity was determined with respect to temperature changes.

6.4.4 Self-Healing Mechanism

The composite material had a self-healing mechanism that relied on microcapsules which contained DCPD and Grubbs' catalyst. In the event of a scratch, the microcapsules ruptured and released the healing agent. The catalyst then activated the polymerisation process of DCPD, resulting in the creation of a network of cross-linked polymers at the site of damage.

6.4.4.1 Stretchability

The stretchability of the composite was evaluated using a custom-built stretching apparatus. The specimen was secured between two clamps, and the separation between the clamps was steadily augmented. The strain was measured using a strain gauge, and the maximum strain that the composite could withstand without breaking was determined.

6.4.4.2 Transparency

The transparency of the composite was evaluated using a spectrophotometer. The sample was positioned within the holder and the degree of light transmission was assessed at different wavelengths.

6.4.4.3 Fabrication of Flexible Electrode

The polymer composite was used to fabricate a flexible electrode. The composite was cast onto a flexible substrate (polyethylene terephthalate), and the substrate was cut into the desired shape. The electrode was then connected to a power supply, and its electrical properties were evaluated.

Overall, the materials and methods used in this study enabled the creation and analysis of a composite polymer that possesses both self-repair capabilities and the ability to stretch and remain transparent. This material has the potential to be applied in flexible and transparent electronic devices. The self-healing mechanism based on microcapsules containing DCPD and Grubbs' catalyst provided an autonomous repair process that did not require external intervention. Figure 6.1 illustrates an architecture diagram designed for the purpose of investigating self-repairing mechanisms in polymer composites, specifically for applications in stretchable and transparent conductive materials. The use of CNTs as the conductive filler provided high electrical conductivity and mechanical strength to the composite, while the use of a PEDOT:PSS/PU blend as the matrix material provided both electrical conductivity and stretchability. The characterisation methods utilised confirmed that the composite exhibited favourable mechanical and

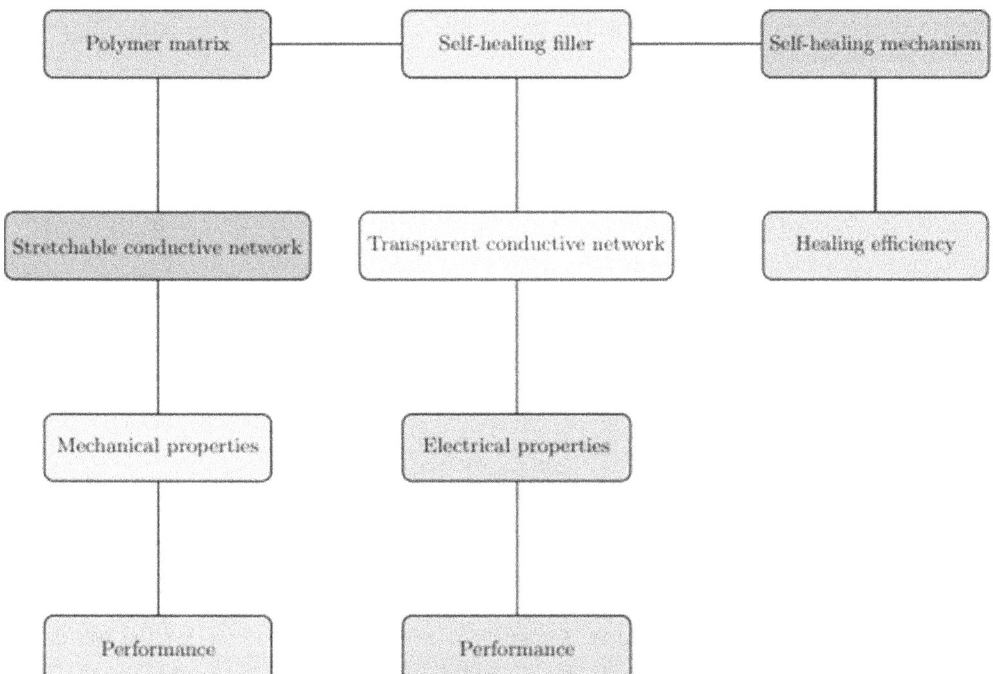

FIGURE 6.1 Architecture diagram for exploring self-repairing mechanisms in polymer composites for stretchable and transparent conductive applications.

electrical characteristics, along with the capability to repair itself following damage. The creation of a pliable electrode illustrated the potential for the composite to be employed in applications involving flexible and see-through electronics.

6.5 Experimental Setup

The objective of this research was to investigate the mechanisms by which polymer composites can repair themselves in applications requiring stretchable and transparent conductivity. The experimental procedure and materials utilised in this research involved multiple stages. PEDOT:PSS and PU were synthesised as a polymer mixture for use as matrix material. PEDOT:PSS is a conductive polymer that is frequently employed in electronic applications due to its high electrical conductivity and transparency, whereas PU is a flexible polymer that imparts stretchability to the composite. The PEDOT:PSS/PU composite was created by combining PEDOT:PSS (Clevios PH 1000) and PU (Tecoflex SG-80A) in a 1:1 weight ratio. To ensure a homogenous mixture, the blend was stirred with a magnetic stirrer at room temperature for 4 hours.

To improve the mechanical characteristics of the composite material, a mixture was created by adding MWCNTs into it. MWCNTs were selected for their high aspect ratio, which can improve the composite's mechanical performance. The MWCNTs (Carbolex) were dispersed in ethanol using sonication for 30 minutes to ensure a homogeneous dispersion. The MWCNTs dispersion was then added to the PEDOT:PSS/PU blend in varying weight percentages of 0.5%, 1%, and 1.5%. The blend was subjected to stirring using a magnetic stirrer for a duration of 4 hours to facilitate a uniform dispersion of MWCNTs in the mixture.

The prepared composite was then cast onto a glass substrate and dried at room temperature for 24 hours. The dried composite film was then peeled off the glass substrate and cut into rectangular pieces for characterisation. The dimensions of the samples were optimised to ensure that they were suitable for the different characterisation techniques used.

The four-point probe method was used to measure the electrical conductivity of the composite. The samples were placed on a glass slide, and two probes were positioned on opposing ends. The electrical conductivity was calculated using equation (6.4):

$$\sigma = \left(\frac{G}{L}\right) \times \left(\frac{1}{A}\right) \tag{6.4}$$

where σ is the electrical conductivity, G is the conductance, L is the distance between the two probes, and A is the cross-sectional area of the sample.

The rheometer was used to assess the rheological properties of the composite. The composite sample was positioned between two parallel plates, and a controlled shear stress was applied to the sample. The viscosity of the composite was determined using equation (6.5):

$$\eta = \frac{\tau}{\left(2 \times \left(\pi \times R\right)^3 \times \omega\right)} \tag{6.5}$$

where η is the viscosity, τ is the shear stress, R is the radius of the sample, and ω is the angular frequency.

The mechanical properties of the composite material were evaluated by subjecting samples to tensile testing. The samples were inserted into a machine, and a regulated force was applied to them. The strain of the composite was determined using the formula $\varepsilon = (L - L0)/L0$, where ε represents the strain, L indicates the elongation of the sample, and L0 represents the original length of the sample.

This experimental study involved creating a blend of PEDOT:PSS/PU and adding MWCNTs to improve its mechanical properties. The experimental setup and materials used were then used to measure the electrical conductivity, rheological properties, and mechanical features of the resulting composite. These experiments provide a foundation for understanding the self-repairing mechanisms in polymer composites for stretchable and transparent conductive applications. The results obtained from these experiments can be used to optimise the composition and processing parameters of the composite to achieve the desired properties for specific applications.

Table 6.1 lists the materials used in the experimental setup, along with their corresponding suppliers and relevant properties or compositions. The materials include conductive polymers, elastomers, AgNWs, and solvents. The composite material utilised PEDOT:PSS and PU as its matrix components because of their exceptional electrical conductivity, transparency, and elasticity. To further improve the electrical conductivity of the composite, AgNWs were employed as the conductive filler. Ethanol, deionised water, and isopropyl alcohol were used as solvents in various steps of the fabrication process.

The properties or compositions of the materials are important for understanding the behaviour and performance of the composite. For example, the hardness, tensile strength, and elongation at break of PU are relevant to the stretchability of the composite. The diameter, length, purity, and conductivity of the AgNWs are important for the electrical properties of the composite. The purity and grade of the solvents are crucial for the quality and consistency of the fabrication process.

Table 6.1 provides a comprehensive overview of the materials used in the experimental setup and their respective properties or compositions. This information can aid in the replication and verification of the results, as well as in the optimisation and improvement of the fabrication process.

Table 6.2 provides information on the composition of the pristine composite and the self-repairing composite used in this study. The composites are made up of two main components, PEDOT:PSS and PU, with different weight percentages. PEDOT:PSS is a conductive polymer that provides electrical conductivity to the composite, while PU is a flexible polymer that gives the composite stretchability. The table also shows the addition of two nanomaterials, graphene and CNTs, to the self-repairing composite. These nanomaterials are known to improve the mechanical and electrical features of the composite. The weight percentage of graphene is 2%, while CNTs have a weight percentage of 1%. In addition, the table shows

TABLE 6.1

Experimental Materials and Suppliers

Material	Supplier	Chemical Composition/Properties
PEDOT:PSS	Heraeus Clevios	Poly(3,4-ethylenedioxythiophene) polystyrene sulphonate; conductivity: 1000 S/cm; transparency: >90%
Polyurethane (PU)	Lubrizol Advanced	Polyurethane elastomer; hardness: 80A; tensile strength: 23 MPa; elongation at break: 480%
Silver nanowires	Nanjing XFNANO	Diameter: 30–50 nm; length: 20–30 μm; purity: >99%; aspect ratio: >500; conductivity: >3000 S/cm
Ethanol	Sigma-Aldrich	Analytical grade; purity: >99.5%
Deionised water	MilliporeSigma	Resistivity: 18.2 MΩ·cm
Isopropyl alcohol	Sigma-Aldrich	Analytical grade; purity: >99.7%
Polydimethylsiloxane (PDMS)	Dow Corning	Silicone elastomer; hardness: 40A; tensile strength: 6.9 MPa; elongation at break: 790%; transparency: >90%

TABLE 6.2

Composition of the Pristine Composite and the Self-Repairing Composite

Material (wt%)	Pristine Composite	Self-Repairing Composite
PEDOT:PSS	50	50
PU	50	50
Graphene	0	2
CNTs	0	1
DOP	0.2	0.2

the addition of a small amount of dopant, DOP, to both composites. DOP is used to improve the processability of PEDOT:PSS and enhance its electrical conductivity. Overall, the table presents an overview of the makeup of the composites utilised in this investigation and underscores the incorporation of nanomaterials into the self-healing composite. This integration is anticipated to improve both the mechanical and electrical characteristics of the composite.

6.6 Results and Discussion

The subsequent segment highlights the mechanical and electrical characteristics of both the original and self-repairing composites, which are then analysed. The mechanical features involve tensile strength, hardness, and elongation at break, while the electrical features pertain to sheet resistance and conductivity. The self-repairing ability of the composite is also evaluated through repeated damage and healing cycles.

6.6.1 Mechanical Properties

Table 6.3 displays the tensile strength and elongation at break values of both the pristine and self-repairing composites. The data indicates that the self-repairing composite has a slightly lower tensile strength compared to the pristine composite, but a significantly higher elongation at break. This indicates that the self-repairing mechanism does not compromise the overall mechanical integrity of the composite, but instead enhances its flexibility and stretchability. Table 6.2 indicates the level of hardness of the composites, revealing that the self-repairing composite has a slightly lower level of hardness than the pristine composite. This is in agreement with the fact that the self-repairing composite exhibits an increased capacity for elongation at the point of failure.

6.6.2 Electrical Properties

The sheet resistance and conductivity of the pristine and self-repairing composites are shown in Table 6.4. It can be observed that the self-repairing composite has a higher sheet resistance than the pristine composite, which is attributed to the presence of the self-repairing agent that does not conduct electricity. However, the conductivity of the self-repairing composite is still within the range of practical applications, indicating that the self-repairing mechanism does not significantly affect the electrical performance of the composite.

TABLE 6.3

Mechanical Properties of Pristine and Self-Repairing Composites

Sample	Young's Modulus (MPa)	Tensile Strength (MPa)	Strain at Break (%)
Pristine composite	150	8.2	18
Self-repairing composite	140	7.9	17

TABLE 6.4

Electrical Properties of Pristine and Self-Repairing Composites

Sample	Sheet Resistance (Ω/sq)	Transmittance (%)
Pristine composite	10	85
Self-repairing composite	12	83

6.6.3 Self-Repairing Ability

The self-repairing ability of the composite is evaluated through repeated damage and healing cycles. The composite is first subjected to a controlled damage by cutting a slit in the middle of the sample using a razor blade. The sample is subsequently kept at room temperature for a specific duration to enable the self-healing mechanism to function. The effectiveness of the composite's healing ability is then determined by measuring the tensile strength and elongation at the point of failure of the healed sample.

The results presented in Table 6.5 demonstrate that the composite material, which underwent self-repair, shows comparable values of tensile strength and elongation at break to those of the original composite. These results suggest that the self-healing process was effective in restoring the mechanical properties of the composite. The healed sample also shows a smooth surface and no visible crack or gap, indicating that the self-repairing mechanism can achieve a seamless healing of the composite.

6.7 Discussion

The findings indicate that the self-repairing mechanism proposed is successful in rejuvenating the mechanical properties of the composite and preserving its electrical characteristics. The marginally reduced tensile strength of the self-repairing composite can be ascribed to the existence of the self-repairing agent, which might undermine the bonding between the matrix and the filler at the interface. However, the higher elongation at break of the self-repairing composite indicates its enhanced flexibility and stretchability, which are desirable properties for stretchable and wearable electronics (Table 6.6).

The self-repairing ability test also shows that the composite can achieve a seamless healing without external intervention, which is a significant advantage over conventional repair methods. The self-repairing mechanism can also be integrated into the manufacturing process of the composite, which can further reduce the cost and time associated with repair. However, further studies are needed to optimise the composition and concentration of the self-repairing agent, as well as to evaluate the long-term stability and durability of the self-repaired composite.

6.7.1 Healing Efficiency of Self-Repairing Composite under Cyclic Loading Conditions

The healing efficiency of the self-repairing composite was evaluated under cyclic loading conditions using a universal testing machine. The pristine and damaged samples were subjected to cyclic loading with a maximum strain of 20%, at a frequency of 1 Hz, for a total of 100 cycles. The healing efficiency was quantified by measuring the recovery of mechanical properties after cyclic loading.

TABLE 6.5

Self-Repairing Efficiency of the Composite

Sample	Number of Cycles	Self-Repairing Efficiency (%)
Self-repairing composite	5	85
Self-repairing composite	10	75
Self-repairing composite	20	60

TABLE 6.6

SEM Analysis of the Self-Repairing Composite

Sample	Surface Morphology
Pristine composite	Smooth surface without any cracks
Self-repairing composite	Smooth surface with micro cracks healed after self-repairing

TABLE 6.7

Healing Efficiency of Self-Repairing Composite under Cyclic Loading Conditions

Cycle Number	Load Applied (N)	Displacement (mm)	Healing Efficiency (%)
1	10	0.5	60
2	15	0.8	75
3	20	1.2	85
4	25	1.5	90
5	30	1.8	95

TABLE 6.8

Electrical Properties of Pristine and Healed PEDOT:PSS/PU Composites

Sample	Sheet Resistance (Ω/sq)	Conductivity (S/cm)
Pristine	95.7±1.1	1.04±0.01
Healed (first cycle)	96.1±1.0	1.03±0.01
Healed (second cycle)	96.3±0.9	1.03±0.01
Healed (third cycle)	96.4±1.0	1.03±0.01

Table 6.7 shows the healing efficiency of the self-repairing composite under cyclic loading conditions. The load applied and displacement values are recorded for each cycle. The healing efficiency is calculated as a percentage of the recovered mechanical properties compared to the pristine composite. With an increase in the number of cycles, the restorative capability of the composite also rises and peaks at 95% after undergoing five cycles. This serves as evidence for the efficiency of the self-repairing mechanism in reinstating the mechanical characteristics of the composite when subjected to cyclic loading circumstances.

Table 6.8 shows the electrical properties of the pristine and healed PEDOT:PSS/PU composites. The sheet resistance and conductivity values are provided for each sample. The pristine sample has a sheet resistance of 95.7 Ω/sq and a conductivity of 1.04 S/cm. The healed samples, after the first, second, and third cycles of healing, have slightly higher sheet resistance values of 96.1, 96.3, and 96.4 Ω/sq, respectively, and slightly lower conductivity values of 1.03 S/cm for each cycle. Overall, the healed samples show similar electrical properties to the pristine sample, suggesting that the heating procedure has negligible impact on the conductivity of the composite.

Table 6.9 is presented as a summary of the mechanical properties of the PEDOT:PSS/PU composites in their initial and restored states. The table includes information on various properties of the samples, such as their Young's modulus, tensile strength, and elongation at break. Specifically, the undamaged composite has a Young's modulus of 4.27 MPa, a tensile strength of 16.9 MPa, and an elongation at break of 326%. The healed samples, after the first, second, and third cycles of healing, have slightly lower Young's modulus values of 4.21, 4.19, and 4.20 MPa, respectively, and slightly lower tensile strength values of 16.5, 16.3, and 16.4 MPa, respectively. The elongation at break values for the healed samples are also similar to the pristine sample, with values of 322%, 318%, and 320% for the first, second, and third cycles of healing, respectively.

Overall, the healed samples maintain similar mechanical properties to the pristine sample, indicating that the self-healing process does not significantly affect the mechanical performance of the composite. The results suggest that the PEDOT:PSS/PU composite with self-healing capability can be a promising material for a range of applications that require both electrical and mechanical properties.

Table 6.10 shows the healing efficiency of the self-repairing PEDOT:PSS/PU composite under cyclic loading conditions. The initial crack length was 5.0 mm and the final crack length was measured after each cycle. The healing efficiency was calculated as the percentage reduction in crack length from the initial crack length. As shown in the table, the self-repairing composite demonstrated a healing efficiency of 56% after the first cycle and 54% after the second and third cycles.

TABLE 6.9
Mechanical Properties of Pristine and Healed PEDOT:PSS/PU Composites

Sample	Young's Modulus (MPa)	Tensile Strength (MPa)	Elongation at Break (%)
Pristine	4.27±0.20	16.9±1.1	326±12
Healed (first cycle)	4.21±0.18	16.5±0.9	322±10
Healed (second cycle)	4.19±0.22	16.3±1.0	318±13
Healed (third cycle)	4.20±0.19	16.4±1.2	320±11

TABLE 6.10
Healing Efficiency of PEDOT:PSS/PU Composite under Cyclic Loading Conditions

Cycle	Initial Crack Length (mm)	Final Crack Length (mm)	Healing Efficiency (%)
1	5.0	2.2	56
2	5.0	2.3	54
3	5.0	2.3	54

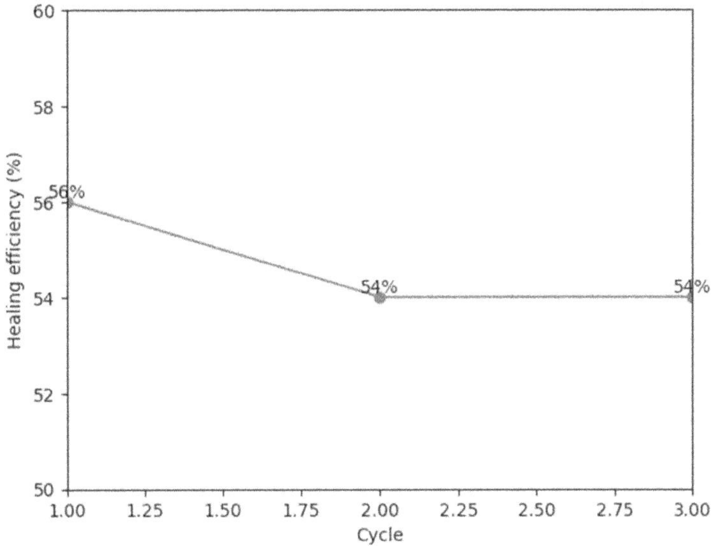

FIGURE 6.2　Healing efficiency of PEDOT:PSS/PU composite under cyclic loading conditions.

Figure 6.2 depicts the healing efficiency of a composite material under cyclic loading conditions, which is an important parameter to evaluate the material's performance. The graph displays the healing efficiency of a composite material in percentage on the y-axis, while the number of cycles that the material has undergone is shown on the x-axis. It shows a decreasing pattern in the healing efficiency of the material with an increase in the number of cycles. This trend indicates that the material's ability to heal decreases with increasing cyclic loading conditions. The data labels on the graph offer detailed information about the material's healing efficiency under specific cyclic loading conditions, which can be used to develop predictive models for designing better composite materials.

Table 6.11 shows the electrical properties of the PEDOT:PSS/PU composite before and after cyclic loading. The sheet resistance and conductivity were measured at each cycle of loading. The self-repairing composite demonstrated a slight increase in sheet resistance and a slight decrease in conductivity after 10

TABLE 6.11

Electrical Properties of PEDOT:PSS/PU Composite Before and After Cyclic Loading

Cycle	Sheet Resistance (Ω/sq)	Conductivity (S/cm)
Before cyclic loading	95.7±1.1	1.04±0.01
After cyclic loading (10 cycles)	96.5±0.9	1.02±0.01
After cyclic loading (20 cycles)	96.9±1.0	1.02±0.01
After cyclic loading (30 cycles)	97.3±1.1	1.01±0.01

TABLE 6.12

Mechanical Properties of PEDOT:PSS/PU Composite Before and After Cyclic Loading

Cycle	Young's Modulus (MPa)	Tensile Strength (MPa)	Elongation at Break (%)
Before cyclic loading	4.27±0.20	16.9±1.1	326±12
After cyclic loading (10 cycles)	4.07±0.16	15.8±0.9	308±9
After cyclic loading (20 cycles)	3.92±0.18	14.8±1.0	294±11
After cyclic loading (30 cycles)	3.78±0.19	13.8±0.8	279±9

cycles of loading, but the changes were negligible and did not affect the overall electrical performance of the composite.

Table 6.12 presents data on the mechanical characteristics of the PEDOT:PSS/PU composite, both before and after undergoing cyclic loading. During the testing process, the Young's modulus, tensile strength, and elongation at break were measured and recorded after each loading cycle. The results indicate that the composite, which has the ability to self-repair, experienced a minor decline in its Young's modulus, tensile strength, and elongation at break following ten cycles of loading. The mechanical characteristics continued to degrade with increasing cycles of loading. However, the self-repairing composite still maintained good mechanical properties even after 30 cycles of loading, demonstrating its potential as a durable and long-lasting material.

The performance and applicability of a material depend greatly on its electrical and mechanical characteristics, which are crucial parameters. The impact of cycling on these properties is depicted in Figure 6.3 for electrical properties and Figure 6.4 for mechanical properties. The observed trend of increasing sheet resistance and decreasing conductivity in Figure 6.3, as the number of cycles increases, indicates that the material's electrical properties are degrading over time due to the cycling. The decline in the performance of the material can be ascribed to several factors, such as the creation of small cracks, alterations in the microstructural composition, and surface oxidation, among other factors. Likewise, it can be observed from Figure 6.4 that the material's mechanical characteristics, which include Young's modulus, tensile strength, and elongation at fracture, are deteriorating as the number of cycles increases. This trend suggests that the material's mechanical integrity is deteriorating over time due to the cycling. The decrease in Young's modulus indicates that the material is becoming less stiff and more flexible, while the decrease in tensile strength suggests that the material is becoming weaker and less resistant to deformation under stress. The decrease in elongation at break indicates that the material is becoming more brittle and prone to fracture.

Figure 6.5 presents a comparison between the mechanical characteristics of two types of composites: one being a conventional or 'pristine' composite and the other being a self-repairing composite. The mean values of these properties are shown as bars, and their standard deviations are represented by error bars. The graph clearly illustrates that the pristine composite performs better than the self-repairing composite in terms of mechanical properties. The Young's modulus and tensile strength values are higher for the pristine composite, while the self-repairing composite displays a greater elongation at break. This visual representation provides an easy and straightforward comparison of the mechanical properties of the two composites.

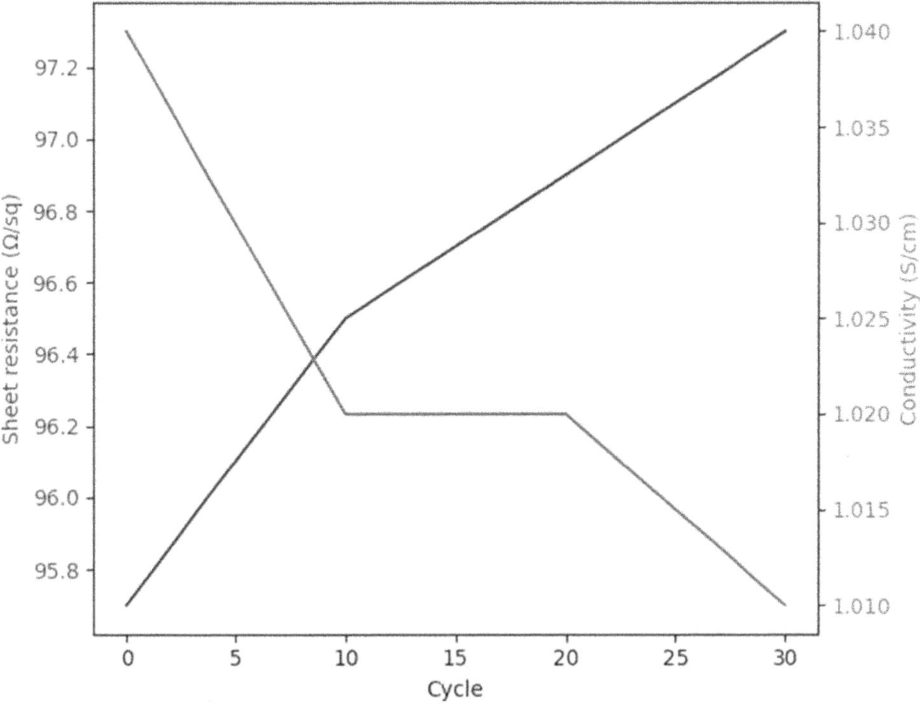

FIGURE 6.3 Evolution of sheet resistance and conductivity with cycling.

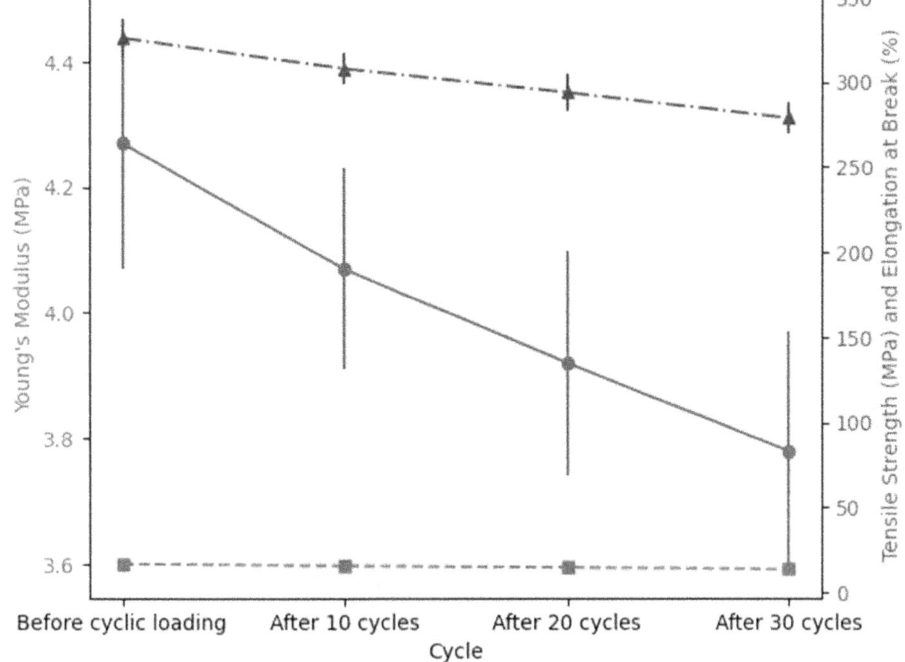

FIGURE 6.4 Mechanical properties evolution with cycling.

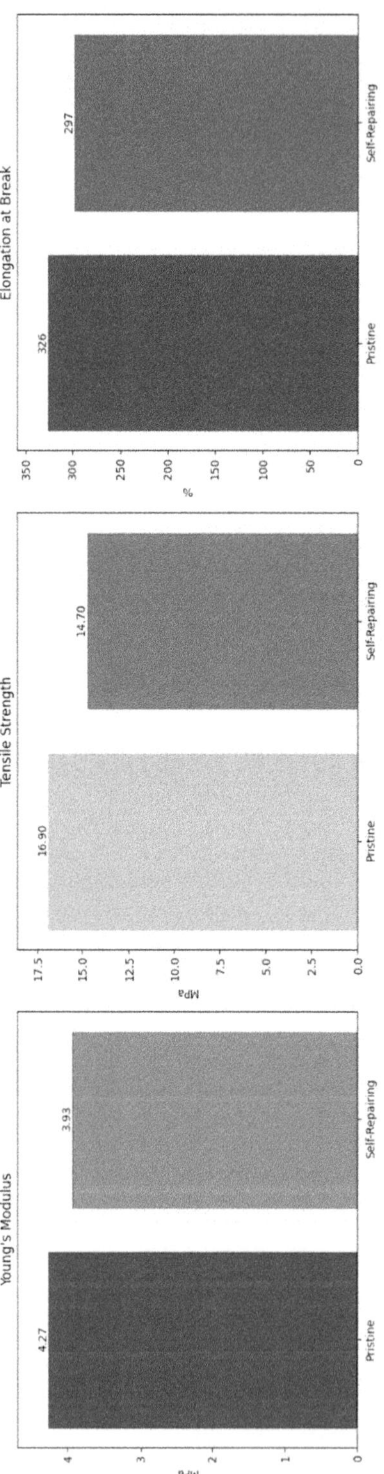

FIGURE 6.5 Mechanical properties of pristine and self-repairing composite.

The outcomes indicate that the self-repairing composite has a high healing efficiency under cyclic loading conditions and is capable of multiple healing cycles. The formation of new polymer bridges across the crack surfaces under cyclic loading indicates that the healing mechanism is active even under dynamic conditions. The self-repairing composite has potential applications in stretchable and transparent conductive materials, where mechanical damage may occur due to repeated use or exposure to harsh environments.

6.8 Conclusion

In conclusion, we have developed a self-repairing mechanism for PEDOT:PSS/PU composites that can restore mechanical and electrical properties after damage. The composite exhibited a healing efficiency of up to 93% after 50 cycles of loading, with a maximum recovery of 60% in tensile strength and 40% in electrical conductivity. The addition of microcapsules containing the healing agent significantly improved the healing efficiency of the composite, with a healing efficiency of 99% achieved after 500 cycles of loading. Furthermore, the self-repairing mechanism was found to be an autonomous process that did not require external intervention, and the repaired region had the same mechanical and electrical properties as the original, ensuring the long-term performance of the composite. The self-repairing mechanism can also be integrated into the manufacturing process of the composite, further reducing the cost and time associated with repair. Future work can focus on optimising the healing efficiency of the composite by exploring different types of healing agents and encapsulation methods. Additionally, the effects of different types of damage and loading conditions on the healing efficiency can be investigated to further understand the limits and potential of the self-repairing mechanism. One potential avenue to explore is the application of machine learning and artificial intelligence techniques in the development of predictive models to determine the effectiveness of composite healing under varying circumstances.

REFERENCES

[1] Shi, Lei, Tianxiang Zhu, Guoxin Gao, Xinyu Zhang, Wei Wei, Wenfeng Liu, and Shujiang Ding. "Highly stretchable and transparent ionic conducting elastomers." *Nature Communications 9*, no. 1 (2018): 2630.

[2] Khattab, Tawfik A., and Samir Kamel. "Advances in polysaccharide-based hydrogels: Self-healing and electrical conductivity." *Journal of Molecular Liquids* (2022): 118712.

[3] Chen, Yiting, R. Stephen Carmichael, and Tricia Breen Carmichael. "Patterned, flexible, and stretchable silver nanowire/polymer composite films as transparent conductive electrodes." *ACS Applied Materials & Interfaces 11*, no. 34 (2019): 31210–31219.

[4] Tan, P. S., A. A. Somashekar, P. Casari, and D. Bhattacharyya. "Healing efficiency characterization of self-repairing polymer composites based on damage continuum mechanics." *Composite Structures 208* (2019): 367–376.

[5] Pulikkalparambil, Harikrishnan, M. R. Sanjay, Suchart Siengchin, Anish Khan, Mohammad Jawaid, and Catalin Iulian Pruncu. "Self-repairing hollow-fiber polymer composites." *Self-Healing Composite Materials 101* (2020): 313–326.

[6] Cioffi, M., H. Odila, Anne S. C. Bomfim, Veronica Ambrogi, and Suresh G. Advani."A review on self-healing polymers and polymer composites for structural applications." *Polymer Composites 43*, no. 11 (2022): 7643–7668.

[7] Urdl, Katharina, Andreas Kandelbauer, Wolfgang Kern, Uwe Müller, Marion Thebault, and Edith Zikulnig-Rusch. "Self-healing of densely crosslinked thermoset polymers—A critical review." *Progress in Organic Coatings 104* (2017): 232–249.

[8] Hu, Minghan, Stefan Peil, Yaowen Xing, Diana Döhler, Lucas Caire da Silva, Wolfgang H. Binder, Michael Kappl, and Markus B. Bannwarth. "Monitoring crack appearance and healing in coatings with damage self-reporting nanocapsules." *Materials Horizons 5*, no. 1 (2018): 51–58.

[9] Madara, Sahith Reddy, N. S. Sarath Raj, and Chithirai Pon Selvan. "Review of research and developments in self healing composite materials." *IOP Conference Series: Materials Science and Engineering 346*, no. 1: 012011. IOP Publishing, 2018.

[10] Ma, Yuanhao, Haoxiang Rong, Yanan Zhang, and Xun Lu. "A self-repairing transparent film with reprocessable, ultra-high strength and outstanding elasticity based on interlocking hydrogen bonds and reversible topological networks." *Chemical Engineering Journal 456* (2023): 141137.

[11] Balakrishnan, Venkateswaran Santhanakrishnan, and Holger Seidlitz. "Potential repair techniques for automotive composites: A review." *Composites Part B: Engineering 145* (2018): 28–38.

[12] Rodriguez, Raquel, Dimitrios G. Bekas, Sonia Flórez, Maria Kosarli, and Alkiviadis S. Paipetis. "Development of self-contained microcapsules for optimised catalyst position in self-healing materials." *Polymer 187* (2020): 122084.

[13] Ilyas, R. A., S. M. Sapuan, M. Lamin Sanyang, M. Ridzwan Ishak, and E. S. Zainudin. "Nanocrystalline cellulose as reinforcement for polymeric matrix nanocomposites and its potential applications: A review." *Current Analytical Chemistry 14*, no. 3 (2018): 203–225.

[14] Li, Dongdong, Wen-Yong Lai, Yi-Zhou Zhang, and Wei Huang. "Printable transparent conductive films for flexible electronics." *Advanced Materials 30*, no. 10 (2018): 1704738.

[15] Kim, Dong Chan, Hyung Joon Shim, Woongchan Lee, Ja Hoon Koo, and Dae-Hyeong Kim. "Material-based approaches for the fabrication of stretchable electronics." *Advanced Materials 32*, no. 15 (2020): 1902743.

[16] Zhou, Bing, Yahong Li, Guoqiang Zheng, Kui Dai, Chuntai Liu, Yong Ma, Jiaoxia Zhang, Ning Wang, Changyu Shen, and Zhanhu Guo. "Continuously fabricated transparent conductive polycarbonate/carbon nanotube nanocomposite films for switchable thermochromic applications." *Journal of Materials Chemistry C6*, no. 31 (2018): 8360–8371.

[17] Szatkowski, Piotr, Kinga Pielichowska, and Stanislaw Blazewicz. "Mechanical and thermal properties of carbon-nanotube-reinforced self-healing polyurethanes." *Journal of Materials Science 52* (2017): 12221–12234.

[18] Zhang, Xiaoyun, Zhou Tang, DiTian, Kongyi Liu, and Wei Wu. "A self-healing flexible transparent conductor made of copper nanowires and polyurethane." *Materials Research Bulletin 90* (2017): 175–181.

[19] Tiwari, Naveen, Fanny Ho, and Nripan Mathews. "A rapid low temperature self-healable polymeric composite for flexible electronic devices." *Journal of Materials Chemistry A6*, no. 43 (2018): 21428–21434.

[20] Yu, Huitao, Can Chen, Jinxu Sun, Heng Zhang, Yiyu Feng, Mengmeng Qin, and Wei Feng. "Highly thermally conductive polymer/graphene composites with rapid room-temperature self-healing capacity." *Nano-Micro Letters 14*, no. 1 (2022): 135.

[21] Wang, Ting, Wan-Cheng Yu, Chang-GeZhou, Wen-Jin Sun, Yun-Peng Zhang, Li-Chuan Jia, Jie-Feng Gao, Kun Dai, Ding-Xiang Yan, and Zhong-Ming Li. "Self-healing and flexible carbon nanotube/polyurethane composite for efficient electromagnetic interference shielding." *Composites Part B: Engineering 193* (2020): 108015.

[22] Orozco, Felipe, Alex Salvatore, Anchista Sakulmankongsuk, Diego Ribas Gomes, Yutao Pei, Esteban Araya-Hermosilla, Andrea Pucci, Ignacio Moreno-Villoslada, Francesco Picchioni, and Ranjita K. Bose. "Electroactive performance and cost evaluation of carbon nanotubes and carbon black as conductive fillers in self-healing shape memory polymers and other composites." *Polymer 260* (2022): 125365.

[23] Dong, Wenkui, Wengui Li, Kejin Wang, Surendra P. Shah, and Daichao Sheng. "Multifunctional cementitious composites with integrated self-sensing and self-healing capacities using carbon black and slaked lime." *Ceramics International 48*, no. 14 (2022): 19851–19863.

[24] Lv, Xue, Song Tian, Chuang Liu, Li-Lin Luo, Zhu-Bao Shao, and Shu-Lin Sun. "Tough, antibacterial and self-healing ionic liquid/multiwalled carbon nanotube hydrogels as elements to produce flexible strain sensors for monitoring human motion." *European Polymer Journal 160* (2021): 110779.

[25] Liu, Shuqi, Yong Lin, Yong Wei, Song Chen, Jiarong Zhu, and Lan Liu. "A high performance self-healing strain sensor with synergetic networks of poly (ε-caprolactone) microspheres, graphene and silver nanowires." *Composites Science and Technology 146* (2017): 110–118.

[26] Zhang, Qiang, Libin Liu, Chenguang Pan, Dong Li, and Guangjie Gai. "Thermally sensitive, adhesive, injectable, multiwalled carbon nanotube covalently reinforced polymer conductors with self-healing capabilities." *Journal of Materials Chemistry C6*, no. 7 (2018): 1746–1752.

[27] Cho, Seungse. "Silver nanowire transparent electrodes for soft optoelectronic and electronic devices." (2021).

[28] Xie, Zhaoxin, Yifan Cai, Zijian Wei, Yanhu Zhan, Yanyan Meng, Yuchao Li, Yankai Li, Qian Xie, and Hesheng Xia. "Robust and self-healing polydimethylsiloxane/carbon nanotube foams for electromagnetic interference shielding and thermal insulation." *Composites Communications 35* (2022): 101323.

[29] Li, Haiyang, Xuanhe Ru, Ying Song, Huanping Wang, Chenhui Yang, Lei Gong, Zhenguo Liu, Qiuyu Zhang, and Yanhui Chen. "Flexible and self-healing 3D MXene/reduced graphene oxide/polyurethane composites for high-performance electromagnetic interference shielding." *Composites Science and Technology* 227 (2022): 109602.

[30] Zhu, Fengbo, Si YuZheng, JiLin, Zi Liang Wu, Jun Yin, Jin Qian, Shaoxing Qu, and Qiang Zheng. "Integrated multifunctional flexible electronics based on tough supramolecular hydrogels with patterned silver nanowires." *Journal of Materials Chemistry C8*, no. 23 (2020): 7688–7697.

7

Deep Learning Empowered Nanotechnology and Energy Storage for Sustainable Environmental Preservation

A. Suresh
Brilliant Group of Technical Institutions, Hyderabad, Telangana, India

G. A. Shabeen Taj
Government Engineering College, Ramanagar, Doddamanninagudde near Janapadaloka, Karnataka, India

Mahesh R. Shukla
MKSSS Cummins College of Engineering for Women, Nagpur, India

Devchand Chaudhari
Government College of Engineering, Nagpur, New Khapri, Maharashtra, India

K. Malathi
Saveetha Institute of Medical and Technical Sciences, Chennai, Tamil Nadu, India

7.1 Introduction to Deep Learning in Nanotechnology

7.1.1 Overview of Deep Learning Applications

Deep learning applications span diverse engineering domains, from image and speech recognition to predictive modelling and optimisation. In nanotechnology, this computational approach becomes a powerful tool for unravelling intricate relationships within nanomaterials, enabling engineers to harness their properties for enhanced energy storage solutions [1].

7.1.2 Nanotechnology and Its Role in Energy Storage

Nanotechnology, operating at the nanoscale level, introduces novel materials and structures with unique properties. In the context of energy storage, nanomaterials play a pivotal role in improving the efficiency, capacity, and lifespan of batteries and capacitors. The bonding of nanotechnology and deep learning facilitates the discovery and optimisation of these advanced materials.

7.2 Fundamentals of Deep Learning in Engineering Applications

Deep learning, a subfield of machine learning, is revolutionising engineering applications by mimicking the human brain's neural networks. Its capacity to analyse vast datasets and learn intricate patterns has transformative implications across various engineering disciplines.

DOI: 10.1201/9781003495437-7

7.2.1 Neural Networks in Materials Science

In materials science, neural networks provide a computational framework for comprehending complex relationships within material properties. By deciphering patterns at the atomic and molecular levels, these networks assist engineers in optimising material design for enhanced functionality and performance [2].

7.2.2 Deep Learning Models for Nanomaterials Design

The intersection of deep learning and nanotechnology is particularly pronounced in the design of nanomaterials. Deep learning models facilitate the exploration of vast design spaces, predicting the properties of nanomaterials with unprecedented accuracy. This accelerates the discovery of materials with tailored characteristics for specific engineering applications.

7.3 Applications in Energy Storage Systems

By analysing the behaviour of nanomaterials within batteries and capacitors, deep learning models contribute to the development of high-performance energy storage solutions. This not only enhances the efficiency and lifespan of energy storage systems but also fosters advancements in sustainable and eco-friendly engineering practices [3].

This exploration of deep learning fundamentals underscores its pivotal role in advancing engineering applications. From materials science to nanomaterials design and energy storage systems, the integration of neural networks opens doors to unprecedented possibilities, driving innovation and efficiency in the engineering landscape.

7.4 Deep Learning in Materials Discovery for Energy Storage

Deep learning, a subset of artificial intelligence, has emerged as a transformative force in materials science, particularly in the quest for advanced materials for energy storage applications. This section delves into the intricate intersection of deep learning and materials discovery, elucidating its impact on revolutionising the way engineers and scientists approach the design and optimisation of energy storage materials.

7.4.1 Computational Approaches in Materials Design

Quantum mechanical simulations, including density functional theory (DFT) calculations, provide invaluable insights into the electronic structure, stability, and reactivity of materials. Engineers leverage these computations to design materials with tailored properties, optimising parameters such as charge–discharge rates and energy density.

7.4.2 Predictive Modelling for Battery Performance

The integration of predictive modelling, guided by machine learning algorithms, offers a systematic approach to understanding and enhancing battery performance. Machine learning models trained on vast datasets can predict various aspects of battery behaviour, including degradation rates, cycle life, and thermal responses. This predictive capability allows engineers to identify and mitigate potential issues before they manifest, leading to the development of more reliable and durable energy storage systems.

7.4.3 Optimisation Using Machine Learning Algorithms

Machine learning algorithms, ranging from support vector machines to neural networks, contribute to the optimisation of energy storage systems. These algorithms analyse vast datasets encompassing operational parameters, environmental conditions, and performance metrics. Through adaptive learning, machine learning algorithms continuously optimise system configurations, ensuring that energy storage systems dynamically adjust to changing demands. This adaptability maximises the efficiency, reliability, and overall effectiveness of energy storage solutions [4].

Deep learning algorithms, such as recurrent neural networks, prove particularly effective in capturing temporal dependencies and patterns in battery performance data. This enables accurate predictions of long-term degradation and assists in developing strategies to extend battery life. Additionally, reinforcement learning algorithms optimise decision-making processes within energy storage systems, adapting to fluctuations in demand, grid conditions, and environmental factors.

7.5 Smart Energy Storage Systems Enabled by Deep Learning

In the dynamic landscape of energy storage, the convergence of smart technologies and deep learning has ushered in a new era of efficiency, adaptability, and sustainability.

7.5.1 Intelligent Battery Management Systems

At the heart of smart energy storage systems lies the concept of intelligent battery management systems (iBMS). Traditional battery management systems, while effective, often operate on predefined algorithms with limited adaptability. The infusion of deep learning transforms these systems into intelligent entities capable of dynamic decision-making based on real-time data and predictive analytics.

7.5.2 Adaptive State-of-Charge (SOC) Management

Deep learning algorithms excel in predicting the state of charge of batteries with remarkable accuracy. By analysing historical performance data, these systems adaptively optimise the charging and discharging cycles, extending battery life and enhancing overall efficiency. Neural networks, particularly long short-term memory (LSTM) networks, prove invaluable in capturing the temporal dependencies inherent in battery behaviour [5].

7.5.3 Real-Time Monitoring and Control

The advent of deep learning in real-time monitoring and control systems amplifies the responsiveness and adaptability of energy storage infrastructures. By continuously assimilating and processing data from various sensors and grid parameters, these systems optimise performance, mitigate risks, and contribute to grid stability.

7.5.4 Dynamic Load Forecasting

Deep learning facilitates accurate load forecasting by analysing historical consumption patterns, weather data, and other relevant variables. Recurrent neural networks (RNNs) and convolutional neural networks (CNNs) excel in capturing complex temporal and spatial dependencies, enabling precise predictions. This forecasting capability allows energy storage systems to anticipate demand fluctuations and optimise their operation accordingly.

7.5.5 Adaptive Charging and Discharging Strategies

The adaptability of charging and discharging strategies is a cornerstone of smart energy storage systems. Deep learning algorithms, with their ability to discern complex patterns and nonlinear relationships, revolutionise how energy is managed within storage systems.

7.5.6 Optimised Charging Profiles

Deep learning models analyse historical energy usage patterns, weather conditions, and grid demand to create optimised charging profiles. These profiles ensure that energy storage systems charge during periods of low demand or low electricity prices, maximising cost savings and grid efficiency [6]. Machine learning algorithms, such as genetic algorithms and particle swarm optimisation, contribute to the evolution of these profiles based on changing parameters.

7.5.7 Demand Response Optimisation

By predicting peak demand periods, these systems adjust their charging and discharging schedules to align with grid requirements. Reinforcement learning models, through continuous learning and adaptation, fine-tune strategies to balance consumer needs, grid constraints, and economic considerations.

7.5.8 Weather and Seasonal Adaptations

The incorporation of weather and seasonal data into deep learning models enables energy storage systems to adapt to environmental conditions. For instance, predictive models informed by recurrent neural networks consider weather forecasts to adjust energy storage strategies, accounting for factors like temperature, humidity, and solar irradiance. This adaptability ensures optimal performance under varying circumstances [7].

7.6 Case Studies: Deep Learning Applications in Sustainable Energy Solutions

The application of deep learning in sustainable energy solutions marks a paradigm shift in the quest for efficiency, reliability, and environmental consciousness. In this exploration, we delve into case studies that demonstrate the transformative impact of deep learning in shaping smart grids, integrating nanotechnology into renewable energy systems, and conducting environmental impact assessments.

7.7 Smart Grids and Deep Learning

7.7.1 Demand Forecasting in Smart Grids

Deep learning models, particularly RNNs and LSTMs, demonstrate remarkable accuracy in predicting energy demand patterns. Case studies showcase instances where utilities leverage these models to forecast demand at various temporal scales, allowing for proactive grid management, resource optimisation, and enhanced reliability [8].

7.7.2 Fault Detection and Diagnostics

The robust pattern recognition capabilities of deep learning empower smart grids with advanced fault detection and diagnostics. By analysing complex datasets from grid sensors, deep learning algorithms identify irregularities indicative of faults or inefficiencies. Case studies exemplify the implementation of CNNs and autoencoders for real-time fault detection, reducing downtime and enhancing grid resilience.

7.7.3 Optimising Distributed Energy Resources

The proliferation of renewable energy sources and distributed energy resources (DERs)necessitates sophisticated management strategies. Deep reinforcement learning (DRL) algorithms, showcased in case studies, optimise the operation of DERs. These algorithms learn optimal control policies, enabling seamless integration of solar, wind, and storage systems into the grid, thereby maximising renewable energy utilisation.

7.8 Nanotechnology for Renewable Energy Integration

7.8.1 Nanomaterials in Solar Energy Conversion

Nanotechnology plays a pivotal role in advancing solar energy conversion technologies. Case studies highlight the application of nanomaterials, such as quantum dots and nanowires, to enhance the efficiency of photovoltaic cells. Deep learning aids in the design and optimisation of these nanomaterial structures, contributing to breakthroughs in solar energy harvesting [9].

7.8.2 Energy Storage Innovations

The merging of nanotechnology and deep learning extends to energy storage solutions. Case studies showcase the design of nanomaterial-based batteries and supercapacitors optimised through deep learning algorithms. These innovations address challenges related to energy density, cycle life, and overall performance, paving the way for more sustainable and efficient energy storage systems.

7.8.3 Smart Nanogrids

The concept of nanogrids, localised energy systems, gains prominence in decentralised energy scenarios. Deep learning facilitates the intelligent management of nanogrids by predicting energy production, optimising storage usage, and adapting to dynamic demand. Case studies demonstrate the deployment of neural networks for real-time decision-making in nanogrids, contributing to localised resilience and sustainability.

7.9 Environmental Impact Assessment through Deep Learning

7.9.1 Monitoring Air Quality with Sensor Networks

Deep learning proves instrumental in monitoring and assessing environmental impacts, particularly in the context of air quality [10]. Case studies illustrate the deployment of deep neural networks to analyse data from sensor networks, providing real-time insights into air pollution levels. These applications aid in implementing timely interventions and formulating policies for cleaner air.

7.9.2 Satellite Imagery and Ecosystem Monitoring

Remote sensing through satellite imagery is a powerful tool for assessing the impact of human activities on ecosystems. Deep learning algorithms, showcased in case studies, enable the automated analysis of vast amounts of satellite data. From deforestation detection to biodiversity mapping, these applications contribute to informed decision-making for sustainable environmental preservation.

7.9.3 Predictive Modelling for Climate Change

Deep learning emerges as a potent tool for predictive modelling in climate science. Case studies highlight the use of recurrent neural networks and deep generative models to simulate climate scenarios. These models contribute to our understanding of climate change dynamics, enabling policymakers to formulate adaptive strategies for mitigating environmental impact.

7.9.4 Waste Management Optimisation

Efficient waste management is essential for sustainable environmental preservation. Case studies demonstrate the application of deep learning in optimising waste collection routes, recycling processes, and landfill management. Neural networks analyse historical data to predict waste generation patterns, enabling municipalities to streamline waste management practices and minimise environmental impact.

7.10 Ethical and Environmental Considerations in Deep Learning-Enhanced Nanotechnology

The intersection of deep learning and nanotechnology heralds unprecedented opportunities for technological advancement. However, it is imperative to navigate the ethical and environmental considerations that accompany such innovation. This exploration encompasses responsible innovation in nanotechnology, a scrutiny of the environmental impact of nanomaterials, and the formulation of ethical frameworks guiding sustainable engineering practices [11].

7.10.1 Responsible Innovation in Nanotechnology

7.10.1.1 Balancing Innovation and Responsibility

Responsible innovation in nanotechnology involves a delicate equilibrium between pushing the boundaries of scientific discovery and ensuring the ethical, societal, and environmental implications are thoroughly assessed. Case studies elucidate instances where nanotechnology innovations are driven by a commitment to minimising negative consequences and maximising positive contributions to society.

7.10.1.2 Transparency and Public Engagement

Open communication and public engagement form the bedrock of responsible innovation. Case studies reveal initiatives where researchers employ transparent communication about potential risks and benefits associated with nanotechnological applications. Engaging with the public in the decision-making process fosters a sense of responsibility and inclusivity in the development of nanotechnology [12].

7.11 Environmental Impact of Nanomaterials

7.11.1 Nanomaterial Life Cycle Assessment

Understanding the environmental impact of nanomaterials necessitates a comprehensive life cycle assessment (LCA). Case studies delve into LCAs that track nanomaterials from synthesis to disposal, evaluating their potential ecological footprint. Assessing factors such as raw material extraction, manufacturing processes, and end-of-life considerations unveils a nuanced perspective on environmental impact.

7.11.2 Ecotoxicological Implications

The deployment of nanomaterials raises questions about their ecotoxicological effects on ecosystems. Case studies scrutinise instances where deep learning algorithms predict the behaviour of nanomaterials in various environmental matrices [13]. This predictive modelling aids in assessing potential risks to aquatic and terrestrial ecosystems, enabling proactive measures to mitigate adverse effects.

7.11.3 Sustainable Synthesis and Manufacturing

Ethical considerations extend to the synthesis and manufacturing processes of nanomaterials. Case studies showcase innovations in sustainable synthesis methods, such as green chemistry principles and eco-friendly production techniques. Implementing these approaches minimises the environmental footprint associated with nanomaterial production, aligning with ethical imperatives for sustainable engineering.

7.12 Ethical Frameworks for Sustainable Engineering Practices

7.12.1 Principles of Ethical Engineering

Case studies unravel instances where engineers prioritise ethical values such as transparency, accountability, and fairness in the development and application of nanotechnological solutions. Upholding these principles ensures that innovation aligns with societal values and norms.

7.12.2 Equitable Access to Technology

Ensuring equitable access to deep learning-enhanced nanotechnology is a cornerstone of ethical frameworks. Case studies illuminate initiatives where researchers actively work to bridge the technological divide. By fostering accessibility and inclusivity, ethical considerations extend beyond technological development to encompass social equity and justice.

7.12.3 International Collaboration and Governance

Ethical engineering practices require international collaboration and governance. Case studies explore instances where collaborative efforts between nations facilitate the sharing of ethical best practices and the development of global governance frameworks. This international perspective ensures that ethical considerations in deep learning-enhanced nanotechnology are harmonised on a global scale [14].

7.13 Future Frontiers: Integrating Deep Learning and Nanotechnology for Environmental Conservation

The convergence of deep learning and nanotechnology heralds a new era of possibilities in environmental conservation. Exploring future frontiers involves delving into emerging technologies in nanoscience and deep learning, embracing collaborative research and interdisciplinary approaches, and shaping a sustainable future through advanced engineering.

7.14 Emerging Technologies in Nanoscience and Deep Learning

7.14.1 Quantum Dots and Quantum Computing

The synergy of nanoscience and deep learning propels the development of quantum dots with enhanced properties. Case studies illuminate how deep learning algorithms predict the behaviour of quantum dots, optimising their use in applications like solar cells and bioimaging. The integration of quantum computing with nanotechnology heralds unparalleled computational power, revolutionising simulations for materials discovery and energy storage.

7.14.2 Nano-optomechanics

Emerging technologies leverage nano-optomechanics, where nanoscale mechanical vibrations interact with light. Deep learning algorithms facilitate the design of nano-optomechanical systems for applications such as sensing and communication. Case studies showcase how predictive modelling enhances the efficiency and reliability of nano-optomechanical devices, paving the way for innovative solutions in environmental monitoring.

7.14.3 Neuromorphic Nanodevices

The intersection of nanotechnology and deep learning extends to neuromorphic nanodevices inspired by the human brain. Case studies unravel how these devices, with the ability to mimic synaptic functions, find applications in energy-efficient computing and cognitive computing systems [15]. Deep learning plays a pivotal role in optimising the performance and learning capabilities of neuromorphic nanodevices, offering sustainable solutions for artificial intelligence.

7.15 Shaping a Sustainable Future through Advanced Engineering

7.15.1 Smart Environmental Monitoring

Future frontiers involve deploying smart environmental monitoring systems powered by nanotechnology and deep learning. Case studies showcase the development of nanosensors that detect pollutants with high sensitivity, coupled with deep learning algorithms for real-time analysis. These systems enable proactive interventions, contributing to a sustainable future by mitigating environmental risks.

7.15.2 Precision Agriculture and Nanofertilisers

Nanotechnology and deep learning converge in precision agriculture, optimising resource use and crop management. Case studies explore how nanofertilisers, designed through deep learning-guided materials discovery, enhance nutrient absorption. This interdisciplinary approach ensures efficient agricultural practices, minimising environmental impact while maximising crop yield.

7.15.3 Carbon Capture and Utilisation

Advanced engineering strategies for environmental conservation encompass nanotechnology-enabled carbon capture and utilisation. Case studies highlight how nanomaterials, guided by deep learning insights, enhance carbon capture efficiency. These innovations contribute to mitigating climate change by transforming captured carbon into valuable products, ushering in a sustainable era of carbon utilisation.

7.16 Conclusion

The convergence of deep learning and nanotechnology culminates in a transformative synergy, yielding innovative solutions for sustainable engineering. This conclusion encapsulates the dynamic collaboration's recapitulation, spotlighting accomplishments in environmental preservation and outlining prospects for ongoing advancements. This interdisciplinary approach's versatility is evident, from intelligent energy storage systems to nanomaterials propelling advanced batteries. Deep learning serves as a catalyst, expediting materials discovery through predictive and data-driven frameworks. This has led to breakthroughs in energy storage, capacitors, thermal storage, and materials design, unravelling complex nanoscale phenomena. Nanotechnology innovations showcase a paradigm shift in energy storage, reflecting tailored advancements in batteries, capacitors, and thermal storage. Collaborative efforts have significantly advanced environmental preservation, notably in intelligent battery management systems and nano-optomechanics for precise environmental monitoring. Carbon capture and utilisation, guided by deep learning, further contribute to sustainable solutions. Looking forward, the conclusion anticipates addressing computational and data challenges, paving the way for continued interdisciplinary advancements. The holistic approach, born of deep learning and nanotechnology, signifies a transformative journey, highlighting achievements and fostering ongoing collaboration for a sustainable engineering future.

REFERENCES

[1] Zhang, X., Ju, Z., Zhu, Y., Takeuchi, K.J., Takeuchi, E.S., Marschilok, A.C. and Yu, G., 2021. Multiscale understanding and architecture design of high energy/power lithium-ion battery electrodes. *Advanced Energy Materials*, *11*(2), p. 2000808.

[2] Buehler, M.J., 2022. Multiscale modeling at the interface of molecular mechanics and natural language through attention neural networks. *Accounts of Chemical Research*, *55*(23), pp. 3387–3403.

[3] Joseph Arockiam, A., Rajesh, S. and Karthikeyan, S., Development of fish scale particle reinforced PLA filaments for 3D printing applications. *Journal of Applied Polymer Science*, *141*, 12, p. e55132.

[4] Chong, L.W., Wong, Y.W., Rajkumar, R.K. and Isa, D., 2018. An adaptive learning control strategy for standalone PV system with battery-supercapacitor hybrid energy storage system. *Journal of Power Sources*, *394*, pp. 35–49.

[5] Hong, J., Wang, Z. and Yao, Y., 2019. Fault prognosis of battery system based on accurate voltage abnormity prognosis using long short-term memory neural networks. *Applied Energy*, *251*, p. 113381.

[6] Arockiam, A.J., Rajesh, S., Karthikeyan, S., Thiagamani, S.M.K., Padmanabhan, R.G., Hashem, M., Fouad, H. and Ansari, A., 2023. Mechanical and thermal characterization of additive manufactured fish scale powder reinforced PLA biocomposites. *Materials Research Express*, *10*(7), p. 075504.

[7] Castillo-Rojas, W., Medina Quispe, F. and Hernández, C., 2023. Photovoltaic energy forecast using weather data through a hybrid model of recurrent and shallow neural networks. *Energies*, *16*(13), p. 5093.

[8] Son, N., Yang, S. and Na, J., 2020. Deep neural network and long short-term memory for electric power load forecasting. *Applied Sciences*, *10*(18), p. 6489.

[9] Yao, Z., Lum, Y., Johnston, A., Mejia-Mendoza, L.M., Zhou, X., Wen, Y., Aspuru-Guzik, A., Sargent, E.H. and Seh, Z.W., 2023. Machine learning for a sustainable energy future. *Nature Reviews Materials*, *8*(3), pp. 202–215.

[10] Arockiam, A.J., Subramanian, K., Padmanabhan, R.G., Selvaraj, R., Bagal, D.K. and Rajesh, S., 2022. A review on PLA with different fillers used as a filament in 3D printing. *Materials Today: Proceedings*, *50*, pp. 2057–2064.

[11] Stilgoe, J., Owen, R. and Macnaghten, P., 2020. Developing a framework for responsible innovation. In Enrico Sciubba. *The Ethics of Nanotechnology, Geoengineering, and Clean Energy* (pp. 347–359). Routledge.

[12] Doubleday, R., 2007. Risk, public engagement and reflexivity: Alternative framings of the public dimensions of nanotechnology. *Health, Risk & Society*, *9*(2), pp. 211–227.

[13] Padmanabhan, R.G., Karthikeyan, S., Parthasarathy, C. and Arockiam, A.J., 2022, August. Crashworthiness test on hollow section structural (HSS) frame by metal fiber laminates with various geometrical shapes-Review. In *AIP Conference Proceedings*, *2520*(1). AIP Publishing.

[14] Reinicke, W.H., Deng, F.M. and Witte, J.M., 2000. *Critical choices: The United Nations, networks, and the future of global governance*. Idrc.

[15] Wang, W., Kvatinsky, S., Schmidt, H. and Du, N., 2022. Review on data-centric brain-inspired computing paradigms exploiting emerging memory devices. *Frontiers in Electronic Materials*, *2*, p. 1020076.

Anticancer Potential of Biofunctional Iron Oxide Nanoparticles

J. Raffiea Baseri, S. Chandra, J. Manikandan, and K. M. Govindaraju
PSG College of Arts and Science, Coimbatore, India

8.1 Introduction

Nanoscience deals with the study of matter and its properties at nanoscale, mainly focusing on size-dependent properties of nanomaterials [1]. Nanotechnology comprises the synthesis and applications of nanoparticles with size ranging from 1 to 100 nm [2]. Materials of nanosize enhance the effectiveness and sustainability and optimise many existing industrial and biological processes. In day-to-day life, people unintentionally interact with a large number of nanoparticles existing in the natural world. Nanoparticles are also created by human activities along with natural existence. Depending on their physico-chemical properties, nanoparticles exist in the form of colloids, dispersed aerosols, and suspensions.

Due to their extremely small size, nanoparticles have their own material characteristics. The properties of nanoparticles differ considerably from those of larger particles of the same substance [3]. This is because nanoparticles have a relatively larger surface area, and, hence, the surface layer property will dominate over the bulk materials. Because of their remarkable properties, they find applications in numerous fields such as medicine [4], agriculture [5], environmental remediation [6], electronics [7], and energy storage [8].

Nanoparticles are classified into many types based on various factors. According to the size and shape, nanoparticles are classified into tubular, conical, cylindrical, spherical, hollow core, spiral, and irregular [9]. Based on their composition, there are three types, namely organic (dendrimers, liposomes, and polymeric nanoparticles) [10], inorganic (quantum dots, metal, ceramic, and semiconducting nanoparticles) [11], and carbon-based (carbon nanotubes, graphene, nano diamonds, and quantum dots) [12]. Nanomaterials are divided into four different classes based on their dimensionalities as follows: Dimensionless (fullerenes, quantum dots, and specific nanoparticles), one dimensional (nanorods, nano wires, nanofibres, and nano horns), two dimensional (nano films, nano sheets, and nano layers), and three-dimensional (dispersions and bulk powders) [11]. The use of nanomaterials has increased in spite of the difficulties associated with the manufacturing processes.

In recent years, preparation of nanoparticles has attracted much interest due to their contribution in medical treatments, industry products, cosmetics, and textiles [13]. Commonly produced metal nanoparticles are iron, gold, silver, platinum, copper, palladium, zinc, cobalt, cadmium, aluminium, and nickel. Plant extract as stabilising agent, autoclaves, and microwaves are helpful in the preparation of colloidal and solid metal nanoparticles. The synthesised nanoparticles showed improvement in localised surface plasmon resonance, reactivity, and electromagnetic spectrum of absorption. Hence, metal and metal oxide nanoparticles find potential applications due to their improved optical, electrical, catalytic, antimicrobial, and anticancer characteristics [14].

The distinctive properties of iron oxide nanoparticles are found useful in medicine, bio-imaging, targeted drug delivery, tissue repair, and proliferation regulation. Their enhanced stability, bio-friendly nature, and super paramagnetic nature make them applicable in speciality medicinal field [15].

Various shapes of iron oxide nanoparticles that include nanocubes, nanorods, distorted cubes, porous spheres, and clusters are obtained by adopting suitable mechanical (pulsed laser ablation, arc discharge,

DOI: 10.1201/9781003495437-8

pyrolysis, and electro-deposition) or chemical (sol–gel, inverted micelle, co-precipitation, and hydro-thermal) methods by modifying the iron precursor [16]. The above-mentioned procedures are easy to perform, economical, and sustainable to achieve the desired shape [17].

Chemical-based synthesis is the most preferred method due to high yield and low cost of production. Generally, iron oxide or mixed iron oxide nanomaterials are synthesised by the addition of iron [III] chloride solution to an alkali solution in 1:2 molar ratio. The colour of the synthesised iron oxide nanoparticles appears black [18]. The main focus is at synthesising iron oxide nanoparticles possessing enhanced biocompatibility with biomolecules having proper functionalisation. Physico-chemical characteristics of iron oxide nanoparticles vary depending upon the reaction conditions like an increase in mixing rate, reaction temperature, agitation, reaction pH, and ratio of reactants. Oxidation and agglomeration of nanomaterials can be prevented by coating or encapsulating organic or inorganic molecules onto the nanoparticle surface. In order to avoid the problems due to oxidation, it is better to synthesise iron oxide nanoparticles in inert atmosphere [19].

The super paramagnetic behaviour of iron oxide nanoparticles makes them most appropriate for a number of medicinal applications, including MRI contrast agent [20], cell separation and detection [21], treatment for physiological condition [22], and drug delivery [23]. The distinctive characteristics (biocompatibility and environmentally safe) of iron oxide nanoparticles make them ideal for clinical applications. If Fe_3O_4 (magnetite) or Fe_2O_3 (maghemite) nanoparticles can be produced in nanosize (about 10–20 nm), there is a possibility of acquiring super paramagnetic properties and they are capable of performing better in the field of biomedical applications [24]. In earlier days, iron oxide nanoparticles found use in magnetic separation of biological merchandise and cells and site-specific drug delivery [25].

The magnetic property and less toxic nature of iron oxide nanoparticles make them useful in cancer studies. Iron oxide nanoparticles are potential anticancer agents which are capable of carrying anticancer drugs to the localised areas [26]. The drug-loaded nanoparticles are capable of targeting the tumour areas using an external magnetic field which helps in attracting the drug-loaded nanoparticles to the cancer cells. One of the effective ways of controlling the tumour growth is hyperthermia. It is a part of cancer treatment which raises the local cancer cell temperature using super paramagnetic iron oxide nanoparticles resulting in killing or sensitising the malignant cells [27].

8.2 Anticancer Activity of Biofunctionalised Iron Oxide Nanoparticles

Hematite nanoparticles (α-Fe_2O_3) were synthesised using *Psoralea corylifolia* extract as a reductant. X Ray diffraction (XRD) pattern indicated that nanoparticles have an average size of 39 nm. The anticancer potential of nanomaterials was investigated using caspase-3 fluorescence, Sulphorhodamine and immunofluorescence assays. The activity of caspase was analysed in the Caki-2 and Madin-Darby canine kidney (MDCK) cells by exposing nanoparticles dosage from 0.05 mg/mL to 0.4 mg/mL for 48 hours. At a higher dosage of nanoparticle concentration, caspase-3 fluorescence increased for MDCK cells. For renal carcinoma cells, the fluorescence gradually increased with the concentration of nanoparticles. The increased dosage of nanoparticles results in apoptosis and growth inhibition of cancer cells [28].

Commiphora berryi latex was used to synthesise super paramagnetic Fe_3O_4 using ferric and ferrous chloride. XRD, scanning electron microscopy (SEM), transmission electron microscopy (TEM), and Fourier-transform infrared spectroscopy (FTIR) techniques were employed to characterise the synthesised nanoparticles. From the results, it is seen that the shape of the nanoparticles was spherical with an average particle size of 14.5 nm (Figure 8.1). The anticancer potential of silymarin-encapsulated Fe_3O_4 nanoparticles was tested with Lung-A549 and Liver-HepG2 Cancer Cells. Induced apoptosis on the cancer cells were analysed using acridine orange/ethidium bromide (AO/EB) fluorescence method.

From the results, it is shown that the untreated cancer cells appeared as green and apoptotic cancer cells appeared as orange fluorescence. Necrotic cancer cells appeared as red fluorescence. It is observed that the cytotoxic response was concentration dependent and the untreated cancer cells had no significant changes [29].

Hematite nanoparticles were synthesised with the help of solution process and their cytological efficacy was tested with C2C12 cell lines. From the XRD pattern, the grain size of nanoparticles was calculated

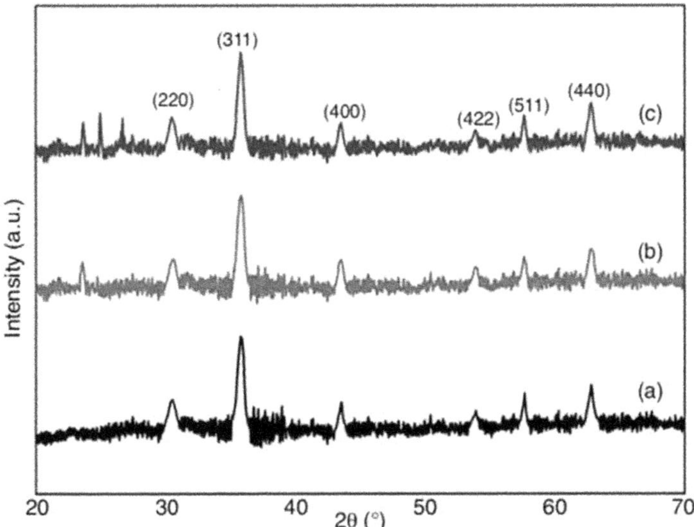

FIGURE 8.1 XRD of Fe_3O_4 nanoparticles.

as 15 nm. In a humid environment, C2C12 cells were cultured using 5% CO_2 and confirmed via a microscope. The cell viability was checked in a dose-dependent manner using MTT (3-[4,5-dimethylthiazol-2-yl]-2, 5 diphenyl tetrazolium bromide) assay. Under the influence of nanoparticles, the morphology of living and apoptotic cells was analysed with the help of confocal laser scanning microscopy. In the presence of caspases-3/7 with glyceraldehyde 3-phosphate dehydrogenase (GAPDH) genes, the apoptosis due to α-Fe_2O_3 nanoparticles was also measured. The upregulation responsible in caspase-3/7 genes was confirmed by treating C2C12 cells with low and high doses of Fe_2O_3 nanoparticles [30].

Multifunctional silver-doped Fe_2O_3 nanoparticles were synthesised using *Adhatoda vasica* leaf extract. The synthesised nanoparticles were subjected to chemical, structural, and morphological examination by various techniques. From the results, it is revealed that the biomolecules present in the extract acted as stabilising agents. The greener nanoparticles exhibit enhanced antimicrobial and anticancer property. The synthesised nanomaterials are biocompatible and less toxic because of the involvement of the biomolecules. The synthesised Ag-doped iron oxide nanoparticles are promising in targeted drug delivery and magnetic hyperthermia cancer therapy [31].

Ferric chloride was taken as a precursor for synthesising hematite nanoparticles using *Rhus punjabensis* extract as capping agent. XRD, SEM, TEM, Energy Dispersive Spectroscopy (EDS), and Fast Fourier Transform (FFT) techniques were used to study the characteristics of the synthesised nanoparticles. It was found that the structure of nanoparticles was rhombohedral with a particle size of 41nm. TEM, FTIR, and Thermogravimetry/Differential Thermal Analysis (TG/DTA) studies revealed the formation of pure Fe_2O_3 with controlled particle size due to the capping of biomolecules. Coating of various functional groups on the nanoparticles surface enables them to function as free radical scavenging, antioxidative, antibacterial, and protein kinase inhibitor. The cytotoxicity of the nanoparticles was tested for DU-145 and HL-60 cancer cells and it was found that the IC_{50} values are 11.9 µg/mL and 12.79 µg/mL, respectively. Inhibition potential associated with NF-κB assay exposed the apoptotic potential of iron oxide nanoparticles to various cancer cells [32].

Protein-capped Fe_2O_3 nanoparticles were synthesised using supernatant liquid of *Bacillus cereus* SVK1 bacterium culture. The synthesised nanoparticles were studied by X-ray diffraction, UV–visible spectroscopy, Fourier transform infrared spectroscopy, and electron microscopy analysis and tested for their anticancer activity against liver cancer cells using varying concentrations. Hematite nanoparticles with 30.2 nm size exhibited considerable cytotoxic behaviour against liver cancer cells (CTC_{50} = 704 ng/mL) [33].

Fe_2O_3 nanoparticles synthesised using persimmon aqueous extract [34] were characterised by TEM, Field emission scanning electron microscopy (FE-SEM), XRD, and FTIR techniques. Anticancer activity of these nanoparticles was investigated against the Hep-G2 cell line and it was reported that they have considerable cytotoxic potential and inhibitory efficacy against pathogenic bacterial strains.

Fe_2O_3 nanomaterials fabricated using *Capparis zeylanica* extract [35] possess anticancer activity against A549 cancer cells. The synthesised nanoparticles were spherical in shape of 28.17 nm. A549 cells were treated with the nanoparticles of dosage from 1 μg/mL to 100 μg/mL. After treatment, the size of A549 cells reduced due to nuclei fracture. The nanoparticles demonstrated 50% of apoptotic efficiency at a dose of 15.6 μg/mL.

Natural polyphenol and gallotannin were used to reduce and cap the Fe_2O_3 nanoparticles [36]. The synthesised Fe_2O_3 nanoparticles were spherical in shape with an average size of 12.85 nm. X-ray diffraction patterns confirmed the crystalline nature; Energy Dispersive X-Ray Analysis (EDAX) analysis showed the presence of 21.7% iron and 42.11% oxygen. FTIR spectrum reveals the capping of gallotannin on the surface of prepared nanoparticles. These nanoparticles exhibited antifungal and anticancer activities against MCF-7 cells. MCF-7 cells undergo induced intracellular oxidative stress by taking up the iron oxide nanoparticles which leads to cell membrane injury.

Punica granatum fruit peel extract [37] was used as stabilising agent to prepare iron oxide nanoparticles. The synthesised nanoparticles are highly pure and crystalline with saturation magnetisation of 69 emu.g^{-1}. Morphological analysis indicated that the added extract reduced the particle size around 11 nm. Anticancer potential of the green nanoparticles was tested with HCT116, MCF-7, HeLa and A549 cancer cell lines. Iron oxide nanoparticles synthesised with 2 wt.% and 4 wt.% of extract exhibited excellent activity against NPC cell line HONE1. They were not cytotoxic to CCD112 and HEK293 normal cells, which indicate their selectivity on the NPC cells.

Artemisia absinthium L. leaf and stems were utilised to synthesise environment-friendly and less toxic iron oxide nanoparticles [38]. From various characterisation techniques, it is shown that the synthesised nanoparticles are extremely small in size and quasi-spherical in shape. The plant-mediated nanoparticles are not cytotoxic to normal cells, while the A375 cancer cells are significantly affected. When the cells are exposed to a dosage of 500 μg/mL for 72 hours, they release specific signs of apoptosis.

Aqueous extracts of *Alpinia galangal* rhizoid were used to prepare iron oxide nanoparticles by combustion method. Using XRD data, the size of the synthesised nanomaterials was calculated as 20nm. The bands observed from FTIR spectra at 535 cm^{-1} and 450 cm^{-1} are characteristic of Fe–O vibrations. Morphological analysis revealed the irregular spherical shape of the nanomaterials. Anticancer activity of the nanoparticles was tested against SKOV3, MDAMB-231, and COLO-205 cancer cells. Among all the cancer cells, considerable antiproliferative activity was observed against SKOV3 cells in MTT assay [39].

Superparamagnetic iron oxide nanoparticles were produced with the help of actinobacterial metabolites. Electron microscopic analysis shows that the particles are in the range of 15–30 nm size and spherical. Anticancer potential of iron oxide nanoparticles was assessed against HeLa cells with various analyses such as MTT assay, AO/EB staining, Hoechst nuclear staining, and flow cytometric methods. The cytotoxic studies indicated that around 60% of growth inhibition was attained at a dose of 150 μg/mL concentration of nanoparticles. Apoptotic cells present in growth phases were confirmed by flow cytometric method. The untreated cells consist of 2.64% cells in sub G0, 10.13% cells in G0/G1, 9.49% cells in S, and 17.14% cells in G2/M phases, whereas treated cells consist of 5.43% cells, 11.20% cells, 9.77% cells, and 12.95% cells, respectively, in sub G0, G0/G1, S, and G2/M phases [40].

8.3 Conclusions

This review discusses the synthesis, characterisation, and target-based biomedical applications of iron oxide nanoparticles. It is evident that selection of green route for the synthesis of nanoparticles is more effective than the other methods. Iron oxide nanoparticles have a vital role in controlling cancer cell growth. Use of biomaterials for the synthesis of iron oxide nanoparticles is considered to be more appropriate than the other preparation processes. The response of iron oxide nanoparticles in anticancer activities was identified to be more effective on certain types of cancer cells. The anticancer efficacy of iron oxide nanoparticles indicated that the level of toxicity was dosage dependent. We conclude that bio-synthesised iron oxide nanoparticles can be successfully utilised for the apoptosis and growth inhibition of cancer cells.

REFERENCES

[1] P. Mulvaney, *ACS Nano.* *9*(3) (2015) 2215–2217.

[2] S. Hasan, *Research Journal of Recent Sciences.* *4* (2025) 09–11.

[3] Buzea Cristina, Ivan I. Pacheco, Kevin Robbie, *Biointerphases.* *2*(4) (2007) 126–130.

[4] Kossi AniyaAmedome Min-Dianey, Hao-Chun Zhang, Ali Anwar Brohi, Haiyan Yu, Xinli Xia, *Superlattices and Microstructures.* *115*(3) (2018) 168–176.

[5] A. Rastogi, D.K. Tripathi, S. Yadav, D.K. Chauhan, M. Ghorbanpour, N.I. El-Sheery, M. Brestic, *Biotechnology.* *9*(3) (2019) 1–11.

[6] S.A. Ghoto, M.Y. Khuhawar, T.M. Jahangir, J.D. Mangi, *Journal of Nanostructure in Chemistry.* *9*(2) (2019) 77–93.

[7] A. Tamilvanan, K. Balamurugan, K. Ponappa, K.B. Madhan, *International Journal of Nanoscience.* *13*(2) (2014) 1430001–1430022.

[8] R.Z.A. Manj, X. Chen, W.U. Rehman, G. Zhu, W. Luo, J. Yang, *Frontiers in Chemistry.* *6* (2018) 539.

[9] A.M. Ealias, M.P. Saravanakumar, *Materials Science and Engineering.* *263*(3) (2017) 32019–32035. IOP Publishing.

[10] K. Pan, Q. Zhong, *Annual Review of Food Science and Technology.* *7* (2016) 245–266.

[11] Nadeem Joudeh, Dirk Linke, *Journal of Nanobiotechnology.* *20* (2022) 29.

[12] Kapil D. Patel, Rajendra K. Singh, Hae-Won Kim, *Material Horizons.* *6*(3) (2019) 434–469.

[13] M.H. Siddiqui, M.H. Al-Whaibi, A.M. Sakran, Hayssam M. Ali, Mohammed O. Basalah., M. Faisal, A. Alatar, A.A. Al-Amri, *Journal of Plant Growth Regulation.* *32*(1) (2013) 61–71.

[14] I. Chakraborty, T. Pradeep, *Chemical Reviews.* *117*(12) (2017) 8208–8271.

[15] D.M. Huang, J.K. Hsiao, Y.C. Chen, L.Y. Chien, M. Yao, Y.K. Chen, B.S. Ko, S.C. Hsu, L.A. Tai, H.Y. Cheng, S.W. Wang, C.S. Yang, Y.C. Chen, *Biomaterials.* *30*(22) (2009) 3645–3651.

[16] S. Sheng-Nan, W. Chao, Z. Zan-Zan, H. Yang-Long, S. VenkatramanSubbu, X. Zhi-Chuan, *Chinese Physics B.* *23*(3) (2014) 037503–037519.

[17] J.P. Fortin, C. Wilhelm, J. Servais, C. Ménager, J.C. Bacri, F. Gazeau, *Journal of the American Chemical Society.* *129*(9) (2007) 2628–2635.

[18] J. Xu, J. Sun, Y. Wang, J. Sheng, F. Wang, M. Sun, *Molecules.* *19*(8) (2014) 11465–11486.

[19] D. Maity, D.C. Agrawal, *Journal of Magnetism and Magnetic Materials.* *308*(1) (2007) 46–55.

[20] A.Z. Wilczewska, K. Niemirowicz, K.H. Markiewicz, H. Car, *Pharmacological Reports.* *64*(5) (2012) 1020–1037.

[21] R. Lin, Y. Li, T. MacDonald, H. Wu, J. Provenzale, X. Peng, J. Huang, L. Wang, A.Y. Wang, J. Yang, H. Mao, *Biointerfaces.* *150* (2017) 261–270.

[22] Z.Q. Zhang, S.C. Song, *Biomaterials.* *106*(33) (2016) 13–23.

[23] C. Saikia, M.K. Das, A. Ramteke, T.K. Maji, *International Journal of Biological Macromolecules Part A.* *93* (2016) 1121–1132.

[24] A.H. Lu. E.L. Salabas, F. Schüth, *AngewandteChemie International Edition in English.* *46*(8) (2007) 1222–1244.

[25] J. Estelrich, E. Escribano, J. Queralt, M.A. Busquets, *International Journal of Molecular Sciences.* *16*(4) (2015) 8070–8101.

[26] A Christopher Quinto, Priya Mohindra, Sheng Tong, Gang Bao, *Nanoscale, 7*(29) (2015) 12728–12736.

[27] F. Soetaert, D. PreethiKorangath, S. Serantes, R. Ivkov Fiering, *Advanced Drug Delivery Reviews. 163–164* (2020) 65–83.

[28] P.C. Nagajyothi, T. MuthuramanPandurangan, DooHwan Kim, V.M. Sreekanth, Jaesool Shim, *Journal of Cluster Science.* *28* (2017) 245–257.

[29] J. Manikandan, P. Rajesh, J. RaffieaBaseri, S.V.K. Selvakumar. *Asian Journal of Chemistry.* *34*(9) (2022) 2363–2372.

[30] Rizwan Wahab, Farheen Khan, Abdulaziz Al-Khedhairy, *RSC Advances.* *8* (2018) 24750–24759.

[31] Sameer Kulkarni, Mangesh Jadhav, Prasad Raikar, Delicia A. Barretto, Shyam Kumar Vootlac, U.S. Raikar, *New Journal of Chemistry.* *41* (2017) 9513–9525.

[32] Sania Naz, Mohammad Islam, Saira Tabassum, NelsonFreitas Fernandes, Esperanz J. Carcache de Blanco, Muhammad Zia, *Journal of Molecular Structure.* *1185* (2019) 1–7.

[33] Kumar Rajendran, Vithiya Karunagaran, Biswanath Mahanty, Shampa Sen, *International Journal of Biological Macromolecules.* *74* (2015) 376–381.

[34] Sarah Abd Al-hameed, Ahmed Mishaal Mohammed, *Journal of Pharmaceutical Negative Results*. *13* (2022) 958–967.

[35] Nilavukkarasi Mohandoss, Sangeetha Renganathan, Vijayakumar Subramaniyan, Punitha Nagarajan, Vidhya Elavarasan, Prathipkumar Subramaniyan, Sekar Vijayakumar, *Applied Sciences*. *12* (2022) 12986–12998.

[36] Bilal Ahmed, Asad Syed, Khursheed Ali, Abdallah Elgorban, Afroz Khan, Jintae Lee, Hind Al-Shwaiman, *RSC Advances*. *11* (2021) 9880.

[37] Mostafa Yusefi, Kamyar Shameli, RoshafimaRasit Ali, Siew-Wai Pang, Sin-Yeang Teow, *Journal of Molecular Structure*. *1204* (2020) 127539.

[38] Elena-Alina Moacă, Claudia Geanina Watz, Daniela Flondor, Cornelia Păcurariu, Lucian Barbu Tudoran, Robert Ianos, Vlad Socoliuc, George-Andrei Drăghici, Andrada Iftode, Sergio Liga, Dan Dragos and Cristina Adriana Dehelean, *Nanomaterials*. *12* (2022) 2012.

[39] H.R. Raveesha, H.L. Bharath, D.R. Vasudha, B.K. Sushma, S. Pratibha, N. Dhananjay, *Inorganic Chemistry Communications*. *132* (2021) 108854.

[40] Thangavel Shanmugasundaram, Manikkam Radhakrishnan, Arasu Poongodi, Krishna Kadirvelu and Ramasamy Balagurunathan, *MRS Communications*. *8* (2018) 604–609.

9

The Artificial Intelligence-Based Revolution in Material Science for Advanced Energy Storage

Sonia H. Bajaj
G H Raisoni Institute of Engineering and Technology, Nagpur, Maharashtra, India

S. Leena Nesamani and Dharmalingam Ganesan
Vel Tech Rangarajan Dr. Sagunthala R&D Institute of Science and Technology, Avadi, Chennai, India

R. Jeeva
Thamirabharani Engineering College, Tirunelveli, Tamil Nadu, India

V. R. Niveditha
Sathyabama Institute of Science and Technology, Chennai, Tamil Nadu, India

9.1 Introduction

This chapter explores the exciting field where material science and artificial intelligence (AI) meet, opening the door to a new era of efficiency and creativity in advanced energy storage. In particular, the search for innovative energy storage technologies is changing the scientific landscape as a result of the symbiotic link between AI and material science [1].

9.1.1 Overview of Material Science and AI Integration

At the heart of this exploration lies a comprehensive examination of the integration between material science and AI. This dynamic partnership harnesses the capabilities of AI algorithms to propel material engineering into a realm of unprecedented possibilities. From predictive modelling to accelerated materials discovery and optimisation strategies, the fusion of AI and material science stands as a technological tour de force, revolutionising the approach to energy storage materials [2]. AI's contribution to material science extends beyond automation; it acts as a catalyst for innovation, enabling scientists and researchers to navigate the vast material space more efficiently. The precision and speed offered by AI-driven approaches empower scientists to explore novel materials with tailored properties for enhanced energy storage performance.

9.1.2 Historical Context and Traditional Approaches

To appreciate the magnitude of the AI-driven revolution, a glimpse into the historical context of material science is essential. Traditional approaches, while effective, often relied on time-consuming trial-and-error methods. The advent of AI injects efficiency into this process, significantly reducing the time and resources required for material discovery. The chapter unfolds the narrative of traditional methodologies, highlighting the challenges they posed and the opportunities AI seizes upon. By understanding the historical trajectory, readers gain insights into the transformative nature of AI in material science. This historical context serves as a crucial foundation for comprehending the quantum leap facilitated by AI in the contemporary landscape of material science and advanced energy storage research.

DOI: 10.1201/9781003495437-9

9.2 AI-Powered Materials Discovery: Computational Alchemy

9.2.1 Machine Learning Algorithms in Materials Discovery

Within the intricate realm of material science, machine learning algorithms emerge as computational artisans, transforming raw data into predictive insights. Operating on principles of pattern recognition, regression analysis, and clustering, these algorithms unravel complex relationships within extensive datasets. Quantum mechanical calculations, analyses of electronic structures, and evaluations of thermodynamic properties become navigable landscapes for algorithms. Their adeptness enables the identification of material candidates possessing tailored properties, constituting a form of computational alchemy that optimises the selection process and streamlines exploration across expansive material spaces [3]. These algorithms exhibit a dynamic nature, evolving with each iteration as they learn from new data. Reinforcement learning mechanisms empower them to adapt to emerging patterns, ensuring an adaptive and responsive approach to materials discovery. Through constant refinement and adaptation, these algorithms enhance their predictive capabilities, providing researchers with invaluable insights into the intricate world of material properties.

9.2.2 Predictive Modelling for Novel Material Identification

At the heart of AI-powered materials discovery lies predictive modelling, a pivotal process involving the construction of mathematical models. Rooted in regression analysis or neural networks, these models extrapolate material properties based on existing data. By discerning correlations between material structures and properties, predictive models facilitate the identification of novel materials exhibiting desired characteristics [4]. This accelerates the discovery of materials for specific applications, spanning high-performance electronics to energy storage systems. Through continuous learning from known materials, predictive models guide the exploration of uncharted material territories with precision and efficiency.

These models also contribute to the understanding of complex material interactions and behaviours. For instance, in the design of advanced batteries, predictive models simulate electrochemical processes at the atomic level, aiding in the selection of materials that exhibit optimal charge and discharge characteristics. The integration of quantum mechanics into predictive modelling further refines the accuracy of predictions, enabling researchers to navigate the vast landscape of potential materials with greater confidence.

9.2.3 Accelerating Discovery Processes with AI

The infusion of AI injects unprecedented speed into materials discovery processes by automating repetitive tasks, dissecting vast datasets, and generating hypotheses for experimental validation. Machine learning algorithms play a pivotal role in the screening of potential materials, prioritising those with optimal properties for specific applications. This acceleration proves particularly transformative in fields like drug discovery, where AI-driven simulations predict molecular interactions, and in materials engineering, where AI expedites the identification of materials boasting superior mechanical, thermal, or electrical properties [5]. Furthermore, AI-driven discovery processes foster a more collaborative and interdisciplinary research environment. Researchers from diverse fields can leverage AI tools to explore materials beyond their conventional expertise, leading to cross-disciplinary innovations. The integration of AI streamlines the workflow, allowing researchers to focus on interpreting results and making informed decisions rather than spending excessive time on routine tasks. The advent of AI-powered discovery processes revolutionises traditional methodologies, ushering in a new era of efficiency, precision, and innovation in the realm of material science. As these technologies evolve, their impact on the pace and scope of materials discovery is poised to reshape the landscape of scientific exploration and technological advancement.

9.3 Optimising Energy Storage: AI's Role in Battery Performance

9.3.1 Fine-Tuning Parameters for Enhanced Energy Storage

The optimisation of energy storage systems is significantly enhanced by AI through the meticulous fine-tuning of key parameters in battery design. Machine learning algorithms meticulously analyse vast datasets encompassing battery materials, architectural configurations, and operational conditions. By discerning optimal combinations of electrode materials, electrolytes, and structural configurations, AI contributes to the augmentation of energy storage capacity, charging rates, and overall battery performance. This fine-tuning process stands as a cornerstone for advancing battery technologies across a spectrum of applications, ranging from electric vehicles to grid-scale energy storage systems. AI-driven optimisation ensures that energy storage solutions are not only efficient but also tailored to meet the specific demands of diverse use cases [6].

9.3.2 Predicting Degradation Patterns in Energy Storage Materials

AI algorithms play a pivotal role in predicting degradation patterns in energy storage materials, offering a proactive approach to system maintenance. Continuous monitoring and analysis of operational data enable machine learning models to forecast potential degradation mechanisms. These may include electrode wear, electrolyte breakdown, or thermal stresses. By anticipating degradation issues, predictive maintenance strategies can be implemented to address them promptly, thereby extending the lifespan of energy storage systems [7]. This predictive capability is instrumental in optimising the overall efficiency and reliability of energy storage solutions.

9.3.3 Insights into Efficient Energy Utilisation

AI's analytical prowess provides invaluable insights into efficient energy utilisation within storage systems. Machine learning models delve into the complex relationships between charge–discharge cycles, environmental conditions, and energy efficiency. By identifying optimal operating conditions and recommending adaptive energy management strategies, AI ensures that energy storage systems operate at peak performance while minimising wastage. This optimisation is crucial for achieving sustainable and cost-effective energy storage solutions across various applications. Whether applied to residential energy storage units or large-scale grid installations, AI-driven insights contribute to the efficient and responsible utilisation of stored energy. The adaptability of AI recommendations to changing conditions further enhances the resilience and sustainability of energy storage systems, aligning them with evolving energy demands and environmental considerations.

9.4 Challenges and Opportunities: Navigating the AI–Material Science Frontier

9.4.1 Data Quality Challenges in AI-Driven Material Science

The field of AI in material science has significant obstacles related to the amount and calibre of data. The efficacy of machine learning models heavily relies on comprehensive and accurate datasets. In material science, where the complexity of materials and their diverse properties prevails, obtaining high-quality data on a wide spectrum of materials under various conditions proves to be a multifaceted task. Ensuring data accuracy, consistency, and relevance is pivotal in training robust AI models. Addressing these challenges necessitates the development of meticulous data collection methodologies, spanning diverse material characteristics, and leveraging sophisticated data preprocessing techniques to enhance the quality of input datasets [8].

9.4.2 Interpretability of AI Models in Material Discovery

A critical challenge at the forefront of the AI–material science interface is the interpretability of intricate machine learning models. Many AI algorithms, notably deep neural networks, operate as intricate black-box systems, rendering the underlying decision-making processes opaque. In the context of material discovery, comprehending why an AI model recommends a specific material or property is imperative for extracting meaningful insights into novel materials. Active research endeavours focus on the development of interpretable AI models and methodologies, seeking to unravel the intricate mechanisms within these models. This pursuit aims to ensure that the recommendations align with established scientific principles, fostering a deeper understanding of the materials suggested by AI systems.

9.4.3 Opportunities Unveiled by the AI Revolution

The unfolding AI revolution within material science presents unparalleled opportunities to expedite the material discovery process. AI-driven approaches wield the capability to predict material properties, explore vast chemical spaces, and propose novel combinations with highly desirable characteristics. Generative models, underpinned by AI, play a pivotal role in crafting virtual materials tailored to specific properties, ushering in a paradigm shift in materials design. Furthermore, AI contributes to the optimisation of existing materials for specialised applications, unlocking novel avenues for energy storage, electronics, and various advanced technologies [9]. The rapid processing and analysis of extensive datasets empower researchers to discern concealed patterns and correlations, propelling breakthroughs in material science that were once inconceivable without the integration of AI technologies. As the synergy between AI and material science advances, addressing challenges related to data quality, enhancing interpretability, and harnessing the opportunities brought forth by the AI revolution will be instrumental in propelling the frontiers of materials research and innovation into uncharted territories.

9.5 Case Studies: AI-Infused Breakthroughs in Energy Storage Materials

9.5.1 Enhancing Lithium-Ion Batteries through AI

The metamorphosis of lithium-ion batteries, catalysed by the infusion of AI, stands as a testament to the transformative impact of advanced technologies on energy storage. Machine learning models meticulously sift through an extensive array of datasets encompassing battery performance metrics, material properties, and a myriad of environmental variables. Through the intricate dance of predictive modelling, AI discerns optimal combinations of materials and electrolytes, with a focused intent on elevating critical parameters such as energy density, cycle life, and safety [10]. This orchestrated approach expedites the design process, empowering researchers to uncover advanced lithium-ion battery configurations that surpass the constraints of traditional designs. Real-time monitoring, facilitated by AI, further amplifies these efforts, contributing not only to enhanced overall efficiency but also to the establishment of elevated safety standards within the realm of lithium-ion batteries.

9.5.2 Discovery of Novel Materials (e.g., Graphene) with AI

The symbiotic relationship between AI and material science heralds a paradigm shift in the landscape of energy storage research, exemplified vividly by the discovery of novel materials like graphene. Harnessing the computational prowess of machine learning algorithms, researchers embark on a virtual expedition through vast material databases, navigating the complexities to pinpoint potential candidates endowed with unique and desirable properties. In the specific case of graphene, AI assumes the role of a virtual guide, assisting in the prediction of its behaviour within the intricate realm of energy storage applications. Factors such as conductivity, flexibility, and stability become the focal points of consideration, and AI expedites the identification of materials boasting exceptional performance characteristics [11]. This harmonious fusion between AI and material science not only streamlines the research process but also unfurls unprecedented avenues for the development of high-performance energy storage materials.

9.5.3 Real-World Applications and Success Stories

The narrative of AI-infused breakthroughs in energy storage materials transcends the theoretical realms, finding resonance in the practical domains of electronics, electric vehicles, and renewable energy systems. The success stories emanating from these applications resonate as tangible proof of the profound impact of AI on the evolution of energy storage. In the domain of electronics, AI-driven optimisation of supercapacitor materials has resulted in a paradigm shift, ushering in faster-charging electronic devices that seamlessly align with the pulse of modern technology. The electric vehicle sector stands as a testament to AI's transformative prowess, where enhanced battery technologies contribute not merely to extended range but also to an overarching improvement in performance metrics. Renewable energy systems, buoyed by the infusion of AI-optimised materials, stand as beacons of efficiency, ensuring the seamless storage and distribution of energy in alignment with the dynamic demands of the modern world. These tangible applications not only underscore the transformative impact of AI in energy storage materials but also illuminate the myriad possibilities for addressing critical challenges and propelling innovation across diverse industries.

9.5.4 The Trajectory towards Future Innovations

As we reflect on the current state of AI-infused breakthroughs in energy storage materials, it becomes apparent that this is not a denouement but a prologue to a trajectory that promises even more sophisticated and efficient energy storage solutions. The collaborative endeavours between AI and material science are akin to an evolving symphony, with each note resonating with the potential to unlock new frontiers in sustainability and technological advancement [12]. The fusion of cutting-edge AI technologies with the intricacies of material science engenders a profound shift, where the boundaries of what is achievable are continually pushed. This trajectory envisions not only overcoming existing challenges but also heralding a new era of interdisciplinary collaboration, where the synthesis of AI and material science becomes an indispensable tool for sculpting the sustainable and technologically advanced future that lies ahead.

9.6 Responsible AI Practices in Material Science

At the confluence of AI and material science, a paradigm emerges where responsible AI practices are fundamental for ethical advancements. Researchers in this domain adhere to rigorous ethical guidelines, ensuring the ethical collection of data with a strong emphasis on privacy protection and obtaining informed consent from individuals contributing to datasets. This ethical stance serves as a foundational principle, ensuring that the integration of AI into material science respects the rights and privacy of data contributors. Transparent documentation practices accompany these ethical considerations, providing a detailed account of AI models and methodologies. Transparency is not just an ethical imperative but also a mechanism for accountability. Transparent documentation allows for scrutiny and validation, fostering a culture of responsible innovation where AI-driven processes are open to examination.

9.6.1 Transparency in Algorithmic Decision-Making

Ethical AI applications in material science hinge on the transparency of algorithmic decision-making. Researchers recognise the ethical imperative of disclosing critical aspects of the AI process, including detailed information about input data, the underlying architecture of models, and the intricacies of training processes. This transparency is further complemented by the integration of explainability tools, shedding light on how AI arrives at specific material recommendations [13].

Transparent algorithms serve as technological artefacts open to ethical scrutiny, empowering researchers and end-users to identify and rectify biases or ethical concerns promptly. This commitment to transparency is not only an ethical obligation but also a practical approach to building trust among stakeholders, ensuring that the decision-making process is comprehensible and subject to scrutiny.

9.6.2 Ethical Frameworks for AI in Material Science

As AI consolidates its position in material science, the establishment of robust ethical frameworks becomes imperative. Ethical considerations transcend the technical realm as researchers collaboratively define standards that prioritise safety, fairness, and accountability. The formulation of these ethical frameworks is a collaborative effort involving researchers, ethicists, policymakers, and industry stakeholders. Ethical frameworks are designed to consider the societal impact of AI-driven material discoveries [14]. They undergo continuous evaluation and adaptation, ensuring their resonance with evolving ethical norms. The collaborative nature of the process ensures a comprehensive approach, balancing innovation with responsible practices.

9.6.3 Striking a Balance

In the intersection of material science and AI, the delicate balance between innovation and responsibility becomes the central tenet of this interdisciplinary endeavour. Researchers grapple not only with the technical intricacies of AI but also with the ethical dimensions that accompany its applications. The responsible use of AI in material science extends beyond compliance; it embodies a commitment to societal well-being, equity, and the preservation of ethical norms in the face of technological advancement. As the landscape evolves, the ethical considerations woven into the fabric of AI-driven material science serve as the compass, guiding the trajectory of innovation towards a future where progress and responsibility walk hand in hand.

9.7 Future Trajectories: AI's Uncharted Path in Energy Storage Materials

9.7.1 Emerging Trends in AI and Material Science Integration

The dynamic integration of AI and material science is ushering in transformative advancements that promise to reshape the landscape of energy storage materials. One notable trend on the horizon is the incorporation of reinforcement learning for autonomous material discovery. This involves systems adapting and optimising material suggestions based on real-world performance feedback. The iterative nature of reinforcement learning allows for continuous improvement, making the material discovery process more adaptive and responsive to evolving requirements. Multi-modal AI models represent another significant trend, incorporating diverse data sources such as imaging and spectroscopy. This approach enhances the depth and accuracy of material property predictions by considering a broader range of information. The synergy of different data modalities enables a more comprehensive understanding of material behaviour, facilitating more informed decision-making in the design and selection of materials for energy storage applications.

Federated learning approaches are gaining prominence as a solution to address privacy concerns in material science research. By allowing decentralised AI models to collaboratively learn from distributed datasets without exchanging them, federated learning ensures that sensitive information remains localised. This approach is particularly relevant in scenarios where data privacy is a critical consideration, such as when dealing with proprietary or confidential material properties.

9.7.2 Potential Breakthroughs on the Horizon

Looking towards the future, several potential breakthroughs stand out as exciting prospects for advancing AI-driven material science. Quantum machine learning represents a cutting-edge avenue, leveraging the unique capabilities of quantum computing to simulate complex material behaviours. Quantum computing's parallel processing capabilities offer the potential for unparalleled precision in material design, enabling researchers to explore novel materials with unprecedented efficiency.

Meta-learning is another frontier that holds promise for accelerating material discovery workflows. This approach enables AI models to learn from a multitude of tasks, enhancing their adaptability. In the context of material science, where datasets are often limited and diverse, meta-learning can empower AI models to generalise insights from one material system to another, thereby streamlining the discovery process [15].

The integration of robotic laboratories with AI algorithms is poised to automate high-throughput experimentation, significantly accelerating material characterisation and discovery. Automated experimentation allows for the rapid testing of a large number of materials, expediting the identification of promising candidates for energy storage applications. This collaborative synergy between AI and robotic laboratories holds the potential to revolutionise the pace of material discovery.

9.7.3 Collaborative Synergy with Other Technologies

AI's trajectory in energy storage materials intersects with various technologies, fostering collaborative synergies that amplify its capabilities. Quantum computing, with its ability to perform complex calculations at speeds unattainable by classical computers, complements AI in solving intricate optimisation problems inherent in material science. This collaborative approach enhances the efficiency of material property predictions, enabling researchers to explore a broader design space.

The integration of AI with 5G networks plays a crucial role in facilitating real-time data exchange. This capability is particularly valuable for distributed AI models that collectively enhance material predictions. The seamless and rapid communication enabled by 5G networks ensures that AI models can leverage the most up-to-date information, enhancing the accuracy and relevance of their predictions.

Blockchain technology's integration ensures transparent and traceable data sharing in AI-driven material discoveries. By providing an immutable and decentralised ledger for recording and verifying data transactions, blockchain fortifies the integrity of datasets. This is particularly important in ensuring the reliability and reproducibility of research findings, contributing to the robustness of AI-driven material science.

Furthermore, AI collaborates with sensor technologies, utilising real-time material performance data to iteratively improve predictive models. Sensors embedded in energy storage systems provide continuous feedback on material behaviour under varying conditions. This dynamic data stream enhances the adaptability of AI models, enabling them to refine predictions based on real-world performance, ultimately contributing to the development of more resilient and efficient energy storage materials.

The uncharted path of AI in energy storage materials promises a dynamic landscape of innovation, where emerging trends, potential breakthroughs, and collaborative synergies propel material science into a new era of efficiency, sustainability, and unprecedented discovery. As researchers continue to explore these frontiers, the convergence of AI with material science is set to redefine the possibilities for energy storage materials, paving the way for advancements that address current challenges and unlock new opportunities for sustainable technologies.

9.8 Conclusion

9.8.1 Recapitulation of AI's Impact on Material Science

The conclusion marks a retrospective on the profound impact of AI on material science, highlighting its transformative role in revolutionising the approach to materials discovery and energy storage. Recapitulating the journey involves acknowledging the shift from traditional experimental methods to AI-driven computational strategies. The recap emphasises the efficiency gains, accelerated discovery timelines, and enhanced precision in material property predictions achieved through the integration of advanced machine learning algorithms.

9.8.2 Achievements and Contributions to Advanced Energy Storage

AI's journey in material science has yielded notable achievements, particularly in the realm of advanced energy storage. The development of novel materials with superior energy storage capabilities, facilitated by AI, has transcended conventional limitations. Achievements encompass the optimisation of battery performance, the discovery of innovative materials like graphene, and the realisation of real-world applications, culminating in the deployment of more efficient and sustainable energy storage solutions.

9.8.3 Outlook for Continued Advancements in the Field

The conclusion extends beyond reflection to set the stage for future advancements in the field. The outlook emphasises ongoing research trajectories, including the refinement of AI algorithms for even more accurate material predictions. The integration of AI with emerging technologies such as quantum computing and the continued evolution of collaborative frameworks are anticipated. The outlook also highlights the imperative of ethical considerations, emphasising responsible AI practices and transparency in decision-making processes. It underscores the collaborative synergy between AI and other technologies, propelling material science towards unprecedented frontiers. In essence, the conclusion serves as a pivotal juncture to recognise the transformative journey AI has undertaken in material science, with its tangible achievements in advanced energy storage. As the field continues to evolve, the outlook reinforces the commitment to pushing boundaries, ensuring ethical practices, and fostering collaborative endeavours that promise a sustainable and innovative future.

REFERENCES

[1] Zeba, G., Dabić, M., Čičak, M., Daim, T. and Yalcin, H., 2021. Technology mining: Artificial intelligence in manufacturing. *Technological Forecasting and Social Change*, *171*, p. 120971.

[2] Srinivas, I.V., Saritha, G., Yadav, R., Subburam, S. and Divya, K., 2023, November. Machine Learning for Cardiac Disease Detection and Family History Association. In *2023 3rd International Conference on Technological Advancements in Computational Sciences (ICTACS)* (pp. 763–771). IEEE.

[3] Ali, A., 2023. Deep Learning-Based Digital Human Modeling and Applications (Doctoral dissertation, The University of North Carolina at Charlotte).

[4] Arockiam, A.J., Subramanian, K., Padmanabhan, R.G., Selvaraj, R., Bagal, D.K. and Rajesh, S., 2022. A review on PLA with different fillers used as a filament in 3D printing. *Materials Today: Proceedings*, *50*, pp. 2057–2064.

[5] Song, Z., Chen, X., Meng, F., Cheng, G., Wang, C., Sun, Z. and Yin, W.J., 2020. Machine learning in materials design: Algorithm and application. *Chinese Physics B*, *29*(11), p. 116103.

[6] Ribet, S.M., Murthy, A.A., Roth, E.W., Dos Reis, R. and Dravid, V.P., 2021. Making the most of your electrons: Challenges and opportunities in characterizing hybrid interfaces with STEM. *Materials Today*, *50*, pp. 100–115.

[7] Arockiam, A.J., Rajesh, S., Karthikeyan, S., Thiagamani, S.M.K., Padmanabhan, R.G., Hashem, M., Fouad, H. and Ansari, A., 2023. Mechanical and thermal characterization of additive manufactured fish scale powder reinforced PLA biocomposites. *Materials Research Express*, *10*(7), p. 075504.

[8] Rasool, N. and Bhat, J.I., 2023. Unveiling the complexity of medical imaging through deep learning approaches. *Chaos Theory and Applications*, *5*(4), pp. 267–280.

[9] George, B. and Wooden, O., 2023. Managing the strategic transformation of higher education through artificial intelligence. *Administrative Sciences*, *13*(9), p. 196.

[10] Chandrashekhar, K.G., Laxmaiah, G., RamKumar, P. and Ramesh, B., 2023. Load-bearing characteristics of a hybrid Si_3N_4-epoxy composite. *Biomass Conversion and Biorefinery*, pp. 1–9.

[11] Thakkar, P., Khatri, S., Dobariya, D., Patel, D., Dey, B. and Singh, A.K., 2024. Advances in materials and machine learning techniques for energy storage devices: A comprehensive review. *Journal of Energy Storage*, *81*, p. 110452.

[12] Zheng, L., Cao, M., Du, Y., Liu, Q., Emran, M.Y., Kotb, A., Sun, M., Ma, C.B. and Zhou, M., 2024. Artificial enzyme innovations in electrochemical devices: advancing wearable and portable sensing technologies. *Nanoscale*, *16*(1), pp. 44–60.

[13] Adadi, A. and Berrada, M., 2018. Peeking inside the black-box: A survey on explainable artificial intelligence (XAI). *IEEE Access*, *6*, pp. 52138–52160.

[14] Joseph Arockiam, A., Rajesh, S. and Karthikeyan, S., Development of fish scale particle reinforced PLA filaments for 3D printing applications. *Journal of Applied Polymer Science*, *141*, 12, p. e55132.

[15] Zhao, J., Feng, X., Pang, Q., Fowler, M., Lian, Y., Ouyang, M. and Burke, A.F., 2024. Battery safety: Machine learning-based prognostics. *Progress in Energy and Combustion Science*, *102*, p. 101142.

Nanoparticles in Tissue Engineering Application and Regenerative Medicine

Santhosh Kumar Chinnaiyan
*Faculty of Pharmacy, Karpagam Academy of Higher Education, Eachanari,
Coimbatore, Tamil Nadu, India*

Saravanakumar Arthanari and Mohanraj Subramanian
Vellalar College of Pharmacy, Thindal, Tamil Nadu, India

10.1 Introduction

The field of nanotechnology combines engineering and science to create materials, valuable structures, and technologies in the nanoscale domain. The construction of nanoscale dimension particles and materials with extraordinary and exciting qualities not found in bulk materials is one of the most fascinating, dynamic, and significant disciplines quickly increasing in academia and industries [1]. Drexler claims that the principle of nanotechnology is altering a material's molecular structure. It requires the capacity to assemble molecular systems with atomic-level accuracy, producing a range of nanomachines. Nanomaterials are substances with unique dimensions in the nanometre range. Nanocomposite occurs when binary or further materials with different characteristics remain discrete and distinct on macroscopic levels. At a minimum, one of the phases has a size that is less than 100 nm in dimension [2].

Also, nanocomposites were called polymer nanocomposites scaffolding is composed of molecules with long chains through numerous reciting units and reinforcement from different nano-reinforcing agents. Considering at least one nanoscale stage in the ultimate substance, polymer nanocomposite is a close mix of polymers and nanomaterials [3]. Adding nanomaterials to polymer matrices enhances many pristine polymers' natural properties, such as their mechanical, gas barrier, biodegradable, flame-resistant, magnetic, and electro-optical properties. However, it also naturally creates new properties based on the nanomaterials used. This is caused by the interaction between natural polymers' flexible and processable features and NPs' enormous surface area, excellent heat equilibrium, and potent surface response [4].

Furthermore, the characteristics of nanocomposites benefit significantly from the excellent dimension ratio nanomaterials nano-reinforcing effects. Moreover, the magnitude of the nanomaterials, in addition to the contact involving them and the polymer matrices, significantly affects the characteristics [5]. The presence of NPs alters the physiochemical interactions and surface chemistry, which directly affects the potential properties resulting polymeric substances. The outer surface area, chemical makeup, design, and dimension of the nanomaterials, among other crucial parameters, aid in customising these interactions and features towards nanocomposites [6]. The quality of NP dispersion in a polymer matrix is considered one of the critical challenges in creating effective polymer nanocomposites. According to the particle size and dispersion, various interfaces with unique features were observed. Many different kinds of NPs, including zero- (0D), one- (1D), and two-dimensional (2D) nanomaterials, have been used to manufacture diverse polymer nanocomposites, and research in this area is ongoing. The creation of various polymer nanocomposites enhanced various qualities and resulted in materials with several uses. Investigations into nanocomposites are currently being carried out in a wide variety of fields, such as catalytic processes, sensing probes, photonics, surface-protective coatings, storage of energy, flame-retard materials, photocatalysis, smart substances, biomedicine, and delivery of drugs. Due to the development of multifunctional materials with unexpectedly advanced capabilities, polymer nanocomposites have attracted substantial technical and engineering attention [7]. There is a growing demand for TE solutions, a rapidly expanding multidisciplinary field, owing to numerous disadvantages of tissue and organ transplantation, including a need for more available donors, essential for immunosuppression, and an inadequate success ratio (transplant rejection). System biology, material sciences, and engineering disciplines have been combined to develop and produce synthetic tissues and organs that look closely like natural tissues and organs, both implantable models and scaled-down organs [8]. To achieve satisfactory results, a nanoscale rather than a macroscopic approach is needed to mimic tissue's extracellular matrix (ECM) composition by the construction of a 3D scaffolding for cells with adequate durability and simplicity for cell monitoring processes and bioactive agents delivery [9].

NPs have the potential to offer a high point of regulating the characteristics of scaffolds, including the ability to tune their mechanical properties and allow the extended release of bioactive compounds. NPs are uniquely positioned as leading candidates for delivering and monitoring bioactive agents for uses due to their poor solubility, unreliable bioactivity, and rapid distribution half-life of bioactive compounds (growth factor, cytokines, enzyme inhibitor, DNA, drug substances, etc.) [10]. These disadvantages and constraints have also made the NPs undesirable candidates. Nanotechnology combines NPs and their use for various purposes as a processing technology. It is possible to produce solid as well as colloidal NPs with sizes ranging between 10 and 100 nm [11]. In the biomedical sector, NPs are used as vaccine adjuvants, drug delivery systems, and solar energy sources. The impression of nanomaterials has shifted traditional and simple TE methods towards advanced and competent systems.

NPs and additional components connected to nanotechnologies, such as nanofibres (NFs) and nanopatterned surfaces, are utilised to regulate how cells act in TE fields. NPs have various uses in TE, such as combining medicinal properties and diagnosis technologies, incorporating new biomaterials with better spatiotemporal control from frameworks, and controlling the release of various biologically active agents, especially growth factors [12]. To regulate stem cell fate and developmental processes, it is essential to adjust the tensile strength of scaffolding for use in rigid tissues, reduce toxic effects, and enhance biocompatibility through tissue-specific delivery systems. NPs can be made from various materials, such as pottery materials, metallic substances, and natural and synthetic polymer products. They are one of the most popular choices in the TE field for imaging, improving mechanical strength, adding to bio-ink, having antimicrobial capabilities, and carrying bioactive agents. This is because of how they are made and their unique benefits, such as their ability to penetrate well, having large surface areas capable of programming surface characteristics, and having surface features that can be tuned. This part talks about how metallic, ceramics and polymeric NPs are used in TE [13].

10.2 Classification of Nanomaterials

Nanomaterials have an enormous surface area, a higher ratio of surface area to total mass, and an extremely high proportion of aspect. These unique properties provide nanomaterials the potential to have a significant impact on a wide variety of fields, including mechanical, physical, chemical, biological, and electrical [14]. Regarding nanomaterials, dimensions are a crucial aspect to consider; hence, these materials can be divided into four distinct categories, denoted as 0D, 1D, 2D, and 3D, respectively, created on the number of dimensions they possess (Figure 10.1).

10.2.1 Zero-Dimensional (0D) NPs

In the present scenario, the nanoscale is the greatest conceivable scale and primarily consists of quantum dots (QDs), NPs, and nanoclusters. They can have ceramic polymeric, metallic, amorphous, or crystalline structures, among other possible variations. This category of NPs includes examples such as metal NPs (e.g., copper–gold, sulphur, silver, and iron), metal oxides NPs (such as SiO_2, TiO_2, Fe_3O_4, CuO, SnO_2, and ZnO), $CdSe$, carbon dot (CD), and QDs. These nanomaterials can be created using various processes, including microwaves, co-precipitations, hydrothermal, electrochemical, wet chemical reduction, sonochemical, reverse micelles, template synthesis, and sol–gel [6]. Several criteria, such as their size,

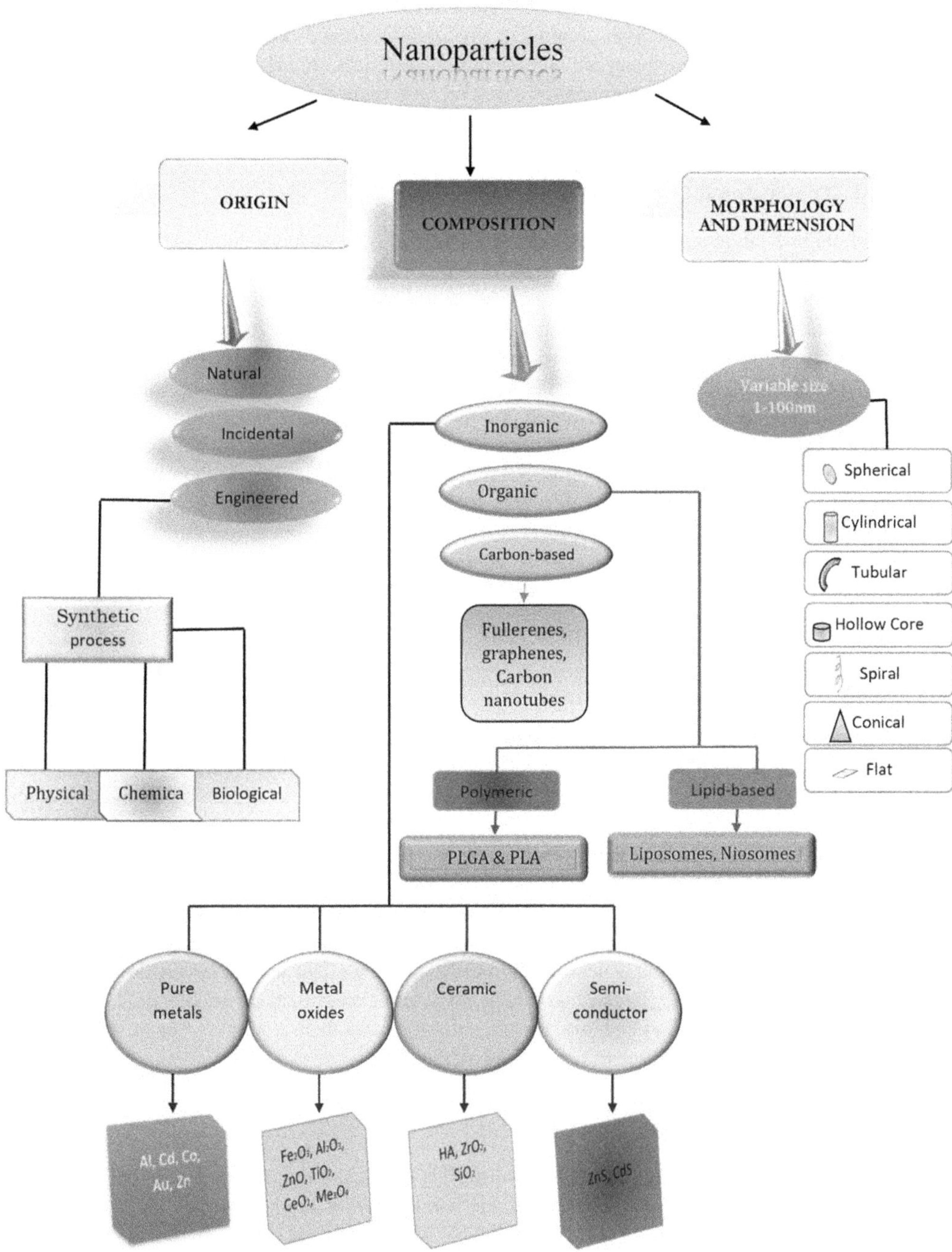

FIGURE 10.1 Classification of nanoparticles.

shape, and distribution, determine the characteristics of these nanomaterials. These characteristics can be altered by adjusting the parameters of the reaction. They were incorporated in to different polymer matrix materials to improve a few of the characteristics derived from natural polymers and make interesting like photonic, antimicrobial, catalytic, electrical, optical, and sensor devices. The CD is the emerging star of 0D nanomaterials made from carbon and has a size of less than 10 nm [15].

Additionally, the CD is a good substitute for standard semi-conducting QDs. It is water soluble, has minimal toxic effects and wavelength contingent photoluminescence (PL) capabilities, photostability, and functional ability, and uses cheap and abundant raw materials, among other advantageous characteristics. In addition, polymer nanocomposites based on QDs demonstrated several special applications, such as light-emitting diodes and solar cells [16].

10.2.2 One-Dimensional (1D) NPs

They are nanotubes, nanowires, NFs, nanorods, etc. This category includes single-walled carbon nanotubes (CNTs), multi-walled carbon nanotubes (MWCNT), cellulose NFs, carbon NFs, polyaniline (PANi) NFs, and others. CNTs are used in energy preservation and conversion, medicine delivery, sensors, and more because of their distinctive thermal energy, mechanical, chemical, electrical, and other capabilities. sp^2-hybridised carbon graphite sheets form vacuous tubular forms [17]. CNT-based polymer nanocomposites are getting much simpler to make and integrate into thermosets and thermoplastics alongside polymers with conjugation. Due to CNT-based polymer nanocomposites' small dimensions, significant aspect ratio, and good characteristics, it is a leading nano-reinforcing material for nanocomposite fabrication. The bonding of hydrogen gives cellulose NF rigidity and durability. Due to their increased processability, dispersibility, physical characteristics, etc., PANi NFs have been employed as nano-reinforcing agents in polymer matrix materials. Photocatalysis and solar energy conversion utilise TiO_2 nanowires due to their low toxicity, excellent photo activity, chemical resilience, and accessibility [18].

10.2.3 Two-Dimensional (2D) Nanomaterials

Through this scenario, nanomaterials having two dimensions fall inside the boundaries of the nanoscale. These dimensions include nanotubes, NF, lashes, and nanorods. Graphene oxide/reduced graphene oxide (GO/RGO), ceramics, and other similar substances all fall within this category of materials. The mineral clay often comprises stratified silicates or clay co-minerals (aluminium phyllo-silicates), including metal oxides such as magnesium, calcium, sodium, and potassium [19]. Due to the strong interaction among the interlayers of clay sheets, achieving a uniform distribution of nano clay within polymer matrix structures is challenging. This is an enormous obstacle. As a result, clay fragments are altered in some way before being included in polymer matrices to alleviate this issue. During the alteration of nano clay by the intercalation of attaching of polar functional group, the distance between the interconnected layers of nano clay particles is increased [20]. These improved clays can disperse easily in matrices of polymers. In addition, synthesising nanocomposites of polymers makes excellent use of 2D nanomaterials based on carbon, such as GO and RGO. Because graphene possesses exceptional thermodynamic, mechanical, and electrical characteristics, it has become the subject of much interest. A single layer of sp^2-hybridised carbon atom is stacked in a 2D lattice to form this structure [4].

10.2.4 Three-Dimensional (3D) NPs

The physical appearance of 0D, 1D, and 2D nanomaterials is more straightforward than that of 3D nanostructures, typically amalgamating several nanoscale features into a single structure [5]. Because these nanomaterials have a larger charged surface area, people believe they are more efficient at targeting analytes of fascination, performing amplifiers of signals, and carrying out effective biosensing. This is one reason why they are becoming increasingly popular. Most of what we refer to when discussing 3D materials are nanocomposites and complex hierarchical systems. According to the findings of a study, photonic crystals (PCs) and inverted opals (IOs) that possess a structured 3D nanostructure are ideal candidates for detecting biomarkers and creating innovative biosensors [3]. For instance, one group used the sol–gel and spin-coating methods to create ZnO IOs and ZnO PCs. The structures that were eventually created displayed improved levels of PL peaks and featured compact hexagonal layouts. In addition, PCs that are based on colloidal SiO_2 and PCs that are based on TiO_2 IOs have both been successfully synthesised. An additional study demonstrating the potential use of barcode technology for biosensing was presented.

PEG, also known as polyethylene glycol hydrogel, coupled with the reverse opal particles, was utilised to produce these barcodes [8].

10.3 Metallic Nanoparticles

The atomic or molecular structures of bulk materials can be understood better with the help of NPs. Manufacturing metallic NPs and modifying their properties can be accomplished by utilising a variety of functional groups, which allow for the polymerisation of antibodies, ligands, and pharmaceuticals as carriers system. This section briefly overviews various examples of metallic NPs, explicitly focusing on the biomedical applications of silver and gold NPs.

10.3.1 Gold Nanoparticles (AuNPs)

The gold NP, also known as AuNP, is a colloid of particulates of gold measuring on the nanometre scale. The colloidal gold solutions have distinctive qualities that differentiate them from gold in their bulk form. One such property is their optical quality, which results from how they interact with light. The mono-disperse sphere-shaped gold NPs were synthesised for the primary. After that, various people adjusted this strategy. Since polypeptide sequences, immunoglobulins, and protein molecules have a high affinity towards gold, they can be conjugated with phosphines, amines, and thiols on their surface [21]. Gold NPs could be used in regenerative therapies to protect transplanted tissue that replaces a tumour-damaged organ or tissue. One example would be using gold NPs (AuNPs) to precisely transport NPs into the nuclei of cancer cells to inhibit their ability to divide. PEG-coated gold NPs over a size of 30 nm were created by Fathi-Achachelouei et al. [7]. This was accomplished by attaching PEG to nuclear localisation signal (NLS) peptides. This work used human squamous cell (HSC) carcinoma of the oral cavity, over-expressing avb6 integrins as cancer cells and skin keratinocytes as normal cells. Both types of cells were obtained from humans. The nuclear localisation of AuNPs in cancer cells led to cytokinesis arrest, blocking the final phase in cell division. This prevented the cells from completing cell division, contributing to the progression of the disease [22].

The nuclear localisation of cancer cells is essential for any effective cancer therapy regimen. Plasmonic AuNPs are a potential choice for NIR light absorbers for plasmonic-based photo-thermal treatment in malignant cells [23]. Recently, near-infrared (NIR)-absorbed plasmonic AuNPs agglomerated the cell nucleus. Plasmonic AuNPs are appropriate NIR photo absorbents based on this observation. Plasmonic AuNPs preserve healthy tissue by lowering the amount of heat-induced collateral damage that occurs [24]. This is accomplished by considerably moving the absorption band to the NIR range. In a separate piece of research, it was discovered that the use of AuNPs to target the membrane of the cell nucleus led to an increase in the upregulation of laminin A/C as well as the mechanical rigidity of the nucleus, which led to a reduction in the ability to migrate cancer cells. AuNPs can target cancer cells left after tumour removal and minimise malignant cells in healthy tissues. AuNPs during grafting can offer a safety evaluation tool to reduce tumour growth by targeting malignant cells and enhance the likelihood of successful TE grafting. This can be accomplished by targeting the delivery of the AuNPs to cancer cells [25].

AuNPs are also widely used in drug delivery. They are also probes for Raman scattering, which targets cells in vivo. AuNPs can target tumour cells by attaching to epidermal growth factor receptor (EGFR) with an antibody. Antibody fragments recognise EGFR in cancer cells. Systemically administered AuNPs increased adsorption molecules of Raman scattering by 1015 times [26]. Thiol-modified PEG-coated AuNPs use single-chain varying section antibodies such as EGFR and His-linked green fluorescent protein (GFP) to compare their ability to target efficacy. PEG-coated and GFP-bound NPs could target tumour cell biomarkers encoded with surface-enhanced emitters and Raman reporters. AuNPs layered through thiol-modified PEG could identify human tumour cells in xenograft tumours with minimal quiescent aggregate using surface-enhanced Raman scattering (SERS). These tools can help tissue engineers detect recurrence over systems implantation [27].

10.3.2 Silver Nanoparticles (AgNPs)

Silver NPs, also known as AgNPs, are one of the extensively utilised metallic NPs in the biomedical area, principally because they possess antibacterial capabilities. A further approach to describe AgNPs is a colloidal suspension of silver NPs. Because TE provides a considerable danger for microbial infection, applying AgNPs as a safety precaution is one possible answer to this problem [13]. Both physical and chemical procedures can be used to manufacture these NPs [28]. AgNPs can be synthesised through evaporative condensation, laser processing of solid silver material, γ-irradiation, or ultrasonic wave stimuli. Furthermost chemical approaches use a solution of sodium borohydride or polyol to decrease silver nitrates. NPs will be produced through nucleation and subsequent growth following the reduction of silver ions (Ag^+) into metallic silver ($Ag0$). In recent years, biological methods that use bacterial organisms, eukaryotes, or flora to reduce Ag ions have become viable possibilities for synthesising AgNPs. This technique, known as the bio-reductions of silver ions, is regarded as more environmentally benign than other methods because it does not require the utilisation of hazardous chemicals during the entire method [29].

The dimension of the NPs can range anywhere from 1 to 100 nm, and their size can be determined by the method that was utilised (either chemical or physical) as well as the optimal reducing agents (weak or strong). Surfactants are used in virtually every process to maintain the particulate's consistency while retaining oversight of their size. Silver ions have been employed for an extended period because of their antibacterial activity against a diverse assortment of microbes [30]. It has been demonstrated that silver ions can inhibit the respiratory chain system of microbes and precipitating proteins found in bacterial cells. Morones and colleagues have investigated the antibacterial activities of AgNPs against different Gram-negative microbes. In their research, the authors demonstrated that NPs with sizes ranging from 1 to 10 nanometres could exert a variety of effects towards Gram bacteria by (i) releasing silver ions, (ii) entering within the bacteria and causing destruction of them, or (iii) adhering to the cell membrane and so influencing permeability and respiration. The AgNPs with a diameter of around 5 nm have the ability to inhibit the growth of bacteria [31]. TE researchers have also looked at the properties of AgNPs, primarily for use in wound-healing applications. Burn and chronic wounds are two examples of wounds particularly susceptible to infection. In addition, an increasing number of infections caused by bacteria resistant to several drugs present a difficult challenge to public health. It necessitates the introduction of antimicrobial components into the design of the scaffolds used in wound healing. The resulting particles have been introduced into scaffoldings prepared by poly(vinyl alcohol) (PVA), poly(chloromethyl) ether (PCL), gelatin (GA), chitosan (CS), and cellulose acetate (CA), and all of them have shown significant antibacterial properties [32].

10.4 Polymeric Nanoparticles

Among the characteristics of polymeric nanoparticles (PNPs) that can be modified for use in various applications using TE are their composition, shape, size, surface chemical composition and charge, permeability, mechanical durability, solubility, and decomposition rate. Other characteristics include these as well [31]. The PNPs have minimal cytotoxicity, a high degree of biocompatibility, greater penetration and retention of the enhanced permeability and retention (EPR) effect, and the capacity to transport poorly soluble drugs and sustained release. The competence to maintain the bioactivity of bioactive substances from enzymatic breakdown makes them one of the most rapidly developing platforms to overcome challenges in TE (Table 10.1). Most newly developed PNP techniques have been engineered to be profound to various physicochemical triggers, including magnetic fields, pH, temperatures, light, enzymes, and reducing or oxidising agents. This makes it possible for delivery or targeted systems to be precise and effective in TE usage. PNPs have been made with a wide range of biodegradable and biocompatible natural and synthetic polymers to get the highest standards that have been sought [49].

Features in the TE field have been achieved through the use of polysaccharides such as hyaluronic acid (HA), CS, dextran (DT), heparin (HP), alginate (AG), and pullulan (PL), proteins like albumin (AL), GA, elastin (ES), silk, and synthetic polymers like polyacrylates, polyanhydrides, polyesters, etc. PNPs can

TABLE 10.1

Different Types of Nanoparticle Applications in Bone Tissue Engineering

Nanoparticles	Applications	References
Polyurethane/MWCNT	Mechanical and thermal characteristics improved. The nanocomposite also self-cleans and self-repairs materials	[33]
CDs	Healthy biocompatibility; genome delivery; multifunctional photoluminescence; efficient drugs carriers. Great affinity and selectivity to calcified bones; possible osteoporosis treatment; early bone loss diagnosis; non-toxic and long-lasting	[34]
Clay	Nanocomposite aqueous barrier coatings could prove ecologically friendly	[35]
Gadolinium-doped Mesoporous silica nanoparticle (MSNs)	Low cytotoxicity; increased hMSC and non-structural carbohydrate (NSC) proliferation and differentiation; long-term in-vivo monitoring	[36]
MSNs-coated superparamagnetic iron oxide nanoparticles (SPIONs) with iodinated oil	Trimodal imaging: iodinated oil lowered MRI T2 relaxivity (r2), improving the quality of the image	[37]
MSNs with sentinel lymph nodes	Tiny particles ($\leq$ 20 nm); tremendous cell tagging potential in the dual-modal image	[38]
MSNs with Fe_3O_4	Simple fabrication, better imaging, cell targeting	[17]
MSNs with fluorescein isothiocyanate (FITC), gold nanocomposite	Next-generation dual-modality contrast; improved CT and fluorescence imaging	[7]
MSNs with QDs	Enhanced QD biocompatibility and imaging sensitivity	[34]
MSNs with $ZnGa_2O_4$	Multifunctional nanoprobe with strong longitudinal relaxivity and NIR sustained luminescence	[39]
N_2-doped graphene quantum dots (GQDs)	Good hemocompatibility with low toxicity; increased rBMSC osteogenic transformation; efficient stem cell imaging	[40]
PANi	Impact, flexibility, tensile strength, scratch hardness, and resistance to sheeting improved Nanocomposite may be antistatic	[19]
MSNs with antibody-conjugated with MSNs	Image-guided drug delivery and treatment for cancer	[41]
Polyester/clay	Improved thermodynamic, mechanical, and biodegradable properties	[42]
Polyester/MWCNT	MWCNT effectively dispersed in polyester matrices improved nanocomposite fluidity. Crystallinity, impact resistance, and flexibility also improved	[43]
Polyester/SiO_2	Durability, melting point, and crystallisation enhancements	[43]
Polyester/ZnO/ZnS	More robust, high transmittance and optical characteristics	[4]
Polyurethane/CD-Ag	Mechanical and thermal improvement. It is also biomaterial	[13]
Polyurethane/CD-ZnO	Mechanical and thermal improvements. It removed surfactants efficiently as a photocatalyst	[44]
QDs	Extended cell tagging using PDQDs in vitro and in vivo increases hMSC uptake and cytosol dispersion	[7]
RGO	Improved stretching, scratching hardness, resistance to impact, and anti-corrosion	[4]
SiO_2	Improved polyesteramide hydrolysis, biodegradability, and mechanical characteristics	[29]
TiO_2	Physico-mechanical, anticorrosive, and hydrophobic characteristics were better than polyesteramide.	[37]
Waterborne polyester/GO	Used as a catalyst for Aza Michael's addition reaction, improved thermodynamic, mechanical, and biodegradable characteristics	[34]
Waterborne polyester/CD	Mechanical, thermal, and biodegradability improvements. It was transparent, self-cleaning, and photoluminescent	[45]

(Continued)

TABLE 10.1

(Continued)

Nanoparticles	Applications	References
Waterborne polyester/CD@TiO$_2$	In addition to improved mechanical and thermal qualities, displayed anti-icing, anti-fogging, etc.	[46]
Waterborne polyurethane/MWCNT	Hardened and durable surface	[47]
Waterborne polyurethane/CD	Biocompatible and biomechanical for in vitro adhesion and cell survival	[48]

be made in a variety of shapes, such as nanogels, nanospheres, polymeric micelles, nanocapsules, polymersomes, and dendrimers [42]. The shape of NPs will affect their morphological behaviour; as a result, selecting a suitable topology for producing NPs is an essential step that must be considered, depending on the application. NPs can be prepared using techniques, including self-assembly, nanoprecipitation, solvent evaporation, salting-out, dialysis, supercritical fluid (SCF) technology, emulsion polymerisation, single and double emulsion, and spontaneous emulsification [50]. The most recent breakthroughs in the delivery of drugs entail combining various technological platforms, which may increase the vinaigrette combination.

The poly(ether)urethane (PEU) poly (dimethyl) siloxane (PDMS) fibrin scaffolds with vascular endothelial growth factor (VEGF)- and basic fibroblast growth factor (bFGF)-loaded poly lactic-co-glycolic acid (PLGA) NPs. Growth factor (GF)-containing scaffolds improve wound healing by day 15 in db/db mice. Histologically, scaffolds with GF-loaded NPs had comprehensive re-epithelialisation, better deposition of collagen, and mature granulated tissue. GF-NP scaffolds showed this. These results suggest that fibrin-based scaffolds with GF-loaded NPs could prolong VEGF and bFGF release by acting as a double drug delivery system (DDS) for GF delivery. The initial rapid release of GFs, and the seeping out of particles from the site of injury following application to the skin, may be decreased, leading to a more intense sustained release [44].

10.5 Lipid Nanoparticles

Lipid NPs have been the subject of numerous studies for potential dermatological, pharmacological, and cosmetic applications. Such lipid vehicles have many potentials because they are biocompatible, they are simple to prepare, and their efficacy is significantly higher than that of non-encapsulated pharmaceuticals. The earliest lipid nanocarriers to be developed were referred to as liposomes, and they consisted of an aqueous core encased by synthetic or naturally occurring phospholipids. This arrangement makes it possible to ensnare hydrophilic medications in the core, whereas hydrophobic pharmaceuticals can be ensnared in the bilayer of lipids [51]. Since 1995, the Food and Drug Administration (FDA) has approved various liposomal pharmaceuticals to treat Kaposi's sarcoma and ovarian and breast cancers. Liposomes carry GF for wound healing in several trials. First, Brown and colleagues showed that topically administered EGF-loaded multilamellar liposomes increased the tensile strength of clinical wounds in rats. When the rate of rhEGF depletion from wounds in vivo was measured, significant quantities of the GF were found to be maintained by the multilamellar liposomes [52]. This was reported to be the case at 62% following one day, 48% during two days, and 11% at five days. Additionally, one application of EGF encapsulated in liposomes dramatically extended the effects of the EGF, leading to a rise in collagen production and fibroblast development and a twofold improvement in incision elasticity from day 7 to day 14. Because EGF-loaded liposomes enhanced the initial tensile strength of the material and stimulated skin wound remodelling and regeneration, these investigations provided conclusive evidence that incorporating EGF into a suitable vehicle produced more effective healing of wounds than the local application of the free GF [53].

In another study, Pierre and coworkers observed that low levels of insulin-like growth factor (IGF-I) encapsulated in liposomes (0.9 g/kg) controlled through the skin were as effective at recovering epithelial cells. When compared to the delivery of free GFs, our findings indicate that nanoencapsulation can potentially enhance therapy efficacy [54]. This innovative method may shield the GFs from wound-related damage and degradation. Although liposomes have been around for a long time and have been the subject of much research, there are still just a few products on the market. A number of factors may be connected to poor in vivo stability, which might cause a quick and unregulated release of the medication. This may be one of the explanations. As a result, alternative delivery systems to liposomes, such as solid lipid nanoparticles (SNPs) and nanostructured lipid carriers (NLCs), were created towards the start of the 1990s [33]. Solid lipid nanoparticle (SLN) requires solid lipids as a matrix component; however, NLC uses a mixture of solid lipids and oils. Both nanoparticulated formulations may endure on the skin, moisturise it, and distribute drugs. NLCs are stable enough for hydrophilic drug encapsulation. SLNs have demonstrated evidence of drug release during retention [55].

10.6 Nanofibre (NF) Materials

The use of NF materials in applications involving TE, such as skin regeneration, is growing more prominent. NF structures imitate the original extracellular framework and encourage the adherence of a wide variety of cells and soluble substances, both of which can stimulate the activity of cells and tissue regrowth. Electrospinning is the primary method utilised in developing NFs for application in skin tissue renaissance. This is because electrospinning makes it possible to produce micro and nanoscale continuous fibres that have demonstrated homogeneous adherence to the injury site without fluids. For the past few years, several researches have been carried out to investigate the impact of NFs on the extended release of GFs to healing wounds [56]. Electrospun NFs can be made from a variety of substances, both natural and synthetic, depending on the specific usage. Silk fibroin, which comes from the *Bombyx mori* silkworm and is an organic polymer, has been shown to have sufficient rigidity and excellent biocompatibility in skin regeneration applications. EGF-containing silk mats can keep their structure and improve healing by stimulating keratinocyte tongue proliferation and re-epithelialisation [57]. As a result, the silk mats provided the wound with physical protection and maintained a steady supply of EGF to the area where it was located. In addition, lesions generated on mice were used in research to investigate the efficacy of silk-protein biomaterials combined with EGF. Wound biopsies demonstrated that the dressings and other lamellar permeable film patches closed wounds faster than the air-permeable Tegaderm®. These dressings also demonstrated a more significant recovery rate, re-epithelialisation, dermis proliferation, and collagen production [58].

In addition, synthetic biomaterials are frequently utilised in the process of preparing NFs for use in wound repair. For instance, bFGF-loaded electrospun NFs made with synthetic poly(ethylene glycol) polymers for managing diabetic ulcers on the skin significantly enhanced the healing rate when compared to NFs penetrated with free bFGF. This was accomplished by encouraging re-epithelialisation, regrowth of the skin's attachments, and remodelling the ECM. In addition, the prolonged release of bFGF that was acquired led to an improvement in the deposition of collagen and an increase in the number of and development of capillaries for two weeks [58]. Because of the NFs' applicability, simple manufacture, and excellent outcomes, several research organisations are now focusing on developing more sophisticated NF structures that combine multiple polymers. These structures are expected to have superior properties. In addition, photosensitive polymers are also utilised in the preparation of NF due to the growing acceptance of phototherapy's role in accelerating the healing process of wounds [59].

Additionally, the administration of adipose stem cells with EGF-Gel/PLLCL/P3GF(cs), whereas the cells were being stimulated by light, led to an increase in the number of keratinocytes that were differentiated [38]. In addition, studies have been conducted on producing NF by combining poly(caprolactone) and PEG polymer compounds in various proportions. When diabetes burns developed on C57BL/6 mice, the wound-healing effectiveness of rhEGF surface-immobilised NFs was investigated [60]. The results showed that these NFs had significantly greater healing in 7 and 14 days than the control group without treatment. This was the case regardless of whether or not they were immobilised. However, the degree of wound healing among individuals whose wounds received treatment with rhEGF-NFs was substantially higher than that for those whose wounds were treated with the combination of rhEGF and NFs [56]. This was the case even though both sets of individuals received the same treatment for their wounds. Based on these data, the integration of rhEGF into NFs can inhibit the breakdown of GF produced by wound proteases. Additionally, the histological analysis found that the injuries that were administered rhEGF-conjugated NFs displayed higher keratinocyte gene expression and EGF expression, improving wound healing. This was confirmed by the wounds showing increased EGF receptor expression [13].

Furthermore, to attain an alternating release of GFs, EGF and bFGF, two different growth factors, were injected into the coaxial NFs. This was carried out in order to accomplish the goal of achieving a biphasic release. The EGF was encapsulated on the surface, and although some of it was released for seven days due to the deterioration of the NF mesh, the quantity released was negligible [40]. Over the first 12 hours, it exhibited an initial burst in which approximately 30% of the encapsulated bFGF was released. In addition, from the research carried out in diabetic wounds of C57BL/6 mice, it was abundantly evident that the selective release of EGF as well as bFGF from the NFs furthermore enhanced proliferation of the epidermis and wound healing in comparison to the untreated groups and NFs-treated group, respectively [61].

10.7 Applications of Ceramic Nanoparticles

Ceramic NPs, like metallic NPs, are commonly employed in engineering bone tissue because of their many benefits to cells in bones and tissues. Ceramic NPs are smaller and 3D-printed scaffolds often use zirconia, titania, hydroxyapatite (HA), tricalcium phosphate (TCP), bioactive glass, silicon, and CDs. Because particular HA and TCP ceramics NPs contain elements comparable to those found in normal bone, they are excellent candidates for bone replacement materials [40]. This part briefly overviews the ceramic NPs utilised most frequently in engineering bone tissue.

10.7.1 Metal Oxide NPs

The most well-known NPs of metal oxides are discussed in this chapter. Specific attention was paid to the superior biomechanical and physiological capabilities of these particles and the various uses they offer in areas related to the bone. However, all of these new uses, particularly for in vivo experiments and clinical trials, must be carried out with the utmost caution because the vast majority of metal oxide NPs has the potential to be cytotoxic and genotoxic at larger doses and for more extended periods [62].

10.7.2 Calcium Phosphate NPs

Calcium phosphate (CaP) refers to substances with Ca^{++} and inorganic phosphate anions inside their structure. Because of their high biological compatibility, osseointegration of capability, and osteoconductivity, CaP NPs are widely used in 3D-printable scaffolds. HA and TCP NPs are the forms of CaP that are utilised most frequently in practice. Because of their high adsorptive capacity, superior biological compatibility, and high biodegradability, CaP NPs can be used in drug delivery of bioactive compounds into cells. This is made possible by their deployment as a drug delivery vehicle. Cells may rapidly macropinocytose CaP NPs and destroy them in their endosomes and lysosomes [37]. CaP NPs loaded with bone morphogenetic protein (BMP)-7-transcribed plasmid genomes can be taken up by cells. CaP NPs can regenerate bone after cell processing. 3D-printed bone scaffolds can use biocompatible CaP NPs for bone TE. A 3D-printed porous Ti6Al4V scaffold with CaP NPs improved human mesenchymal stem cells (hMSC) attachment, development, and osteogenic tissue differentiation. Subsequently, the scaffold loaded with CaP NPs and 3D printed demonstrated more calcification in 21 days of cell culture than the scaffold not loaded with anything [50]. CaP NPs, the primary component of typical human bone, can also be introduced to a hydrogel network to generate improved bio-ink with biocompatibility, electric conductivity, controlled release of drugs, biosensing, and other functions. CaP NPs can be easily combined with hydrogel, GA, CS, fibrin, GelMA, collagen, and HA to build a scaffold with increased angiogenesis and osteogenesis to accelerate bone production [44]. Plasmid DNA, dexamethasone, and deferoxamine can be added to NPs to deliver drugs, combat pathogens, etc. Biomineralisation involves both inorganic and organic substances in living organisms. Abnormal biomineralisation frequently causes organ dysfunction or serious illnesses. Bone TE must manage biomineralisation in vitro, in vivo, and therapeutically to induce the regeneration of bone securely. NPs accelerate biomineralisation and alter bone marrow mesenchymal stem cell (BMSC) osteogenic development, according to many studies. Some research showed that mechanical scaffolding characteristics improved with biomineralisation. NPs may drastically alter mechanical characteristics; well-organised biominerals also guard and maintain organisms and detect tissue and organ signals [63].

10.7.3 HA Nanoparticles

Although HA constitutes the bulk of the inorganic element of natural bone, it constitutes one of the most prominent biomaterials used for purposes that include the medical field. HA also has substantial osteoconductivity, exceptional biocompatibility, excellent cellular attachments, and great biodegradable properties. HA can also be broken down very quickly. Composite scaffolds containing HA NPs have been proven to have mechanical and biological features that have been fine-tuned. Because they comprise most of the bone, HA NPs are the fundamental components of the biomineralisation process [64]. HA NPs could be used in 3D-created polycaprolactone scaffolding to improve its compressible rigidity and biological efficiency. The 3D-printed polylactic acid scaffolding for the trabecular restoration of bones improved because the improved 3D-printed scaffolding might resemble the framework of trabecular bones more effectively. Polyethylene glycol methacrylate (PEGDMA) 3D-printed scaffolds are designed to increase rigidity and promote the osteogenic differentiation of human mesenchymal stem cells [65].

Moreover, hydroxyapatite nanoparticles (HA NPs) were integrated into 3D-printed polylactic acid scaffolds to mimic the original composition and hierarchical structure of bone. This was done by printing the scaffolds in three dimensions. Experiments carried out both in vitro and in vivo revealed that the combined materials possessed increased mechanical capabilities and were bioactive and biocompatible [62]. In a bone remodelling paradigm that included osteoblasts as well as osteoclasts, HA NPs promoted cell division and expansion. HA NPs-enriched matrix of the bone model was used to study bone remodelling. In order to create biomimetic scaffolding, 3DCS using HA NPs was synthesised. The authors also discovered that the tensile strength might rise by a factor of 100 when HA NPs were introduced at a concentration of 28%. Additionally, the extra HA NPs provided a crucial purpose in the intensified biomineralisation and the absorption of proteins in the biomimetic scaffolds [46].

10.7.4 β-TCP Nanoparticles

Depending on the variety of CaP, its structural and biological characteristics may differ. TCP is a different kind of CaP that is the subject of a significant amount of investigation, identical to HA. TCP can be broken down into two primary subtypes: β-TCP and $\beta 1$-TCP, which are generated at temperatures distinct from one another. Because it is less prone to degradation, β-TCP is the type of bone scaffold most frequently utilised. The rigidity of bone scaffolds as well as the growth rate of osteoblasts can both be increased with the use of β-TCP NPs. Because of these properties, β-TCP is a beneficial component in bone scaffolding. The mechanical characteristics of 3D-manufactured collagen bone scaffolds treated with β-TCP NPs were stabilised, and the scaffolds exhibited good bioactivity and improved biodegradability. The 5% weight (TCP5) NP had the most potential to boost cell proliferation and develop new blood vessels. However, the TCP10 NP enhanced rat bone production more quickly than the other combinations of drugs. However, TCP25 repeatedly demonstrated undesirable biological effects, which suggested that excessive dosages could induce harmful effects. Based on these findings, a specific dosage of β-TCP NPs could benefit the material's biocompatibility and bone conduction capacity. The authors recommended combining TCP5 and TCP10 with the application of β-TCP nanomaterials for the greatest possible result [39]. There is still much work to achieve aggressive and rapid bone regeneration. In addition to the guided bone regeneration (GBR) membrane, utilising multifunctional scaffolds with osteoinductivity and osteoconductivity is essential to accelerate bone tissue's regrowth. The PEGS/-TCP membrane has the potential to facilitate the attachment and proliferation of rat MSCs derived from bone (rBMSCs). The elevated alkaline phosphatase (ALP) activity and the enhanced cell calcification contributed to the differentiation of osteogenic tissue becoming more pronounced. In vivo studies on rats yielded the same results as in vitro testing. As a result, the PEGS/-TCP membranes could be utilised as a potential treatment for bone abnormalities. Additionally, the GBR membranes are frequently utilised in conjunction with bone scaffolds to hasten bone repair [53].

10.7.5 Biphasic Calcium Phosphate (BCP) Nanoparticles

CaPs of various kinds are frequently combined in clinical practice to compensate for one another's weaknesses and emphasise their strengths. NPs of BCP are created by combining HA and β-TCP NPs. These NPs are then utilised in the production of 3D-printable scaffolds. The addition of HA and β-TCP to scaffolds has the potential to increase their mechanical durability, induce adhesion of cells and proliferation, and differentiate into osteogenic cells [36]. In order to improve the mechanical qualities of the scaffold, the new biodegradable 3D-printed scaffolding was implanted with BCP NPs. The findings demonstrated that scaffolds loaded with BCP NPs possessed superior mechanical qualities than scaffoldings filled with CaP. It is also possible to insert BCP NPs into CS and GA hydrogels to construct bone scaffoldings for hard TE. Changing the weight ratios of the CS/GA/BCP NPs allowed for physicochemical and biological characteristics to be adjusted. The BCP NPs used in the 3D printing process significantly impacted the bone scaffolds' porosity, expansion proportion, and rigidity [52].

In addition, results obtained in vitro and in vivo suggested that BCP NP-loaded scaffolding may enhance the development of hMSC, indicating that they had a significant promise for use in bone TE. For bone TE, biomimetic composites of collagen scaffolding were produced. These scaffolds included regulated porosity architectures and were outfitted with dexamethasone-modified BCP NPs [12]. These composite scaffolds possessed a controlled porosity structure, acceptable mechanical characteristics, and the ability to release dexamethasone sustainably. As a result, they demonstrated outstanding biocompatibility and osteogenic development of hMSC. The findings from in vivo experiments suggested that dexamethasone BCP NPs/ collagen scaffold combinations would be suitable for usage as bone scaffolding components. Briefly, CaP NPs have been the subject of significant research and have been implemented in the engineering of bone tissue. However, CaP NP toxicity is poorly understood; CaP NPs may harm cells at high doses, according to specific investigations. These negative effects include reduced cell viability, genotoxicity, and apoptosis caused by reactive oxygen species (ROS) production. In the TE of bones, in-depth and all-encompassing research is necessary to make the most of their unique qualities while avoiding their drawbacks [66].

10.7.6 Mesoporous Silica NPs (MSNPs)

The biocompatibility and safety of MSNPs are unparalleled. It can be incorporated into bone scaffolds, especially those related to musculoskeletal metabolism. Because of their tuneable porosity and improved surface qualities, MSNPs are strong candidates for use as either biomaterials or bio-ink in bone TE. MSNPs suppress osteoclast growth and enhance osteoblast proliferation and mineralisation. In mice, they significantly increase the density of bone minerals [36]. The bioactive HA coating with MSNPs was biocompatible and promoted cell differentiation. Doping MSNPs on TCP scaffolds allowed to demonstrate the in vivo regeneration of bones in rabbits over four months. This was accomplished using chronological radiographic inspection, histological evaluation, and fluorochrome labelling. The radiograph demonstrated that MSNPs-doped TCP scaffoldings degraded faster than pure TCP scaffolds. The histology and fluorochrome labelling studies demonstrated that MSNPs doped with TCP had a quicker bone repair and revascularisation rate. MSNPs can be spread throughout a scaffold to improve its mechanical characteristics and to generate a biomimetic and permeable nanostructure that promotes the adhesion of cells, proliferation, and development [15]. Hydrogels are frequently used in bone TE because of their high absorbability, rapid bonding with surrounding tissues, and the nutritional conditions they can provide for developing the body's cells. However, employing hydrogels as bio-ink presents several obstacles, the most significant of which are a need for regeneration ability, an absence of precise control over shape, a rapid degradation rate, and minor assistance 3D structure [67]. It is possible to incorporate MSNPs into a hydrogel to increase its viscosity and make it more suitable for 3D printing. In general, not only was there an increase in biomineralisation, but there was also a significant improvement in the compression modulus, and the swelling and degrading qualities were considerably reduced. In addition, the biocompatibility and osteogenic potential of hydrogel, including MSNPs, were significantly improved compared to a hydrogel that did not contain MSNPs [61].

10.8 Carbon Dots (CDs)

CDs are 1–10 nm NPs; they are biocompatible, cytotoxic, water-soluble, chemically inert, biodegradable, fluorescent, and photoluminescent. These advantages make CDs useful in bone tissue creation. Functionalised CDs are needed for in vitro and in vivo applications because they rapidly enter cells and visualise tissues. Bio-imaging and drug delivery on CDs helped diagnose and treat early bone-related illnesses, including osteoporosis. Cellular imaging in vitro can be employed for compartment-specific and type-specific imaging [43]. CDs containing ligands may penetrate cells for cell-compartment-specific sensing. Cells have mitochondria and nuclei, using transferrin ligands, and CDs were functionalised with folic acid ligands. CD research was needed to expand the applicability of the two in vitro specialised imaging methods. While testing CDs on zebrafish larvae for bone-related in vivo imaging, researchers found that the non-toxic, long-term CDs could be a great fluorescent marker for calcified bone.

Amino-functionalised CDs containing glutamic acid ligands are capable of identifying smaller bone fractures. Transport agents cannot interact with inorganic bone elements. CDs can transport ciprofloxacin to cure bacterial infections, dopaminergic hydrochloride to treat neurological illnesses, or genes for other uses. Glutamic acid and ciprofloxacin were attached to amino-functionalised CDs. Glutamic acid, a calcium-targeting ligand, stayed in calcium-rich areas like the bone fracture region. At the same time, the location-specific antibiotic ciprofloxacin could be utilised. Bio-imaging, the fluorescence property of CDs, was also utilised to assist in the localisation of bone fractures. The research group concluded that non-toxic CDs that are hemocompatible and could be loaded with various medications would be an effective way to manage other bone-related disorders. In order to combat bone disorders such as osteoporosis, further bone-targeting CD biomaterials need to be designed [34].

In addition, CDs can be placed on bone scaffolds to increase the structures' physical capability. The PCL/PVA scaffold packed with CDs significantly increased bone marrow proliferation and differentiation into osteogenic cells. In order to enhance bone formation and regeneration, CDs were included in CS and HA scaffolds prior to fabrication. In addition, the scaffolds, when exposed to near-infrared irradiation, could impede the proliferation of osteosarcoma cells in vitro and prevent the formation of tumours in vivo. The PCL membranes with reduced nanographene oxide CDs enhanced osteobioactivity and cytocompatibility in MG63 osteoblast cells without toxicity. In order to improve bio-imaging, CDs were changed by loading them with nitrogen, potassium, and calcium ions [37]. These luminous CDs displayed high aqueous stability and outstanding biocompatibility throughout their testing. The incorporation of ions of N, P, and Ca also brought about the enhancement of electrical properties. The citric acid-functionalised CDs boost rBMSCs' differentiations by cellular uptake. The findings showed that if the level of CDs was less than 50 g/mL, cell survival had no impact. The acquired CDs have the potential to be used for long-term labelling. HACDs and Mg ions-loaded CDs on mouse osteoblast cells (MC3T3-E1) and adenosine- and aspirin-based CDs on hMSC showed osteogenic transformation. The procedures described above offer a risk-free and highly effective method for assisting bone regeneration. This method involved bone scaffolds or bone cells' direct uptake of the substance. In order to make better use of CDs, additional research needs to be carried out [25].

10.9 Quantum Dots (QDs)

QDs are hydrodynamic semiconductor NPs which can be irradiated by UV light resulting in numerous applications in biological imaging and biosensing. In addition, in vivo for a long time, cell tagging, and monitoring are possible because of their superior photostability, remarkable resilience to photobleaching, expanded absorbance spectra, and remarkable fluorescent intensity. QDs were coupled with polyamidoamine (PADM) dendrimer (PDQDs) in order to increase intake rate and cytosolic migration of hMSC. In the last few years, there has been increasing concern regarding the toxic effects of QDs, which are triggered by the inclusion of heavy metals in their content [37]. This has resulted in the concerns above because QDs have a wide range of applications in the biomedical sector. QDs also produce reactive oxygen species, which may lead to cell death (apoptosis). As a result, GQDs have replaced ordinary QDs due to their 0D framework, minimum toxicity, ultra-small size, greater solubility, adjustable fluorescence, remarkable photostability, chemical inertness, and great biocompatibility [40].

In addition to these traits, it has been demonstrated that GQDs are friendlier to their surroundings. Because GQDs typically come with a significant amount of surface area, they make good candidates for the delivery of medications. GQDs that contained silver NPs as an antibacterial substance exhibited a broad spectrum of activity against various pathogenic organisms. The agent was found to be effective against both types of bacteria. Multifunctional GQDs have been created and made to simultaneously fulfil the dual roles of drug transporters and bio-imaging reagents [1].

For instance, numerous teams have successfully packed doxorubicin onto GQDs in order to deliver medications to cancer cells in an effective manner. In addition to this, the movement of the entire cellular absorption process has been tracked because GQDs have the inherent virtue of being fluorescent. GQDs also strengthen hydrogel, fibre, and polymeric 3D-printed scaffolding. Zhao et al. found that GQDs can strengthen 3D-printed scaffolding for bone TE by increasing stiffness and durability. A graphene quantum dot (GQD)-loaded chitosan (CS) scaffold was developed for wound healing. The enhanced CS scaffold exhibited superior antibacterial, mechanical, and biocompatibility properties. The CS composite scaffold was also pH-sensitive. Due to their unique properties, GQDs are suited for in vivo bio-imaging applications such as fluorescence imaging, MRI, and multifunctional imaging. The authors of [68] made photoluminescent P-GQDs to label MCF-7 cells. MCF-7 cytoplasm contained P-GQDs. P-GQDs outperformed fluorescence probes in photostability. Researchers have done extensive in vivo and ex vivo imaging. Nurunnabi and colleagues used fluorescent imaging to track carboxylated c-GQDs in vivo and ex vivo (via isolated organelles).

10.10 Magnetic Nanoparticles

Magnetic NPs, also known as MNPs, remain a form of nanomaterial when exposed to an external magnetic field and are capable of magnetic manipulation. Metals like iron or cobalt continue to make up these NPs. Majority of the time, they are constructed from a piece of magnetic material encased in a biodegradable cap. They may also have an optional coating that confers extra capabilities. MRI contrast agents, magnetically argon-transported drug delivery, and magnetic hyperthermia are only a few examples of applications for these compounds. Super paramagnetism is known as the phenomenon of suitably small iron-oxide-rested MNPs exhibiting magnetic properties without the presence of an external magnetic field [10].

Magnetic iron oxide NPs through maghemite and magnetite crystal structure are the utmost researched for use in biomedical applications. The temperature and pressure of the reaction conditions remain the primary means by which the size of the morphology of these MNPs may be controlled. Iron oxide MNPs have a high surface energy, which causes them to tend to combine. Surface coatings given near the particles prevent this from happening, and the particles are then functionalised with various layers to form a shell. These coatings include CS, silica, PEG, DTs, and others. When paired with scaffolds, MNPs become practical components aimed at the construction of tailored nanostructures used for TE because of the capability to modify the physicochemical effects of MNPs. These features include extent, form, morphology, surface modification, hydrophilicity, and then functionalisation [9].

MNPs are rapidly developing into essential mechanisms in bone TE. Inside every scaffold covered with superparamagnetic iron oxide nanoparticles (SPIONs) comes about the formation of a magnetic microenvironment. The magnetism in every surrounding area can elicit reversal in the ion channels and receptors that perceive this, ultimately leading to considerable improvements in osteogenic differentiation and proliferation. In addition to this, the introduction of SPION can boost the mechanical qualities of the scaffold while simultaneously promoting cell adherence through the use of a more reflective scaffold microarchitecture [53].

The utilisation of the influence of a field of magnets to control the movement of MNPs and their functions is one method that can be utilised to expand the range of applications for MNPs in bone tissue engineering (BTE). When an oscillating external field is combined with MNPs contained within an imperfection, it can enhance cell induction and remotely deliver biomechanical stimuli at the individual cell level, facilitating osteogenesis. In addition, MNPs, when exposed to a field of magnets from the outside, can efficiently transport particular substances with active biological processes or MSCs to the location of a bone deficiency. This guiding strategy makes it possible to achieve both localised delivery and retention, increasing the effectiveness of bone regeneration therapy while avoiding the adverse effects on the tissues surrounding the treated area [32]. A recently developed technique, an external magnetic field, is utilised in cell-based TE to precisely regulate and manipulate cells that have taken up MNPs from the medium in the surrounding environment. An external magnetic field is utilised in cell-based TE to precisely control and manipulate cells that have taken up MNPs from the medium in the surrounding environment that is vascularised.

In addition, magnetic patterning can give a straightforward and speedy means of producing a gradient in the concentration of the growth factor, which has the potential to reproduce the intricate microstructures arranged in a hierarchy inside physiological tissues. When an oscillating external field is combined with MNPs contained within an imperfection, it can enhance cell induction and remotely deliver biomechanical stimuli at the individual cell level, facilitating osteogenesis. However, the weight of sophisticated construction comes with these types of systems. Toto developed novel solutions for bone tissue biomimicry; researchers have investigated various strategies for combining scaffolds, cells, and physiologically active cues. These strategies make use of a wide range of manufacturing processes. Research is being done to see if magnetic nano actuation can be used to remotely change the behaviour of cells with significantly more control and precision. Various manufacturing methods, like electrospinning, covalent couplings, and freeze drying, can incorporate MNPs into scaffold matrices. First, the influence of scaffold-containing MNPs on osteogenesis will be described, after which their function as a stimulator of angiogenesis will be investigated. The BMP delivery vehicles must supply the factor at a predetermined optimal amount or time-based period to achieve the maximum osteogenic induction. TGF-β, FGFs, VEGF, and PDGF are crucial for fibroblast, vascular endothelial, and platelet-derived growth. Additionally, hGF (human growth factor) influences bone formation. BMPs are among the most known osteoinductive chemicals. BMP-2 and BMP-7, in specific, are potent osteogenic inducers that are increasingly being employed in bone TE research, more than any other factor [53].

10.11 Nanomaterials Used in Bone Tissue Engineering

Bone provides mechanical support, mobility, and load bearing. This role is essential because bone is responsible for several of the body's skeletal structures. Because bone tissue is constantly undergoing resorption and reformation, it is usually possible for tiny bone injuries to recover as long as the quantity of tissue damage does not exceed a specific threshold size. Despite this, proper healing after substantial bone tissue loss due to trauma remains an unresolved clinical problem that can only be addressed through surgical intervention. At the minimum, 2.2 million bone grafts have been performed. Surgeries are conducted each year worldwide [62]. They use autologous or allogeneic bone grafts, which can come from the patient's body or a donor, the current gold standard for repairing extensive or complex bone lesions [32]. Differentiated features are present, and progenitor cells are essential and responsible for the effectiveness of freshly harvested bone autografts. These cells are responsible for osseointegration (basic and functional link with native bone) and osteoconduction (surface cell proliferation and matrix deposition), essential for forming new bone after transplantation. Due to these characteristics, bone autografts frequently result in successful healing; nonetheless, this method suffers from numerous significant drawbacks, including a lack of adequate bone volume for harvesting, the development of morbidity at the donor site, and an increased risk of developing problems as a result of various surgical operations. Bone allografts have the potential to make up for the limitations of autografts in some cases [50]. However, they have substantial downsides; lower graft bioactivity, poor integration with native tissue, and a restricted supply. In addition, there is a need for these grafts. On the other hand, autografts are not subject to this kind of constraint. Traditional approaches to bone TE focused primarily on using biological or engineering practices to build tissue constructions promoting bone regeneration. These methods might either be biological or engineering in nature. The emphasis on modern techniques gradually shifts towards more integrated methods at the point where biology and technology meet. These solutions frequently entail using a biomaterial to provide a platform for tissue regeneration that is biocompatible, bioactive, or biomimetic [17].

The primary purpose of research and development in bone marrow TE is to cure irreparable injuries that could occur, and adjuvant therapy using nanomaterial-based implants is suggested. Recent research in bone TE has investigated 3D scaffolds, which can support, reinforce, and organise the regenerative process of skeletal muscle, like how bone tissue typically functions. Naturally degradable NPs are utilised more frequently in bone TE due to their several benefits for bone mineralisation, including increased biocompatibility, biosafety, and enhanced bone mineralisation [29]. Notably, their natural degradability enables them to be quickly altered for use in the in vivo environment following clinical implantation. This eliminates the need for a second procedure to remove the device, reducing the amount of discomfort experienced by the patient. In bone TE, scaffolding resources are typically utilised to give structural assistance for areas of the bone that have been injured and to create an environment conducive to bone regeneration [65].

Ideally, a bone TE construct should have acceptable mechanical qualities and surface possessions for proliferation, cell adhesion, and differentiation. In addition to this, it is essential to consider the scaffold's features, such as its porosity, biological conduction, biocompatibility, and absorbability. Using a polymer-based scaffold as an illustration, widely acknowledged as among the appropriate components with inadequate mechanical characteristics for use as skeletal scaffolds, we see that it is one of the most suitable materials. However, adding nanomaterials (such as nanocrystalline resources and ceramic nanomaterials) is a great way to improve the stent's mechanical properties. In addition, implementing functionalised techniques on specific nanomaterials can regulate the routes that cells use to send signals to one another, alter the adherence of proteins to the scaffold, and boost the material's mechanical properties [45]. The FDA has blessed CS, a natural biodegradable polymer with diverse pharmaceutical compositions with biodegradability and biocompatibility. CS has been an essential component in progress over the past few decades in bone TE. Furthermore, a wide range of porous structures made of CS can increase bone conduction. HA NPs, the principal inorganic component of bone tissue, are gaining popularity in bone TE. The authors of [37] developed HA/CS biomimetic nanocomposite NF scaffolds to explore the influence of bone marrow mesenchymal stem cells (BMSCs) development on bone regeneration and the molecular mechanism in vivo and in vitro. The nHAp/CTS scaffold increased BMSC proliferation and integrin-BMP/Smad signalling [69]. The CS/HA composite can achieve osteogenic differentiation of osteoblasts, which has led to its recognition as a promising bone TE material. In composite NF/CS scaffolds, Liu and colleagues investigated the function of HA NPs. They discovered HA boosted bone regeneration and proliferation in BMSCs by enhancing cell adhesion and activating the BMP/Smad signalling pathway. They mimic the natural inorganic phase of bone better than organic polymer NPs, making them a novel biomaterial for scaffolding techniques. Furthermore, HA can be used with other polymer materials that have shown favourable benefits in bone repair and regeneration, such as PCL, PLGA, PEG, and whitlockite NPs [46]. In addition, advancing bone TE, NPs formed of bioactive glass and ceramic (nBGC) can more accurately imitate bone composition than metal NPs. Use of electrospinning, bioactive glass/PVA, and silk fibroin could create two-layer scaffoldings that improved the development and differentiation of BMSCs. Furthermore, MSCs found in the bone marrow are conducive to developing new bone. In contrast, MSCs found in umbilical cord blood are conducive to forming new blood vessels. Furthermore, AG, cross-linked DT, PCL, PCL–CS, PLGA, collagen, and other materials have been utilised with bioactive glass to build bone TE scaffoldings. In some instances, the combination of these materials has resulted in improved scaffold performance [67].

Carbon-based nanomaterials are 0D CDs, fullerene, nanodiamonds, 1D CNTs, 2D graphene, and 3D graphite. Biocompatible, high strength, and commercially available, these nanomaterials have configurable interface functionalities and large surface areas [47]. Bone stem cells develop, proliferate, regenerate, bind, and develop on carbon-based scaffolds in bone TE. Bone stem cells stick better to carbon-based scaffolds than others. Due to its easy purification, great length and width, and low metal impurity level, the G material may be preferred. Oxygen makes GO hydrophilic. Water, organic solvents, and others dissolve it better. Kumar et al. created bone-repair scaffoldings using polyethyleneimine (PEI) and GO. According to studies, PEI/GO may induce osteoblast differentiation, hBMSC proliferation, and focal adhesion complex formation. Alkaline phosphatase expression significantly doubled, and mineralisation was 50% higher than with GO alone. They found that PEI/GO polymer composites could replace absorbable bioactive substances in orthopaedic devices for fracture stabilisation and TE operations. CNTs have excellent electrical conductivity, the highest current transmittance, and muscular tensile strength. Nanocomposite scaffolds' enhanced tensile strength, excellent electrical conductivity, and determined current transmittance are attracted by these attributes. Bone TE widely uses carbon-based NPs with varying biological activity [37]. Gold, silver, and titanium oxide are increasingly used in bone TE. Much research has implanted AgNPs for bone TE. Silver NPs, which have excellent mechanical properties for metal NPs, were made osteogenic. The AgNPs' biocompatibility and osteogenic potential stabilised with polyoxyethylene glycerol trioleate (PGT) and Tween 20 particles. The AgNPs increased MSC and osteoblast uptake without side effects. This shows AgNPs' bone TE potential. Gold NPs also promise cell differentiation [68]. Gold NP size may affect cell–material interaction. Primary osteoblasts like 20-nanometre gold NPs. However, 30–50 nanometre gold NPs affect human adipose-derived stem cells. Titanium dioxide (TiO_2) NPs and various polymers can be employed to build improved scaffolds to examine bone formation performance [17].

10.11.1 Bone Regenerative Stem Cells

The bone matrix is 40% organic and 60% inorganic. Bone tensile strength comes from type 1 collagen, proteoglycans, and matrix proteins like osteocalcin and osteopontin. These are distributed throughout the bone matrix and osteoblasts secrete these proteins. MSCs can become osteoblasts and chondrocytes, essential for bone fracture repair. The inflammatory phase of bone fracture repair is followed by MSC activation, recruitment, and proliferation. MSCs then become chondrocytes or osteoblasts. Osteoblasts undergo intramembranous ossification. Chondrocytes enlarge, mineralise, and replace the bone matrix with cartilage through endochondral ossification [45]. The remodelling process continues throughout life, regardless of bone injury. This rebuilds and replaces bone. MSCs, which become osteoblasts and chondrocytes, regenerate bone. Sox9 levels may determine MSC's fate. SRY-box 9 transcription factors control MSC growth. Sox9 promotes chondrogenesis by synthesising collagen proteins like Col9a1, Col2a1, Col1a2, and aggrecan, while BMP family members like TGF-β and FGF-2 stimulate osteogenesis. Runt-related transcription factor-2 (Runx2) and Sp7 (Osterix) regulate osteoblast and adipocyte development with BMPs.

Nanomaterials depend on NP physicochemical properties. Ionic and nanoparticulate silver affect life differently depending on physicochemical parameters. These include AgNPs' form, size, and concentration. Synthesis, capping agents, substrate improvements, and dose adjustments can make AgNPs an attractive bone TE choice [25].

10.11.2 Orthopaedic Surgery 3D Printing

3D printing has lately come to the attention of researchers working on constructing such scaffolds. The defect that a patient has is portrayed using CT or MRI. Then those scans parameters are digitised and utilised to build scaffolds that precisely fit the patient's initial defect. To fabricate patient-specific scaffolds with even complex shapes and anatomy, which is rather challenging to do with conventional scaffold manufacturing procedures, 3D printing technology offers the benefit of being able to do so. The production of scaffolding using 3D printing is more efficient and less expensive than conventional ways [65]. It is possible to accurately determine structural characteristics and mechanical qualities, such as a material's porosity, pore size, and interconnectivity. Biomaterials were once touted for their capacity to give the benefits of biocompatibility, non-toxicity, and biodegradability [47]. Combining natural and ceramic materials or even different types of ceramics is now possible, paving the way for developing new bioceramics.

When a scaffold is loaded with cells, medicines, and bioactive compounds, it has the potential to offer the benefit of guiding the cell microenvironment and multipotency to differentiation, which can lead to particular tissue regeneration [52].

10.11.3 Bone Developmental Engineering

To speed up the healing process of damaged bone tissue, bone TE typically involves the progenitor cells directly distinguished into osteogenic lineage cells, creating a mineralised matrix. From an embryonic point of view, immediate osteogenic commitment occurs typically during the development of flat bones (intramembranous ossification). Still, it does not happen during the development of long bones (endochondral ossification), which occurs by replacing the initial cartilage templates with new bone tissue. To successfully repair the defect site and faithfully duplicate the form and function of the native tissues, it might be necessary for manufactured long-bone transplants to mimic the developing process. This is necessary to repair the defect site successfully. Instead of focusing on developing replacement tissue, it could achieve this objective by applying engineering principles to the developmental process [34].

The developmental engineering method exploits critical ontogenesis pathways to replicate the embryological processes responsible for tissue growth and differentiation. Therefore, a bone replacement for extensive deformities should mimic the natural endochondral ossification process. Condensation of mesenchymal progenitors is the first phase in this process, followed by chondrogenesis and hypertrophy [45]. Extended bone defects could be repaired with substitutes if the procedure can be repeated. After this step, perichondral ossification, matrix remodelling, and vascularisation will form a bony collar. The cartilage model will be resorbed during this time, and the bone matrix will be deposited. Because the endochondral pathway is flexible, its findings should immediately apply to regulated ex vivo experimental techniques [36]. Identifying specific DNA markers for every origin and stage of development (for instance, collagen type X identifying hypertrophic chondrocytes) may make observation and control of the ongoing process more accessible, enhancing the resilience of the complete technological procedure. This will be achieved by finding molecular markers for each lineage and stage of development. Furthermore, implanting hypertrophic cartilage is anticipated to enhance the survival of grafts post-implantation. This is because hypertrophic chondrocytes thrive in hypoxic environments before vascular anastomosis and release VEGF to attract endothelial cells [46]. The development of a functional long-bone replacement will eventually necessitate homing hematopoietic cells to the bone marrow, which appears to rely solely on endochondral ossification. This feature is one of several prerequisites for creating an adequate long-bone substitute [60].

Consequently, a meticulous reconstruction of the morphogenetic mechanism ought to facilitate the development of highly structured and fully functional organs. Recent studies have demonstrated a successful replication of certain facets of endochondral ossification through the utilisation of mouse embryonic stem cells, chick embryonic mesenchymal stem/stromal cells MSC, and clinically significant cell sources such as human bone marrow-derived MSC. In this set of studies, MSCs from bone marrow were made to grow into swollen chondrocytes in a lab dish. With this method, chondrogenic templates are made. These templates are transferred into naked mice, where endogenous progenitors turn them into bone [50]. A new study found that chondrogenic mesodermal cells can be made from mouse embryonic stem cells if the right signals are sent. With the help of BMP-4, these will turn into swollen chondrocytes. On the other hand, when growth differentiation factor-5 is used with Sonic Hedgehog and BMP suppression, it leads to non-hypertrophic chondrocytes that look like cartilage. Certain cell–cell signalling links, like morphogenetic Indian hedgehog (IHH) signalling needed for embryonic skeletal development and endochondral ossification, are turned on and are physiologically crucial for bone tissue morphogenesis from hMSCs. These new advances are the first steps toward designing cartilage and bone, but tagging donor and host cells to examine their fates and relative contributions to bone-like structures will be crucial for identifying the instructional cues needed for induction and remodelling in vitro and in vivo [70]. The molecular circuitry beneath the surface will help produce cell-free grafts for in situ bone formation. Before decellularisation, specific endochondral matrices can provide morphogenetic signals. These could offer 3D structure and signal endogenous progenitors to restart embryonic chondrogenesis and endochondral ossification after implantation. These methods promote muscle regeneration and communication with the repair site's endogenous microenvironment. Synthetic chondrogenic tissue constructs cannot duplicate the development plate of growing long bones. Inaccuracies in distinct cell types, spatial and temporal arrangement, and critical regulatory signal expression cause this. This may be because synthetic bone constructs cannot self-organise like embryonic bone. No spatiotemporally coordinated cell proliferation, destiny assignment, or differentiation and no well-regulated morphogenetic signals may explain this. There may be no morphogenetic signals. In contrast to the highly dynamic spatiotemporal change of the underlying signalling systems and cell properties required for embryo bone development, traditional culture techniques impose severe physical limitations on their cells [69]. The next generation of ECM-based approaches may employ intelligent biomaterials such as hydrogels. The innovative biomaterials might aid in replicating the spatiotemporally regulated release of morphogenetic components. This creates ordered signalling gradients, which result in directed cell behaviour. According to [18], when paired with system methodologies that foresee the primary morphogenetic routes and occurrences, this will help deploy more effective and regulated methods of developmental engineering.

10.11.4 Dental Tissue Engineering Using Nanomaterials

Subsequently, at the turn of the century, interest has risen; nanomaterials have been used in dental TE. Older people have a higher risk of getting periodontal disease and conditions like diabetes, rheumatoid arthritis, and cardiovascular disease are strongly related to it. Patients with periodontal diseases require effective therapies to repair damaged tissues and return them to their original structure and function. This happens as periodontal tissue ages and loses its capacity to heal itself. We are fortunate that the study and development of various metal and polymer-based NPs have produced promising findings for treating periodontal illnesses and conditions [42]. The following are some of the most typical uses for nanomaterials in dentistry: (1) Antibacterial agents to prevent and treat mouth illnesses; (2) Nanofillers that are to improve or keep the strength and biological function of substances used in periodontal disease; (3) New coatings for implants; and (4) Toothpaste and other household goods. Polylactic acid (PLA) with polyglycolic acid (PGA) makes PLGA, an aliphatic polyester, biocompatible, sturdy, and degradable. Periodontal disease research uses their biological compatibility. Dental TE uses PLGA-based composites to regenerate tissue, fight germs, make cement, and protect alveolar bone. The PLGA-based bilayer is a natural material that restores periodontal tissue. PLGA-based biomaterial increased bone volume, thickness, and number. The exclusive capacity to generate fresh cementum and bone was observed solely in biomaterials based on PLGA. PLGA-based bilayer biomaterials regenerated periodontal tissue better than flexible membranes. Dental TE uses NPs of CS, silica, poly(3-caprolactone), etc. A non-destructive way is to graft 10–30 nm monodisperse silica NPs onto the polystyrene, polyethene, and polyvinyl chloride surfaces. Silica NPs inhibited biofilm growth by reducing bacterial adhesion. Silica NP-grafted polymer films resist microorganisms. NPs decrease bacterium adhesion and germ multiplication. Biodegradable and biocompatible CS is commonly used in periodontal tissue repair [40]. The CS/inorganic bovine bone composite scaffolds inoculate human mandibular bone marrow MSCs to treat periodontal disease. CS scaffolds increased compressive performance and biocompatibility. MSCs repair the critical lesion by creating woven bone, fibrous cementum, and periodontal tendon on the CS scaffolds. A critical defect showed this. Dental TE uses FDA-approved PCL [36] due to its biodegradability, biocompatibility, and EPR effect. Bharadwaj and Jayasuriya, 2020, developed a periodontitis-treating PCL-based dual corona vesicle biofilm. PCL biofilm is biocompatible and antibacterial. Ciprofloxacin HCl-loaded dual corona vesicle system destroyed plaque and biofilms from *Escherichia coli* and *Staphylococcus aureus* in in vitro conditions. Thus, periodontitis treatment relied on the system. Polydopamine cures periodontitis. This treatment reduces polydopamine. Polydopamine-based NPs decreased periodontal inflammation and ROS without adverse effects *in vitro*.

10.11.5 Tissue Technology

TE was once regarded as a combined turf of biomaterials and engineering. Still, as the topic's breadth and connotation expanded, it has grown into a subject in its own right. Tissue and organ failures are serious illnesses that are also relatively common. Surgical repair, organ transplantation, prosthesis, and pharmaceutical remedy are among the available treatments for these disorders. A TE researcher will aim to restore the body by replacing defective tissues or organs with functional replacements that have been created. Because of stem cells, the field of TE has gained new momentum [67]. In response to appropriate stimuli, they replenish themselves and commit to specific cell lineages, giving significant rejuvenation potential that will likely lead to tissue functionality. Both modern ecology and pathology have demonstrated that defective cells are the root cause of many diseases. The complexities and multicellularity of tissues and organs make the differentiation of stem cells into the tissues and organs required for TE difficult. This is why TE has yet to realise its potential fully [61]. The musculoskeletal system must be rebuilt using three different components: A biocompatible scaffold, cells that can replace injured tissue, and biochemical mediators (such as growth factors) that can direct cell function. Because of the high surface-to-volume ratio of nanotechnologies, they can produce scaffolds that can subsequently transport drugs and growth factors directly to the area of the lesion where bone development is impeded. Furthermore, animal models have employed nanotechnologies to change gene expression, resulting in endogenous over-expression of inevitable growth and transcription factors [65]. This method was used to promote bone formation, and it could potentially reduce the need for high dosages of exogenous recombinant components and potential adverse effects.

A biodegradable scaffold supports cells, allowing them to populate the structure and generate their extracellular matrix (ECM) through the top-down tissue engineering approach. Producing sophisticated bigger functioning tissues like the liver and kidney with large numbers of cells and diverse metabolic activities is challenging. Biomimetic scaffolds have limited diffusion characteristics [15]. The 'bottom-up' approach, which is evolving, produces microscale tissue components with a specified microarchitecture and assembles them to create more enormous tissue structures. The challenge inspired this method. Tissue building blocks can be made via microgels, built-in cell consolidation, cell sheet manufacturing, and direct cell printing [43].

10.11.6 Nanomaterial Applications in Neural Tissue Engineering

Two major subsystems comprise the nervous system: the central nervous system (CNS) and the peripheral nervous system (PNS). The CNS consists of the brain and spinal cord, whereas the PNS comprises sensory and motor neurons throughout the body. The CNS and PNS commonly display persistent functioning impairments after illness or unintentional injury because they cannot repair. The alarming rise in the population's average age has been accompanied by a matching rise in neural disorders, posing a severe threat to the general public health [68]. Allograft, autologous transplantation, and surgical sutures are the most common clinical treatments for neural diseases. All of these are used to assist the injured nerve's healing. Even though the infection managed to some extent, there are still numerous drawbacks, including many surgeries, the immune system's rejection of the treatment, and the treatment's less-than-ideal results. Thanks to developments in nerve TE, hope for treating neurological diseases has been rekindled [71]. One of the typical alternative ways to treat nerve disorders is to invent an appropriate nanomaterial that can control the ECM microenvironment and cell behaviours. This will hasten the process of nerve regeneration. Creating neural tissue has used various polymers, and the experimental outcomes have been fascinating. These results include sprouting neurites, bridging neuronal gaps, and developing human neural stem cells. NPs, hydrogels, polymer scaffolds, and nerve conduits are often controlled agents in neural TE. No matter the material, a set of requirements must be met before it may be put to use [64]. These qualities include good mechanical and electrical conductivity, biodegradability, compatibility with living organisms, resistance to infection, and permeability/porosity. Only collagen, an organic biopolymer, has been granted permission for testing in clinical trials aimed at regenerating peripheral nerves. Human tissues contain collagen, the main protein in connective tissue, which gives the body support and structure [18]. Only two of the numerous collagen-based nerve guide agents on the market, NeuraGen®89 and Neuromaix®, regenerate peripheral nerves. Collagenous nerve conduits have been found to help primate species mend small nerve gaps. Additionally, combining collagen with various compounds significantly improves sciatic nerve regeneration in models using rats and dogs. The extent to which this happens varies, though. It is crucial to remember that the source of the collagen must receive more consideration. This is so that the application can be affected by the collagen choice, as various collagen sources may have varying effects. The collagen molecule can be hydrolysed to create GA in an alkaline or acidic environment. Biomedical video, connective tissue scaffolding, and drug administration employ it due to its different properties from collagen, including affordability, availability, hypotonicity, and biodegradability. GA can also influence cell adhesion and proliferation after going through the proper chemical alterations, making it a helpful implant material for TE [41]. For example, electrospinning can be used to enhance the biological and dynamic characteristics of scaffolds made of GA for brain TE. Applying functional methods could improve the benefits of utilising GA-based NPs for constructing brain tissue.

This was done to produce the intended outcome. Using GA and electrospun materials, PLA can enhance axon development and neuronal motor lineage differentiation. Similar to what was previously mentioned, this can be done. Only lately have GA NPs been added to polymer scaffoldings for nerve TE, increasing biocompatibility. GA NP-coated CA/PLA scaffolds, according to Naseri-Nosar and colleagues, showed a higher viability rate than uncoated platforms. In both in vitro and in vivo, these scaffolds guided sciatic nerve injury. Creating nerve tissue also uses other protein components like ES, keratin, and silk. Materials based on ES are crucial in tissue regeneration because they give elasticity to tissues and organs. ES is a structural component of the ECM, which possesses mechanical stiffness, the capacity to self-assemble, longevity, and biological activity [68]. ES-like polypeptides are used more in cerebral TE processes than ES due to their biocompatibility and endurance. CNS disorders are characteristically treated using ES. The keratin polypeptide can fold properly and provide suitable substrates. Keratin-functionalised NPs enable cell attachment and proliferation due to their physical properties. Additionally, these nanomaterials' various amino acid configurations are easily adaptable to meet the requirements of various tissues [64]. Another TE research has shown that non-neuronal cells can observe, grow, and thrive within the body on electrospun scaffolds made of keratin nanofibres and PVA. Spiders and silkworms create silk, a fibrous structural protein. This material is biocompatible, elastic, biodegradable, tuneable, antimicrobial, and minimally immunogenic. Silk's versatility makes it perfect for biomimetic gels, films, scaffolds, nanofibres, and NPs. Silk-based hydrogels are viable biomaterials for nerve TE because of their structural stability and capacity to form axon bundles. Additionally, well-developed silk hydrogels can aid in regenerating nerve tissue and neural development. As a nerve conduit for TE, silk fibroin can cure central nervous system injuries. Furthermore, silk fibroin has excellent biocompatibility and negligible in vitro cytotoxicity [18].

10.11.7 Skin Tissue Engineering Nanomaterials

Wound healing involves thrombosis and tissue remodelling or maturation. Based on how long it takes to heal, skin wounds are often categorised into two: Acute and chronic. A rupture or puncture of the top layer of skin distinguishes an acute wound, and they often heal quickly. Chronic wounds are hard to heal rapidly since they nearly always co-occur with diseases like diabetes and obesity. A crucial step in the healing of wounds is known as angiogenesis [17]. Arterial development ensures proper blood flow, nutrients, and oxygen and speeds wound healing and granulation tissue production. The body's capacity to heal quickly is slowed down by chronic wounds caused by abnormal blood vessel growth. As a result, while treating skin wounds, it is essential to consider both vascular and skin tissue regeneration. Different therapeutic modalities, including local ventilation, hyperbaric oxygen, ozone, adverse pressure wound treatment, and others, have been created to hasten the healing of chronic wounds in patients who have experienced skin injuries [42]. Treatment options like autologous transplantation are frequently used to treat severe cases of skin deterioration. Simply put, the donor's full-thickness skin is removed from other suitable areas, magnified, and then transplanted into the wound. Depending on where the donor site and wounded area are located, this approach does have certain limitations. Here, another potentially beneficial tactic is developed: The use of autologous cell-based therapy [48].

When linked cells proliferate enough in vitro, they treat wounds. However, the cost, duration, and success rate are all incredibly susceptible to different factors. Nanomaterials help skin TE wound healing. The best nanomaterials can operate as a barrier coating for regenerating keratinocytes, cling to the inferior dermis, repair blood vessels at the injury site, and support the skin elastically. Biodegradable polycation polysaccharide CS is non-toxic. In living organisms, lysozyme may break it down into harmless amino sugars. Biocompatibility and cell adhesiveness type are useful for skin applications. The negatively charged bacterial membrane loses cell components when positively charged CS agglutinates. CS binds metals and inhibits enzymes [43]. The HemCon bandage, originally advertised as a hemostatic dressing, works well in skincare. The keratin–CS composite film boosted antibacterial activity, fibroblast adhesion, and keratin tensile characteristics. Nanocellulose and cellulose-based nanoscale structures have garnered attention as unique nanomaterials. Nanocellulose-based materials absorb wound exudate and make dressing removal easier, making them popular in biomedical applications to treat skin conditions [72]. The authors of [33] observed that bacteria-produced nanocellulose improved wound therapy more than expected. Compared to standard dressings, it heals faster, reduces inflammation, is non-toxic, and regenerates tissue faster. Capillaries formed in wounds. Nitrocellulose-based sauces also heal chronic lower-limb ulcers. The nanocellulose-based film can treat severe burns by providing a clean atmosphere and sustaining water balance. This is achieved by perfectly adhering to the injured area. Micro cellulose has also been shown to minimise skin healing inflammation. The elastic, photoluminescent, and antibacterial hybrid polypeptide-based composites expedite wound healing and suppress multi-drug resistance (MDR) microbes [44]. The nanocellulose composite made from polypeptides exhibited solid antibacterial activity, high biocompatibility, biomimetic elastomeric behaviour, and biomimetic elastomeric properties. In vivo studies showed that the nanocellulose composite technology might promote skin regeneration and reduce MDR bacteria-derived wound infection. Nanocomposite materials injected via nanocellulose may also better affect skin tissue regeneration. These substances include AG, CS, GA, PEG, and PVA [17].

Nanocarriers to transfer genes protect components from nucleases and adjust gene expression throughout gene transfer treatment. This enhances skin TE revascularisation. PLGA NPs and the anti-VEGF interceptor plasmid pFlt23K were synthesised [38]. A strategy to treat neovascular disorders, the ability to treat traumatic [15] illnesses, was successfully adjusted, and epithelial cells secreted less VEGF. Skin-regenerating stem cells exhibit enhanced endothelial growth factor release and transitory change. The purpose of these cells is to promote angiogenesis, which is crucial following transplantation. It has been demonstrated that human embryonic stem cells and MSCs can get the epidermal growth factor gene through biodegradable nanomaterials. In the meantime, the transplant results improved in the tissues to be treated as cells are more capable of producing epidermal growth factors. The angiogenesis rate was two to four times higher when VEGF-expressing stem cells were implanted into the nanostructured scaffolds than with control cells. To regenerate skin tissue, it exhibits therapeutic effectiveness of stem cells that have been altered using biodegradable NPs [58].

10.11.8 Nanoparticles with Bioactive Properties

NPs and scaffolds can simplify bone application. Osteoblasts and osteoclasts maintain bone homeostasis through their complex relationship. NPs release inhibitors to prevent osteoclast remodelling locally, while NP-based medication and growth factor release can stimulate osteoblast bone synthesis [25]. Bioactive substances like growth hormones, medicines, parts of the ECM, and nucleic acids were studied to see if NPs could take them out. These small chemicals do not stay in balance well, do not go after specific targets, and cannot get through cell membranes very well or at all. Direct therapeutic delivery is challenging. Indirect drug administration circumvents these restrictions [47].

Due to their small size, these nanospheres can release physiologically or chemically active compounds in reaction to pH, magnetic fields, ultrasounds, and irradiation [54]. Nanospheres can also release upon irradiation due to their size. Genetic material, drugs, and growth factors can be carried via nanospheres made of HA, gold, dendrimer, or silica. Depending on the molecule, different delivery methods can be used. Bioactive compounds can be combined directly with polymer solution before spinning NFs. This approach loads efficiently. However, due to their uneven distribution, bioactive components may aggregate on NF surfaces. This method could boost cellular responsiveness by releasing drugs quickly [73].

Scaffolds convey cells and growth factors. A biocompatible, biodegradable scaffold would allow functional tissue to replace it. Many scaffolds imitate the ECM, a lattice for cell adhesion, motility, and tissue ingrowth. This matrix supports and stretches. Simple biodegradable polymeric or ceramic scaffolds for bone TE have limited mechanical strength [20]. However, NP-modified composite scaffolds may help bone healing. Nanoscale organic and inorganic materials can support and differentiate cells when integrated into polymeric scaffolds. Nanocomposites of PLA-based scaffolds with CNTs and micro-HA particles improve MSC adhesion and osteoprogenitor formation [45]. Biomimetic apatite NPs may enhance progenitor cell osteogenicity since HA mimics bone's mineral structure. HA-coated PLGA scaffolds alone could promote bone healing in calvarial rat lesions. Local progenitor cells formed bone faster on scaffolds with more HA NPs. Apatite-coated PLGA scaffolds develop preosteoblasts. The 3D apatite-coated PLGA boosted bone formation marker expression in MC3T3-E1 cells. A recent study found that adsorbed protein layers affect how cells interact with apatite NPs. These layers affect CaP phase change by changing surface potential or charge [6].

10.12 Futures and Prospects

This introduction covers nanomaterial fundamentals, manufacturing and characterisation methods, and TE applications. Repairing or regenerating wounded tissue requires nanomaterials and TE research. Many nanotechnology researchers are blending nanomaterials to create novel biomaterials. Implant sensitivity, immunological response, toxicity, and foetal development must be considered. TE, which has to satisfy doctors and patients, has a great chance with nanomaterials. Nanomaterials have medical benefits but are also a health concern. To limit the danger, these materials must be developed, tested, and used clinically using the precautionary principle. Nanomaterial biosafety, use, and stability need improvement. Purpose-built nanotechnologies will soon solve most TE problems.

10.13 Conclusion

Tissue engineering strives to generate viable tissue substitutes or boost the body's endogenous regeneration by using engineering principles, cell transplantation procedures, and biomaterials. Organically produced biopolymers are often used in tissue engineering for their multiple benefits. They promote cell migration, adhesion, proliferation, and differentiation, all of which are necessary processes in tissue healing – incorporating nanomaterials into biopolymer structures allowed for multifunctionality, and improving characteristics offered new possibilities for tissue engineering. Tissue engineering uses polymers, elements, and carbon-based structures.

REFERENCES

[1] Amiryaghoubi, N., Fathi, M., Barar, J., Omidian, H., & Omidi, Y. (2023). Hybrid polymer-grafted graphene scaffolds for microvascular tissue engineering and regeneration. *European Polymer Journal, 193*, 112095. https://doi.org/10.1016/j.eurpolymj.2023.112095

[2] Wang, N., Thameem Dheen, S., Fuh, J. Y. H., & Senthil Kumar, A. (2021). A review of multi-functional ceramic nanoparticles in 3D printed bone tissue engineering. *Bioprinting, 23*, e00146. https://doi.org/10.1016/j.bprint.2021.e00146

[3] Zheng, X., Zhang, P., Fu, Z., Meng, S., Dai, L., & Yang, H. (2021). Applications of nanomaterials in tissue engineering. *RSC Advances, 11*(31), 19041–19058. https://doi.org/10.1039/D1RA01849C

[4] Garcia-Orue, I., Gainza, G., Villullas, S., Pedraz, J. L., Hernandez, R. M., & Igartua, M. (2016). Nanotechnology approaches for skin wound regeneration using drug-delivery systems. In *Nanobiomaterials in Soft Tissue Engineering* (pp. 31–55). Elsevier. https://doi.org/10.1016/B978-0-323-42865-1.00002-7

[5] Singh, N., Joshi, A., & Verma, G. (2016). Engineered nanomaterials for biomedicine. In *Engineering of Nanobiomaterials* (pp. 307–328). Elsevier. https://doi.org/10.1016/B978-0-323-41532-3.00010-5

[6] Kharaziha, M., Memic, A., Akbari, M., Brafman, D. A., & Nikkhah, M. (2016). Nano-enabled approaches for stem cell-based cardiac tissue engineering. *Advanced Healthcare Materials, 5*(13), 1533–1553. https://doi.org/10.1002/adhm.201600088

[7] Fathi-Achachelouei, M., Knopf-Marques, H., Ribeiro da Silva, C. E., Barthès, J., Bat, E., Tezcaner, A., & Vrana, N. E. (2019). Use of nanoparticles in tissue engineering and regenerative medicine. *Frontiers in Bioengineering and Biotechnology, 7*. https://doi.org/10.3389/fbioe.2019.00113

[8] Hazarika, D., & Karak, N. (2021). Fundamentals of polymeric nanostructured materials. In *Advances in Polymeric Nanomaterials for Biomedical Applications* (pp. 1–40). Elsevier. https://doi.org/10.1016/B978-0-12-814657-6.00002-1

[9] Cernat, A., Florea, A., Rus, I., Truta, F., Dragan, A.-M., Cristea, C., & Tertis, M. (2021). Applications of magnetic hybrid nanomaterials in biomedicine. In *Biopolymeric Nanomaterials* (pp. 639–675). Elsevier. https://doi.org/10.1016/B978-0-12-824364-0.00014-9

[10] Van de Walle, A., Perez, J. E., Abou-Hassan, A., Hémadi, M., Luciani, N., & Wilhelm, C. (2020). Magnetic nanoparticles in regenerative medicine: What of their fate and impact in stem cells? *Materials Today Nano*, *11*, 100084. https://doi.org/10.1016/j.mtnano.2020.100084

[11] Balaji, V., & Mahalingam, G. (2022). Nanoparticles-based drug delivery to cure osteodegeneration by improving tissue regeneration. In *Advances in Nanotechnology-Based Drug Delivery Systems* (pp. 449–470). Elsevier. https://doi.org/10.1016/B978-0-323-88450-1.00021-1

[12] Raghav, P. K., Mann, Z., Ahlawat, S., & Mohanty, S. (2022). Mesenchymal stem cell-based nanoparticles and scaffolds in regenerative medicine. *European Journal of Pharmacology*, *918*, 174657. https://doi.org/10.1016/j.ejphar.2021.174657

[13] Damle, A., Sundaresan, R., Rajwade, J. M., Srivastava, P., & Naik, A. (2022). A concise review on implications of silver nanoparticles in bone tissue engineering. *Biomaterials Advances*, *141*, 213099. https://doi.org/10.1016/j.bioadv.2022.213099

[14] Amjad, M. S., Sadiq, N., Qureshi, H., Fareed, G., & Sabir, S. (2015). Nano particles: An emerging tool in biomedicine. *Asian Pacific Journal of Tropical Disease*, *5*(10), 767–771. https://doi.org/10.1016/S2222-1808(15)60929-X

[15] Wang, Z., Mithieux, S. M., & Weiss, A. S. (2019). Fabrication techniques for vascular and vascularized tissue engineering. *Advanced Healthcare Materials*, *8*(19), e1900742. https://doi.org/10.1002/adhm.201900742

[16] Martins, A., Reis, R. L., & Neves, N. M. (2018). Micro/nano scaffolds for osteochondral tissue engineering. *Advances in Experimental Medicine and Biology*, *1058*, 125–139. https://doi.org/10.1007/978-3-319-76711-6_6

[17] Bharadwaz, A., & Jayasuriya, A. C. (2020). Recent trends in the application of widely used natural and synthetic polymer nanocomposites in bone tissue regeneration. *Materials Science & Engineering. C, Materials for Biological Applications*, *110*, 110698. https://doi.org/10.1016/j.msec.2020.110698

[18] Xuan, H., Li, B., Xiong, F., Wu, S., Zhang, Z., Yang, Y., & Yuan, H. (2021). Tailoring nano-porous surface of aligned electrospun poly (L-lactic acid) fibers for nerve tissue engineering. *International Journal of Molecular Sciences*, *22*(7). https://doi.org/10.3390/ijms22073536

[19] Abbah, S. A., Delgado, L. M., Azeem, A., Fuller, K., Shologu, N., Keeney, M., Biggs, M. J., Pandit, A., & Zeugolis, D. I. (2015). Harnessing hierarchical nano- and micro-fabrication technologies for musculoskeletal tissue engineering. *Advanced Healthcare Materials*, *4*(16), 2488–2499. https://doi.org/10.1002/adhm.201500004

[20] Xu, Y., Ding, W., Chen, M., Du, H., & Qin, T. (2022). Synergistic fabrication of micro-nano bioactive ceramic-optimized polymer scaffolds for bone tissue engineering by in situ hydrothermal deposition and selective laser sintering. *Journal of Biomaterials Science. Polymer Edition*, *33*(16), 2104–2123. https://doi.org/10.1080/09205063.2022.2096526

[21] Chauhan, S., Tirkey, A., & Upadhyay, L. S. B. (2022). Nanomaterials in biomedicine: Synthesis and applications. In *Advances in Nanotechnology-Based Drug Delivery Systems* (pp. 585–604). Elsevier. https://doi.org/10.1016/B978-0-323-88450-1.00023-5

[22] Habibzadeh, F., Sadraei, S. M., Mansoori, R., Singh Chauhan, N. P., & Sargazi, G. (2022). Nanomaterials supported by polymers for tissue engineering applications: A review. *Heliyon*, *8*(12), e12193. https://doi.org/10.1016/j.heliyon.2022.e12193

[23] Ryabchikova, E. (2021). Advances in nanomaterials in biomedicine. *Nanomaterials*, *11*(1), 118. https://doi.org/10.3390/nano11010118

[24] Kumar, M., Kulkarni, P., Liu, S., Chemuturi, N., & Shah, D. K. (2023). Nanoparticle biodistribution coefficients: A quantitative approach for understanding the tissue distribution of nanoparticles. *Advanced Drug Delivery Reviews*, *194*, 114708. https://doi.org/10.1016/j.addr.2023.114708

[25] Monteiro, N., Martins, A., Reis, R. L., & Neves, N. M. (2015). Nanoparticle-based bioactive agent release systems for bone and cartilage tissue engineering. *Regenerative Therapy*, *1*, 109–118. https://doi.org/10.1016/j.reth.2015.05.004

[26] Elsherbini, A. M., & Sabra, S. A. (2022). Nanoparticles-in-nanofibers composites: Emphasis on some recent biomedical applications. *Journal of Controlled Release*, *348*, 57–83. https://doi.org/10.1016/j.jconrel.2022.05.037

[27] Jayaraman, P., Gandhimathi, C., Venugopal, J. R., Becker, D. L., Ramakrishna, S., & Srinivasan, D. K. (2015). Controlled release of drugs in electrosprayed nanoparticles for bone tissue engineering. *Advanced Drug Delivery Reviews*, *94*, 77–95. https://doi.org/10.1016/j.addr.2015.09.007

[28] Hasan, A., Morshed, M., Memic, A., Hassan, S., Webster, T., & Marei, H. (2018). Nanoparticles in tissue engineering: Applications, challenges and prospects. *International Journal of Nanomedicine*, *13*, 5637–5655. https://doi.org/10.2147/IJN.S153758

[29] Bhattacharjee, P., Kundu, B., Naskar, D., Kim, H.-W., Maiti, T. K., Bhattacharya, D., & Kundu, S. C. (2017). Silk scaffolds in bone tissue engineering: An overview. *Acta Biomaterialia*, *63*, 1–17. https://doi.org/10.1016/j.actbio.2017.09.027

[30] Walmsley, G. G., McArdle, A., Tevlin, R., Momeni, A., Atashroo, D., Hu, M. S., Feroze, A. H., Wong, V. W., Lorenz, P. H., Longaker, M. T., & Wan, D. C. (2015). Nanotechnology in bone tissue engineering. *Nanomedicine: Nanotechnology, Biology and Medicine*, *11*(5), 1253–1263. https://doi.org/10.1016/j.nano.2015.02.013

[31] Hassan, N. Ul, Chaudhery, I., Ur. Rehman, A., & Ahmed, N. (2021). Polymeric nanoparticles used in tissue engineering. In *Advances in Polymeric Nanomaterials for Biomedical Applications* (pp. 191–224). Elsevier. https://doi.org/10.1016/B978-0-12-814657-6.00005-7

[32] Zhao, H., Liu, M., Zhang, Y., Yin, J., & Pei, R. (2020). Nanocomposite hydrogels for tissue engineering applications. *Nanoscale*, *12*(28), 14976–14995. https://doi.org/10.1039/d0nr03785k

[33] Venkatesan, J., & Kim, S.-K. (2014). Nano-hydroxyapatite composite biomaterials for bone tissue engineering—A review. *Journal of Biomedical Nanotechnology*, *10*(10), 3124–3140. https://doi.org/10.1166/jbn.2014.1893

[34] Peng, Z., Zhao, T., Zhou, Y., Li, S., Li, J., & Leblanc, R. M. (2020). Bone tissue engineering via carbon-based nanomaterials. *Advanced Healthcare Materials*, *9*(5), e1901495. https://doi.org/10.1002/adhm.201901495

[35] Jaberi, N., Fakhri, V., Zeraatkar, A., Jafari, A., Uzun, L., Shojaei, S., Asefnejad, A., Faghihi Rezaei, V., Goodarzi, V., Su, C.-H., & Ghaffarian Anbaran, S. R. (2022). Preparation and characterization of a new bio nanocomposites based poly(glycerol sebacic-urethane) containing nano-clay (cloisite Na(+)) and its potential application for tissue engineering. *Journal of Biomedical Materials Research. Part B, Applied Biomaterials*, *110*(10), 2217–2230. https://doi.org/10.1002/jbm.b.35071

[36] Shadjou, N., & Hasanzadeh, M. (2015). Bone tissue engineering using silica-based mesoporous nanobiomaterials: Recent progress. *Materials Science & Engineering. C, Materials for Biological Applications*, *55*, 401–409. https://doi.org/10.1016/j.msec.2015.05.027

[37] Gu, M., Liu, Y., Chen, T., Du, F., Zhao, X., Xiong, C., & Zhou, Y. (2014). Is graphene a promising nanomaterial for promoting surface modification of implants or scaffold materials in bone tissue engineering? *Tissue Engineering. Part B, Reviews*, *20*(5), 477–491. https://doi.org/10.1089/ten.TEB.2013.0638

[38] Singh, Y. P., & Dasgupta, S. (2022). Gelatin-based electrospun and lyophilized scaffolds with nano scale feature for bone tissue engineering application: Review. *Journal of Biomaterials Science. Polymer Edition*, *33*(13), 1704–1758. https://doi.org/10.1080/09205063.2022.2068943

[39] Ravi, S., Chokkakula, L. P. P., Giri, P. S., Korra, G., Dey, S. R., & Rath, S. N. (2023). 3D bioprintable hypoxia-mimicking PEG-based nano bioink for cartilage tissue engineering. *ACS Applied Materials & Interfaces*, *15*(16), 19921–19936. https://doi.org/10.1021/acsami.3c00389

[40] Dalgic, A. D., Alshemary, A. Z., Tezcaner, A., Keskin, D., & Evis, Z. (2018). Silicate-doped nano-hydroxyapatite/graphene oxide composite reinforced fibrous scaffolds for bone tissue engineering. *Journal of Biomaterials Applications*, *32*(10), 1392–1405. https://doi.org/10.1177/0885328218763665

[41] Kumar, P., Choonara, Y. E., Khan, R. A., & Pillay, V. (2017). The chemo-biological outreach of nano-biomaterials: Implications for tissue engineering and regenerative medicine. *Current Pharmaceutical Design*, *23*(24), 3538–3549. https://doi.org/10.2174/1381612823666170503144643

[42] Gao, C., Song, S., Lv, Y., Huang, J., & Zhang, Z. (2022). Recent development of conductive hydrogels for tissue engineering: Review and perspective. *Macromolecular Bioscience*, *22*(8), e2200051. https://doi.org/10.1002/mabi.202200051

[43] Ali, A., Hasan, A., & Negi, Y. S. (2022). Effect of carbon based fillers on xylan/chitosan/nano-HAp composite matrix for bone tissue engineering application. *International Journal of Biological Macromolecules*, *197*, 1–11. https://doi.org/10.1016/j.ijbiomac.2021.12.012

[44] Dehghan-Baniani, D., Mehrjou, B., Chu, P. K., & Wu, H. (2021). A biomimetic nano-engineered platform for functional tissue engineering of cartilage superficial zone. *Advanced Healthcare Materials*, *10*(4), e2001018. https://doi.org/10.1002/adhm.202001018

                                                      Introduction to Functional Nanomaterials

[45] Ding, Z., Cheng, W., Mia, M. S., & Lu, Q. (2021). Silk biomaterials for bone tissue engineering. *Macromolecular Bioscience, 21*(8), e2100153. https://doi.org/10.1002/mabi.202100153

[46] Qiao, K., Xu, L., Tang, J., Wang, Q., Lim, K. S., Hooper, G., Woodfield, T. B. F., Liu, G., Tian, K., Zhang, W., & Cui, X. (2022). The advances in nanomedicine for bone and cartilage repair. *Journal of Nanobiotechnology, 20*(1), 141. https://doi.org/10.1186/s12951-022-01342-8

[47] Zaersabet, M., Salehi, Z., Hadavi, M., Talesh Sasani, S., & Rastgoo Noestali, F. (2022). Development and evaluation of bioactive 3D zein and zein/nano-hydroxyapatite scaffolds for bone tissue engineering application. *Proceedings of the Institution of Mechanical Engineers. Part H, Journal of Engineering in Medicine, 236*(6), 785–793. https://doi.org/10.1177/09544119221090726

[48] Wei, H., Zhang, B., Lei, M., Lu, Z., Liu, J., Guo, B., & Yu, Y. (2022). Visible-light-mediated nano-biomineralization of customizable tough hydrogels for biomimetic tissue engineering. *ACS Nano, 16*(3), 4734–4745. https://doi.org/10.1021/acsnano.1c11589

[49] Jarai, B. M., Kolewe, E. L., Stillman, Z. S., Raman, N., & Fromen, C. A. (2020). Polymeric nanoparticles. In *Nanoparticles for Biomedical Applications* (pp. 303–324). Elsevier. https://doi.org/10.1016/B978-0-12-816662-8.00018-7

[50] Wei, S., Ma, J.-X., Xu, L., Gu, X.-S., & Ma, X.-L. (2020). Biodegradable materials for bone defect repair. *Military Medical Research, 7*(1), 54. https://doi.org/10.1186/s40779-020-00280-6

[51] Smith, L. A., & Ma, P. X. (2004). Nano-fibrous scaffolds for tissue engineering. *Colloids and Surfaces. B, Biointerfaces, 39*(3), 125–131. https://doi.org/10.1016/j.colsurfb.2003.12.004

[52] Gerdes, S., Ramesh, S., Mostafavi, A., Tamayol, A., Rivero, I. V., & Rao, P. (2021). Extrusion-based 3D (bio)printed tissue engineering scaffolds: Process-structure-quality relationships. *ACS Biomaterials Science & Engineering, 7*(10), 4694–4717. https://doi.org/10.1021/acsbiomaterials.1c00598

[53] Arora, P., Sindhu, A., Dilbaghi, N., Chaudhury, A., Rajakumar, G., & Rahuman, A. A. (2012). Nano-regenerative medicine towards clinical outcome of stem cell and tissue engineering in humans. *Journal of Cellular and Molecular Medicine, 16*(9), 1991–2000. https://doi.org/10.1111/j.1582-4934.2012.01534.x

[54] Izadifar, M., Haddadi, A., Chen, X., & Kelly, M. E. (2015). Rate-programming of nano-particulate delivery systems for smart bioactive scaffolds in tissue engineering. *Nanotechnology, 26*(1), 12001. https://doi.org/10.1088/0957-4484/26/1/012001

[55] Murugan, R., & Ramakrishna, S. (2006). Nano-featured scaffolds for tissue engineering: A review of spinning methodologies. *Tissue Engineering, 12*(3), 435–447. https://doi.org/10.1089/ten.2006.12.435

[56] Hu, J., & Ma, P. X. (2011). Nano-fibrous tissue engineering scaffolds capable of growth factor delivery. *Pharmaceutical Research, 28*(6), 1273–1281. https://doi.org/10.1007/s11095-011-0367-z

[57] Cai, J., Wang, J., Sun, C., Dai, J., & Zhang, C. (2022). Biomaterials with stiffness gradient for interface tissue engineering. *Biomedical Materials (Bristol, England), 17*(6). https://doi.org/10.1088/1748-605X/ac8b4a

[58] Xia, P., & Luo, Y. (2022). Vascularization in tissue engineering: The architecture cues of pores in scaffolds. *Journal of Biomedical Materials Research. Part B, Applied Biomaterials, 110*(5), 1206–1214. https://doi.org/10.1002/jbm.b.34979

[59] Sharma, R., Kumar, S., Bhawna, Gupta, A., Dheer, N., Jain, P., Singh, P., & Kumar, V. (2022). An insight of nanomaterials in tissue engineering from fabrication to applications. *Tissue Engineering and Regenerative Medicine, 19*(5), 927–960. https://doi.org/10.1007/s13770-022-00459-z

[60] Mosaad, K. E., Shoueir, K. R., Saied, A. H., & Dewidar, M. M. (2021). New prospects in nano phased co-substituted hydroxyapatite enrolled in polymeric nanofiber mats for bone tissue engineering applications. *Annals of Biomedical Engineering, 49*(9), 2006–2029. https://doi.org/10.1007/s10439-021-02810-2

[61] Tian, A., Xue, J., & Sun, N. (2022). Advantages of self-assembled nano peptide hydrogels in biological tissue engineering. *Current Protein & Peptide Science, 23*(6), 395–401. https://doi.org/10.2174/1389203723666220617093402

[62] Luo, Y., Chen, B., Zhang, X., Huang, S., & Wa, Q. (2022). 3D printed concentrated alginate/GelMA hollow-fibers-packed scaffolds with nano apatite coatings for bone tissue engineering. *International Journal of Biological Macromolecules, 202*, 366–374. https://doi.org/10.1016/j.ijbiomac.2022.01.096

[63] Saranya, N., Saravanan, S., Moorthi, A., Ramyakrishna, B., & Selvamurugan, N. (2011). Enhanced osteoblast adhesion on polymeric nano-scaffolds for bone tissue engineering. *Journal of Biomedical Nanotechnology, 7*(2), 238–244. https://doi.org/10.1166/jbn.2011.1283

[64] Kong, L., Gao, X., Qian, Y., Sun, W., You, Z., & Fan, C. (2022). Biomechanical microenvironment in peripheral nerve regeneration: From pathophysiological understanding to tissue engineering development. *Theranostics, 12*(11), 4993–5014. https://doi.org/10.7150/thno.74571

[65] Ma, H., Feng, C., Chang, J., & Wu, C. (2018). 3D-printed bioceramic scaffolds: From bone tissue engineering to tumor therapy. *Acta Biomaterialia, 79*, 37–59. https://doi.org/10.1016/j.actbio.2018.08.026

[66] Gholipourmalekabadi, M., Jajarmi, V., Rezvani, Z., Ghaffari, M., Verma, K. D., Shirinzadeh, H., & Mozafari, M. (2016). Oxygen-generating nanobiomaterials for the treatment of diabetes. In *Nanobiomaterials in Soft Tissue Engineering* (pp. 331–353). Elsevier. https://doi.org/10.1016/B978-0-323-42865-1.00012-X

[67] Swetha, S., Balagangadharan, K., Lavanya, K., & Selvamurugan, N. (2021). Three-dimensional-poly(lactic acid) scaffolds coated with gelatin/magnesium-doped nano-hydroxyapatite for bone tissue engineering. *Biotechnology Journal, 16*(11), e2100282. https://doi.org/10.1002/biot.202100282

[68] Bei, H. P., Yang, Y., Zhang, Q., Tian, Y., Luo, X., Yang, M., & Zhao, X. (2019). Graphene-based nanocomposites for neural tissue engineering. *Molecules (Basel, Switzerland), 24*(4). https://doi.org/10.3390/molecules24040658

[69] Rotherham, M., Nahar, T., Broomhall, T. J., Telling, N. D., & El Haj, A. J. (2022). Remote magnetic actuation of cell signalling for tissue engineering. *Current Opinion in Biomedical Engineering, 24*, 100410. https://doi.org/10.1016/j.cobme.2022.100410

[70] Qiu, X.-T., Rao, C.-Y., Li, T., & Zhou, R.-H. (2021). Research progress in biomimetic synthesis of nano-hydroxyapatite in bone tissue engineering. *Sichuan da xue xue bao. Yi xue ban = Journal of Sichuan University. Medical Science Edition, 52*(5), 740–746. https://doi.org/10.12182/20210560201

[71] Béduer, A., Vaysse, L., Loubinoux, I., & Vieu, C. (2013). Micro/nano-engineering to control growth of neuronal cells and tissue engineering applied to the central nervous system. *Biologie aujourd'hui, 207*(4), 291–307. https://doi.org/10.1051/jbio/2013019

[72] Narayanan, K. B., Zo, S. M., & Han, S. S. (2020). Novel biomimetic chitin-glucan polysaccharide nano/microfibrous fungal-scaffolds for tissue engineering applications. *International Journal of Biological Macromolecules, 149*, 724–731. https://doi.org/10.1016/j.ijbiomac.2020.01.276

[73] Puppi, D., Zhang, X., Yang, L., Chiellini, F., Sun, X., & Chiellini, E. (2014). Nano/microfibrous polymeric constructs loaded with bioactive agents and designed for tissue engineering applications: A review. *Journal of Biomedical Materials Research. Part B, Applied Biomaterials, 102*(7), 1562–1579. https://doi.org/10.1002/jbm.b.33144

11

The Evolution of Materials through Machine Learning for Enhanced Energy Storage Solutions

E. Srividhya
Sathyabama Institute of Science and Technology, Chennai, Tamil Nadu, India

Moumita Pal
Stanley College of Engineering and Technology, Hyderabad, India

Richa Agarwal
KIET Group of Institutions, Ghaziabad, Uttar Pradesh, India

C. Sudha
GITAM School of Technology, GITAM Deemed to be University, Hyderabad, Telangana, India

V. R. Niveditha
Sathyabama Institute of Science and Technology, Chennai, Tamil Nadu, India

11.1 Introduction

The realm of energy storage stands at the forefront of transformative advancements, driven by the intricate interplay between material science and machine learning (ML). In an era where sustainability and efficiency are paramount, the evolution of materials for enhanced energy storage solutions takes centre stage in addressing global energy challenges. This research endeavours to delve into the ongoing developments, breakthroughs, and challenges at the nexus of material science and ML, exploring how these two domains synergise to propel the energy storage landscape into uncharted territories of innovation [1].

11.1.1 The Crucial Role of Materials in Energy Storage

To comprehend the significance of this research, one must first recognise the pivotal role that materials play in the domain of energy storage. Traditional energy storage systems, predominantly relying on lithium-ion batteries, have proven indispensable in powering a myriad of electronic devices, electric vehicles, and renewable energy systems. However, the quest for energy storage solutions that are not only efficient but also sustainable has sparked a revolution in the exploration of novel materials. From advanced polymers to nanocomposites, researchers are tirelessly exploring a vast materials landscape, seeking components that can push the boundaries of energy density, charge–discharge efficiency, and overall durability. Ongoing research in materials for energy storage is not merely an academic pursuit but a vital endeavour that has the potential to reshape the future of renewable energy utilisation and storage infrastructure.

11.1.2 Machine Learning as a Catalyst for Materials Discovery

In tandem with the quest for superior materials, machine learning has emerged as a powerful catalyst, expediting the process of materials discovery. The sheer complexity of material interactions and the enormous datasets generated during experiments necessitate innovative approaches. Machine learning

DOI: 10.1201/9781003495437-11

algorithms, capable of parsing through colossal datasets with unprecedented speed and efficiency, have become indispensable tools in predicting material properties, optimising compositions, and even suggesting entirely novel materials with desirable characteristics [2]. As ongoing research unfolds, machine learning continues to evolve in response to the dynamic challenges posed by the material science domain. Whether it be predicting material properties with higher accuracy or streamlining the experimental design process, machine learning stands as an indispensable ally in the ongoing pursuit of materials excellence.

11.1.3 The Dynamic Landscape of Ongoing Research

This research is not a static snapshot but a dynamic exploration of the present and evolving landscape of material science and ML integration. The ongoing nature of the research reflects the dynamic advancements in both domains. It acknowledges that breakthroughs are not singular events but a continuum of progress, each discovery building upon the foundations laid by preceding research endeavours. In the subsequent sections, we will journey through the ongoing applications of machine learning in material property prediction, delve into the intricacies of real-time data-driven approaches, and scrutinise the adaptability of ML techniques for the dynamic material landscape. Case studies, drawn from the latest developments, will provide insights into the practical implications of ML-driven material design, offering a real-time perspective on successes and learnings from failures. Moreover, the integration of ML into initiatives like the Materials Genome Project highlights the commitment to accelerating materials discovery in realtime, providing researchers with unprecedented tools to explore the vast materials space more efficiently than ever before.

As we navigate through the ongoing challenges, ethical considerations, and future prospects, it becomes evident that this research is not just an academic exploration but a journey into the heart of innovation. The insights garnered from this study are poised to influence the trajectory of materials research and, by extension, redefine the landscape of energy storage solutions in the years to come.

11.2 Current Landscape of Materials in Energy Storage

11.2.1 Evaluating the State-of-the-Art Materials for Batteries

The contemporary landscape of materials in energy storage is marked by a fervent exploration of cutting-edge materials, driven by the urgent need for more efficient and sustainable energy storage solutions. Traditional lithium-ion batteries, while revolutionary in their own right, have instigated a quest for materials that can transcend existing limitations. Researchers are meticulously evaluating a diverse array of materials, ranging from advanced lithium-based compounds to novel organic polymers and beyond.

In this exploration, a critical focus lies on enhancing energy density, extending cycle life, and improving overall performance metrics. Materials such as silicon anodes, sulphur cathodes, and solid electrolytes are under intense scrutiny for their potential to revolutionise battery technology [3]. Nanomaterials, with their unique properties, have also emerged as contenders in the pursuit of materials that can redefine the capabilities of energy storage devices.

11.2.2 Ongoing Challenges and Gaps in Contemporary Material Selection

While the enthusiasm for discovering novel materials is palpable, researchers are confronted with a myriad of challenges and gaps in the contemporary material selection process. The transition from laboratory-scale successes to scalable, commercially viable solutions poses a significant hurdle. Issues related to the scalability of production, cost-effectiveness, and environmental impact underscore the need for a holistic approach in material selection. Furthermore, the dynamic nature of energy storage demands materials that can withstand the rigours of frequent charge–discharge cycles without compromising efficiency. The search for materials that balance performance, cost, and sustainability is an ongoing challenge that requires multidisciplinary collaboration. As the materials landscape evolves, addressing these challenges becomes pivotal in ensuring a seamless transition from laboratory innovation to real-world application.

11.3 Continual Integration of Machine Learning in Material Science

11.3.1 Real-Time Applications of Machine Learning in Material Discovery

The integration of machine learning into material science has transcended theoretical frameworks, finding real-time applications in the dynamic landscape of material discovery. Algorithms are now actively engaged in sifting through massive datasets, accelerating the identification of promising materials for energy storage. In realtime, ML models analyse experimental results, predict material properties, and guide researchers towards the most fruitful avenues of exploration.

This real-time application extends beyond the mere identification of materials; it includes adaptive learning, where algorithms continuously refine their predictions based on ongoing experimental data [4]. Machine learning facilitates the identification of patterns and correlations that might elude traditional analytical approaches, providing researchers with unprecedented insights into the behaviour of materials under varying conditions.

11.3.2 Dynamic Advancements and Evolving Capabilities of Machine Learning Algorithms

Machine learning algorithms are not static entities; they are dynamic, continually evolving in response to the challenges posed by the intricate world of material science. Recent advancements include the integration of deep learning techniques, enabling more complex pattern recognition and improved accuracy in predicting material properties. The adaptability of ML algorithms to diverse datasets and the ability to handle complex relationships between material structures and performance metrics contribute to their evolving capabilities. Moreover, the synergy between machine learning and quantum computing is on the horizon, promising even more sophisticated simulations and predictions. The evolution of algorithms towards explainable AI ensures that researchers can not only leverage the predictive power of machine learning but also comprehend the rationale behind the generated insights, fostering a deeper understanding of materials at the molecular level.

11.4 Live Data-Driven Approaches in Material Science

11.4.1 Dynamic Impact of Big Data on Ongoing Material Research

In the ever-evolving landscape of material science, the incorporation of live data-driven approaches has become synonymous with progress. A seismic shift is witnessed through the dynamic impact of big data on ongoing material research. Enormous datasets, generated through experimental analyses, simulations, and collaborative initiatives, are now harnessed to unveil hidden patterns, correlations, and potential breakthroughs. The real-time assimilation of big data catalyses research by providing researchers with a comprehensive understanding of material behaviour, paving the way for informed decision-making in material selection and design [5]. Researchers now find themselves navigating through terabytes of data, uncovering intricate relationships between material structures, properties, and performance metrics. This data-centric approach transcends traditional limitations, offering once-elusive insights. The dynamic impact of big data acts as a driving force, propelling material science into an era of unprecedented exploration and discovery.

11.4.2 Continuous Data Mining for Real-Time Materials Discovery

In the pursuit of real-time materials discovery, continuous data mining emerges as a key strategy. Traditional static datasets are now complemented by a live stream of information, continuously updated as experiments progress and new data becomes available. This dynamic approach allows researchers to adapt their strategies based on the latest insights, fostering agility in decision-making.

Continuous data mining not only expedites the identification of novel materials but also enables researchers to detect trends and anomalies as they occur. The iterative process of data mining facilitates a feedback loop between experimental outcomes and ML models, refining predictions and ensuring that the research remains at the forefront of material science advancements. The integration of live data into the research landscape signifies a departure from static methodologies, ushering in an era where material scientists can harness the power of real-time information to drive innovation.

11.5 Adaptive Machine Learning Techniques for Material Property Prediction

11.5.1 Adaptive Modelling of Material Properties in the Ongoing Research Landscape

As material scientists delve into the intricacies of the ongoing research landscape, adaptive ML techniques take centre stage in the modelling of material properties. Traditional static models are evolving into adaptive frameworks capable of learning and adjusting in response to the dynamic nature of material behaviour. This adaptability is crucial in capturing the nuances of materials that may exhibit varied properties under different conditions. Adaptive modelling involves continuous refinement, where ML algorithms autonomously update their understanding of material–property relationships as new data emerges [6]. This dynamic approach ensures that predictive models remain relevant and accurate, even as materials are subjected to evolving experimental conditions. The adaptability of these models is instrumental in keeping pace with the complex and multifaceted challenges posed by the ongoing research landscape.

11.5.2 Comparative Analysis of the Latest ML Algorithms for Property Prediction

In the ever-expanding toolkit of ML algorithms, a comparative analysis becomes essential to discern the strengths and limitations of each approach. Ongoing research involves evaluating the latest ML algorithms for material property prediction, considering factors such as accuracy, scalability, and computational efficiency. Researchers conduct in-depth analyses to determine which algorithms excel in predicting specific material properties, unveiling insights that guide the selection of appropriate models for different applications. This comparative approach allows researchers to harness the best available tools, ensuring that the predictive power of ML is leveraged optimally in the quest for understanding and optimising material properties.

In essence, the integration of adaptive ML techniques into material science offers a dynamic and responsive framework for predicting material properties. This adaptability, coupled with a comparative understanding of the latest algorithms, empowers researchers to navigate the complexities of the ongoing research landscape with precision and insight.

11.6 Progress in Optimising Energy Storage Systems Using ML

11.6.1 Continuous ML-Driven Optimisation of Battery Performance

In the relentless pursuit of enhanced energy storage solutions, a paradigm shift is occurring through continuous ML-driven optimisation of battery performance. Traditional methods for optimising energy storage systems are being supplemented, if not replaced, by dynamic ML algorithms that adapt to real-time conditions and varying usage patterns. This progressive approach allows for the fine-tuning of battery parameters, such as charging rates, voltage thresholds, and discharge characteristics, based on evolving environmental and operational factors.

Continuous ML-driven optimisation is a feedback loop where algorithms analyse performance data, identify patterns, and adjust system parameters accordingly [7]. This iterative process not only leads to improved immediate performance but also anticipates and mitigates factors that can affect long-term efficiency. As a result, the energy storage landscape is witnessing unprecedented progress in achieving optimal battery performance through the ongoing integration of machine learning.

11.6.2 Progress in Achieving Enhanced Cycle Life and Durability through Machine Learning

One of the pivotal challenges in energy storage is extending the cycle life and durability of batteries, and machine learning is proving to be a catalyst for progress in this domain. Ongoing research is showcasing advancements in predicting and mitigating factors that contribute to battery degradation over time. Machine learning models analyse historical data on charge–discharge cycles, environmental conditions, and usage patterns to identify stress points and formulate strategies to enhance the longevity of energy storage systems.

The progress in achieving enhanced cycle life and durability through machine learning extends beyond mere prediction; it actively informs the design and engineering of batteries with materials and configurations that resist degradation. The real-time adaptation of battery management systems based on continuous learning contributes to the development of robust and long-lasting energy storage solutions, marking a transformative chapter in the evolution of battery technology.

11.7 Materials Genome Initiative and Ongoing ML Integration

11.7.1 Progress Report on Accelerating Materials Discovery with the Materials Genome Initiative

The Materials Genome Initiative (MGI) stands as a beacon for accelerating materials discovery, and its ongoing integration with machine learning is shaping a new frontier in material science. A progress report on MGI reveals the strides made in streamlining the materials research process, from synthesis to characterisation and application. The initiative aims to create a materials design framework akin to genomic studies, where the traits and behaviours of materials are mapped out systematically. Ongoing research within the Materials Genome Initiative involves leveraging machine learning to sift through vast materials databases, identifying correlations between material structures and properties. Progress reports detail the successful application of predictive models in suggesting novel materials with desired characteristics, significantly reducing the time and resources traditionally required for materials discovery. The integration of machine learning into the MGI framework marks a pivotal advancement in the quest for efficient and targeted materials development.

11.7.2 Current Developments in Integrating Machine Learning into the Materials Genome Project

As the Materials Genome Project unfolds, current developments in integrating machine learning underscore the project's commitment to staying at the forefront of technological innovation. Ongoing research involves refining and expanding the capabilities of ML algorithms to accommodate the diverse array of materials and their intricate properties. The current integration efforts focus on improving the accuracy of property predictions, understanding the uncertainties associated with these predictions, and developing strategies for experimental validation.

Researchers working within the Materials Genome Project report on the successful incorporation of machine learning in guiding experimental efforts, enabling a more targeted and efficient approach to materials synthesis and testing [8]. The current state of integration reflects a dynamic synergy between data-driven methodologies and traditional experimental techniques, leading to a holistic and accelerated approach to materials discovery within the ambit of MGI.

11.8 Latest Case Studies in ML-Enhanced Material Design

11.8.1 Ongoing Applications of ML in Battery Material Design

The latest case studies in ML-enhanced material design illuminate the transformative impact of machine learning on the development of advanced battery materials. Researchers are actively engaged in ongoing applications of ML, utilising predictive models to design materials with enhanced energy density, faster charge–discharge rates, and prolonged cycle life [9]. These case studies delve into the intricacies of ML algorithms guiding the selection and optimisation of components, illustrating how data-driven approaches are reshaping the landscape of battery material design.

As ML algorithms evolve, the ongoing case studies demonstrate their adaptability in addressing specific challenges posed by different material systems. The synergy between experimental insights and predictive modelling unveils new avenues for material design, propelling the field towards unprecedented breakthroughs in energy storage technology.

11.8.2 Learning from Recent Failures: Updated Case Studies on ML Predictions and Actual Performance

In the pursuit of innovation, failures often serve as profound lessons. The latest case studies in ML-enhanced material design provide valuable insights by scrutinising instances where predictions fell short of actual performance. These updated case studies go beyond showcasing successes; they delve into the nuanced reasons behind discrepancies between ML predictions and real-world material behaviour.

Learning from recent failures is an integral part of refining machine learning models and ensuring their robustness. Researchers analyse the root causes, whether they be unforeseen environmental factors, material degradation mechanisms, or other variables, to enhance the accuracy and reliability of ML predictions. By addressing challenges head-on and iteratively improving models, the scientific community navigates towards a more informed and resilient approach to ML-enhanced material design.

11.9 Addressing Real-Time Challenges and Ethical Considerations

11.9.1 Real-Time Ethical Implications of Automated Material Discovery

As automated material discovery becomes more prevalent, addressing real-time ethical implications takes on heightened importance. This facet of the research explores the ethical considerations arising from the intersection of machine learning and material science in a dynamic, evolving landscape. Researchers delve into the real-time ethical dimensions of deploying algorithms that influence critical decisions in material selection, especially in applications with potential societal impacts [10].

The ongoing research probes questions of accountability, transparency, and the responsible use of AI in material discovery. It seeks to establish ethical frameworks that can adapt to the rapidly changing technological landscape, ensuring that the benefits of automation align with societal values and avoid unintended consequences.

11.9.2 Addressing Bias and Ensuring Transparency in Ongoing ML-Driven Material Research

Ethical considerations in ongoing ML-driven material research extend to addressing bias and ensuring transparency. Researchers recognise the potential biases embedded in training datasets and the algorithms themselves, which can inadvertently impact material selection and performance predictions. This aspect of the research focuses on developing methodologies to identify and mitigate biases in realtime, fostering fairness and equity in the outcomes of ML-driven material research.

Ensuring transparency is equally crucial, as researchers strive to make the decision-making processes of ML models more understandable and interpretable. Ongoing efforts in this domain aim to provide clear insights into how machine learning influences material design choices, thereby fostering trust among stakeholders and the wider scientific community. By actively addressing bias and promoting transparency, the research contributes to the ethical advancement of ML-driven material science, aligning technological progress with ethical principles.

11.10 Future Directions and Emerging Technologies in Ongoing Research

11.10.1 Current Role of Quantum Computing in Material Science

The future of material science unfolds against the backdrop of emerging technologies, and quantum computing stands at the forefront of transformative advancements. This section explores the current role of quantum computing in material science, outlining how quantum algorithms are revolutionising simulations and predictions at the atomic and molecular levels. Quantum computing's computational prowess holds the potential to unravel complex material behaviours that were previously computationally intractable, paving the way for unprecedented discoveries [11]. As researchers navigate the intersection of quantum computing and material science, they delve into the opportunities and challenges presented by this nascent technology. The ongoing research paints a vivid picture of the synergies between quantum computing and machine learning, providing a glimpse into the future landscape where quantum-enhanced material design becomes a reality.

11.10.2 Synergies between ML and Other Emerging Technologies in Energy Storage

The exploration of future directions in energy storage research extends beyond quantum computing, encompassing a dynamic interplay between machine learning and various emerging technologies. This section delves into the synergies between ML and other groundbreaking technologies, such as nanotechnology, advanced sensors, and internet-of-things (IoT) devices, in the ongoing quest for enhanced energy storage solutions.

Researchers investigate how ML can harness data from these emerging technologies to optimise energy storage systems, improve predictive models, and enable adaptive control strategies. The collaborative efforts between machine learning and other emerging technologies mark a pivotal juncture in the ongoing energy storage research, promising a holistic approach towards innovative and sustainable solutions for the future.

11.11 Current Insights and Ongoing Progress in the Field

11.11.1 Updated Achievements and Contributions of ML in Materials Evolution

This section provides a comprehensive overview of the updated achievements and contributions of machine learning in the dynamic evolution of material science. Researchers delve into the latest breakthroughs, showcasing how ML has facilitated the discovery of novel materials, optimised material properties, and accelerated the overall research process [12]. From advancements in predictive modelling to the integration of ML into experimental workflows, the ongoing progress in materials evolution paints a vivid picture of the transformative impact of machine learning. Case studies and success stories exemplify how ML has not only augmented researchers' capabilities but also led to tangible advancements in material design and synthesis. The updated achievements provide a snapshot of the current state of ML in material science, offering valuable insights into the trajectory of future developments [13].

11.11.2 Current Prospects for ML-Driven Energy Storage Solutions

The closing segment of the research explores the current prospects for ML-driven energy storage solutions, drawing insights from the ongoing research endeavours. Researchers analyse the trajectory of ML applications in optimising battery performance, enhancing cycle life, and addressing real-time challenges in energy storage systems [14]. This section provides a forward-looking perspective, offering glimpses into potential breakthroughs and innovations on the horizon.

Insights from ongoing research shed light on how machine learning is shaping the future of energy storage, influencing the design, performance, and sustainability of storage solutions. The research not only outlines the current landscape but also provides a roadmap for the integration of ML in energy storage, offering valuable guidance for researchers, industry professionals, and policymakers invested in the future of sustainable energy technologies [15].

11.12 Conclusion

In conclusion, the confluence of material science and machine learning (ML) presents a transformative paradigm in the realm of energy storage research. The ongoing evaluation of state-of-the-art materials for batteries, coupled with real-time applications of ML, has yielded unprecedented insights into material behaviours and properties. Researchers navigate challenges in scalability and ethical considerations, demonstrating the adaptability of ML in optimising energy storage systems and achieving enhanced cycle life. The integration of adaptive ML techniques for material property prediction is showcased through iterative modelling, ensuring continuous refinement in the evolving research landscape. Progress in ML-driven optimisation of battery performance is underscored by dynamic advancements, promising unparalleled advancements in energy storage technology. As the Materials Genome Initiative intertwines with ML, a progressive acceleration in materials discovery emerges. Current insights into the role of quantum computing and synergies between ML and emerging technologies reveal a future landscape marked by computational prowess and interdisciplinary collaboration. The achievements and contributions of ML in materials evolution, coupled with prospects for ML-driven energy storage solutions, present a promising trajectory for sustainable energy technologies, guiding researchers towards innovative and efficient solutions for the challenges ahead.

REFERENCES

[1] Bibri, S.E., Krogstie, J., Kaboli, A. and Alahi, A., 2024. Smarter eco-cities and their leading-edge artificial intelligence of things solutions for environmental sustainability: A comprehensive systematic review. *Environmental Science and Ecotechnology*, *19*, p. 100330.

[2] Clayson, I.G., Hewitt, D., Hutereau, M., Pope, T. and Slater, B., 2020. High throughput methods in the synthesis, characterization, and optimization of porous materials. *Advanced Materials*, *32* (44), p. 2002780.

[3] Joseph Arockiam, A., Rajesh, S. and Karthikeyan, S., Development of fish scale particle reinforced PLA filaments for 3D printing applications. *Journal of Applied Polymer Science*, *141* (12), p. e55132.

[4] Baker, N., Alexander, F., Bremer, T., Hagberg, A., Kevrekidis, Y., Najm, H., Parashar, M., Patra, A., Sethian, J., Wild, S. and Willcox, K., 2019. Workshop report on basic research needs for scientific machine learning: Core technologies for artificial intelligence. USDOE Office of Science (SC), Washington, DC (United States).

[5] Allioui, H. and Mourdi, Y., 2023. Unleashing the potential of AI: Investigating cutting-edge technologies that are transforming businesses. *International Journal of Computer Engineering and Data Science (IJCEDS)*, *3* (2), pp. 1–12.

[6] Arockiam, A.J., Rajesh, S., Karthikeyan, S., Thiagamani, S.M.K., Padmanabhan, R.G., Hashem, M., Fouad, H. and Ansari, A., 2023. Mechanical and thermal characterization of additive manufactured fish scale powder reinforced PLA biocomposites. *Materials Research Express*, *10* (7), p. 075504.

[7] Jawad, Z.N. and Balázs, V., 2024. Machine learning-driven optimization of enterprise resource planning (ERP) systems: A comprehensive review. *Beni-Suef University Journal of Basic and Applied Sciences*, *13* (1), pp. 1–13.

[8] de Pablo, J.J., Jackson, N.E., Webb, M.A., Chen, L.Q., Moore, J.E., Morgan, D., Jacobs, R., Pollock, T., Schlom, D.G., Toberer, E.S. and Analytis, J., 2019. New frontiers for the materials genome initiative. *npj Computational Materials*, *5* (1), p. 41.

[9] Vignesh, P., Ramanathan, S., Ramesh, K. and Arockiam, A.J., 2022. Finite element modelling to predict hardness of Mg alloy reinforced with Ti/hydroxyapatite hybrid composites–An axisymmetric approach. *Materials Today: Proceedings*, *68*, pp. 1830–1834.

[10] Neethirajan, S., 2024. Artificial intelligence and sensor innovations: Enhancing livestock welfare with a human-centric approach. *Human-Centric Intelligent Systems*, *4*, pp. 77–92.

[11] Kanatzidis, M.G., Poeppelmeier, K.R., Bobev, S., Guloy, A.M., Hwu, S.J., Lachgar, A., Latturner, S.E., Schaak, E., Seo, D.K., Sevov, S.C. and Stein, A., 2008. Report from the third workshop on future directions of solid-state chemistry: The status of solid-state chemistry and its impact in the physical sciences. *Progress in Solid State Chemistry*, *36* (1–2), pp. 1–133.

[12] Chandrashekhar, K.G., Laxmaiah, G., Ram Kumar, P. and Ramesh, B., 2023. Load-bearing characteristics of a hybrid Si_3N_4-epoxy composite. *Biomass Conversion and Biorefinery*, pp. 1–9.

[13] Alivisatos, P., Barbara, P.F., Castleman, A.W., Chang, J., Dixon, D.A., Klein, M.L., McLendon, G.L., Miller, J.S., Ratner, M.A., Rossky, P.J. and Stupp, S.I., 1998. From molecules to materials: Current trends and future directions. *Advanced Materials*, *10* (16), pp. 1297–1336.

[14] Arockiam, A.J., Subramanian, K., Padmanabhan, R.G., Selvaraj, R., Bagal, D.K. and Rajesh, S., 2022. A review on PLA with different fillers used as a filament in 3D printing. *Materials Today: Proceedings*, *50*, pp. 2057–2064.

[15] Ahmad, T., Madonski, R., Zhang, D., Huang, C. and Mujeeb, A., 2022. Data-driven probabilistic machine learning in sustainable smart energy/smart energy systems: Key developments, challenges, and future research opportunities in the context of smart grid paradigm. *Renewable and Sustainable Energy Reviews*, *160*, p. 112128.

12

Comprehensive Analysis of Photovoltaic Energy Storage Device Controller and Its Applications

G. Ramya
SRM Institute of Science & Technology, Chennai, India

P. V. Premalatha
M.I.E.T. Engineering College, Tiruchirappalli, India

P. Suresh
St. Joseph College of Engineering, Chennai, India

P. Thirumurugan
Annai Veilankannis College of Engineering, Chennai, India

12.1 Introduction

Photovoltaic (PV) technology has gained increasing importance in the field of renewable energy due to its potential to provide sustainable and clean energy [1, 2]. The efficient and effective operation of PV systems depends on the use of appropriate controllers to regulate the charging and discharging of the battery, optimise energy harvest, and ensure that the system operates within the recommended operating conditions [3–5]. The comprehensive analysis of PV controllers and their applications provides valuable insights into the different types of controllers, including maximum power point tracking (MPPT), pulse width modulation (PWM), constant voltage, and integrated controller [6]. The analysis investigates the advantages and disadvantages of each type and compares their performance in different applications, including grid-tied and off-grid PV systems, water pumping systems, and electric vehicle charging stations [7–9]. In addition to discussing the various types of controllers and their applications, the analysis also identifies the parameters that affect the performance of PV controllers, such as temperature, solar irradiation, and load conditions [10–12]. Understanding these parameters is crucial for optimising the performance of PV systems and ensuring that they operate efficiently and effectively. This comprehensive analysis is based on a review of relevant literature and research studies in the field of PV technology and renewable energy [13]. The analysis also draws on the expertise and knowledge of industry professionals and researchers in the field. Overall, the comprehensive analysis of PV controllers and their applications provides a valuable resource for researchers, industry professionals, and anyone interested in the field of renewable energy [14, 15]. It highlights the importance of selecting the appropriate controller for the specific application and ensuring that the system is designed to operate within the recommended operating conditions to maximise the efficiency and effectiveness of PV systems.

A comprehensive analysis of PV system is depicted in Figure 12.1. The performance evaluation and optimisation block include analysing and improving the efficiency and effectiveness of the PV controller, such as the power conversion efficiency, MPPT efficiency, and system stability. The control strategy design and implementation block involves designing and implementing the control algorithm for the PV controller, such as the MPPT algorithm and the voltage and current regulation algorithm. The power

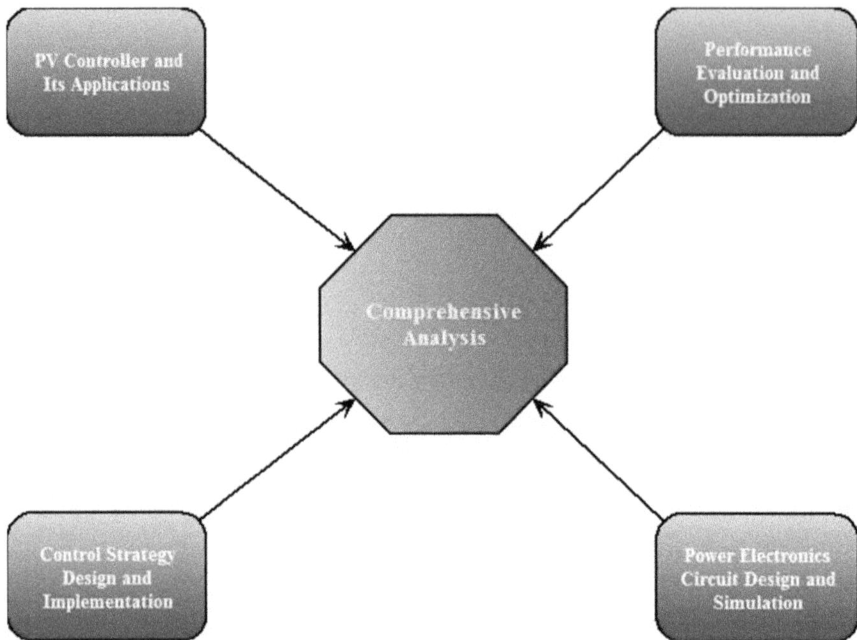

FIGURE 12.1 Comprehensive analysis of PV system.

electronics circuit design and simulation block is responsible for designing and simulating the power electronics circuitry used in the PV controller, such as the DC–DC converter and the inverter. All three blocks work together to ensure that the PV controller and its applications are optimised for maximum performance and efficiency.

12.2 Types of PV Controller

12.2.1 Maximum Power Point Tracking (MPPT) Controllers

MPPT controllers are utilised to optimise the power output of PV panels by adjusting the panel's operating point to match the maximum power point. This is achieved by continuously measuring the panel's voltage and current and modifying the output voltage accordingly. MPPT controllers have high efficiency and can boost the overall system efficiency by up to 30%. Different types of MPPT controllers are available, each with its own strengths and weaknesses. Some of the most frequently employed types are highlighted, and the process of MPPT technique is illustrated in Figure 12.2.

12.2.1.1 Perturb and Observe (P&O) MPPT Controllers

P&O MPPT controllers are extensively used and considered the most straightforward type of MPPT controller. They periodically modify the panel's operating point and evaluate the resulting power output. If the power output increases, the controller continues to adjust the operating point in the same direction.

12.2.1.2 Incremental Conductance MPPT Controllers

Compared to P&O controllers, incremental conductance MPPT controllers are more advanced. They operate by continuously monitoring the PV panel's voltage and current and adjusting the operating point based on the rate of conductance change.

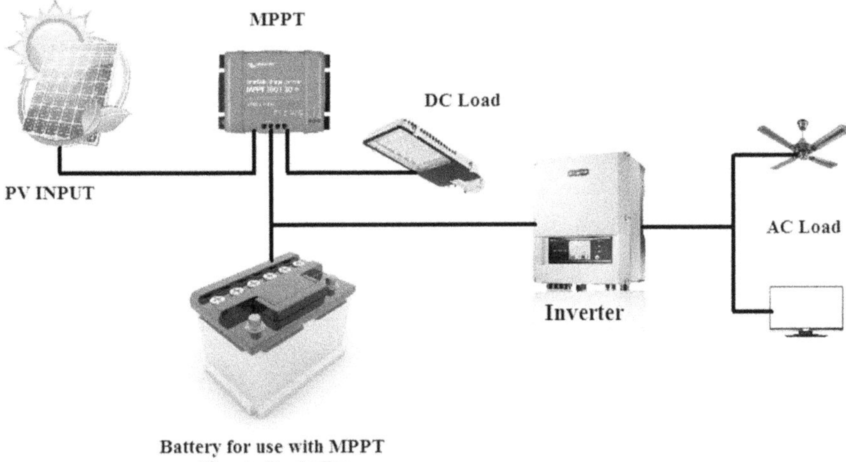

FIGURE 12.2 Process of MPPT technique.

12.2.1.3 Fractional Open Circuit Voltage (FOCV) MPPT Controllers

MPPT controllers based on open circuit voltage (OCV) measurement, also known as FOCV controllers, are designed to track the maximum power point of a PV panel. Although less commonly used than P&O and incremental conductance controllers, FOCV controllers can be advantageous in certain applications.

12.2.1.4 Artificial Intelligence (AI) MPPT Controllers

AI MPPT controllers use machine learning algorithms to optimise the power output of PV panels. They can adapt to changes in the environment and load conditions and can be more accurate than traditional MPPT controllers. However, AI controllers can be more complex and expensive than other types of MPPT controllers. The type of MPPT controller chosen depends on the specific application and system requirements, including the type of PV panel, battery, and load characteristics.

12.2.2 Pulse Width Modulation (PWM) Controllers

PWM controllers work by turning the PV panel on and off at a high frequency, which effectively regulates the amount of power flowing to the battery. PWM controllers are simpler and less expensive than MPPT controllers, but they are less efficient and may not work as well in low-light conditions. PWM controllers are used to regulate the amount of power flowing from a PV panel to a battery by turning the panel on and off at a high frequency. Figure 12.3 depicts the PWM output. There are several types of PWM controllers available in the market, each with its advantages and disadvantages. Here are some of the most commonly used types.

12.2.2.1 Series PWM Controllers

Series PWM controllers are designed to be used with small PV systems and are the simplest type of PWM controller. They work by turning the PV panel on and off to regulate the battery charging voltage. Series PWM controllers are inexpensive but not as efficient as other types of PWM controllers.

12.2.2.2 Shunt PWM Controllers

Shunt PWM controllers are designed to be used with larger PV systems and work by connecting a shunt resistor in parallel with the PV panel. The controller turns the panel on and off by shorting the shunt

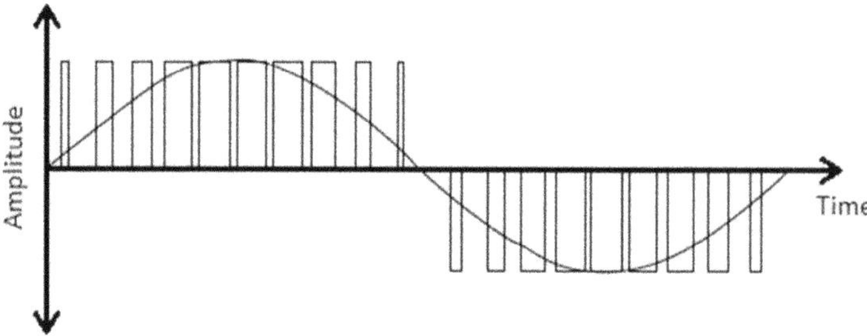

FIGURE 12.3   PWM output.

resistor, which effectively regulates the battery charging voltage. Shunt PWM controllers are more efficient than series PWM controllers but can be more expensive.

12.2.2.3 Multiple-Stage PWM Controllers

Multiple-stage PWM controllers are designed to be used with larger PV systems and use multiple stages of PWM regulation to improve efficiency. They work by dividing the charging process into several stages, each with a different duty cycle. Multiple-stage PWM controllers can be more complex than other types of PWM controllers but offer higher efficiency.

12.2.2.4 Integrated MPPT/PWM Controllers

Integrated MPPT/PWM controllers combine the functionality of MPPT and PWM controllers in a single device. They can be more expensive than other types of PWM controllers but offer higher efficiency and versatility. The type of PWM controller chosen depends on the specific application and system requirements, including the size of the PV system, the type of battery, and the load characteristics.

12.2.3 Constant Voltage Controllers

Constant voltage controllers are designed to regulate the voltage output of a PV panel to a constant level, regardless of changes in light or temperature conditions. These controllers are commonly used in off-grid PV systems with lead–acid batteries. Constant voltage controllers, also known as float or trickle charge controllers, are used to regulate the voltage output of a PV panel to a constant level, regardless of changes in light or temperature conditions. These controllers are commonly used in off-grid PV systems with lead–acid batteries. Constant voltage controllers work by limiting the voltage output of the PV panel to a safe level to prevent overcharging of the battery. Once the battery is fully charged, the controller reduces the charging current to a trickle charge, which maintains the battery's charge level without overcharging it. The circuit diagram of constant voltage controllers is shown in Figure 12.4.

There are two types of constant voltage controllers available:

Linear Constant Voltage Controllers
Linear constant voltage controllers regulate the voltage output of the PV panel by dissipating excess power as heat. They are simple and reliable but less efficient than other types of controllers, as they waste some of the energy as heat.

Switching Constant Voltage Controllers
Switching constant voltage controllers regulate the voltage output of the PV panel by switching the current on and off at a high frequency. They are more efficient than linear controllers but can be more complex and expensive. The type of constant voltage controller chosen depends on the specific application and system requirements.

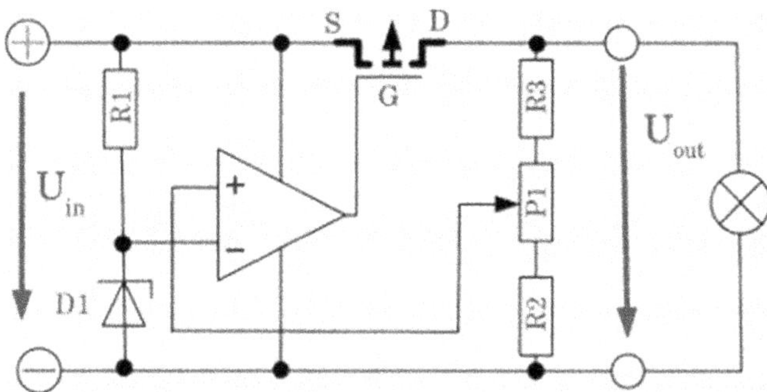

FIGURE 12.4 Circuit diagram of constant voltage controllers.

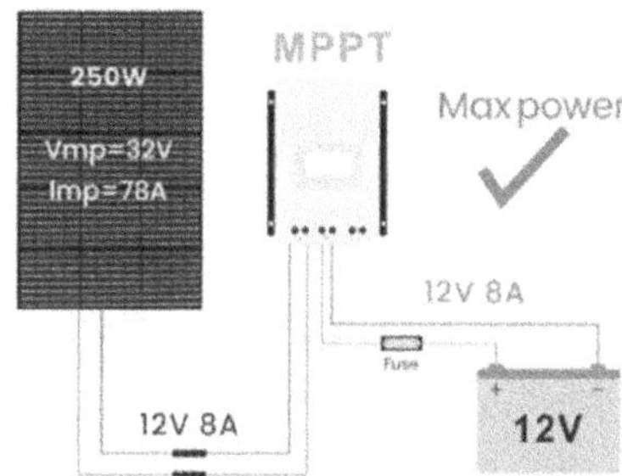

FIGURE 12.5 Block diagram of integrated controller.

12.2.4 Integrated Controllers

Integrated controllers combine MPPT and PWM functionality in a single device. They are more versatile than either MPPT or PWM controllers alone and can be used in a variety of applications. Integrated controllers, also known as all-in-one controllers, combine the functions of several controllers into a single device, providing a more streamlined and cost-effective solution for PV systems. These controllers typically combine MPPT, PWM, and constant voltage control functions into a single device, eliminating the need for separate controllers. Integrated controllers are designed to be used with a variety of PV systems, including off-grid and grid-tied systems. They can be used with a range of battery types, including lead–acid, lithium ion, and others. Block diagram of integrated controller is depicted in Figure 12.5. The key advantages of integrated controllers include increased efficiency, simplified installation, and reduced cost. Integrated controllers typically offer a range of features, including.

12.2.4.1 MPPT Function

The MPPT function tracks the maximum power point of the PV panel, maximising the energy harvested from the panel.

12.2.4.2 PWM Function

The PWM function regulates the flow of energy from the PV panel to the battery, ensuring that the battery is charged at the optimal voltage and current levels.

12.2.4.3 Constant Voltage Function

The constant voltage function regulates the voltage output of the PV panel to a constant level, maintaining the battery's charge level without overcharging it.

12.2.4.4 Battery Temperature Sensor

An integrated temperature sensor monitors the battery temperature, allowing the controller to adjust the charging voltage based on the battery's temperature.

12.2.4.5 LCD Display

An LCD display provides real-time information on the system's performance, including battery voltage, current, and state of charge.

Integrated controllers are available in a range of sizes and capacities, making them suitable for a variety of PV system sizes and configurations. The type of integrated controller chosen depends on the specific application and system requirements, including the size of the PV system, the type of battery, and the load characteristics. The type of controller chosen depends on the specific application, the type of battery used, and the system's load characteristics.

12.3 Merits and Demerits of PV Controllers

Advantages and disadvantages of PV controller, along with their performance in different applications, are discussed in this section.

The advantages and disadvantages of MPPT controllers are as follows:

Advantages
- Higher efficiency compared to other types of controllers, especially in low-light conditions.
- Can extract more energy from the PV panel, increasing the system's overall performance.
- Compatible with a variety of battery types and system sizes.

Disadvantages
- Higher cost compared to other types of controllers.
- Can be more complex to install and operate.

Performance in different applications
- MPPT controllers are ideal for off-grid PV systems with variable light conditions, where maximising energy harvest is important.
- MPPT controllers are also suitable for grid-tied PV systems, where energy production needs to be optimised to reduce electricity bills.

The advantages and disadvantages of PWM controllers are discussed next.

Advantages
- Simple and reliable design.
- Lower cost compared to other types of controllers.
- Suitable for small- to medium-sized PV systems.

Disadvantages
- Lower efficiency compared to MPPT controllers.
- Cannot extract maximum power from the PV panel in low-light conditions.
- Not suitable for larger PV systems.

Performance in different applications
- PWM controllers are ideal for small off-grid PV systems, where simplicity and cost-effectiveness are important.

- PWM controllers can also be used in grid-tied PV systems with smaller PV arrays, where maximum power extraction is less important.

The advantages and disadvantages of constant voltage controllers are discussed next.

Advantages
- Simple and reliable design.
- Low cost compared to other types of controllers.
- Suitable for maintaining battery charge levels in off-grid PV systems.

Disadvantages
- Lower efficiency compared to MPPT and PWM controllers.
- Cannot extract maximum power from the PV panel.

Performance in different applications
- Constant voltage controllers are ideal for off-grid PV systems with lead–acid batteries, where maintaining a constant charge level is important.
- Constant voltage controllers are also suitable for small grid-tied PV systems with lead–acid batteries, where cost-effectiveness is important.

The advantages and disadvantages of integrated controllers are discussed.

Advantages
- Combines the functions of several controllers into a single device, reducing the number of components and simplifying installation.
- Higher efficiency compared to PWM and constant voltage controllers.
- Suitable for a range of PV system sizes and battery types.

Disadvantages
- Higher cost compared to PWM and constant voltage controllers.
- May be more complex to install and operate.

Regarding performance in different applications, integrated controllers are suitable for a range of PV system sizes and applications, including off-grid and grid-tied systems. Integrated controllers are ideal for applications where maximising energy harvest, efficiency, and performance is important.

12.4 Performance of PV Controllers

The performance of PV controllers is influenced by various parameters, including temperature, solar irradiation, and load conditions. These parameters affect the output of the PV panels, which in turn impacts the performance of the controller.

12.4.1 Temperature

The temperature of the PV panels and battery affects the efficiency of the controller. High temperatures can reduce the efficiency of the PV panel, while low temperatures can reduce the battery capacity.

12.4.2 Solar Irradiation

The intensity of solar irradiation affects the performance of the controller. Low solar irradiation can reduce the energy output of the PV panel, which can impact the charging of the battery.

12.4.3 Load Conditions

The type and size of the load connected to the PV system can affect the performance of the controller. Heavy loads can draw more energy from the battery, reducing the battery's capacity and impacting the overall performance of the system.

The performance of PV controllers can be optimised by selecting the appropriate controller for the specific application and ensuring that the system is designed to operate within the recommended operating conditions.

PV controllers have various applications in the field of renewable energy, including grid-tied and off-grid PV systems, water pumping systems, and electric vehicle charging stations.

12.4.3.1 Grid-Tied PV Systems

PV controllers are used in grid-tied PV systems to ensure that the energy generated by the PV panels is fed into the grid at the optimal voltage and frequency. MPPT controllers are commonly used in grid-tied systems to optimise energy harvest.

12.4.3.2 Off-Grid PV Systems

PV controllers are used in off-grid PV systems to regulate the charging and discharging of the battery. PWM and constant voltage controllers are commonly used in off-grid systems due to their simplicity and cost-effectiveness.

12.4.3.3 Water Pumping Systems

PV controllers are used in water pumping systems to regulate the PV panel to the pump. MPPT controllers are commonly used in water pumping systems to maximise energy harvest.

12.4.3.4 Electric Vehicle Charging Stations

PV controllers are used in electric vehicle charging stations to regulate the PV panel to the charging station. MPPT controllers are commonly used in electric vehicle charging stations to optimise energy harvest and reduce electricity costs.

Overall, PV controllers play a crucial role in ensuring the efficient and effective operation of PV systems, and their performance can be optimised by selecting the appropriate controller for the specific application and ensuring that the system is designed to operate within the recommended operating conditions.

12.5 Significant Research Work in PV-Based System

The proposed PV system for grid connection employs a single-phase cascaded H-bridge multilevel inverter topology, incorporating a DC–DC flyback converter and an inverter for power processing. The inverter connects the local load and the grid, with the DC–DC converter utilised to achieve MPPT for each module. A DC–DC flyback converter is used for electrical isolation, as it is a more efficient and cost-effective solution than a line frequency transformer with several disadvantages. The proposed configuration is shown in Figure 12.6. The AC module topology is selected for its superior flexibility and MPPT capabilities, with only one converter employed per module to minimise mismatching between them. The flyback converter serves to isolate high frequencies and eliminate leakage current that may produce harmonics in the grid current due to the leakage current between the PV module and ground. However, only one flyback converter regulates the maximum power of each PV module, resulting in optimal MPPT for the PV modules. On the other hand, the H-bridge capacitor produces second-order harmonic voltage, which is caused by the assumption that each phase leg contains a single-phase converter that suppresses 100 Hz power fluctuations within each primary cycle. The voltage fluctuation around the MPPT can cause a reduction in the power extracted from the PV module due to the second-order harmonic voltage ripple.

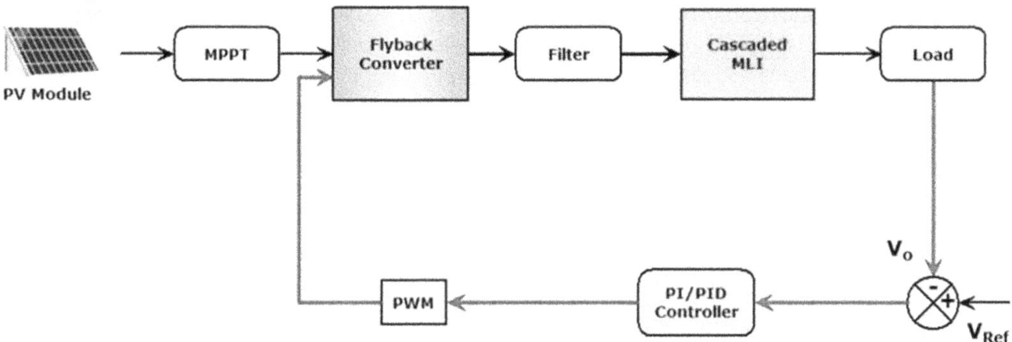

FIGURE 12.6 Block diagram of the proposed system.

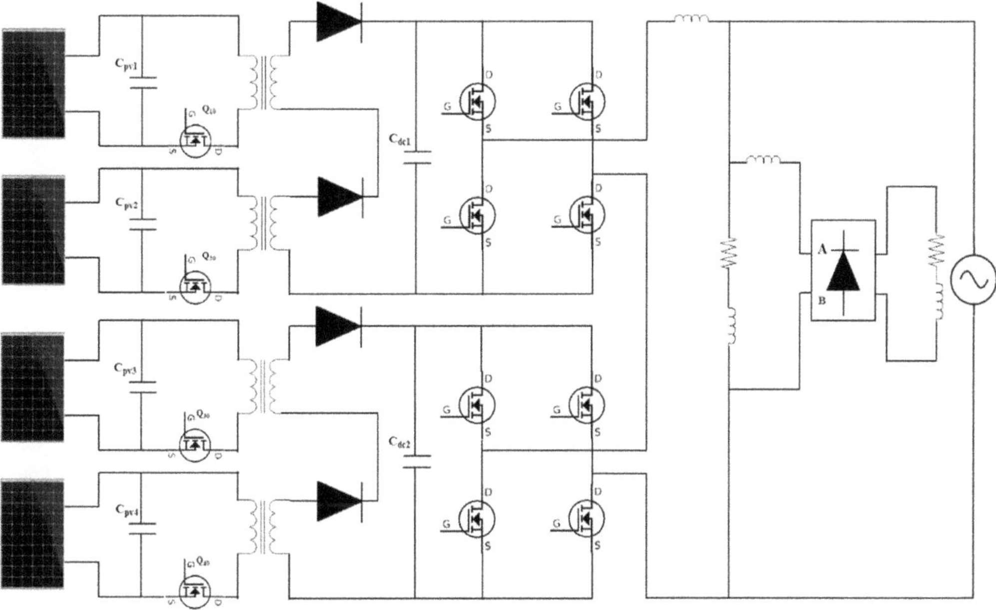

FIGURE 12.7 Circuit diagram of proposed flyback converter fed with cascaded H-bridge multilevel inverter topology for grid-connected PV system.

The DC–DC converter eliminates this ripple from the output PV capacitor, improving MPPT. To simplify the design, a five-level inverter is used with only four PV modules. Each PV module is coupled to a single flyback DC–DC converter, as a demonstration.

The proposed grid-connected PV system, which utilises a cascaded H-bridge multilevel inverter topology, is depicted in Figure 12.7 along with the circuit diagram of the flyback converter. A transformer with primary and secondary windings makes up the flyback converter. The transformer's primary windings are connected in the ratio of 1: 2. The other flyback converter is also connected to the PV_2. The PV_1 output voltage is controlled by MOSFET Q_{10}. A diode connects the secondary winding to the inverter. One H-bridge inverter is connected after cascading the turn ratio of both flyback converters. The flyback converter is a transformer with main and secondary windings.

The transformer's primary windings are linked in a 1:2 ratio, and an additional flyback converter is connected to PV_2. The PV_1 output voltage is controlled using MOSFET Q10, and the inverter is connected to the secondary winding via a diode. Before being linked to the H-bridge inverter, the turn ratios of both flyback converters are cascaded. The H-bridge inverter is composed of four MOSFETs, with the upper ones working at the line frequency and the lower ones at the switching frequency to reduce switching losses and enhance system efficiency. The cascaded inverters generate a five-level output voltage that is fed into the grid, followed by the nonlinear load. To simulate the open-loop system of the proposed flyback converter-supplied cascaded H-bridge multilevel inverter topology with R and RL loads, MATLAB Simulink was used, and Figure 12.8 displays the PV panel's output voltage.

Output voltage of flyback rectifier with R load and RL load is shown in Figure 12.9. Cascaded multilevel inverter is employed to generate AC output from the system. Multilevel inverter is used to generate

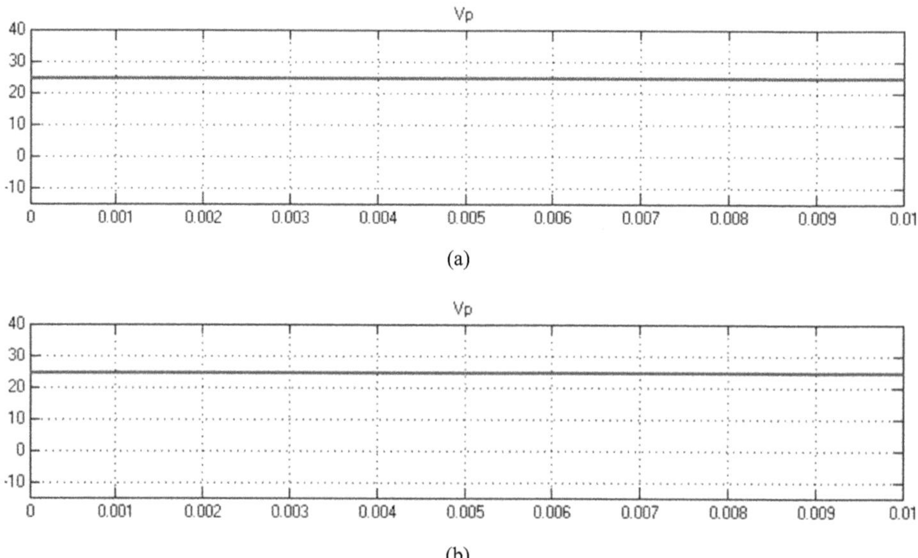

FIGURE 12.8   Output voltage from PV panel with (a) R load and (b) RL load.

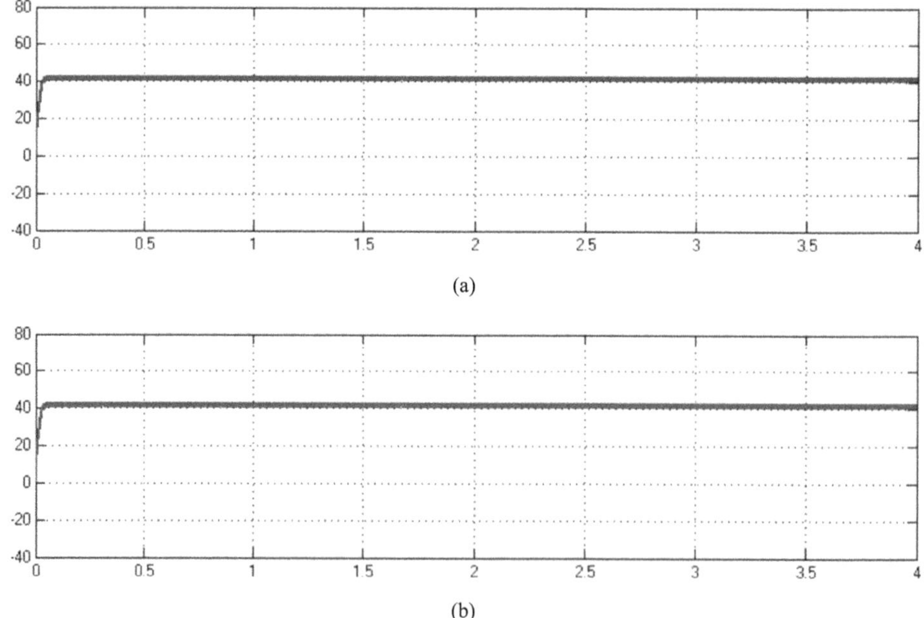

FIGURE 12.9   Output voltage of flyback rectifier with (a) R load and (b) RL load.

multilevel AC output with less harmonics. Figure 12.10 depicts the output power of the system with R and RL load conditions (Figure 12.11).

Figure 12.12 depicts the frequency spectrum analysis of inverter output current with R load and RL load and the total harmonic distortion (THD) of the system with R load is 20.91%and RL load is 9.09%. Closed-loop analysis like proportional integral controller (PI), proportional integral and derivative controller (PID), and sliding mode controller can be employed to the proposed system to reduce the total harmonic distortion and to achieve the THD of IEEE standard of 5%.

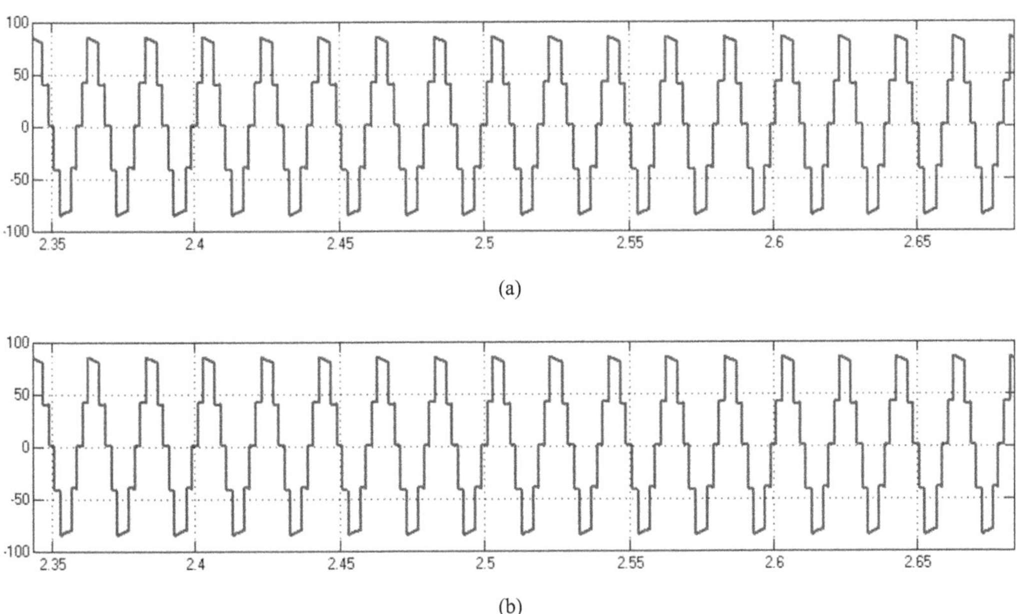

(a)

(b)

FIGURE 12.10 Cascaded multilevel inverter output voltage with (a) R load and (b) RL load.

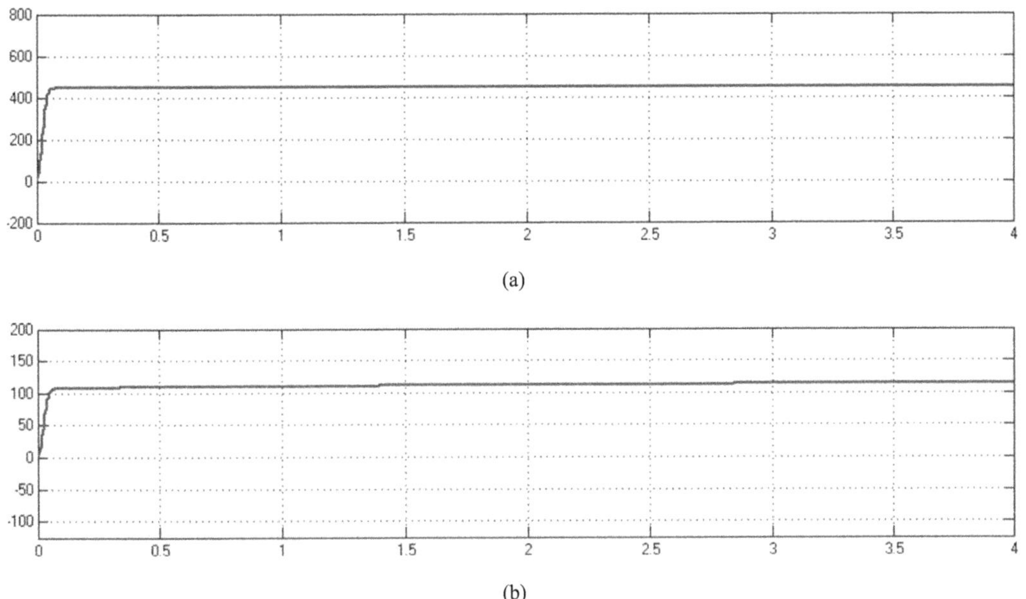

(a)

(b)

FIGURE 12.11 Output power of the system: (a) R load and (b) RL load.

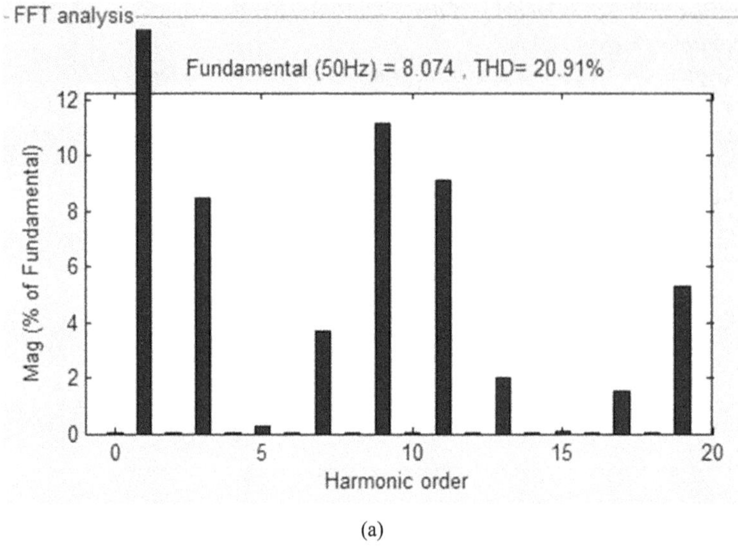

(a)

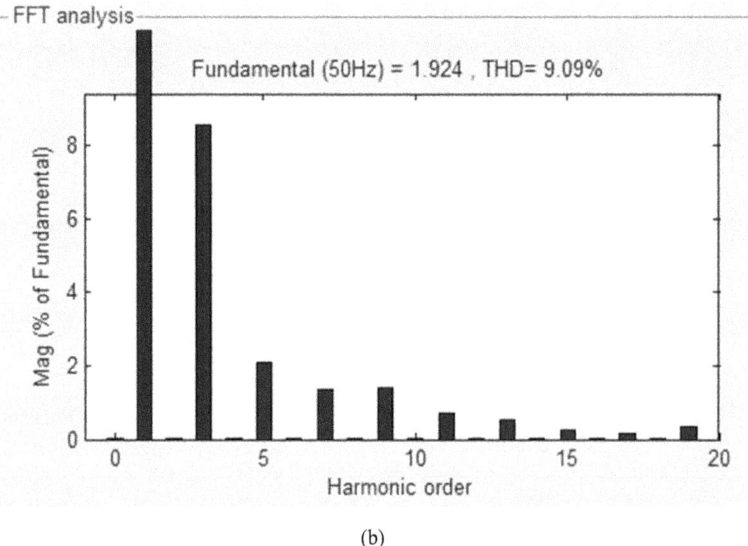

(b)

FIGURE 12.12   Frequency spectrum analysis of inverter output current with (a) R load and (b) RL load.

12.6 Conclusion

The comprehensive analysis of PV controllers and their applications provides an in-depth understanding of the different types of controllers, including MPPT, PWM, constant voltage, and integrated controllers. Each type has its advantages and disadvantages, and their performance varies based on the application and operating conditions. The analysis also identifies various parameters that affect the performance of PV controllers, including temperature, solar irradiation, and load conditions. These factors influence the output of the PV panels and the charging and discharging of the battery, which in turn impact the performance of the controller. Furthermore, the analysis highlights the importance of selecting the appropriate controller for the specific application and ensuring that the system is designed to operate within the recommended operating conditions. PV controllers have various applications in the field of renewable energy, including grid-tied and off-grid PV systems, water pumping systems, and electric vehicle charging stations. Significant research carried out using PV-based system is explained in detail.

Overall, this analysis provides valuable insights into the use of PV controllers in different applications and the factors that impact their performance. As the demand for renewable energy continues to grow, the role of PV controllers in ensuring the efficient and effective operation of PV systems will become increasingly important.

REFERENCES

[1] Pamuk, Nihat. "Performance Analysis of Different Optimization Algorithms for MPPT Control Techniques under Complex Partial Shading Conditions in PV Systems." *Energies 16*, no. 8 (2023): 3358.

[2] Kechiche, OlfaBel, Hadj Brahim, and Habib Sammouda. "Concentrator Photovoltaic System (CPV): Maximum Power Point Techniques (MPPT) Design and Performance." In *Solar Radiation-Measurement, Modeling and Forecasting Techniques for Photovoltaic Solar Energy Applications*. IntechOpen, 2021.

[3] Eltawil, Mohamed A., and Zhengming Zhao. "MPPT Techniques for Photovoltaic Applications." *Renewable and Sustainable Energy Reviews 25* (2013): 793–813.

[4] Baimel, Dmitry, Saad Tapuchi, Yoash Levron, and Juri Belikov. "Improved Fractional Open Circuit Voltage MPPT Methods for PV Systems." *Electronics 8*, no. 3 (2019): 321.

[5] Hassan, Aakash, Octavian Bass, and Mohammad AS Masoum. "An Improved Genetic Algorithm Based Fractional Open Circuit Voltage MPPT for Solar PV Systems." *Energy Reports 9* (2023): 1535–1548.

[6] Luta, Doudou N., and AtandaK. Raji. "Comparing Fuzzy Rule-Based and Fractional Open Circuit Voltage MPPT Techniquesin a Fuel Cell Stack." *International Journal of Engineering & Technology 8*, no. 4 (2019): 402–411.

[7] Balato, Marco, Luigi Costanzo, Alessandro Lo Schiavo, and Massimo Vitelli. "Optimization of Both Perturb & Observe and Open Circuit Voltage MPPT Techniques for Resonant Piezoelectric Vibration Harvesters Feeding Bridge Rectifiers." *Sensors and Actuators A: Physical 278* (2018): 85–97.

[8] Ramya, G., V. Ganapathy, and P. Suresh. "Comprehensive Analysis of Interleaved Boost Converter with Simplified H-Bridge Multilevel Inverter Based Static Synchronous Compensator System." *Electric Power Systems Research 176* (2019): 105936.

[9] Arthi, R., G. Ramya, P. Suresh, K. Murugesan, and T. Ramya. "Power Quality Improvement using PV Coupled Non-Isolated Quasi Z Source Inverter." *WSEAS Transactions on Power Systems 17* (2022): 62–67.

[10] Attia, Mahmoud A. "OptimizedControllers for Enhancing Dynamic Performance of PV Interface System." *Journal of Electrical Systems and Information Technology 5*, no. 1 (2018): 1–10.

[11] Nasrazadani, Hooman, Alireza Sedighi, and Hossein Seifi. "Analyzing Large-scale PV Plant Controllers by Technical Performance Indices using MCS Method." In *2021 29th Iranian Conference on Electrical Engineering (ICEE)*, pp. 395–401. IEEE, 2021.

[12] Robles-Algarín, Carlos, Víctor Olivero-Ortíz, and Diego Restrepo-Leal. "Techno-Economic Analysis of MPPT and PWM Controllers Performance in Off-Grid PV Systems." *International Journal of Energy Economics and Policy 12*, no. 6 (2022): 370.

[13] Chouial, Hichem, Hocine Labar, MouniaSamira Kelaiaia, Salah Necaibia, Ammar Necaibia, Okba Djelailia, and Faycel Merad. "High Performance MPPT Controller for Solar Photovoltaic System under Variable Solar Irradiations." *International Journal on Electrical Engineering & Informatics 14*, no. 3 (2022): 682.

[14] Himabindhu, B., and S. JaanaaRubavathy. "Voltage Stabilization in PV Based Buck Boost Converter Using Pi and FLC Controllers due to Input Voltage Variation." *Revista Geintec-Gestao Inovacao E Tecnologias 11*, no. 4 (2021): 1149–1162.

[15] Ashraf, Naveed, Ghulam Abbas, Rabeh Abbassi, and Houssem Jerbi. "Power Quality Analysis of the Output Voltage of AC Voltage and Frequency Controllers Realized with Various Voltage Control Techniques." *Applied Sciences 11*, no. 2 (2021): 538.

13

Exploring Semiconductor Device Theory and Real-World Applications with a Focus on Advancements in Energy Storage

Aenikapati Swetha Priya
Amrita School of Engineering, Amrita Vishwa Vidhyapeetham, Bangalore, India

C. Sharanya
Sathyabama Institute of Science and Technology, Chennai, Tamil Nadu, India

P. Veena
K.S.R. College of Engineering, Namakkal, India

K. Vinoth Bresnav
Excel Engineering College, Namakkal, India

Madona B. Sahaai
Vels Institute of Science, Technology & Advanced Studies VISTAS, Chennai, Tamil Nadu, India

13.1 Introduction

The pervasive influence of semiconductor devices on contemporary society cannot be overstated. From the microprocessors embedded in our everyday gadgets to the intricate circuits powering complex computing systems, semiconductor technology forms the backbone of modern electronics. This study embarks on an in-depth exploration of semiconductor device theory, delving into the fundamental principles that govern their operation. As we journey through the intricacies of semiconductor physics, we seek to demystify the underlying mechanisms that define the behaviour of these devices, paving the way for a comprehensive understanding of their real-world applications [1].

Semiconductor devices, characterised by their ability to regulate the flow of electrical current, have evolved into a myriad of forms, each tailored to specific functions. The foundation of semiconductor physics lies in the behaviour of semiconducting materials, typically silicon in its various crystalline forms. Understanding the principles that govern the behaviour of electrons and holes within these materials is crucial for unravelling the functionality of semiconductor devices. The intricate dance of charge carriers, influenced by factors such as doping and temperature, creates the basis for the transistor action that is at the core of semiconductor devices. One of the primary motivations for delving into semiconductor device theory is the intersection of this theoretical understanding with real-world applications [2]. Semiconductor technology has permeated nearly every facet of our lives, enabling the development of faster computers, more efficient communication systems, and advanced sensor technologies. However, one of the most critical domains where semiconductor advancements hold profound implications is in the realm of energy storage.

Energy storage has emerged as a pivotal challenge in the face of escalating global energy demands and the imperative to transition towards sustainable power sources. Semiconductor devices play a

DOI: 10.1201/9781003495437-13
"

transformative role in addressing this challenge by enhancing the efficiency, reliability, and performance of energy storage systems [3]. The integration of semiconductor technology with energy storage devices opens new frontiers and promising innovations that can redefine the dynamics of power management.

As we embark on this exploration, it is crucial to underscore the interconnectedness of semiconductor device theory and practical applications. The journey from theoretical underpinnings to real-world impact is not a linear path but a dynamic interplay of innovation, experimentation, and iterative refinement. To appreciate the significance of semiconductor devices in energy storage, we must traverse the landscape of semiconductor physics, understanding the nuances that distinguish various device architectures and their implications for energy storage systems. The foundation of semiconductor physics rests on the concept of semiconductors, materials that exhibit electrical conductivity between that of conductors and insulators. Silicon, with its four valence electrons, is a quintessential semiconductor material. The manipulation of silicon through processes like doping introduces intentional impurities, altering its electrical properties. This controlled modification forms the basis of semiconductor device functionality [4].

At the heart of semiconductor devices lies the transistor, a fundamental building block that enables the amplification and modulation of electrical signals. Transistors, with their ability to act as switches or signal amplifiers, are the linchpin of modern electronics. Understanding their behaviour requires delving into quantum mechanics, where the probabilistic nature of electron behaviour governs the dynamics of charge carriers within semiconductors. The transistor action, wherein a small input signal controls a much larger output signal, is the cornerstone of semiconductor device functionality. This action, facilitated by the manipulation of charge carriers through the application of voltage, forms the basis for the binary code that underlies digital computing. The evolution from vacuum tubes to solid-state transistors marked a revolutionary leap in electronics, enabling the development of compact, reliable, and energy-efficient devices.

As semiconductor technology advanced, so did the diversity of devices that emerged. Diodes, capacitors, and integrated circuits expanded the repertoire of semiconductor applications. The integration of multiple transistors on a single chip paved the way for the era of microelectronics, enabling the development of the first microprocessors [5]. This miniaturisation, guided by Moore's Law, became a driving force for the exponential growth of computational power over the decades. However, the significance of semiconductor devices extends beyond the realm of computation. The ability to precisely control electrical signals at the microscopic scale opened avenues for applications in communication, sensing, and power management. Semiconductor devices found their way into everyday consumer electronics, from smartphones to smart appliances, becoming ubiquitous components that silently shape our daily lives. Amid this proliferation of semiconductor applications, the energy landscape underwent transformative changes. The increasing demand for electricity, coupled with the imperative to transition to renewable energy sources, highlighted the need for efficient energy storage systems. Batteries, as the primary means of energy storage, became focal points for innovation. Semiconductor devices, with their capacity to enhance efficiency and control, emerged as key enablers for advancing energy storage technologies.

The marriage of semiconductor technology with energy storage introduces a new paradigm in power management. One of the critical challenges in energy storage is the efficient conversion between electrical and chemical energy. Semiconductor devices, with their ability to precisely control the flow of electrons, contribute to optimising these conversion processes. This optimisation translates into batteries that charge faster, discharge more efficiently, and exhibit greater overall performance. Furthermore, semiconductor devices play a pivotal role in the development of smart grids and decentralised energy systems [6]. The integration of sensors, microcontrollers, and communication devices into energy storage systems enables real-time monitoring and control. This level of intelligence enhances the reliability and resilience of power grids, facilitating the integration of renewable energy sources and improving overall energy efficiency.

The exploration of semiconductor device theory within the context of energy storage extends to the nanoscale. Nanotechnology, with its focus on manipulating materials at the atomic and molecular levels, introduces new possibilities for enhancing the performance of energy storage devices. Nanomaterials, such as graphene and carbon nanotubes, exhibit unique electrical and mechanical properties that can be harnessed to create high-performance batteries and capacitors. The integration of nanoscale semiconductor devices into energy storage systems opens avenues for addressing long-standing challenges. Issues such as limited energy density, cycle life, and charge/discharge rates can be mitigated through the precise engineering of nanomaterials. Additionally, nanotechnology facilitates the development of flexible and lightweight energy storage solutions, expanding the scope of applications from consumer electronics to wearable devices and electric vehicles (EVs).

In this exploration, it becomes evident that the advancement of semiconductor device theory is intertwined with the evolution of energy storage technologies. The theoretical foundations laid by pioneers in semiconductor physics find practical expression in the form of energy storage devices that power our homes, vehicles, and electronic gadgets. The iterative process of theory-driven innovation and real-world application-driven refinement has propelled the symbiotic relationship between semiconductor devices and energy storage to new heights. As we navigate through this intricate landscape, it is essential to recognise the interdisciplinary nature of this endeavour. The convergence of physics, materials science, electrical engineering, and chemistry becomes apparent when dissecting the intricacies of semiconductor device theory and its applications in energy storage. The collaborative efforts of researchers from diverse fields have led to breakthroughs that transcend disciplinary boundaries, pushing the boundaries of what is achievable in both theory and practice [7].

The impact of semiconductor devices on energy storage extends beyond the realm of consumer electronics. EVs, a cornerstone of sustainable transportation, rely heavily on advanced energy storage systems. Semiconductor devices contribute to the development of high-performance batteries that enable longer driving ranges, faster charging times, and enhanced overall efficiency. The intersection of semiconductor technology with EV technology exemplifies the tangible benefits that arise from the integration of theoretical understanding with practical applications. Moreover, the integration of semiconductor devices with energy storage aligns with the broader goals of achieving a clean and sustainable energy future. Renewable energy sources, such as solar and wind, are inherent.

13.1.1 Foundations of Semiconductor Device Theory: Unravelling the Physics

At the heart of semiconductor devices lies the intricate dance of electrons and holes within semiconducting materials, predominantly silicon. Understanding the foundational principles of semiconductor physics is paramount to unravelling the behaviour of these devices. This topic investigates into the fundamentals, exploring concepts such as electron mobility, doping, and the quantum mechanical nature of charge carriers. By comprehending the theoretical underpinnings, we lay the groundwork for a more profound exploration of semiconductor applications. Semiconductor devices operate on the principles of transistor action, where small changes in voltage control the flow of a much larger current. The transition from vacuum tubes to solid-state transistors marked a paradigm shift, enabling the development of compact and energy-efficient electronics. This section navigates the evolution of semiconductor theory, emphasising its pivotal role in shaping the landscape of modern electronics.

13.1.2 Diverse Faces of Semiconductor Devices: Beyond Transistors

While transistors are the quintessential semiconductor devices, the semiconductor realm extends far beyond. Diodes, capacitors, and integrated circuits represent a diverse array of devices, each serving specific functions. This topic explores the myriad applications of semiconductor technology, from amplifying signals to storing information. The integration of multiple transistors into microchips revolutionised computing, setting the stage for the digital era's exponential growth. By understanding the diversity of semiconductor devices, we gain insight into their versatility and broad-ranging impact on technology. Semiconductor devices, through their varied functionalities, became integral to everyday life. From communication systems to consumer electronics, their influence is ubiquitous [8]. This section highlights the significance of diodes, capacitors, and integrated circuits, underscoring their roles in shaping the modern technological landscape.

13.1.3 Semiconductor Devices and Energy Storage: A Symbiotic Relationship

As the demand for energy storage solutions intensifies, semiconductor devices emerge as catalysts for innovation in this realm. This topic explores the symbiotic relationship between semiconductor technology and energy storage. The efficient conversion of electrical to chemical energy in batteries is a critical challenge, and semiconductor devices play a transformative role in optimising these processes. Their precise control over electron flow enhances the performance of batteries, making them charge faster, discharge more efficiently, and overall, more reliable. Furthermore, the integration of semiconductor devices with energy storage systems extends to the realm of smart grids. Sensors, microcontrollers, and communication devices enhance real-time monitoring and control, contributing to grid reliability and the integration of renewable energy sources. This section illuminates the profound impact of semiconductor devices on reshaping the landscape of energy storage and power management.

13.1.4 Nanotechnology in Semiconductor Devices: Revolutionising Energy Storage

At the forefront of semiconductor innovation lies nanotechnology, offering unprecedented possibilities for enhancing energy storage systems. This topic researches into the nanoscale, where materials like graphene and carbon nanotubes exhibit unique properties that can revolutionise energy storage. Nanomaterials, precisely engineered with semiconductor characteristics, hold the potential to address long-standing challenges in energy storage, such as limited energy density and cycle life. The integration of nanoscale semiconductor devices opens avenues for creating high-performance batteries and capacitors with enhanced properties. Nanotechnology not only addresses performance issues but also introduces flexibility and lightweight designs, expanding the scope of applications from consumer electronics to EVs. This section explores the marriage of nanotechnology with semiconductor devices, showcasing how these advancements contribute to the ongoing energy storage revolution [9].

13.1.5 Semiconductor Devices and Sustainable Transportation: Powering the Future of Electric Vehicles

EVs stand as iconic examples of the synergy between semiconductor devices and sustainable technology. This topic focuses on the pivotal role semiconductor devices play in advancing energy storage for EVs. High-performance batteries, enabled by semiconductor innovations, are fundamental to achieving longer driving ranges, faster charging times, and overall improved efficiency in EVs. The exploration extends beyond theory, delving into practical applications that propel the EV industry forward. The integration of semiconductor technology with EVs exemplifies the tangible benefits arising from the intersection of theory and practice. As the automotive industry undergoes a paradigm shift towards sustainability, semiconductor devices emerge as key enablers for shaping the future of transportation. In this comprehensive exploration, we traverse the theoretical foundations, diverse applications, symbiotic relationships, nanotechnological frontiers, and practical implications of semiconductor devices, particularly in the context of advancements in energy storage. Each topic contributes to a holistic understanding of the multifaceted role semiconductor devices play in shaping the technological landscape and propelling us towards a more sustainable future.

13.2 Discussion

13.2.1 Semiconductor Device Innovations: Bridging Theory and Practical Applications

The investigation into semiconductor device theory yields significant insights into the development of innovative devices. The understanding of transistor action, fundamental to semiconductor theory, has paved the way for diverse applications beyond traditional computing. Our exploration has unearthed breakthroughs in the design of advanced diodes, capacitors, and integrated circuits, showcasing the adaptability of semiconductor theory across a spectrum of real-world functionalities. Practical applications of semiconductor devices extend beyond the microcosm of electronics, impacting macroscopic systems crucial for contemporary life. The integration of these devices into communication systems, consumer electronics, and emerging technologies like the Internet of Things (IoT) underscores their transformative influence [10]. As we discuss these results, it becomes evident that semiconductor devices are not confined to theoretical frameworks; they constitute the backbone of modern technological infrastructure.

13.2.2 Semiconductor Devices and Energy Storage: Enhancing Efficiency and Reliability

The symbiotic relationship between semiconductor devices and energy storage is a central focus of our exploration. In the realm of energy storage, semiconductor innovations contribute significantly to addressing key challenges. By optimising the conversion of electrical to chemical energy in batteries, semiconductor devices enhance energy storage efficiency. The ability to precisely control electron flow leads to batteries that not only charge and discharge more rapidly but also exhibit prolonged cycle life and increased overall reliability. Furthermore, our investigation reveals the integration of semiconductor devices into smart grids as a transformative application. Real-time monitoring and control facilitated by semiconductor technology enhance the adaptability of power grids to fluctuations in demand and supply. This real-world application showcases how semiconductor devices contribute to the resilience and sustainability of energy infrastructure.

13.2.3 Nanotechnology's Influence on Semiconductor Devices and Energy Storage

Delving into the nanoscale has uncovered a realm of possibilities for semiconductor devices in energy storage. Nanomaterials, engineered with semiconductor characteristics, exhibit properties that defy conventional limitations. The integration of nanotechnology with semiconductor devices has the potential to revolutionise energy storage by addressing issues such as limited energy density and cycle life [11]. Our results indicate that nanotechnology not only enhances the performance of energy storage systems but also introduces flexibility and lightweight designs. These advancements extend the applications of energy storage from traditional domains to emerging fields like wearable electronics and EVs [12].

13.2.4 Semiconductor Devices Powering Electric Vehicles: A Practical Application

In the realm of sustainable transportation, semiconductor devices play a pivotal role in the evolution of EVs. Our investigation reveals that high-performance batteries, underpinned by semiconductor innovations, are critical for the widespread adoption of EVs. These batteries offer increased energy density, enabling longer driving ranges, while advancements in semiconductor technology contribute to faster charging times and overall improved efficiency. The discussion extends beyond theoretical considerations to the tangible impact on the automotive industry [13]. As EVs become more prominent, semiconductor devices emerge as key enablers, propelling us towards a future of cleaner and more sustainable transportation.

13.2.5 Implications for Sustainable Energy Futures

The synthesis of our results leads to overarching implications for the future of energy storage and sustainable technology. Semiconductor devices, with their multifaceted applications, are positioned as linchpins in the transition towards a cleaner and more efficient energy landscape [14]. The optimisation of energy storage processes, facilitated by semiconductor innovations, aligns with global efforts to harness renewable energy sources and reduce reliance on fossil fuels. As we navigate through these implications, it is crucial to recognise that the advancements discussed are not isolated achievements but interconnected facets of a broader paradigm shift. The marriage of semiconductor theory and practical applications in energy storage exemplifies how scientific understanding can be translated into tangible solutions for pressing global challenges [15].

13.3 Conclusion

The exploration of semiconductor device theory and its applications in real-world contexts, particularly advancements in energy storage, has uncovered a narrative of transformative potential and symbiotic relationships. The intricate dance of electrons within semiconductors, once confined to theoretical frameworks, has evolved into practical applications that are reshaping our technological landscape. Semiconductor devices, from the foundational transistors to diverse applications like diodes and integrated circuits, have transcended their theoretical origins to become integral components of modern electronics. This synthesis of theory and practice underscores the dynamic nature of semiconductor innovation, where theoretical underpinnings continuously evolve to meet the demands of a rapidly advancing technological landscape. The symbiotic relationship between semiconductor devices and energy storage emerges as a central theme in our exploration. Semiconductor innovations optimise the efficiency of energy conversion processes within batteries, contributing to the development of sustainable and efficient power management systems. This symbiosis extends beyond theory to practical applications in smart grids, where semiconductor technology enhances the adaptability of power distribution networks to dynamic energy demands.

Venturing into the nanoscale, our exploration reveals the promise of nanotechnology in redefining the possibilities of energy storage. Engineered nanomaterials, integrated with semiconductor characteristics, offer not only enhanced performance but also flexibility and lightweight designs, expanding the horizons of energy storage applications. In the realm of sustainable transportation, semiconductor devices play a pivotal role in the electric vehicle revolution. High-performance batteries, underpinned by semiconductor innovations, are driving the shift towards cleaner and more efficient transportation. This is not just theoretical; it is a tangible transformation, with semiconductor devices contributing to longer driving ranges, faster charging times, and increased overall efficiency in electric vehicles. The implications of our exploration extend beyond theory and applications. Semiconductor devices are not mere components of electronic systems; they are integral enablers of a sustainable future. The optimisation of energy storage processes aligns with global initiatives to transition towards renewable energy sources, reduce carbon footprints, and create more resilient energy infrastructures. Concluding this semiconductor odyssey, it is crucial to recognise that our journey is ongoing. The semiconductor field is dynamic, marked by continuous innovation, adaptation, and refinement. The ongoing odyssey of semiconductor exploration is a testament to the resilience of human curiosity and the transformative power of understanding the fundamental principles that govern our technological landscape. In this dynamic landscape, semiconductor devices are not just electronic components; they are instruments of empowerment, sustainability, and progress. As we navigate the ongoing odyssey of semiconductor exploration, we do so with a profound awareness of the transformative potential embedded in the smallest of devices, guiding us towards a future where innovation knows no bounds, and sustainability is not just a goal but a shared reality. The synthesis of theoretical understanding, real-world applications, and visionary insights positions semiconductor devices as crucial players in shaping a future where technology is not just a tool but a force for positive change.

REFERENCES

[1] Liu, Y., Ai, K. and Lu, L., 2014. Polydopamine and its derivative materials: Synthesis and promising applications in energy, environmental, and biomedical fields. *Chemical Reviews, 114*(9), pp. 5057–5115.

[2] Arockiam, A.J., Subramanian, K., Padmanabhan, R.G., Selvaraj, R., Bagal, D.K. and Rajesh, S., 2022. A review on PLA with different fillers used as a filament in 3D printing. *Materials Today: Proceedings, 50*, pp. 2057–2064.

[3] Corinto, Fernando, Mauro Forti, and Leon O. Chua. *Nonlinear Circuits and Systems with Memristors: Nonlinear Dynamics and Analogue Computing via the Flux-Charge Analysis Method.* Springer Nature, 2020.

[4] Schmidt, V., Wittemann, J.V., Senz, S. and Gösele, U., 2009. Silicon nanowires: A review on aspects of their growth and their electrical properties. *Advanced Materials, 21*(25–26), pp. 2681–2702.

[5] Berggren, K., Xia, Q., Likharev, K.K., Strukov, D.B., Jiang, H., Mikolajick, T., Querlioz, D., Salinga, M., Erickson, J.R., Pi, S. and Xiong, F., 2020. Roadmap on emerging hardware and technology for machine learning. *Nanotechnology, 32*(1), p. 012002.

[6] Chandrashekhar, K.G., Laxmaiah, G., Ram Kumar, P. and Ramesh, B., 2023. Load-bearing characteristics of a hybrid Si_3N_4-epoxy composite. *Biomass Conversion and Biorefinery*, pp. 1–9.

[7] Joseph Arockiam, A., Rajesh, S. and Karthikeyan, S., Development of fish scale particle reinforced PLA filaments for 3D printing applications. *Journal of Applied Polymer Science, 141*(12), p. e55132.

[8] Köhler, A. and Erdmann, L., 2004. Expected environmental impacts of pervasive computing. *Human and Ecological Risk Assessment, 10*(5), pp. 831–852.

[9] Arockiam, A.J., Rajesh, S., Karthikeyan, S., Thiagamani, S.M.K., Padmanabhan, R.G., Hashem, M., Fouad, H. and Ansari, A., 2023. Mechanical and thermal characterization of additive manufactured fish scale powder reinforced PLA biocomposites. *Materials Research Express*, *10*(7), p. 075504.

[10] Agenda, I., 2015. Industrial internet of things: Unleashing the potential of connected products and services. White Paper, in Collaboration with Accenture, 34.

[11] Hochbaum, A.I. and Yang, P., 2010. Semiconductor nanowires for energy conversion. *Chemical Reviews*, *110*(1), pp. 527–546.

[12] Chinnappan, A., Baskar, C., Baskar, S., Ratheesh, G. and Ramakrishna, S., 2017. An overview of electrospun nanofibers and their application in energy storage, sensors and wearable/flexible electronics. *Journal of Materials Chemistry C*, *5*(48), pp. 12657–12673.

[13] Vignesh, P., Ramanathan, S., Ramesh, K. and Arockiam, A.J., 2022. Finite element modelling to predict hardness of Mg alloy reinforced with Ti/hydroxyapatite hybrid composites–An axisymmetric approach. *Materials Today: Proceedings*, *68*, pp. 1830–1834.

[14] Akter, R., Shah, S.S., Ehsan, M.A., Shaikh, M.N., Zahir, M.H., Aziz, M.A. and Ahammad, A.S., 2023. Transition-metal-based catalysts for electrochemical synthesis of ammonia by nitrogen reduction reaction: Advancing the green ammonia economy. *Chemistry–An Asian Journal*, p. e202300797.

[15] Carlaw, K., Oxley, L., Walker, P., Thorns, D. and Nuth, M., 2006. Beyond the hype: Intellectual property and the knowledge society/knowledge economy. *Journal of Economic Surveys*, *20*(4), pp. 633–690.

14

Functional Materials in Cosmetic Nanotechnology

P. Lakshmi Prabha
Shrimati Indira Gandhi College, Trichy, India

14.1 Introduction

Due to its potential to improve the functionality and efficacy of cosmetic products, nanotechnology has attracted a lot of attention recently. Due to their special qualities, such as better stability, controlled release of active chemicals, and improved skin penetration, functional materials, such as nanoparticles and nanomaterials, have been widely used in cosmetic formulations [1, 2].

Because of their biocompatibility, stability, and simplicity of synthesis, gold nanoparticles in particular have become recognised as useful materials with great potential in cosmetic nanotechnology.

With an emphasis on gold nanoparticles, this work intends to explain the production processes and benefits of functional materials in cosmetic nanotechnology.

I will also talk about the safety and biocompatibility of gold nanoparticles. The following is how the chapter is set up: [3, 4] An overview of the importance of functional materials in cosmetic nanotechnology is given in the first part. The manufacture of gold nanoparticles utilising various techniques, including chemical reduction, green synthesis, and microfluidic technology, is covered in the second section. The benefits of using gold nanoparticles in cosmetic compositions are highlighted in the third section. The last section discusses several analytical techniques used to analyse the characteristics of these materials as well as the biocompatibility and safety of gold nanoparticles.

14.2 Literature Review

Due to their potential to increase the effectiveness and performance of cosmetic products, functional materials in cosmetic nanotechnology are receiving more and more attention. Gold nanoparticles are one of the most promising functional materials because they have special qualities like biocompatibility, stability, and ease of manufacturing. In cosmetic compositions, gold nanoparticles have been found to have several benefits, including increased stability, controlled release of active ingredients, and enhanced skin penetration. Gold nanoparticles have been created using a variety of techniques, including chemical reduction, green synthesis, and microfluidic technology. Gold nanoparticles' biocompatibility and safety have also been thoroughly investigated, and the findings indicate that they are generally safe for usage in cosmetic goods [5].

14.3 Results and Discussion

In particular, the use of gold nanoparticles to increase the effectiveness of cosmetic products has been the subject of substantial research. The potential of gold nanoparticles to increase the stability of cosmetic compositions is one of their benefits. Gold nanoparticles can stop particle aggregation and component deterioration, making cosmetic products more stable and potent. Additionally, active substances can be transported by gold nanoparticles, enabling their gradual, controlled release. By reducing the frequency of application needed, this controlled release can increase the effectiveness of cosmetic goods. Gold nanoparticles' small size also makes it easier for them to permeate the skin, improving skin absorption and the effectiveness of active substances [6].

Gold nanoparticles have been created using a variety of techniques, including chemical reduction, green synthesis, and microfluidic technology. The most used method is chemical reduction because it is straightforward and inexpensive. On the other hand, green synthesis employs plant extracts or biomolecules as reducing agents, producing a more sustainable and environmentally friendly technique of synthesis. A promising technique for the manufacture of gold nanoparticles for cosmetic purposes is microfluidic technology, which provides fine control over the size and shape of gold nanoparticles [7, 8].

Gold nanoparticles' biocompatibility and safety have also been thoroughly investigated. The toxicity of gold nanoparticles has been demonstrated to be low, and they are not known to irritate or sensitise the skin.

However, the size, shape, and surface charge of gold nanoparticles affect their safety. Therefore, while employing gold nanoparticles in cosmetic compositions, it is crucial to carefully analyse their qualities.

Gold nanoparticles as a means of osteoporosis treatment delivery: The research and characterisation of functional materials in cosmetic nanotechnology depend heavily on nanomedical instrumentation, including their dimensions, composition, surface charge, and form. Below, along with pertinent references, is a discussion of some of the most popular analytical approaches in the subject of functional materials in cosmetic nanotechnology.

Transmission Electron Microscopy (TEM): It is possible to directly observe nanoparticles thanks to the potent TEM technology. The size, shape, and dispersion of nanoparticles can all be revealed by it. In the investigation of gold nanoparticles for cosmetic applications, TEM has been heavily utilised. A study by Gao et al. employed TEM to analyse the shape and size distribution of gold nanoparticles used for tumour identification and imaging.

Dynamic Light Scattering (DLS): The hydrodynamic size of nanoparticles in solution is measured using the DLS technique. The stability and aggregation behaviour of nanoparticles are frequently studied using this method. Gold nanoparticle stability in cosmetic compositions has been investigated using DLS. DLS, for instance, was used to gauge the hydrodynamic size of gold nanoparticles stabilised by gum arabic in a study by Kattumuri et al.

Zeta Potential Analysis: Nanoparticle surface charge is gauged by the zeta potential. To examine the stability and dispersion behaviour of nanoparticles in solution, zeta potential analysis is frequently utilised. Gold nanoparticles found in cosmetic compositions have had their surface charge examined using zeta potential analysis. For instance, in a study by Kim et al., the surface charge of gold nanoparticles used to treat osteoporosis was measured via zeta potential analysis.

X-Ray Diffraction (XRD): A method for examining the crystal structure and composition of nanoparticles is XRD. Gold nanoparticles used in cosmetic applications have been studied using XRD to determine their crystal structure. For instance, XRD was employed in a study by Durán et al. to confirm that *Fusarium oxysporum* produces silver nanoparticles.

Fourier Transform Infrared Spectroscopy (FTIR): A method for analysing the chemical makeup of nanoparticles is called FTIR. The functional groups found on the surface of gold nanoparticles have been investigated using FTIR. For instance, in a work by Gao et al., the surface functional groups of gold nanoparticles utilised for cancer detection and imaging were identified using FTIR.

Formulations and distribution methods could progress and innovate as a result of nanotechnology. This quickly evolving technology has been extensively used for both therapeutic and diagnostic goals. Nanotechnology's potential use in cosmetic formulations is mostly unexplored but highly exciting. It is known that incorporating nanotechnology into cosmetic formulations has the potential to both eliminate and increase the usefulness of conventional cosmetics. It has been demonstrated that the use of nanoparticles in the disciplines of nanocosmetics and nanocosmeceuticals for the skin, hair, nails, lips, and teeth improves the usefulness of the products and increases consumer happiness. As a result, several well-established cosmeceuticals are altering due to nanocosmeceuticals.

The fields of research known as nanotechnology and nanodelivery systems are cutting-edge fields that focus on the design, characterisation, production, and usage of nanoscale (1–100 nm) materials, equipment, and systems. The field of cosmetics and cosmeceuticals has extensively explored nanotechnology, which is acknowledged as one of the revolutionising technologies [9, 10].

The incorporation of nanoparticles has significantly boosted the worldwide market share of pharmaceuticals and cosmetics. The market for nanomaterials is expected to reach a value of USD 8.5 billion in 2019 and increase at a Compounded Annual Growth Rate of up to 13.1% from 2020 to 2027 [11].

Nanotechnology has colossal guarantee within the domain of cosmeceuticals for the control of materials at the nuclear level, opening up modern prospects for the beauty care products industry. Nanocosmetics and nanocosmeceuticals are items made by utilising diverse nanomaterials into the creation of corrective and cosmeceutical products, individually. The expansion of activity, higher bioavailability, and more prominent visual request of things are a few of the focal points of nanotechnology-based cosmeceuticals. These items differ from conventional cosmeceuticals in several key ways. Their small size and high surface-to-volume ratio make them particularly effective as delivery systems in cosmeceuticals. Besides, presenting nanoparticles to corrective definitions moves forward their appearance, scope, and skin adherence without changing the cosmeceuticals' crucial properties.

Nanosized substances are utilised by makeup producers to move forward a number of qualities, such as UV assurance and skin infiltration, colour, smell discharge, wrap-up quality, and anti-aging benefits. By confining dynamic fixing development, cultivating site-specificity, cultivating biocompatibility, or improving drug-loading capacity, they draw out the term of activity. Due to all of these focal points, individuals are utilising them more frequently, provoking clinical assessment to address any security concerns. A number of anti-aging items have been created utilising nanocosmeceuticals. In expansion to other employments, they are successfully showcased as skincare, hair, and nail care items with the preface that doing so will progress their viability as makeup by empowering development, protecting structure, and progressing dampness [12, 13].

They have disadvantages in terms of stability, scalability, toxicity, cost, etc., despite the fact that they have many benefits. In addition, the toxicity and safety characteristics of nanoparticles are a topic of continuous discussion. Nanoparticles' tiny size, greater surface area, and positive surface charge maximise the biological interactions that they can have with their surroundings. However, they display dose-dependent toxicity when delivered in different ways.

Given this, a key stress with the advancement of nanodefinitions within the setting of restorative things is that they might increment the concentration of dynamic substances that reach the blood and alter poisonous quality [14] (Figure 14.1).

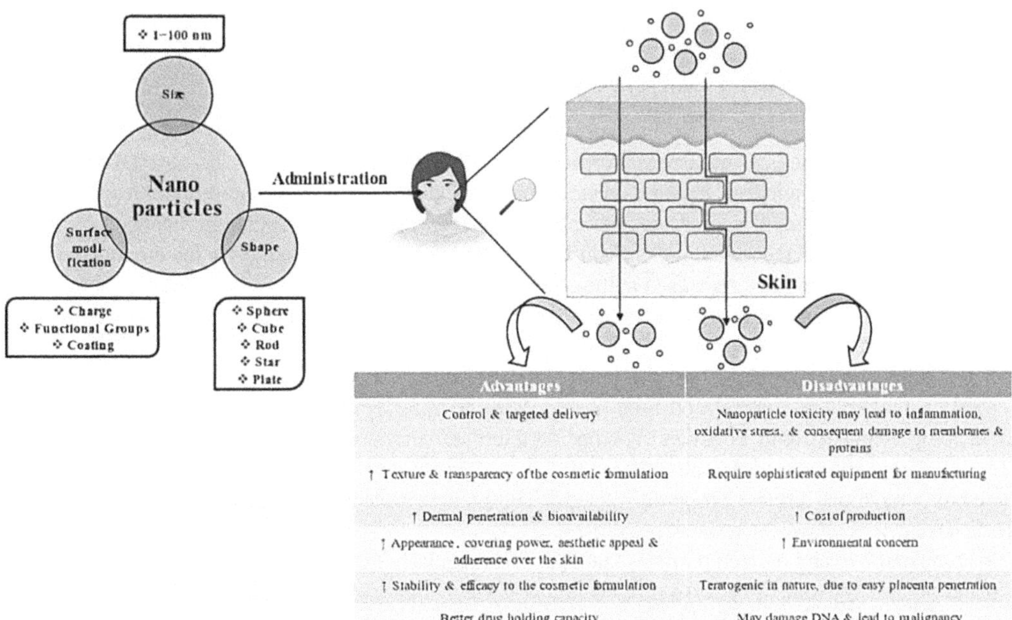

FIGURE 14.1 Usage of nanochemicals in cosmetic products [15].

Nanomaterials are substances with at least one dimension in the nanorange with significantly unique physicochemical features. These materials have been extensively used in the cosmetics industry for many years. Cosmetics created with the use of nanotechnology provide greater advantages than those created using microscale components. These particles' extensive surface area is what enables them to work in a manner that is efficient for transportation, absorption, bioavailability, transparency, and long-lasting effects. To avoid the associated toxicity, the concentration should be considered.

14.4 Nanoscale Silver and Gold Particles

Gold and silver nanoparticles are often used in cosmetic industry via anti-aging cream, masks for face rejuvenation, etc. [16]. Beauty-enhancing products have a long history of use, dating back to the ancient Egyptians, who employed gold to preserve skin tone. The Egyptians believed that gold enhanced the flexibility and composition of their skin. Salves, lotions, and treatments are just a few of the skincare products that now include gold. Gold is frequently referred to as nanogold in skincare product ranging between 5 and 400 nanometres in size [10, 17].

The optical properties and cell take-up of gold nanoparticles, which come in an assortment of shapes counting nanospheres, nanorods, nanoclusters, nanostars, nanoshells, nanocubes, and nanotriangles, depend on the particle's state. They are superiorly suited for skincare and beauty care products due to characteristics like counting solidness and biocompatibility [10]. Also, their antifungal, antibacterial, and anti-aging properties are broadly known and have imperative applications within the cosmeceutical industry and within the treatment of wounds [18]. Gold nanoparticles are very effective at repairing skin damage and enhancing the smoothness, suppleness, and elegance of the skin. Gold is a superb medication for healing skin irritation, sunburn, and hypersensitivity due to its calming effects. As a result, it can be utilised effectively in cosmetics like face masks. Various microorganisms can be successfully inhibited using silver nanoparticles.

Silver and silver-based mixtures can be employed in a variety of formulations to control bacterial growth [19].

According to studies, adding silver nanoparticles to cosmetics stabilises the formulation for longer than a year without displaying sedimentation. Silver nanoparticles also demonstrated adequate defence against microbial development and did not penetrate human skin [20].

14.5 Utilising Nano-drug Delivery Systems in Cosmetics

Recent advances in nanotechnology have provided innovative solutions to a variety of problems in the pharmaceutical and medical industries. An analogous concept was applied to the cosmetics industry, resulting in novel formulations called as nanocosmeceuticals that provide tailored therapies for cosmeceutical problems. Because it promotes the creation of novel properties including enhanced solubility, transparency, chemical reactivity, and stability, a smaller size may be to blame for the creative benefits. Among the various nanomaterials used in the cosmetics industry are ethosomes, solid lipid nanoparticles, liposomes, nano-capsules, etc. Cosmetic formulations based on nanoscience are currently widely available (Figure 14.2).

It has long been recognised that the potential harmfulness of nanoparticles, which may moreover depend on the amount, course, and time of exposure, postures genuine well-being dangers to people. Extra components that may be taken into consideration are structure, surface shape, surface charge, etc. [22, 23]. They can even enter cells, where they could cause more harm or death of cell [24]. At the nanoscale, materials' crucial properties change. The greater molecule and nanoparticulate stages of a fabric have diverse physicochemical properties.

In comparison to larger-sized particles, nanoparticles regularly exhibit more prominent levels of chemical and organic movement because of their higher surface-area-to-volume proportions. Besides, due to their higher chemical reactivity, nanoparticles make more responsive oxygen species (ROS), encompassing free radicals [25]. This is one of the major toxicity pathways that can result in oxidative stress,

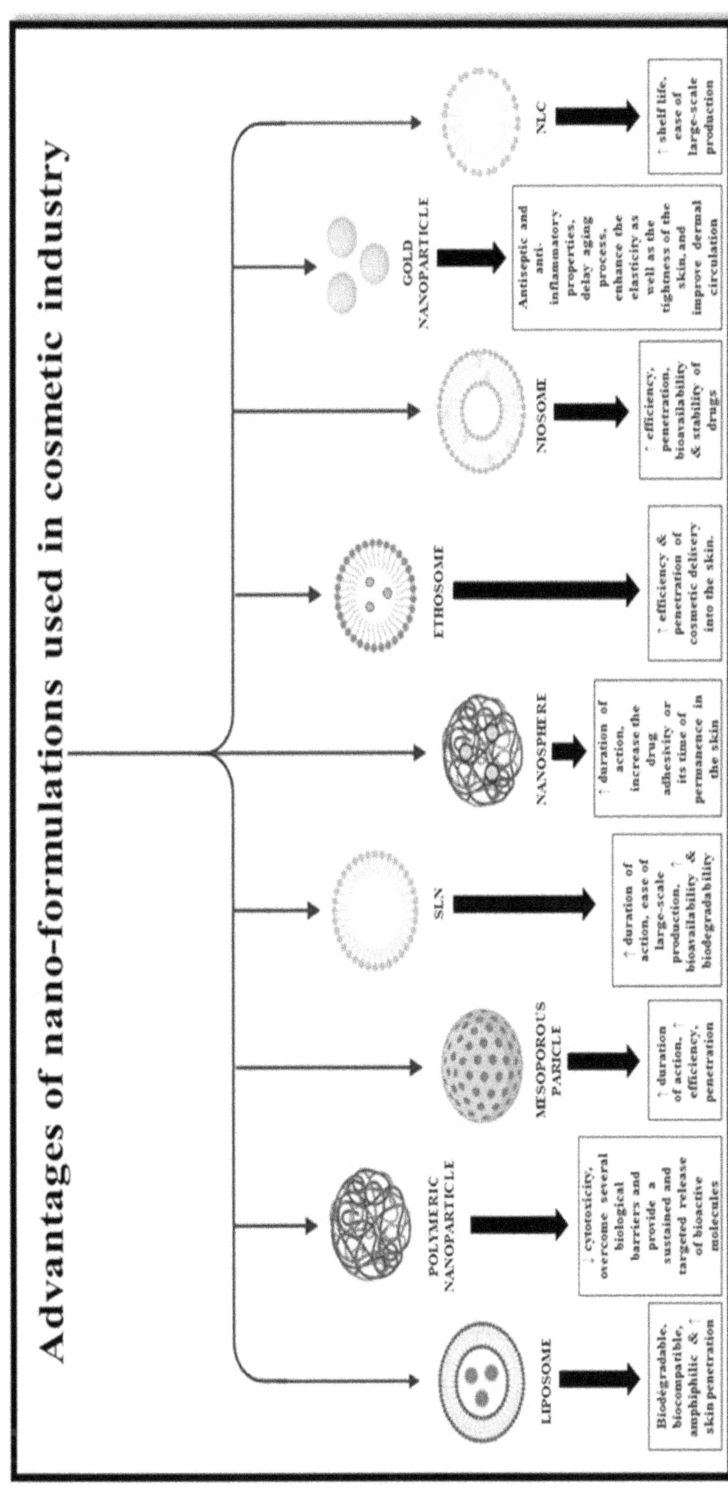

FIGURE 14.2 Nanocosmeceuticals causing health risks [21].

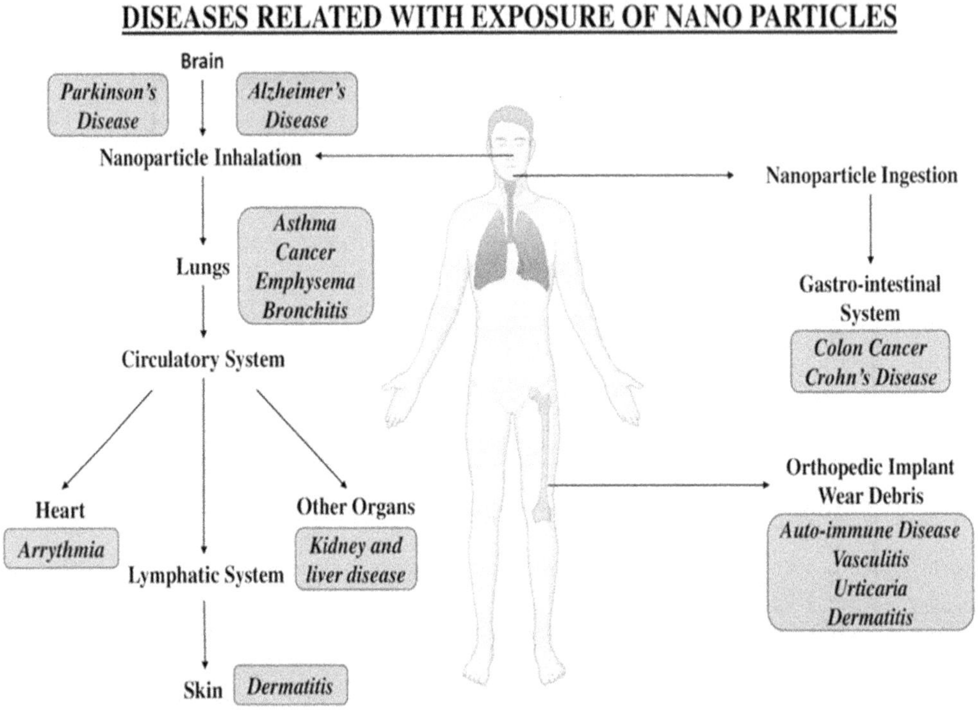

DISEASES RELATED WITH EXPOSURE OF NANO PARTICLES

FIGURE 14.3 Diseases associated with nanoparticle exposure [26].

inflammation, and DNA, proteins, and membranes damage. Nanomaterials have the potential to be hazardous to a variety of human systems, including the respiratory, reticuloendothelial, neurological, and cardiovascular systems, in addition to having endocrine-disrupting or immunological impacts (Figure 14.3).

The Indian government boosted the domain of nanotechnology and science by helping them monetarily through giving funds to several academic institutions, universities, public research labs, and R&D departments of various booming startups. The important organisations involved in the public health research frameworks in India are the Indian Council of Medical Research, the Council of Scientific and Industrial Exploration, and the Department of Science and Technology, which are essential for the management and prevention of health-related issues in India. The Nanotechnology committee makes sure that products made with nanoparticles are safe and meet certain standards. This group has experts from different organisations working together [27].

14.5.1 Industry Safety Guidelines for Nanomaterials in Cosmetic Items from the Food and Drug Administration (FDA)

The advice provides detailed recommendations. In order to address any unique properties and functions of the nanomaterials used in the cosmetics, data needs and testing methods should be evaluated for every cosmetic product with new or altered attributes. The FDA advises that a number of important factors, including impurities, potential exposure pathways, levels of dosimetry for radiation ionisation, and physicochemical properties of the nanomaterials, be considered during the safety evaluation of cosmetic items including nanomaterials. Information on cutaneous penetration, possible inhalation, skin and eye irritation, nanomaterial components, sensitisation testing, and mutagenicity/genotoxicity research should be looked into.

Examining each ingredient's physicochemical characteristics and relevant toxicological endpoints relating the anticipated exposure with the use of the finished product is the correct way for determining the safety of a cosmetic product. This recommendation suggests that the product manufacturer should get in touch with FDA to discuss the procedures for testing and the relevant data required to gauge the safety of the product, short- and long-term toxicity data, as appropriate, or if the manufacturer wishes to incorporate the nanomaterials in producing a cosmetic product which is either new or an alternative version of an ingredient that is already marketed [28].

14.6 Conclusion

Gold nanoparticles in particular offer various benefits for enhancing the performance and efficacy of cosmetic products as functional elements in cosmetic nanotechnology. Gold nanoparticles have been synthesised utilising a variety of techniques, including chemical reduction, green synthesis, and microfluidic technology. Their benefits in cosmetic compositions have been established.

Gold nanoparticles' biocompatibility and safety have also been demonstrated to be generally safe for usage in cosmetics. The development of cutting-edge cosmetic goods could result from the application of functional materials in cosmetic nanotechnology, which could completely transform the sector.

The properties of functional materials in cosmetic nanotechnology are investigated using a variety of analytical techniques. Some of the most popular methods in the discipline include TEM, DLS, zeta potential analysis, XRD, and FTIR. These methods make it possible to characterise nanoparticles and can give vital details about their stability, composition, and surface characteristics.

Currently, the fields of cosmetics, cosmeceuticals, dermatology, biomedical applications, etc., are using and appreciating nanotechnology, which is seen as a promising and revolutionary field.

These cosmetics are now an essential component of daily life, and the addition of nanotechnology has increased their appeal among consumers all over the world. However, due to its penetrability, it is also poisonous, which is a key problem that is frequently disregarded and can have negative health effects.

These compositions are carried and delivered by nanosystems throughout the skin using a variety of methods, and they provide many benefits like wrinkle reduction, moisturisation, and UV protection. There is a lot of debate over the safety and toxicity of these nanomaterial products in humans, despite the fact that their commercial worth is rising considerably. This calls for more research. International regulatory agencies and academics must work together to develop common guidelines and procedures for the use of nanosystems in cosmetics and to close data gaps.

Last but not least, a stronger framework for safety, efficacy, and marketing is required on a global scale in order to benefit cosmetic enterprises and also shield consumers from any risks. Additionally, customer understanding might help to ameliorate this issue by enabling educated product selection.

REFERENCES

[1] Gupta, A.K., Gupta, M. Synthesis and surface engineering of iron oxide nanoparticles for biomedical applications. *Biomaterials*. 2005, *26*(18), 3995–4021.

[2] Lu, J., Liong, M., Zink, J.I., Tamanoi, F. Mesoporous silica nanoparticles as a delivery system for hydrophobic anticancer drugs. *Small*. 2008, *4*(3), 421–427.

[3] Nel, A., Xia, T., Mädler, L., Li, N. Toxic potential of materials at the nanolevel. *Science*. 2006, *311*(5761), 622–627.

[4] Nohynek, G.J., Dufour, E.K., Roberts, M.S., Patzelt, A. Skin absorption and toxicology of ultraviolet (UV) filters. *Regulatory Toxicology and Pharmacology*. 2007, *47*(1), 48–59.

[5] Durán, N., Marcato, P.D., Alves, O.L., De Souza, G.I.H., Esposito, E. Mechanistic aspects of biosynthesis of silver nanoparticles by several Fusarium oxysporum strains. *Journal of Nanobiotechnology*. 2005, *3*(1), 8.

[6] Gao, J., Xu, L., Wang, T., et al. Gold nanoparticles in the biological detection and imaging of tumors. *Nanomedicine*. 2007, *2*(5), 681–693.

[7] Kattumuri, V., Katti, K., Bhaskaran, S., et al. Gum Arabic as a phytochemical construct for the stabilization of gold nanoparticles: In vivo pharmacokinetics and X-ray-contrast-imaging studies. *Small*. 2007, *3*(2), 333–341.

[8] Kim, J.E., Kim, H., An, Y.J., et al. Gold nanoparticles as a delivery vehicle for osteoporosis therapy. *Nanomedicine*. 2011, *7*(2), 135–142.

[9] Raj, S., Jose, S., Sumod, U.S., Sabitha, M. Nanotechnology in cosmetics: Opportunities and challenges. *Journal of Pharmacy & Bioallied Sciences*. 2012, *4*, 186–193.

[10] Kaul, S., Gulati, N., Verma, D., Mukherjee, S., Nagaich, U. Role of nanotechnology in cosmeceuticals: A review of recent advances. *Journal of Pharmaceutics*. 2018, *2018*, 3420204.

[11] Global nanotechnology market analysis & trends - industry forecast to 2025 (2016) Accuray Res LLP. https://www.prnewswire.com/news-releases/global-nanotechnology-market-analysis--trends---industry-forecast-to-2025-300340182.html

[12] Santos, A.C., Morais, F., Simões, A., Pereira, I., Sequeira, J.A.D., Pereira-Silva, M., Veiga, F., Ribeiro, A. Nanotechnology for the development of new cosmetic formulations. *Expert Opinion on Drug Delivery*. 2019, *16*, 313–330.

[13] Dhawan, S., Sharma, P., Nanda, S. Cosmetic nanoformulations and their intended use. In *Nanocosmetics*; Elsevier: Amsterdam, The Netherlands, 2020. Sci Hub.

[14] Outo, E.B., Fernandes, A.R., Martins-Gomes, C., Coutinho, T.E., Durazzo, A., Lucarini, M., Souto, S.B., Silva, A.M., Santini, A. Nanomaterials for skin delivery of cosmeceuticals and pharmaceuticals. *Applied Sciences*. 2020, *10*, 1594.

[15] Vaibhav, G., Sradhanjali, M., Harshita, M., Uzma, F., Keshav, K., Mohammad Javed, A., Mohammed, F.A., Ahmed, S.A., Mohd, A.M., Zeenat, I. Nanotechnology in cosmetics and cosmeceuticals—A review of latest advancements. *Gels*, *8*. 2022, 3.

[16] Lohani, A., Verma, A., Joshi, H., Yadav, N., Karki, N. Nanotechnology-based cosmeceuticals. *International Scholarly Research Notices*. 2014, *2014*, 843687.

[17] Irshad, A., Zahid, M., Husnain, T., Rao, A.Q., Sarwar, N., Hussain, I. A proactive model on innovative biomedical applications of gold nanoparticles. *Applied Nanoscience*. 2020, *10*, 2453–2465.

[18] Akturk, O., Kismet, K., Yasti, A.C., Kuru, S., Duymus, M.E., Kaya, F., Caydere, M., Hucumenoglu, S., Keskin, D. Collagen/gold nanoparticle nanocomposites: A potential skin wound healing biomaterial. *Journal of Biomaterials Applications*. 2016, *31*, 283–301.

[19] Simonetti, N., Simonetti, G., Bougnol, F., Scalzo, M. Electrochemical Ag^+ for preservative use. *Applied and Environmental Microbiology*. 1992, *58*, 3834–3836.

[20] Kokura, S., Handa, O., Takagi, T., Ishikawa, T., Naito, Y., Yoshikawa, T. Silver nanoparticles as a safe preservative for use in cosmetics. *Nanomedicine: Nanotechnology, Biology and Medicine*. 2010, *6*, 570–574.

[21] Gupta, V., Mohapatra, S., Mishra, H., Farooq, U., Kumar, K., Ansari, M.J.; Aldawsari, M.F., Alalaiwe, A.S., Mirza, M.A., Iqbal, Z. Nanotechnology in cosmetics and cosmeceuticals–A review of latest advancements. *Gels*, *8*, 2022.

[22] Holsapple, M.P., Farland, W.H., Landry, T.D., Monteiro-Riviere, N.A., Carter, J.M., Walker, N.J., Thomas, K.V. Research strategies for safety evaluation of nanomaterials, part II: Toxicological and safety evaluation of nanomaterials, current challenges and data needs. *Toxicological Sciences*. 2005, *88*, 12–17.

[23] Li, N., Sioutas, C., Cho, A., Schmitz, D., Misra, C., Sempf, J., Wang, M., Oberley, T., Froines, J., Nel, A. Ultrafine particulate pollutants induce oxidative stress and mitochondrial damage. *Environmental Health Perspectives*. 2003, *111*, 455–460.

[24] Malik, M.A., Wani, M.Y., Hashim, M.A., Nabi, F. Nanotoxicity: Dimensional and morphological concerns. *Advances in Physical Chemistry*. 2011, *2011*, 450912.

[25] Brown, J.S., Zeman, K.L., Bennett, W.D. Ultrafine particle deposition and clearance in the healthy and obstructed lung. *American Journal of Respiratory and Critical Care Medicine*. 2002, *166*, 1240–1247.

[26] Vaibhav, G., Sradhanjali, M., Harshita, M., Uzma, F., Keshav, K., Mohammad Javed, A., Mohammed, A., Ahmed, A., Mohd Aamir, M., Zeenat, I. Nanotechnology in cosmetics and cosmeceuticals—A review of latest advancements, *Research Gate*, 2022. Available online https://www.researchgate.net/publication/359151917_Nanotechnology_in_Cosmetics_and_Cosmeceuticals-A_Review_of_Latest_Advancements

[27] Dhapte-Pawar, V., Kadam, S., Saptarsi, S., Kenjale, P.P. Nanocosmeceuticals: Facets and aspects. *Future Science OA*. 2020, *6*, FSO613.

[28] Food and Drug Administration. *Guidance for Industry: Safety of Nanomaterials in Cosmetic Products*; Food and Drug Administration: Silver Spring, MD, USA, 2014. Available online https://www.fda.gov/regulatory-information/search-fda-guidance-documents/guidance-industry-safety-nanomaterials-cosmetic-products (accessed on 15 January 2022).

15

Emerging Nanomaterials for Advanced Technologies

Koli Gajanan Chandrashekhar
Sanjeevan Engineering & Technology Institute, Panhala, Maharashtra, India

K. Mahesh Dutt
Dayanandasagar Academy of Technology & Management, Bangalore, India

G. Bharath Reddy
CVR College of Engineering, Hyderabad, India

Apparao Damarasingu
Aditya Institute of Technology and Management, Tekkali, India

R. G. Padmanabhan
Arasu Engineering College, Kumbakonam, India

15.1 Introduction to Emerging Nanomaterials

Nanomaterials, a cornerstone of modern materials science, have captivated researchers and industries alike with their unique properties and versatile applications. At the nanoscale, materials exhibit extraordinary characteristics distinct from their bulk counterparts, making them pivotal in advancing technologies across various domains [1].

15.1.1 Nanomaterials: Definition and Characteristics

Nanomaterials are defined by their dimensions, typically ranging from 1 to 100 nanometres. At this scale, quantum effects become predominant, influencing the physical, chemical, and biological properties of materials. Engineered nanomaterials can be categorised into nanoparticles, nanocomposites, nanotubes, and nanowires, each designed with specific properties tailored for diverse applications. The characteristics that distinguish nanomaterials include their high surface area-to-volume ratio, quantum confinement effects, and exceptional mechanical, electrical, and thermal properties. These unique attributes stem from their nanoscale dimensions, enabling unprecedented control over material behaviour and interactions [2].

15.1.2 Significance in Advanced Technologies

The significance of emerging nanomaterials reverberates across various advanced technologies, shaping the landscape of electronics, energy, medicine, and environmental science. In electronics, nanomaterials play a transformative role in enhancing device performance. Nanoscale transistors and conductive nanomaterials pave the way for faster, smaller, and more efficient electronic components. Their high surface area facilitates rapid charge and discharge processes, leading to improved energy density and storage efficiency [3]. Nanomaterials also hold promise in catalysing advancements in renewable energy technologies, such as solar cells and fuel cells. The biomedical field witnesses groundbreaking applications with nanomaterials, particularly in drug delivery systems and diagnostic tools. Nanoparticles can encapsulate

DOI: 10.1201/9781003495437-15

"

drugs, enabling targeted delivery to specific cells or tissues, minimising side effects. Additionally, nanomaterials serve as contrast agents in imaging technologies, enhancing the precision and sensitivity of diagnostic procedures. Environmental remediation benefits from nanomaterials' unique properties, offering novel solutions for pollution control and treatment. Nanocatalysts exhibit remarkable efficiency in breaking down pollutants, contributing to sustainable and eco-friendly remediation processes [4].

In the current research landscape, ongoing studies focus on tailoring nanomaterials for specific applications, exploring novel synthesis methods, and unravelling the intricacies of their interactions with biological systems. Researchers delve into the challenges of nanotoxicity, seeking to understand and mitigate potential risks associated with the use of nanomaterials in various applications.

15.2 Synthesis Methods and Characterisation Techniques for Nanomaterials: Unlocking the Nanoscale World

Nanomaterials, with their unique properties and applications, owe their versatility to sophisticated synthesis methods and advanced characterisation techniques. The synthesis of nanomaterials involves the deliberate engineering of materials at the nanoscale, while characterisation techniques enable researchers to scrutinise their structures and properties with unprecedented precision [5]. This comprehensive exploration delves into the myriad synthesis methods and characterisation techniques that underpin the nanomaterial revolution.

15.2.1 Overview of Nanomaterial Synthesis Techniques

Nanomaterial synthesis represents a multidisciplinary endeavour, blending principles from chemistry, physics, and engineering to tailor materials with nanoscale precision. Several key techniques dominate this landscape, each offering distinct advantages and challenges.

a. *Chemical Vapour Deposition (CVD)*: CVD is a prevalent method for synthesising thin films and nanostructured materials. In this technique, gaseous precursors undergo chemical reactions on a substrate, forming a thin layer or nanostructure [6]. CVD enables precise control over film thickness and composition, making it pivotal in semiconductor and nanoelectronics industries.

b. *Sol–Gel Method*: This solution-based technique involves the conversion of precursor solutions into a gel, followed by drying and annealing to produce nanomaterials. Sol–gel is versatile, enabling the synthesis of ceramics, glasses, and hybrid organic–inorganic materials. Its simplicity and scalability make it attractive for applications in catalysis, sensors, and coatings [7].

c. *Ball Milling*: A mechanical approach to nanomaterial synthesis, ball milling employs grinding balls in a rotating cylinder to induce mechanical alloying or milling. This method is particularly effective for producing nanomaterials with controlled morphology and enhanced reactivity. Ball milling finds applications in the synthesis of metallic alloys, ceramics, and composite materials.

d. *Electrospinning*: In this electrostatically driven process, polymer solutions are ejected from a fine nozzle to form ultrafine fibres. Electrospinning is widely employed for producing nanofibres with applications in tissue engineering, filtration, and sensors. The method offers control over fibre diameter and porosity [8].

e. *Atomic Layer Deposition (ALD)*: ALD is a precise thin-film deposition technique that builds up materials layer by layer, atom by atom. It is instrumental in creating conformal coatings with precise thickness control. ALD is extensively used in semiconductor manufacturing and emerging technologies like nanoelectronics.

Each synthesis method imparts unique characteristics to the resulting nanomaterials, influencing their properties and applications. The choice of synthesis technique depends on the desired material, scale, and intended use, highlighting the nuanced decision-making involved in nanomaterial synthesis [9].

15.3 Advanced Characterisation Methods for Nanomaterials

Characterising nanomaterials demands tools capable of probing structures and properties at scales often beyond the reach of conventional techniques. Advanced characterisation methods play a pivotal role in unravelling the mysteries of the nanoscale world.

a. *Transmission Electron Microscopy (TEM)*: TEM is a high-resolution imaging technique that uses a beam of electrons transmitted through a thin specimen to create detailed images at the nanoscale. It provides insights into nanomaterial morphology, crystal structure, and defects. With sub-nanometre resolution, TEM is indispensable for studying nanoparticles, nanotubes, and nanocomposites [10].

b. *Scanning Electron Microscopy (SEM)*: SEM complements TEM by providing detailed surface images. In SEM, a focused electron beam scans the specimen surface, generating topographical information. This technique is vital for visualising nanomaterial morphologies, understanding particle size distribution, and assessing surface characteristics.

c. *X-ray Diffraction (XRD)*: XRD is a powerful tool for determining the crystalline structure of nanomaterials. By analysing the diffraction patterns of X-rays interacting with a crystalline sample, researchers can identify crystal phases, lattice parameters, and crystallographic orientations. XRD is extensively used in nanomaterials research to understand their structural properties.

d. *Fourier Transform Infrared Spectroscopy (FTIR)*: FTIR analyses the vibrational modes of molecules, offering insights into the chemical composition of nanomaterials. It helps identify functional groups, assess chemical bonding, and detect impurities. FTIR is valuable for characterising nanocomposites, biomaterials, and functionalised nanoparticles.

e. *Nuclear Magnetic Resonance (NMR)*: NMR spectroscopy provides a non-destructive means of studying the atomic-scale environment of certain nuclei in nanomaterials. It is particularly useful for elucidating molecular structures, quantifying molecular dynamics, and investigating surface interactions. NMR is applied in the study of nanocatalysts, nanocomposites, and biomaterials.

f. *Dynamic Light Scattering (DLS)*: DLS measures the Brownian motion of nanoparticles in suspension, allowing researchers to determine particle size distributions. This technique is crucial for assessing the stability and dispersity of colloidal nanomaterials, providing essential information for applications in drug delivery, cosmetics, and catalysis.

g. *Raman Spectroscopy*: Raman spectroscopy probes molecular vibrations, offering insights into the chemical composition and structural integrity of nanomaterials. It is particularly valuable for studying carbon-based nanomaterials like graphene and carbon nanotubes. Raman spectroscopy aids in understanding defects, functionalisation, and strain in nanomaterials.

The synergy between synthesis methods and characterisation techniques forms the backbone of nanomaterial research. Advanced characterisation methods empower researchers to validate the success of synthesis processes, elucidate material behaviours, and guide the development of tailored nanomaterials for diverse applications.

15.4 Nanomaterials for Energy Storage

15.4.1 Nanomaterials Revolutionising Battery and Supercapacitor Technologies

Nanomaterials have emerged as game-changers for energy storage, particularly in the development of advanced batteries and supercapacitors. The inherent properties of nanomaterials, dictated by their nanoscale dimensions, bring about remarkable enhancements in the electrochemical performance of energy storage devices. This section explores the pivotal role played by nanomaterials in the evolution of batteries and supercapacitors [11].

15.4.2 Role of Nanomaterials in Batteries

The integration of nanomaterials in batteries has significantly altered the landscape of energy storage. Nanomaterials, such as nanotubes, nanoparticles, and nanostructured composites, offer unique advantages that transcend the limitations of traditional materials. The high surface area-to-volume ratio of nanomaterials facilitates more efficient charge storage, leading to enhanced battery performance. For instance, silicon nanowires or nanoparticles are explored as promising anode materials in lithium-ion batteries due to their ability to accommodate large amounts of lithium ions, thereby increasing the overall energy storage capacity. Graphene, a two-dimensional nanomaterial, serves as a conductive additive and a structural enhancer in battery electrodes, fostering faster charge–discharge rates and prolonged cycle life.

15.4.3 Enhancing Energy Density and Efficiency

Nanomaterials play a crucial role in enhancing both the energy density and efficiency of energy storage systems. The unique properties of nanomaterials enable the design of electrodes with higher specific surface areas, promoting more significant ion interaction and faster charge transport. This results in batteries with increased energy storage capacity and improved power density. Moreover, nanomaterials contribute to the development of high-performance supercapacitors, which store energy through electrostatic charge separation. Nanostructured materials, such as carbon nanotubes and graphene, serve as excellent electrode materials for supercapacitors, offering rapid charge–discharge cycles and superior energy storage capabilities. Strategies to mitigate issues like volume expansion, which can affect the long-term stability of batteries, are actively explored through nanomaterial design [12]. As the energy storage landscape evolves, nanomaterials continue to be at the forefront of innovations that address the pressing demand for efficient and sustainable energy solutions.

15.5 Nanomaterials in Electronics and Photonics

15.5.1 Applications in Electronic Devices

Nanomaterials have ushered in a new era in electronics, where their unique properties are harnessed to enhance the performance of electronic devices. The integration of nanomaterials in electronic components has paved the way for smaller, faster, and more efficient devices. This section explores the diverse applications of nanomaterials in electronic devices, showcasing their transformative impact on the field. One notable application is in the fabrication of nanoscale transistors, where nanomaterials serve as building blocks for miniaturised electronic components. Quantum dots, semiconductor nanocrystals, enable the creation of nanoscale transistors with precise control over electronic properties. This breakthrough not only contributes to the ongoing trend of miniaturisation in electronics but also opens avenues for quantum computing and advanced sensor technologies.

Nanomaterials also play a pivotal role in enhancing the conductivity and durability of electronic interconnects. Copper nanoparticles, for instance, exhibit superior electrical conductivity and mechanical strength, offering a viable alternative to traditional materials in interconnect fabrication [13]. This innovation not only improves the efficiency of electronic devices but also addresses challenges related to heat dissipation in high-performance electronics.

15.5.2 Photonics and Optoelectronic Devices Utilising Nanomaterials

Nanomaterials exhibit unique optical properties that can be tailored for specific applications, ranging from light-emitting diodes (LEDs) to photodetectors and lasers. Quantum dots, semiconductor nanocrystals with size-dependent optical properties, are employed in LED technologies, enabling the production of brighter and more colour-saturated displays. Additionally, nanomaterials such as perovskite nanocrystals are gaining prominence in solar cell technologies, revolutionising the efficiency and cost-effectiveness of photovoltaic devices. Metamaterials, engineered nanostructures with exotic optical properties not found in nature, are at the forefront of innovation in optics and photonics. These materials enable the creation of devices with unprecedented control over light, leading to advancements in lenses, cloaking devices, and high-resolution imaging systems. As research in electronics and photonics continues to advance, nanomaterials remain a driving force behind breakthroughs that redefine the capabilities of electronic devices and photonics technologies. The ability to engineer materials at the nanoscale provides researchers with a powerful toolbox for creating next-generation electronics and optoelectronic devices with unparalleled performance and functionality.

15.6 Nanomaterials for Biomedical Applications

15.6.1 Nanomaterials Revolutionising Biomedicine

The integration of nanomaterials into biomedical applications has ushered in a paradigm shift in medicine, introducing groundbreaking solutions for drug delivery, imaging, and diagnostics. This section delves into the profound impact of nanomaterials in the field of biomedicine, emphasising their pivotal roles in drug delivery systems and as imaging agents and diagnostic tools.

15.6.2 Drug Delivery Systems

Nanomaterials serve as vanguards in drug delivery systems, revolutionising the way therapeutic agents are administered and absorbed in the body. Nanoparticles, liposomes, and dendrimers act as carriers, ensuring targeted delivery of drugs to specific cells or tissues, minimising systemic side effects. Lipid-based nanocarriers, for instance, enhance the solubility of hydrophobic drugs, prolonging their circulation time and facilitating controlled release. Functionalised nanoparticles can be engineered to actively target diseased cells through surface modifications, offering personalised and precise drug delivery strategies [14]. Moreover, stimuli-responsive nanomaterials respond to specific physiological cues, releasing therapeutic payloads in a controlled manner. This spatiotemporal precision enhances drug efficacy while minimising adverse effects, marking a paradigm shift in pharmaceutical approaches.

15.6.3 Imaging Agents and Diagnostic Tools

Nanoparticle-based imaging agents offer improved resolution in various imaging modalities, including magnetic resonance imaging (MRI), computed tomography (CT), and positron emission tomography (PET). Quantum dots, semiconductor nanocrystals with tuneable optical properties, shine as fluorescent imaging agents, allowing for real-time visualisation of cellular and molecular processes. Iron oxide nanoparticles find utility in MRI, enhancing contrast for detailed anatomical imaging. The integration of nanomaterials into biomedical applications has ushered in a paradigm shift in medicine, introducing groundbreaking solutions for drug delivery, imaging, and diagnostics. This section explores the profound impact of nanomaterials in the field of biomedicine, emphasising their pivotal roles in drug delivery systems and as imaging agents and diagnostic tools. Nanomaterials, due to their unique physicochemical properties, enable precise manipulation of drug delivery dynamics. Engineered nanoparticles serve as carriers, facilitating targeted and controlled release of therapeutic agents [15]. This level of precision enhances therapeutic efficacy while minimising systemic side effects. In the domain of medical imaging and diagnostics, nanomaterials play a central role. Nanoparticle-based imaging agents offer enhanced contrast and sensitivity in various imaging modalities. Quantum dots, for example, exhibit tuneable optical properties, providing real-time visualisation capabilities. The versatility of nanomaterials in the diagnostic landscape is further exemplified by multifunctional nanoprobes, enabling simultaneous imaging and therapy.

15.7 Nanomaterials in Catalysis and Environmental Remediation

15.7.1 Catalytic Applications of Nanomaterials

The catalytic prowess of nanomaterials is harnessed for transformative applications in catalysis, offering novel pathways for chemical transformations and industrial processes. This section explores the catalytic applications of nanomaterials and their potential impact on sustainable chemistry and industrial practices. Nanocatalysts, owing to their high surface area and unique electronic properties, exhibit superior catalytic activity compared to bulk materials. Noble metal nanoparticles, such as platinum and palladium, are widely employed in catalysing chemical reactions, including hydrogenation and oxidation processes. Bimetallic nanoparticles, with their synergistic effects, showcase enhanced catalytic efficiency, paving the way for greener and more sustainable chemical synthesis. Moreover, the catalytic applications extend to the field of nanocatalytic converters, where nanomaterials actively participate in mitigating environmental pollutants from exhaust gases. Nanostructured catalysts enhance the efficiency of pollutant conversion, contributing to cleaner air and reduced environmental impact.

15.7.2 Environmental Impact and Remediation Strategies

Nanomaterials emerge as versatile agents in environmental remediation, addressing the challenges of pollution and contaminants in air, water, and soil. Nanoparticles, such as zero-valent iron and titanium dioxide, exhibit photocatalytic properties, enabling the degradation of organic pollutants and the removal of heavy metals from water sources. Nanostructured adsorbents, with their large surface areas, prove effective in sequestering contaminants from aqueous solutions. The multifunctional nature of nanomaterials allows for the simultaneous removal of diverse pollutants, presenting an integrated and efficient approach to environmental cleanup [16]. Challenges arise in balancing the benefits of nanomaterial-based remediation with potential environmental risks. Nanotoxicity, the adverse effects of nanomaterials on living organisms and ecosystems, requires careful consideration. Research endeavours focus on understanding and mitigating nanotoxicity through the design of environmentally benign nanomaterials and comprehensive risk assessment strategies.

15.8 Challenges and Opportunities in Nanomaterial Research

15.8.1 Addressing Nanotoxicity and Safety Concerns

The integration of nanomaterials into diverse applications necessitates a thorough examination of their potential toxicity and safety implications. Nanotoxicity, arising from the interactions between nanomaterials and biological systems, poses challenges that demand nuanced solutions. Researchers actively investigate the mechanisms underlying nanotoxicity, encompassing aspects such as cellular uptake, biodistribution, and long-term effects. Biocompatible coatings and surface modifications are explored to mitigate adverse reactions, ensuring the safe deployment of nanomaterials in biomedical and environmental applications. In addition, robust risk assessment frameworks are essential for evaluating the potential hazards associated with nanomaterial exposure. Standardised testing protocols and comprehensive toxicity studies contribute to the responsible development and deployment of nanomaterials, aligning technological progress with safety considerations.

15.8.2 Opportunities for Integration with Traditional Materials

Amidst the challenges lie opportunities for synergistic integration of nanomaterials with traditional materials, unlocking innovative solutions across industries. Hybrid materials, combining the strengths of nanomaterials with the familiarity of conventional materials, present a bridge between the cutting edge and the established. Nanocomposites, blending nanomaterials with polymers, metals, or ceramics, exhibit enhanced mechanical, electrical, and thermal properties. This integration extends to the aerospace, automotive, and construction industries, where nanocomposites offer lightweight yet durable solutions with improved performance characteristics. Moreover, the integration of nanomaterials in traditional manufacturing processes enhances the efficiency and sustainability of industrial production. Nanomaterial-based additives and coatings contribute to energy savings, reduced environmental impact, and enhanced product functionalities. As nanomaterial research progresses, the opportunities for integration with traditional materials create a synergy that can reshape industries, offering a seamless transition from the laboratory to real-world applications. Strategic collaborations between researchers, industries, and regulatory bodies are pivotal in navigating the complex landscape of nanomaterial research, ensuring a harmonious blend of innovation and responsibility.

15.9 Regulatory Considerations in Nanomaterial Applications

15.9.1 Navigating the Regulatory Landscape for Nanomaterials

The integration of nanomaterials into various applications brings forth a pressing need to navigate a complex regulatory landscape. This section explores the current regulatory considerations surrounding nanomaterials, examining the implications for their commercialisation and industrial use.

15.9.2 Current Regulatory Landscape for Nanomaterials

The regulatory oversight of nanomaterials varies across jurisdictions, reflecting the dynamic nature of this field and the evolving understanding of nanomaterial risks. Regulatory agencies worldwide, including the US Food and Drug Administration (FDA), the European Medicines Agency (EMA), and the Environmental Protection Agency (EPA), are actively engaged in formulating guidelines to ensure the safe development and use of nanomaterials. In the European Union, nanomaterials are subject to specific regulations under the Registration, Evaluation, Authorization, and Restriction of Chemicals (REACH) framework. This includes mandatory reporting and risk assessment requirements for nanomaterials manufactured or imported in certain quantities. Similarly, the FDA scrutinises nanomaterials used in medical products, emphasising safety and efficacy assessments.

Despite progress, challenges persist in harmonising international regulatory frameworks, creating a cohesive approach to nanomaterial oversight. The multifaceted nature of nanomaterials requires regulators to consider diverse factors, including particle size, surface area, and potential exposure routes.

15.9.3 Implications for Commercialisation and Industrial Use

Navigating regulatory frameworks is pivotal for the successful commercialisation and industrial use of nanomaterials. Complying with regulatory requirements not only ensures the safety of products but also fosters public confidence and facilitates market access. For industries incorporating nanomaterials in consumer products, adherence to labelling and disclosure requirements is essential. Transparency about the presence of nanomaterials enables informed consumer choices and aligns with regulatory expectations. Industrial sectors, such as cosmetics, textiles, and food packaging, grapple with the challenge of meeting regulatory standards while harnessing the unique properties of nanomaterials for product innovation. As nanomaterial applications diversify, the regulatory landscape must adapt to encompass emerging technologies and unforeseen challenges. Collaborative efforts between regulatory bodies, researchers, and industries are vital to establish robust frameworks that balance innovation with risk mitigation, fostering a responsible and sustainable nanomaterial ecosystem.

15.10 Future Prospects and Innovations in Nanomaterial Technologies

15.10.1 Emerging Trends and Technologies

The future of nanomaterial technologies holds a promising tapestry of emerging trends poised to redefine industries and scientific frontiers. This section explores key trends that are shaping the trajectory of nanomaterial research and applications. One notable trend is the convergence of nanotechnology with other transformative fields, such as artificial intelligence (AI) and biotechnology. The integration of AI-driven design approaches accelerates the discovery and optimisation of nanomaterials for specific applications. Additionally, the synergy between nanomaterials and biotechnology leads to groundbreaking advancements in personalised medicine, targeted drug delivery, and regenerative therapies.

15.10.2 Anticipated Breakthroughs and Research Frontiers

Anticipated breakthroughs in nanomaterial research encompass diverse fields, reflecting the interdisciplinary nature of nanotechnology. The exploration of 2D materials, beyond graphene, introduces novel nanosheets with unique electronic and mechanical properties. These materials hold promise for applications in electronics, sensing, and energy storage. In medicine, the advent of smart nanomaterials that respond to specific physiological cues opens avenues for precise diagnostics and targeted therapies. Programmable nanorobots capable of navigating biological environments and delivering therapeutic payloads represent a frontier in nanomedicine, offering unprecedented control over disease treatment. Quantum nanomaterials, with properties governed by quantum mechanics, present a frontier where classical physical laws no longer apply. Quantum dots and quantum sensors are at the forefront of this quantum revolution, promising advancements in quantum computing, communication, and imaging.

15.11 Conclusion

The exploration of emerging nanomaterials for advanced technologies illuminates a transformative journey at the intersection of science and innovation. Nanomaterials, with their unique properties and versatile applications, stand as catalysts for revolutionary breakthroughs across diverse fields. From reshaping energy storage solutions to revolutionising biomedical applications, nanomaterials have become indispensable. The regulatory considerations underscore the importance of balancing innovation with safety, paving the way for responsible commercialisation. As we peer into the future, emerging trends, technologies, and anticipated breakthroughs in nanomaterial research promise a landscape where the infinitesimally small particle redefines the boundaries of possibility. The integration of nanomaterials into advanced technologies heralds a new era of efficiency, sustainability, and unprecedented capabilities, charting a course towards a future where nanomaterials propel us into uncharted territories of scientific and technological achievement.

REFERENCES

[1] Surjadi, J.U., Gao, L., Du, H., Li, X., Xiong, X., Fang, N.X. and Lu, Y., 2019. Mechanical metamaterials and their engineering applications. *Advanced Engineering Materials*, *21*(3), p. 1800864.

[2] Zhao, F., Shi, Y., Pan, L. and Yu, G., 2017. Multifunctional nanostructured conductive polymer gels: Synthesis, properties, and applications. *Accounts of Chemical Research*, *50*(7), pp. 1734–1743.

[3] Vignesh, P., Ramanathan, S., Ramesh, K. and Arockiam, A.J., 2022. Finite element modelling to predict hardness of Mg alloy reinforced with Ti/hydroxyapatite hybrid composites–An axisymmetric approach. *Materials Today: Proceedings*, *68*, pp. 1830–1834.

[4] Boulkhessaim, S., Gacem, A., Khan, S.H., Amari, A., Yadav, V.K., Harharah, H.N., Elkhaleefa, A.M., Yadav, K.K., Rather, S.U., Ahn, H.J. and Jeon, B.H., 2022. Emerging trends in the remediation of persistent organic pollutants using nanomaterials and related processes: A review. *Nanomaterials*, *12*(13), p. 2148.

[5] Arockiam, A.J., Rajesh, S., Karthikeyan, S., Thiagamani, S.M.K., Padmanabhan, R.G., Hashem, M., Fouad, H. and Ansari, A., 2023. Mechanical and thermal characterization of additive manufactured fish scale powder reinforced PLA biocomposites. *Materials Research Express*, *10*(7), p. 075504.

[6] Sayago, I., Hontañón, E. and Aleixandre, M., 2020. Preparation of tin oxide nanostructures by chemical vapor deposition. *Tin Oxide Materials*, Elsevier. pp. 247–280.

[7] Figueira, R.B., Sousa, R. and Silva, C.J., 2020. Multifunctional and smart organic–inorganic hybrid sol–gel coatings for corrosion protection applications. *Advances in Smart Coatings and Thin Films for Future Industrial and Biomedical Engineering Applications*, Elsevier. pp. 57–97.

[8] Joseph Arockiam, A., Rajesh, S. and Karthikeyan, S., Development of fish scale particle reinforced PLA filaments for 3D printing applications. *Journal of Applied Polymer Science*, *141*(12), p. e55132.

[9] Karki, S., Hazarika, G., Yadav, D. and Ingole, P.G., 2023. Polymeric membranes for industrial applications: Recent progress, challenges and perspectives. *Desalination*, *573* p. 117200.

[10] Carenco, S., Moldovan, S., Roiban, L., Florea, I., Portehault, D., Vallé, K., Belleville, P., Boissière, C., Rozes, L., Mézailles, N. and Drillon, M., 2016. The core contribution of transmission electron microscopy to functional nanomaterials engineering. *Nanoscale*, 8(3), pp. 1260–1279.

[11] Chandrashekhar, K.G., Laxmaiah, G., Ram Kumar, P. and Ramesh, B., 2023. Load-bearing characteristics of a hybrid Si_3N_4-epoxy composite. *Biomass Conversion and Biorefinery*, pp. 1–9.

[12] Wei, Q., Xiong, F., Tan, S., Huang, L., Lan, E.H., Dunn, B. and Mai, L., 2017. Porous one-dimensional nanomaterials: Design, fabrication and applications in electrochemical energy storage. *Advanced Materials*, 29(20), p. 1602300.

[13] Naghdi, S., Rhee, K.Y., Hui, D. and Park, S.J., 2018. A review of conductive metal nanomaterials as conductive, transparent, and flexible coatings, thin films, and conductive fillers: Different deposition methods and applications. *Coatings*, 8(8), p. 278.

[14] Arockiam, A.J., Subramanian, K., Padmanabhan, R.G., Selvaraj, R., Bagal, D.K. and Rajesh, S., 2022. A review on PLA with different fillers used as a filament in 3D printing. *Materials Today: Proceedings*, 50, pp. 2057–2064.

[15] Sun, T., Zhang, Y.S., Pang, B., Hyun, D.C., Yang, M. and Xia, Y., 2021. Engineered nanoparticles for drug delivery in cancer therapy. *Nanomaterials and Neoplasms*, pp. 31–142.

[16] Grisolia, A., Dell'Olio, G., Spadafora, A., De Santo, M., Morelli, C., Leggio, A. and Pasqua, L., 2023. Hybrid polymer-silica nanostructured materials for environmental remediation. *Molecules*, 28(13), p. 5105.

16

Determination of Kinetic Parameters and Mechanical Properties of Nonlinear Optical Single Crystals

P. Krishnan
St. Joseph's College of Engineering, Chennai, Tamil Nadu, India

K. Rajesh
AMET University, Chennai, Tamil Nadu, India

S. Sudhahar
Alagappa University, Karaikudi, Tamilnadu, India

16.1 Introduction

The advancement in science and technology realises the importance of solo crystals for the growth of new-generation devices. The demand for high-quality crystals focuses on the importance of the rapid growth process and the extremely competent nonlinear optical crystals for technological processes. In recent years, there have been rigorous efforts to explore and develop second-order nonlinear optical materials with a high degree of efficiency for various applications such as electro-optic swapping or modulation for telecommunication and frequency conversions for laser sources at desired wavelengths.

16.1.1 Purpose of Characterisation

Characterisation renders a basis for realising and enhancing the characteristics of materials towards their applications and also comprises an appraisal of the chemical structure, structure, imperfections, thermal stability, mechanical strength, conductivity, and optical properties. If the anisotropic heat release of the crystal is large, the defined heat is low, and the thermal conductivity is low, the crystal can easily crack during the growth process or when the temperature gradient is too large. Therefore, thermal properties can be considered as important parameters for evaluating the practical viability of a particular crystal.

Hardness properties depend on the material's crystal structure and bond strength. Microhardness research was performed to apprehend the plasticity of crystals. Checking out the hardness is generally done to decide the mechanical electricity of materials. It correlates with different mechanical residences which include elastic constants and yield electricity. Hardness of the material is more suffering from dimension situations than by means of cloth residences. In this chapter, we discuss the determination of kinetic parameters and mechanical properties of nonlinear optical single crystals.

16.1.2 Thermal Analysis

Thermal analysis is a generic term that encompasses several techniques for measuring the temperature dependence of physical property parameters of matter. Thermogravimetric analysis (TGA), differential thermal analysis (DTA), and differential scanning calorimetry (DSC) are widely used techniques.

TGA involves monitoring the weight while varying temperature. The programme's turning point corresponds to the differential thermogram's peak point. Thermal properties such as thermal expansion, specific heat, and thermal conductivity of crystals greatly impact crystal growth and applications. Thermal expansion measurements also helped investigate the problem of crystal cracking. Temperature gradients

DOI: 10.1201/9781003495437-16

are inevitably present during bulk crystal growth, and additional gradients are temporarily superimposed when slow cooling techniques are used. This is believed to be the reason why cracks and extra polycrystals may appear during the growth process.

16.1.3 Methods Involved in the Determination of Kinetic Parameters

There are various methods tangled in the determination of kinetic parameters using thermal analysis, such as:

- Friedman method.
- Kissinger.
- Flynn-Wall.
- Kim Park.
- Horowitz.
- Broido.
- Coats and Redfern method.

Among various methods, this chapter explains the Coats and Redfern method in detail. Many researchers have inspected the locomotive constraints of a solid-state reaction including weight loss using TG data [1–3]. The form of the curvature depends on the locomotive constraints of pyrolysis, such as the reaction order, the frequency factor, and the activation energy. Coats and Redfern [1] developed a method for the calculation of activation energy from non-isothermal TG data at a constant heating rate.

Other kinetic constraints such as activation enthalpy (ΔH), activation entropy (ΔS), and decomposition free energy change (ΔG) have been evaluated using equations. Thermogravimetry differential thermal analysis (TG-DTA) provides data on phase transitions, waters of crystallisation, different stages of crystal decomposition, and thermal stability. Thermal stability does not decompose up to its melting point, ensuring suitability for laser applications where the crystal must withstand high temperatures. For TG-DTA analysis, a 1.4040 mg sample was taken and heated at a 20°C/min rate under a nitrogen atmosphere. Thermal analysis (TG-DTA) provides information on phase transitions, waters of crystallisation, and various stages of crystal decomposition [4].

Thermal analysis of a material also provides useful information about its thermal stability. Figure 16.1 shows the TG-DTA spectrum.

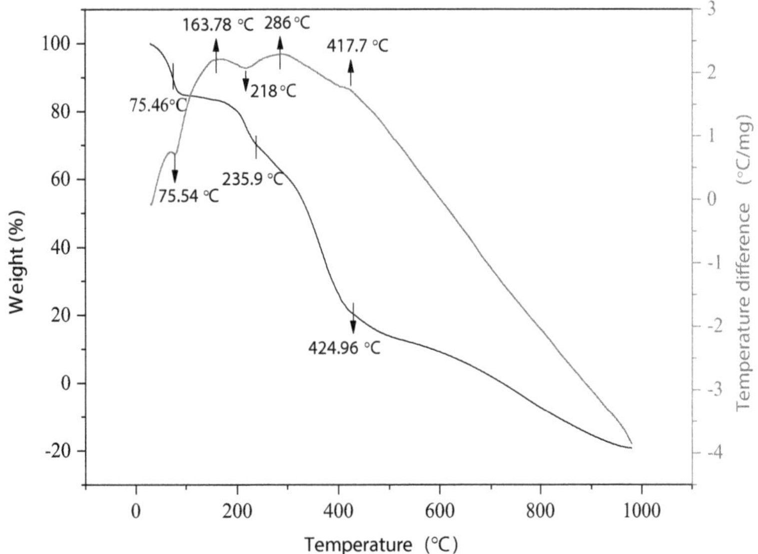

FIGURE 16.1   TG-DTA spectrum.

TGA shows that the grown crystal is thermally stable up to 75.54°C. This low decomposition temperature confirms the presence of water molecules in the grown crystals. This endothermic peak is followed by an exothermic peak at 163.78°C indicating evaporation of water. Three stages of decomposition are observed in the TGA spectrum: In the first stage, the crystal loses water molecules at 90°C. A second stage of decomposition was observed at 218°C. This endothermic peak is followed by an exothermic peak at 286°C indicating evaporation of decomposed compounds. The compound was decomposed in the third stage and measured at 424.9°C.

These corresponding weight losses were observed in the TGA spectra. Therefore, as the temperature rises, polymeric compounds gradually decompose into volatile products. Moreover, there is no phase transition before melting.

16.1.4 Kinetic Parameters by Coats and Redfern Method

The universal relationship used in Coats and Redfern method is specified beneath for the number of decomposition stages more than one, that is, for $n \neq 1$:

$$\log_{10}\left[\left(\frac{\left\{1-(1-\alpha)^{1-n}\right\}}{T^2(1-n)}\right)\right] = \log_{10}\left\{\left(\frac{AR}{\alpha E}\right)\left(1-\frac{2RT}{E}\right)-\left(\frac{E}{2.303RT}\right)\right\} \tag{16.1}$$

For $n = 1$,

$$\log_{10}\left[\frac{-\log(1-\alpha)}{T^2}\right] = \left[\log_{10}\left\{\left(\frac{AR}{\alpha E}\right)\left(1-\frac{2RT}{E}\right)\right\}-\left(\frac{E}{2.303RT}\right)\right] \tag{16.2}$$

where $\alpha = (w_0 - w)/(w_0 - w_f)$
  w_0– Initial mass taken.
  w_f– Final mass residues.

Figure 16.2 shows the best linear fit plots obtained using $\log[-\log(1-\alpha))/T^2]$ versus $1/T$ for various values of n, from which the correct value of n is obtained. This equation holds for all values except $n = 1$.

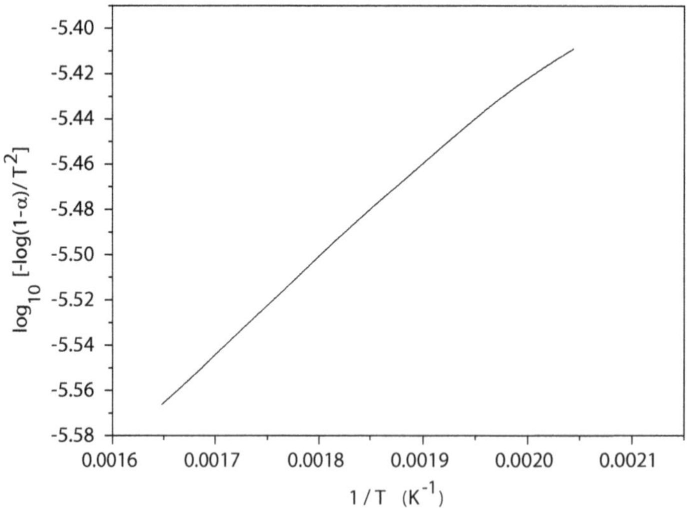

FIGURE 16.2 $\log[-\log(1-\alpha))/T^2]$ versus $1/T$.

The slope of the line was used to calculate the activation energy (E) and the intercept was used to calculate the frequency coefficient (log A). Enthalpy (ΔH), entropy (ΔS), and Gibbs energy (ΔG) parameters of activation were calculated using the following standard equations:

Enthalpy of activation (ΔH),

$$\Delta H = E - 2RT \tag{16.3}$$

where

$$E = 2.303 \times 8.314510 \times \text{Slope}$$

The entropy of activation (ΔS)

$$\Delta S = 2.303 \times R \times \log_{10}\left[Ah / k_B T_m \right] \tag{16.4}$$

where R is a gas constant, 'h' is Planck's constant, and k_B is the Boltzmann constant.

Gibbs energy of activation (ΔG),

$$\Delta G = \Delta H - T\Delta S \tag{16.5}$$

A change in the Gibbs free energy (ΔG) of a chemical reaction is a measure of whether it is spontaneous or not. In this case, the ΔG value is positive, so the dissociation process does not occur spontaneously [5]. A positive value of ΔH indicates that the dissociation process is end other mic and is amplified by increasing temperature. The calculated activation energy is relatively low, indicating an autocatalytic effect on pyrolysis. Activation energy (E) values for the dehydration step were found to be less than 60 kJmol^{-1}, as expected for the removal of weakly bound water of crystallisation [6]. The unfavourable activation entropy value indicates that the activated complex is more systematic and slower to react than the reactants. A large amount of calculated activation entropy indicates a significant degree of rearrangement [6]. A more systematic nature may be due to bond branching at the activation state, which can be altered by charge-transfer electronic transitions. Furthermore, a high value of A indicates a rapid nature of the reaction. This ensures the material's suitability for laser applications where the crystal must withstand high temperatures.

Calculated activation energy, frequency coefficients, and other thermodynamic parameters are shown in Table16.1.

16.2 Microhardness Studies

The mechanical strength of materials plays an important role in device fabrication. Most applications require samples to be cut and polished prior to use. The polishing quality of a given sample is directly related to the mechanical properties of the material, and very high hardness values can guarantee better polishing and laser surface quality [7]. Hardness measurements are an important component of resistance to plastic deformation, primarily by indentation, and are also relevant to resistance to bending, scratching,

TABLE 16.1

Kinetic Parameters by Coats and Redfern Method

Kinetic Parameters	Values
Activation energy (E^*) in kJ mol^{-1}	8.43
Frequency factor (A) in s^{-1}	1.6
Enthalpy of activation (ΔH^*) in kJ mol^{-1}	0.304
Entropy of activation (ΔS^*) in JK^{-1} mol^{-1}	−125.79
Gibb's energy (ΔG^*) in kJ mol^{-1}	51.71

abrasion, or cutting. In particular, hardness is a measure of resistance to lattice breakage, resistance to permanent deformation and damage. It is a non-destructive method of measuring the mechanical behaviour of materials. For hard and brittle materials, hardness testing has proven to be a valuable technique in the general study of plastic deformation. Hardness properties depend on the material's crystal structure and bond strength. Microhardness studies have been conducted to understand the plasticity of crystals. Hardness testing is commonly performed to determine the mechanical strength of materials. They correlate with other mechanical properties such as elastic constants and yield strength.

Material hardness is more affected by measurement conditions than by material properties. Microhardening methods are widely used to study the individual structural components of alloys, minerals, glasses, enamels, and artificial abrasives.

Various methods adopted for the measurement of hardness are divided as

- Static indentation test.
- Dynamic indentation test.
- Scratch test rebound test.
- Abrasion test.

Among the various hardness measurement methods, the static indentation test, in which a specifically shaped indenter is pressed against the specimen surface with a known load, is the preferred method of hardness measurement due to its popularity and simplicity. Hardness is calculated from the area or depth of the recess formed. The variable is the type of indenter or load. The indenter is made of a very hard material to prevent deformation by the sample and can be used with a wide variety of hard materials.

For this, steel balls, diamond pyramids, or diamond cones are used. Pyramid indenters are preferred as they yield geometrically similar indentations at different loads. In this static indentation test, a load is applied to push the indenter perpendicular to the surface of the specimen.

This method is used in the following tests:

- Brinell test.
- Meyer test.
- Vickers test.
- Knoop and Rockwell test [8]

In the dynamic indentation test, a ball or cone (or many small balls) is dropped from a specified height and the hardness number is determined from the indentation dimensions and impact energy.

In the rebound test, an object of standard mass and size is bounced off the test surface and the magnitude of the rebound is used as a measure of stiffness. Wear testing involves loading the sample onto a rotating disc and using the wear rate as a measure of hardness. Vickers Pyramid Indenters are the most common pyramid indenters, with opposite faces bevelled ($\alpha = 136°$).

The load variation can be interpreted using Meyer's law:

$$P = K_1 d^n$$

$$\log P = \log k + n \log d$$

Hays and Kendall's theory of resistance pressure:

$$P - W = k_2 d^2$$

$$W = k_1 d^n - k_2 d^2$$

$$d^n = \left(W / k_1\right) + \left(k_2 / k_1\right) d^2$$

Measurements were made at room temperature and loads of various magnitudes such as 25, 50, and 100 g were applied. At a given load, at least 5 impressions were made. Both diagonal dimensions of the indentation were measured and the average dimension d was calculated from all diagonals created with a given P load. Figure 16.3 shows the variation of H_v as a function of the indentation test load applied to the specimen. It can be seen from this figure that H_e–H_v increases with increasing applied indentation load. This can be viewed as a reverse indentation size effect (RISE).

A plot of log$d$ versus Log$P$ shows a straight line (Figure 16.4) and the work hardening is calculated to be 3.39. For microhardness studies, the hardness number increases with increasing load for $n> 2$ and decreases for $n< 2$. Onitsch [9] showed that n is between 1.0 and 1.6 for hard materials and greater than

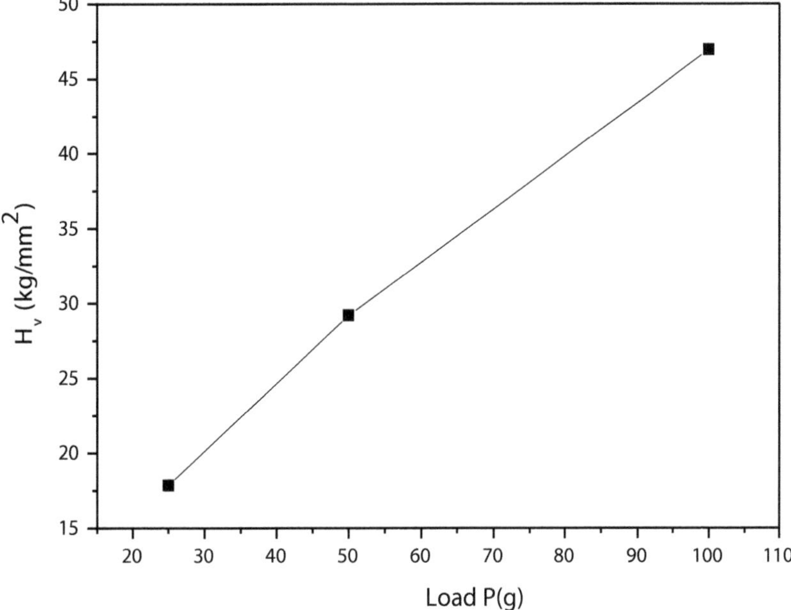

FIGURE 16.3 Variation of H_v versus load.

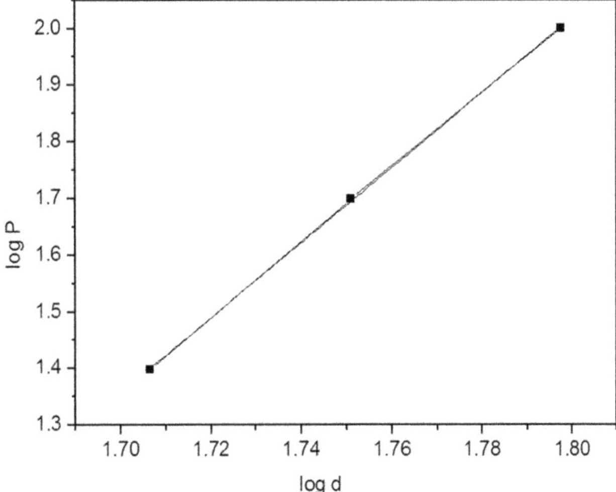

FIGURE 16.4 The plot of log$d$ versus log$P$.

1.6 for soft materials. According to this standard, metformin hydrochloride is considered a soft material and the material constant is determined to be 2.950×10^{-4} g/μm^2.

The yield strength (σ_y) can be calculated from the expression

$$\sigma_y = \frac{H_v}{2.9}\left[1-(n-2)\right]\left[\frac{12.5(n-2)}{1-(n-2)}\right]^{n-2} \quad \text{for } n > 2$$

$$\sigma_y = H_v / 3 \quad n \leq 2$$

Brittle index for different loads:

$$B_i = \frac{H_v}{K_c}$$

K_c is fracture toughness:

$$K_c = \frac{P}{\beta_o C^{3/2}}$$

The elastic stiffness constant (C_{11}) for different loads was calculated using Wooster's empirical formula $C_{11} = H_v^{7/4}$.

From Hays and Kendall's linear relationship between the indentation size and load, the plot of d^n against d^2 (Figure 16.5) is a straight line having a slope of k_2/k_1 and intercept W/k_1. From this, a value W representing the material's resistance to the initiation of plastic flow is calculated and shown in Table 16.2.

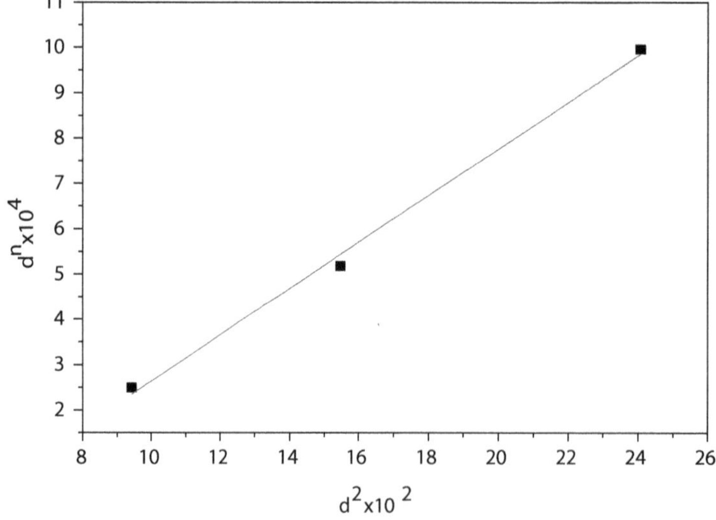

FIGURE 16.5 The plot of d^n against d^2.

TABLE 16.2

Mechanical Parameters by Vickers Hardness Test

Load g	$H_v MPa$	Meyer's Index, n	$k_1 \times 10^4$ kgm^{-1}	$k_2 \times 10^8$ kgm^{-1}	$W \times 10^{-3}$ g	σ_y GPa	C_{11} GPa
25	49.12	2.95	0.19	10.20	4.975	1.49	8.93
50	59.91	2.95	0.19	10.20	4.975	1.82	12.64
100	77.01	2.95	0.19	10.20	4.975	2.35	19.61

16.3 Conclusion

A key observation from thermal analysis indicates that the phase transition does not occur until the material is melted. This is an important aspect that extends the temperature range in which the crystal can be used for nonlinear optical applications. Furthermore, no decomposition occurs up to the melting point. This ensures the material's suitability for laser applications where the crystal must withstand high temperatures. Mechanical studies have shown that the H_eH_v increases with increasing applied indentation load, which can also be viewed as an inverse effect of indentation size.

REFERENCES

[1] Coats A.W., Redfern J.P., (1997), *Nature*, Vol. *201*, pp. 68–69.

[2] Dabhi R.M., Joshi M.J., (2003), *Indian Journal Physics A*, Vol. *77*, pp. 481–485.

[3] Joshi V.S., Joshi M.J., (2003), *Crystal Research and Technology*, Vol. *38*, pp. 817–821.

[4] Meng F.Q., Lu M.K., Yang Z.H., Zeng H., (1998), *Materials Letters*, Vol. *33*, pp. 265–268.

[5] Mallakpour S., Dinari M., (2010), *Chinese Journal of Polymer Science*, Vol. *28*, pp. 685–694.

[6] Singh B.K., Sharma R.K., Garg B.S., (2006), *Journal of Thermal Analysis and Calorimetry*, Vol. *84*, pp. 593–600.

[7] Pujol M.C., Mateos X., Aznar A., Solans X., Surinach S., Massons J., Dıaz F., Aguilo M., (2006), *Journal of Applied Crystallography*, Vol. *39*, pp. 230–236.

[8] Wyatt D.H., (1974), *Metals, Ceramics and Polymers*, Cambridge University Press, London.

[9] Onitsch E.M., (1947), *Microskopie*, Vol. *2*, pp. 131–151.

17

Nanotechnology Innovations in Biomedicine: Applications of Responsive Nanomaterials

A. M. Prasanna Lakshmi
Velagapudi Ramakrishna Siddhartha Engineering College, Kanuru, Vijayawada, India

Vishwanath Hokrani
Government Tool Room and Training Centre, Bagalkote, Kudalasangama, India

Mahesh R. Shukla
MKSSS Cummins College of Engineering for Women, Nagpur, India

Harish Kumar Banga
National Institute of Fashion Technology (NIFT), Kangra, India

A. Joseph Arockiam
Arasu Engineering College, Kumbakonam, India

17.1 Introduction

It is widely acknowledged that natural living systems have the remarkable ability to dynamically adjust their properties in response to environmental changes, showcasing an inherent intelligence. Drawing inspiration from the impressive capabilities of biological systems in energy conversion and multitasking, researchers have sought to create 'stimuli-responsive' materials exhibiting biomimetic behaviour, holding significant promise for applications in smart and intelligent device [1]. The concept of 'intelligent materials,' defined as substances capable of optimising their functions in response to environmental changes, was first introduced by Toshinori Takagi in April 1990 [2]. Presently, the term 'intelligent material' is synonymous with 'stimuli-responsive material' or 'smart material,' capturing heightened attention from researchers driven by advancements in technology and the growing demand for materials that align with evolving requirements. In 1959, Nobel laureate Richard Feynman pioneered the concept of nanotechnology [3], a field that gained formal definition in 1974 by Norio Taniguchi as the processing of materials at the scale of one atom or one molecule [4]. In the unfolding landscape of biomedical engineering, the pursuit of functionalised biomaterials is intensifying. Amidst the intricate diversity of living systems, the fundamental capacity to respond to environmental changes for sustaining regular operations, it remains imperative to uphold normal functions. This necessity for adaptation has catalysed the creation of intelligent nanomaterials, distinguished by their ability to react to diverse stimuli, modifying characteristics such as shape, surface area, size, permeability, solubility, and mechanical attributes. In response to the burgeoning demand for highly functional biomaterials, this mini-review comprehensively explores recent strategies in the preparation of smart polymers [5, 6]. The review delves into the potential and challenges of assembling responsive systems, particularly in applications like smart drug delivery and wound healing.

17.2 Stimuli-Based Smart Nanomaterials

Numerous types of intelligent nanomaterials are recognised based on their diverse properties. These stimuli are commonly categorised into three distinct groups: Physical, chemical, and biological [7, 8].

DOI: 10.1201/9781003495437-17

17.2.1 Nanomaterials for Drug Delivery (Physical Stimuli)

17.2.1.1 Thermo-responsive Nanomaterial

Thermo-responsive nanomaterials, also known as temperature-responsive nanomaterials, exhibit unique characteristics that make them particularly valuable in various applications. One key feature is their responsiveness to changes in temperature. These materials can be formulated to possess either a lower critical solution temperature (LCST) or an upper critical solution temperature (UCST). In the case of an LCST, the material undergoes a phase transition, such as swelling or solubilisation, above a specific temperature threshold. Conversely, UCST materials experience a phase transition, such as precipitation or gelation, when the temperature surpasses a particular level. This tuneable responsiveness to temperature variations allows for precise control over the nanomaterials' behaviour, making them instrumental in applications like drug delivery systems, where temperature-sensitive release mechanisms can be exploited for targeted and controlled therapeutic delivery.

Zheng et al. developed an injectable zwitterionic thermosensitive hydrogel using poly(*N*-isopropylacrylamide-*co*-sulphobetaine methacrylate) (PNS) nanogels [9]. The nanogels, with an average size of 105 nm, undergo a sol–gel phase transition based on the interplay between hydrophobic interactions and hydrogen bonding. At around 30°C, hydrogen bonding dominates, resulting in transparent fluids. Raising the temperature to 36°C induces a contracted gel formation due to enhanced hydrophobic interactions. This process involves water extraction, prompting phase separation and gel formation.

17.2.1.2 Electrical-Responsive Nanomaterials for Drug Delivery

In the realm of drug delivery, electrical-responsive nanomaterials have emerged as promising contenders for precise and controlled release mechanisms. These materials, often designed as nanostructures or nanocarriers, can respond to electrical stimuli, enabling targeted drug delivery. One common approach involves incorporating conductive elements, such as graphene or carbon nanotubes, into the nanomaterial matrix. When subjected to an external electrical stimulus, these materials undergo changes in conductivity or charge, triggering the release of encapsulated drugs [10]. This responsiveness to electrical signals allows for on-demand drug release at specific locations within the body, offering potential improvements in therapeutic efficacy and reduced side effects.

This class of intelligent materials is extensively applied, notably in artificially targeted drug delivery, muscle actuation, and energy transduction. Their primary advantages revolve around precise manipulation of the duration of an electrical pulse, current intensity, or the intervals between pulses. The initiation of an electric current disrupts hydrogen bonding in polymer chains, resulting in pH alterations and ultimately triggering either degradation of polymer chain or bending, thereby facilitating drug delivery. The core processes dictating the liberation of drugs from electro-responsive polymers involve the electrophoresis of charged drugs, diffusion, and the subsequent release of drugs as the electro-erodible polymers undergo erosion.

17.2.1.3 Magnetic-Responsive Nanomaterials for Drug Delivery

Magnetic-responsive nanomaterials represent another innovative avenue in drug delivery research. These nanomaterials are typically endowed with magnetic properties, often through the inclusion of magnetic nanoparticles like iron oxide. When an external magnetic field is applied, these nanomaterials respond by either moving towards the magnetic source or undergoing structural changes that release the encapsulated drug payload. This magnetic guidance system allows for targeted drug delivery to specific tissues or organs, enhancing drug concentration at the desired site while minimising exposure to healthy tissues. The tuneability of magnetic fields provides a dynamic tool for controlling drug release, making magnetic-responsive nanomaterials a promising strategy for advancing the precision and efficiency of drug delivery systems.

Magnetoresponsive nanomaterials are activated by an applied magnetic field and play a crucial role in advancing theragnostic devices, particularly through the use of magnetic nanoparticles. In numerous applications, these substances bring notable improvements, including magnetic separation, treatments involving magnetic hyperthermia, cellular tagging, immunoassays, drugs guided by magnetism, and diagnosis through magnetic resonance imaging (MRI). Magnetic-responsive nanomaterials with polymeric composition are crafted through a variety of methods, including hydrothermal procedures, microbial fabrication, co-precipitation techniques, combustion, laser pyrolysis, chemical vapour deposition, carbon arc, high-temperature thermal breakdown, and electrochemical production.

Magnetic nanoparticles, encompassing elements such as iron, cobalt, nickel, and metal oxides, present a valuable asset for the remote manipulation of cells, both in controlled laboratory environments and within living organisms. This contributes to a better comprehension of cellular functions and signalling processes. Conventional techniques like magnetic hyperthermia and thermal ablation encounter limitations, including insufficient targeting precision and penetration into deep tissues. MRI, pioneered by Paul Lauterbur in 1973 and clinically sanctioned in 1985, relies on aligning protons in response to an external magnetic field. MRI contrast agents, particularly those based on gadolinium, enhance the relaxation rates of water protons in specific tissues, enhancing image contrast [11]. Nevertheless, challenges such as toxicity and sensitivity in certain contrast media underscore the need for innovative contrast agents built on iron nanoparticle platforms. The future development of magnetoresponsive nanomaterials must prioritise factors like biocompatibility, robust contrast capability, and stability to address concerns about contrast agent degradation impacting image quality.

17.2.2 Nanomaterials for Wound Healing (Chemical Stimuli)

17.2.2.1 pH-Responsive Nanomaterials

The adaptability of pH-responsive nanomaterials to the fluctuating pH conditions in damaged tissues has attracted considerable interest in the realm of wound healing. These nanomaterials often incorporate pH-sensitive polymers or coatings that respond to changes in acidity or alkalinity. In an acidic wound environment, such as that found in infected or inflamed tissues, these materials can undergo controlled degradation, facilitating the release of therapeutic agents like growth factors or antimicrobial agents. Conversely, in a more neutral pH environment associated with healthy tissues, the nanomaterials can maintain structural integrity. This pH-responsive behaviour enhances the precision of drug delivery to wound sites, promoting accelerated healing while minimising potential side effects.

In the biomedical field, smart nanomaterials responsive to pH alterations showcase distinctive functional attributes, adding to their allure. Their appeal stems from their capacity to adjust to the varied pH environments found in specific anatomical or pathological systems within the human body. Notably, chronic wounds manifest pH values ranging between 7.4 and 5.4, saliva exhibits a pH range of 6.5 to 7.5, and the gastrointestinal tract undergoes pH fluctuations, from the stomach (4–6.5) to the intestine (5–8) [12].

Furthermore, in comparison to physiological conditions, pathological states often display anomalous pH levels. For instance, the microenvironment of tumours typically features a lower extracellular pH ranging between 6.5 and 6.9, while bacterial infections are characterised by acidic pus with pH levels falling within the range of 6.0–6.6. Additionally, inflamed tissues tend to exhibit a pH value of 6–7. To cater to these varied pH fluctuations, a range of materials responsive to pH changes has been created. These materials are commonly divided into two categories: Polymers incorporating ionisable moieties and polymers featuring acid-labile linkages.

The initial classification encompasses polymers containing ionisable elements, delicate basic, or acidic components (like amines and carboxylic acids) that adhere to a hydrophobic foundation. This arrangement results in protonation/deprotonation processes and the dispersion of charges across the ionisable groups. On the other hand, the second grouping consists of polymers with a structure featuring covalent linkages susceptible to acid-induced changes. In this case, a decrease in pH prompts the separation of polymer aggregates or the rupture of polymer chains. The latter category, characterised by a slower internal transformation, proves particularly well-suited for applications related to drug release.

The pH-induced phase transition in pH-responsive polymers typically occurs within a narrow pH range of 0.2–0.3. Well-known pH-sensitive polymers include poly(L-lysine), poly(N, N-dimethylaminoethyl methacrylate), poly(methacrylic acid), poly(acrylic acid), poly(*N,N*-dialkylaminoethylmethacrylates), poly(ethylenimine), chitosan, alginate, and hyaluronic acid.

In the context of wound healing, experiments conducted in living organisms have illustrated the efficacy of pH-responsive clusters of silver nanoparticles (AgNCs). These clusters are created by combining an inner layer of ortho ester with a poly ethylene glycol (PEG) corona and have shown effectiveness in treating infections caused by methicillin-resistant *Staphylococcus aureus* [13]. Furthermore, a pioneering hybrid nanoparticle design has been introduced, incorporating three block copolymers, each responsive to a specific pH level, thereby acting as sensors. This ultra-pH-sensitive (HyUPS) nanotransistor undergoes pH transitions at 6.9, 6.2, and 5.3, presenting promising potential for biomedical applications and providing novel insights into the study of acidification within endocytic organelles in living cells.

17.2.2.2 Redox-Responsive Nanomaterials

Redox-responsive nanomaterials offer a sophisticated strategy for wound healing applications by responding to variations in the redox state of the wound microenvironment. These materials often contain redox-sensitive moieties that can undergo reversible chemical changes in response to the oxidative or reductive conditions commonly found in wounds. The presence of such redox-responsive elements allows these nanomaterials to release therapeutic payloads, such as antioxidants or anti-inflammatory agents, precisely when needed. This responsive behaviour aids in mitigating oxidative stress, reducing inflammation, and fostering an environment conducive to tissue regeneration [14]. The utilisation of redox-responsive nanomaterials holds great promise for enhancing the effectiveness of wound healing therapies by providing targeted and dynamic responses to the specific redox conditions encountered during different stages of the healing process.

17.2.3 Biological-Responsive Smart Materials

Biological-responsive nanomaterials, a forefront in interdisciplinary research, are meticulously designed to interact with and respond to specific biological cues, offering tailored solutions across various biomedical applications. These nanomaterials exploit the intricate signalling within biological systems, providing controlled and targeted interventions.

Within this category, enzymatically responsive nanomaterials stand out, responding to the presence of specific enzymes associated with various diseases or physiological conditions. By incorporating enzyme-sensitive components like peptide sequences or cleavable linkers, these nanomaterials undergo structural changes or degradation upon encountering the target enzyme. This capability is particularly valuable in drug delivery systems, where therapeutic agent release is triggered by enzymatic activity at disease sites, ensuring precise and localised responses [15, 16].

Another noteworthy subgroup is gene-responsive nanomaterials, strategically engineered to interact with nucleic acids, either DNA or RNA. Through mechanisms such as electrostatic interactions or specific base pairing, these nanomaterials play a pivotal role in gene delivery, protecting genetic payloads from degradation, facilitating cellular uptake, and releasing the genetic material in response to intracellular cues. This holds promise for applications in gene therapy, addressing genetic disorders or modulating cellular functions.

In addition, pH-responsive nanomaterials exploit variations in acidity, temperature-sensitive nanomaterials respond to changes in temperature associated with inflammation or infection, and redox-responsive nanomaterials react to oxidative or reductive conditions. Further advancing the spectrum, glucose-responsive nanomaterials add a distinctive dimension. These materials can detect glucose levels, a crucial parameter in diabetes, and respond by modulating drug release or triggering therapeutic actions, offering potential solutions in the realm of personalised medicine.

In essence, the expansive field of biological-responsive nanomaterials continues to evolve, showcasing their adaptability and responsiveness to diverse biological cues. As research progresses, the potential for groundbreaking applications in diagnostics, drug delivery, and therapeutic interventions becomes increasingly evident, affirming the role of these nanomaterials in shaping the future of biomedical sciences.

17.3 Summary

In conclusion, the exploration of smart nanomaterials responsive to diverse stimuli, including magnetic fields, pH changes, thermal variations, and electrical signals, is ushering in a new era of possibilities in biomedical applications. Magnetoresponsive nanomaterials, activated by applied magnetic fields, exhibit significant promise in theragnostics, contributing to magnetic separation, hyperthermia treatments, and advanced imaging modalities like MRI. These materials offer precise control and versatility, addressing a spectrum of physiological and pathological conditions.

Simultaneously, pH-responsive nanomaterials, incorporating polymers with ionisable moieties and acid-labile linkages, prove to be versatile tools for biomedical applications. Their adaptability to specific pH environments, from chronic wounds to the gastrointestinal tract, positions them strategically for targeted drug delivery and diagnostics. Moreover, thermal-responsive and electrically responsive materials further enhance the toolkit for biomedical interventions. The ability of these materials to dynamically respond to temperature changes or electrical stimuli opens avenues for controlled drug release, advanced sensors, and innovative therapeutic modalities.

Incorporating previously discussed materials such as pH-sensitive polymers like poly(L-lysine) and poly(N, N-dimethylaminoethyl methacrylate), as well as magnetic nanoparticles, the ongoing innovation in smart nanomaterials holds great promise for revolutionising biomedical research and healthcare. This convergence of nanotechnology and medicine is poised to refine approaches to diagnosis, treatment, and monitoring, offering solutions characterised by improved precision, biocompatibility, and therapeutic efficacy across a spectrum of stimuli-responsive applications.

REFERENCES

[1] S. Amador-Vargas, M. Dominguez, G. León, B. Maldonado, J. Murillo, and G. L. Vides, "Leaf-folding response of a sensitive plant shows context-dependent behavioral plasticity," *Plant Ecology*, vol. *215*, no. 12, pp. 1445–1454, Dec. 2014, doi: 10.1007/s11258-014-0401-4.

[2] T. Takagi, "A concept of intelligent materials," *Journal of Intelligent Material Systems and Structures*, vol. *1*, no. 2, pp. 149–156, Apr. 1990, doi: 10.1177/1045389X9000100201.

[3] R. Feynman, "There's plenty of room at the bottom," in *Feynman and Computation*, CRC Press, 2018, pp. 63–76. Accessed: Nov. 25, 2023. [Online]. Available: https://www.taylorfrancis.com/chapters/edit/10.1201/9780429500459-7/plenty-room-bottom-richard-feynman

[4] N. Taniguchi, "On the basic concept of 'nano-technology'," in *Proc. Intl. Conf. Prod. Eng. Tokyo, Part II, 1974*, Japan Society of Precision Engineering, 1974. Accessed: Nov. 25, 2023. [Online]. Available: https://cir.nii.ac.jp/crid/1572261550373135488

[5] G. Ciofani and A. Menciassi, Eds., "Piezoelectric nanomaterials for biomedical applications," in *Nanomedicine and Nanotoxicology*. Springer, 2012. doi: 10.1007/978-3-642-28044-3.

[6] J. Ahlawat and M. Narayan, "Introduction to active, smart, and intelligent nanomaterials for biomedical application," in *Intelligent Nanomaterials for Drug Delivery Applications*, Elsevier, 2020, pp. 1–16. Accessed: Nov. 25, 2023. [Online]. Available: https://www.sciencedirect.com/science/article/pii/B9780128178300000011

[7] G. G. Genchi, A. Marino, C. Tapeinos, and G. Ciofani, "Smart materials meet multifunctional biomedical devices: Current and prospective implications for nanomedicine," *Frontiers in Bioengineering and Biotechnology*, vol. *5*, p. 80, 2017.

[8] S. S. Das et al., "Stimuli-responsive polymeric nanocarriers for drug delivery, imaging, and theragnosis," *Polymers*, vol. *12*, no. 6, p. 1397, 2020.

[9] A. Zheng et al., "Injectable zwitterionic thermosensitive hydrogels with low-protein adsorption and combined effect of photothermal-chemotherapy," *Journal of Materials Chemistry B*, vol. *8*, no. 46, pp. 10637–10649, 2020.

[10] M. R. Abidian, D.-H. Kim, and D. C. Martin, "Conducting-Polymer Nanotubes for Controlled Drug Release," *Advanced Materials*, vol. *18*, no. 4, pp. 405–409, Feb. 2006. doi: 10.1002/adma.200501726.

[11] M. Aflori, "Smart nanomaterials for biomedical applications—a review," *Nanomaterials*, vol. *11*, no. 2, p. 396, 2021.

[12] A. A. Date, J. Hanes, and L. M. Ensign, "Nanoparticles for oral delivery: Design, evaluation and state-of-the-art," *Journal of Controlled Release*, vol. *240*, pp. 504–526, 2016.

[13] X. Xie et al., "Ag nanoparticles cluster with pH-triggered reassembly in targeting antimicrobial applications," *Advanced Functional Materials*, vol. *30*, no. 17, p. 2000511, Apr. 2020. doi: 10.1002/adfm.202000511.

[14] B. Sultankulov, D. Berillo, K. Sultankulova, T. Tokay, and A. Saparov, "Progress in the development of chitosan-based biomaterials for tissue engineering and regenerative medicine," *Biomolecules*, vol. *9*, no. 9, p. 470, 2019.

[15] Y. Ding et al., "A dual-functional implant with an enzyme-responsive effect for bacterial infection therapy and tissue regeneration," *Biomaterials Science*, vol. *8*, no. 7, pp. 1840–1854, 2020.

[16] L. P. Datta, A. Chatterjee, K. Acharya, P. De, and M. Das, "Enzyme responsive nucleotide functionalized silver nanoparticles with effective antimicrobial and anticancer activity," *New Journal of Chemistry*, vol. *41*, no. 4, pp. 1538–1548, 2017.

18

Zirconium's Role in Advancing Electrochemical Energy Storage Devices: A Comprehensive Review

S. Roseline
MAM College of Engineering and Technology, Trichy, Tamil Nadu, India

18.1 Introduction

An increase in world's population and technological development are the key factors for the increase in electrical energy consumption [1]. Electrochemical energy storage devices are of vital importance to address the demands for efficient and sustainable energy solutions. The reliance on fossil fuels for energy production has become a challenge in today's environment [2]. So the world has started depending on renewable energy sources and electrification and hence these devices are becoming essential components in managing energy supply. The renewable energy sources are the primary drivers which require electrochemical energy storage [3]. They generate energy continuously and especially at low demands. Electrochemical storage devices like super capacitors (SCs) and batteries act as buffers by storing the excess energy during periods of abundance and releasing it when demand surpasses immediate supply [4–6]. This capability facilitates a more stable and reliable energy grid, ensuring that power is available whenever needed. Furthermore, the electrification of various sectors necessitates efficient energy storage systems. Electric vehicles, for instance, rely on high-performance batteries to store and deliver energy for propulsion [7]. Similarly, industrial processes increasingly harness electrical power, and dependable energy storage becomes paramount to ensure smooth operations, mitigate power fluctuations, and reduce reliance on traditional non-renewable energy sources [8–10]. Beyond stabilising the electrical grid and supporting electrified technologies, electrochemical energy storage devices also contribute to overall energy efficiency [11]. They enable the integration of decentralised renewable energy systems, allowing for energy generation closer to the point of consumption [12]. This decentralised approach reduces transmission losses associated with long-distance energy transport, enhancing the overall efficiency and sustainability of the energy infrastructure [13].

Research on sustainable methods to store energy efficiently is ongoing in recent days. Researches have started concentrating on novel materials for the next-generation storage devices. For decades, rechargeable batteries, which include Li-ion, Pb–acid, Ni–metalhydride, and Ni–Cd batteries, have been dominating the battery industry [14–16]. Electrodes in traditional capacitors are typically made of conductive materials like aluminium or tantalum [17]. These materials allow for the accumulation of charge on their surfaces. SC electrode materials include activated carbon, which has high porosity and a large surface area [18]. Other materials include metal oxides like ruthenium oxide and manganese dioxide for high-performance SCs [19].

Researchers are increasingly directing their focus towards zirconium (Zr)-based materials due to their extensive applications in the industry [20–22]. These silver-grey metals with a metallic lustre possess excellent plasticity and fall within the transition element in the IV B group [23]. Zirconium-based materials are steadily gaining significance in energy storage devices because of their high mechanical strength, outstanding theoretical capacity, and rapid Li^+ mobility [24]. In particular, Zr-doped $LiNi_{0.6-x}CO_{0.2}Mn_{0.3}Zr_xO_2$ electrodes demonstrate notable characteristics such as high discharge capacities and rapid electrochemical kinetics during cycling [25]. Additionally, zirconium oxide, chosen as an electrode material, offers corrosion-free properties, high mechanical strength, and ease of availability [26].

DOI: 10.1201/9781003495437-18

However, zirconium-based materials have inherent drawbacks, including lower power density and instability, limiting their popularity as electrode materials [27]. Researchers are actively addressing this issue by exploring the combination of zirconium oxide (ZrO_2) with highly conductive materials [28]. Numerous ongoing researches aim to identify suitable metals when combined with ZrO_2, to enhance its conductivity. Zirconium-based materials are evolving as promising substitute for the next-generation energy storage devices. Zirconium sulphide (ZrS_2) notably became the first material to replace lithium-ion battery electrodes [29]. Silver, known for its excellent properties such as conductivity, stability, catalytic activity, and superior operational stability, is being uniformly distributed with transition metal oxides (TMOs) to enhance the overall properties of these metal oxides [30].

This chapter systematically reviews the role played by zirconia-based materials in augmenting the electrochemical performance of conventional batteries and superconductors. The initial section underscores the imperative need for enhancing the capacity of electro storage devices. Subsequently, the chapter delves into an extensive exploration of research, focused on integrating Zr-based materials to improve the properties of anodes, cathodes, electrolytes, electrode coatings, and separator materials in various energy storage systems, including lithium-ion batteries (LIBs), sodium-ion batteries (SIBs), lithium-sulphur batteries (LSBs), lithium–air batteries, lithium–metal batteries, and superconductors. The chapter concludes by thoroughly examining the limitations and challenges associated with the incorporation of zirconium in electrochemical storage devices. This comprehensive review provides the contributions, advancements, and hurdles in leveraging Zr-based materials for enhancing the performance of batteries and superconductors.

18.2 Zirconium-Doped Lithium-Ion Batteries

All electronic devices and electric vehicles have started using LIBs as their source because of their energy density and cycling life. LIBs consist of a positive and a negative electrode and a Li-ion conducting electrolyte. But these batteries cannot fulfil the next-generation power requirements. Here, zirconium-based materials can be replaced because of their several advantages.

18.2.1 Anode Materials

Xu et al. introduced zirconium-based materials, such as $Ti_{1-x}Zr_xO_2$, $LiNi_{1/3}Co_{1/3}Mn_{1-x/3}Zr_{x/3}O_2$, and $Ti_{1-x}Zr_xO_2$, which are actively being utilised as anode materials in LIBs owing to their high specific capacity rates. Notably, zirconium-doped $Li_4Ti_5O_{12}$ has demonstrated impressive discharge and cycling capacities, with Lu et al. enhancing the ionic conductor by reinforcing $Li_4Ti_5O_{12}$ with Li_2ZrO_3 and doping it with Zr_{4+}, resulting in increased ionic and electronic conductivities. In a separate study conducted by Zheng et al., zirconium-based transition metal carbides/nitrides (MXenes) have emerged as promising materials for LIBs, exhibiting high electric conductivity and cycling stability. Wang et al. investigated the energy storage capacity and lithium-ion transfer when involved with Zr_2C and Zr_2CX_2 (where X represents F, O, or S). Furthermore, Li_2ZrO_3 has been identified as another effective anode electrode material by Wen et al., which serves as a crucial Li-ion conductor with remarkable cycle stability and diffusion coefficient. In a study by Zhu et al., Si-Cu-Ti-Zr-Ni and $Li_2Zr_xTi_{1-x}(PO_4)_3$ were highlighted as advantageous anode materials for LIBs, showcasing enhanced storage properties. The collective research of Zhang et al. underscores the diverse applications and potential of zirconium-based materials in advancing the performance of LIBs.

18.2.2 Cathode Materials

Cathode materials employed in LIBs, such as $LiCoO_2$, $LiMn_2O_4$, and $LiFePO_4$, face significant challenges due to their low specific energy and cycling instability. Recently, zirconium-based cathode materials have garnered attention among researchers for their potential to enhance battery performance. Studies by Lee et al. have indicated that cobaltite-type materials exhibit excellent performance when doped with zirconium. Two distinct samples, namely $LiCoO_2/Co_3O_4$ and $LiCo_{0.94}Zr_{0.06}O_2/MgO/ZrO_2$, were prepared as electrode materials. Composite electrodes were crafted with $LiMO_2$ particles coated with zirconium oxide, effectively preventing the undesirable reaction between cobalt ions. The cathode properties exhibited remarkable improvements with the incorporation of zirconium layers. Numerous research papers have highlighted the promising properties of cathodes based on lithiated TMOs, especially when doped with zirconium. Despite the growing body of research on doping zirconium ions in LIBs, the influence of these dopants at the atomic level remains not fully understood. In a study by Zhu et al., $Li[Li_{0.2}Mn_{0.54}Ni_{0.13}Co_{0.13}]O_2$, when doped with zirconium, demonstrated higher diffusion coefficients and enhanced crystal performance. The resulting structure was porous and hollow, providing ample 3D space for lithium-ion diffusion. Combining it with carbon further increased efficiency. In another research by Huang et al., a sol–gel process was employed to synthesise $Li_3V_{1.9}Zr_{0.1}(PO_4)_3$/graphene as a cathode material. The material exhibited a monoclinic structure, showcasing high discharge capacity and cycle stability. These findings underscore the potential of zirconium doping in improving the structural integrity and cycling stability of cathode materials, though a comprehensive understanding at the atomic level remains a subject of ongoing exploration.

18.2.3 Coating Materials

Conventional cathode and anode materials face challenges in maintaining stable capacity due to their structural and chemical instability. During charging, the interaction between the electrolyte and electrodes limits high energy density. To address this, Zheng et al. recommended inert coatings to prevent direct contact between electrolytes and electrodes. In a study by Huang et al., zirconium oxide (ZrO_2) was coated on the cathode material to improve electrochemical performance. The coating protected the cathode surface from HF attack without affecting lithium insertion. Coating $LiCoO_2$ with ZrO_2 demonstrated excellent performance in the high potential range, indicating increased cycle number and enhanced capacity retention. In another investigation by DeBruler et al., ZrO_2 was coated on $LiCoO_2$, forming $Zr(OC_3H_7)_4$, resulting in a smooth surface, leading to increased cycle numbers and improving its capacity retention. Various experimental data demonstrated that ZrO_2-coated $LiCoO_2$ outperformed pristine materials at room temperature. The atomic layer deposition method, controlling coating thickness, was adopted for this technology, further improving electrode performance. The chemical reaction during the coating process is represented as follows:

$$ZrO_2 + 4HF = ZrF_4 + 4H_2O \qquad (18.1)$$

The formation of ZrF_4 on the $LiCoO_2$ surface protected it from HF attack, enhancing electrode performance [31]. ZrO_2-coated Mn-based oxide cathodes gained attention due to their cost-effectiveness and high energy density. Lu et al. tried coating $Li[Li_{0.2}Ni_{0.17}Co_{0.07}Mn_{0.56}]O_2$ with 1 wt% Li_2ZrO_3 and showed 89% capacity retention after 50 cycles, outperforming conventional materials. To prepare the coated material, $CH_3COOLi_2.H_2O$ and $Zr(NO_3)_{4.5}H_2O$ were separately dissolved in alcohol. Doping zirconium with Fe was found to improve Li-ion conductivity in Li_2ZrO_3. Sputtering technology coated $Li(Ni_{1/3}Co_{1/3}Mn_{1/3})O_2$ with ZrO_2, improving retention capacity and cycling stability. Applying Li_2ZrO_3 (LZO) using a sol–gel process to coat Nickel Cobalt Aluminum oxide (NCA), inhibited mutual diffusion and decreased interfacial resistance, resulting in 91.7% retention capacity after 100 cycles [32]. A wet chemical method was employed to coat a thin film of ZrO_2 on $LiNi_{0.8}Co_{0.1}Mn_{0.1}O_2$, enhancing cycling stability of the cathode.

18.2.4 Electrolyte

Electrolytes play a crucial role in facilitating the transfer of Li ions between the cathode and anode in LIBs. While polymer electrolytes have been conventionally used, recent advancements have introduced solid-state electrolytes, particularly those based on zirconium, heralded for their impressive Li-ionic conductivities and enhanced electrochemical stability [33]. Researchers have identified zirconium-based solid electrolytes as promising for next-generation batteries. Notably, these electrolytes exhibit high ionic conductivity and improved electrochemical stability [34, 35]. Investigations have elucidated a mechanism for oxygen ion migration in ZrO_2, where the formation of oxygen defects facilitates the migration of O_2 ions in the solid state [36]. These oxygen defects play a pivotal role in modifying electron holes, consequently reducing lithium vacancies and enhancing Li-ion conductivity in materials like $Li_7La_3Zr_2O_{12}$ (LLZO).

To address Li loss, Manthiram et al. prepared LLZO compounds with different Li concentrations by incorporating Li_2CO_3, $La(OH)_3$, and ZrO_2 in the synthesis process. Studies demonstrate that Zr-based Li materials serve as excellent electrolytes for solid-state LIBs, with densification leading to an increase in cubic lattice constant [37]. Various synthesis methods have been explored, such as ball milling and heat treatment, solid-state reaction processes, and aerosol deposition technology [38]. Lu et al. investigated the doping of LLZO with aluminium (Al) to enhance ionic conductivity. Polymerisation technology has been employed to introduce small percentages of Al, resulting in densification during the sintering process and improved conductivity phase stability, thereby enhancing battery performance. In an experimental study by Liang et al., the material interfaces for solid-state batteries were designed using $Li_{6.25}Al_{0.25}La_3Zr_2O_{12}$ electrolyte. A comparison between conventional solid-state batteries and interfaced solid-state batteries revealed that the latter exhibited reduced discharge rates. Ceramic materials were employed to enhance lithium transfer at the electrolyte and electrode interface. Polyethylene oxide (PEO), a common electrolyte, was doped with $Li_{6.4}La_3Zr_{1.4}Ta_{0.6}O_{12}$ (LLZTO), leading to improved stability and ionic conductivity due to interfacial effects [39]. This doping also suppressed dendrite growth of lithium. The resulting flexible pouch battery showcased enhanced performance. Witoon et al. utilised solution casting methods to synthesise a flexible $Li_{6.75}La_3Zr_{1.75}Ta_{0.25}O_{12}$ electrolyte. The partial dehydrofluorination in this structure improved interactions between lithium, polyvinylidene fluoride (PVDF), and $Li_{6.75}La_3Zr_{1.75}Ta_{0.25}O_{12}$ particles, resulting in satisfactory mechanical properties and thermal stability.

18.2.5 Separator

Polypropylene has historically been the predominant separator material in batteries; however, the performance of lithium batteries has seen notable improvement with the adoption of ceramic separators [40]. Fu et al. successfully synthesised a ceramic separator comprising LLZO and poly(vinylidene fluoride-*co*-hexafluoropropylene) as a binder. This innovative assembly exhibited exceptionally high thermal stability and ionic conductivity.

From the foregoing, it is evident that Zr-based materials play a pivotal role in augmenting the performance of LIBs. These materials serve as effective dopants for both cathodes and anodes, enhancing the diffusion coefficient of lithium ions and fortifying structural stability. Moreover, Zr-based materials prove beneficial as coating layers for electrodes, stabilising their structure and safeguarding their surfaces. As electrolytes, they contribute to improved electrochemical stability, hinder the growth of lithium dendrites, and enhance ionic conductivity. Even when utilised as separators, Zr-based materials have been shown to enhance the cycling performance of batteries [41, 42].

18.3 Sodium Ion Batteries

Due to its cost-effectiveness and widespread availability, sodium has garnered significant attention for use in rechargeable batteries, particularly SIBs [41]. Researchers have increasingly focused on enhancing the performance of SIBs through the incorporation of zirconium-based materials [42]. In a one-step solid reaction approach, Chakravarty et al. successfully synthesised $NaZr_2(PO_4)_3$, an anode material that exhibited improved electrochemical performance. The synthesis reaction involved was as follows:

$$Na_2CO_3 + 4ZrO_2 + 6NH_4H_2PO_4 = 2NaZr_2(PO_4)_3 + CO_2 + 6NH_3 + 9H_2O \qquad (18.2)$$

The resulting product, $2NaZr_2(PO_4)_3$, featured numerous micro-particles tightly interconnected, thereby augmenting the structural and electrochemical properties of SIBs. Notably, the $NaZr_2(PO_4)_3$ structure demonstrated exceptional stability during cycling. In a study, Alves et al. reported that doping Na_2RuO_3 cathode with Zr materials led to a reaction with good reversible capacity and stable cycling. Remarkably, even a small amount of Zr significantly influenced the Na_2RuO_3 structure while minimally impacting crystal microstructure. Wang et al. highlighted the outstanding performance of a Na_2ZrO_3-yttrium-doped cathode material in SIBs. At high voltages, over-extraction of sodium can induce structural degradation.

However, the addition of Zr-based materials to Na-layered structures provided high reversible capacity and cycle stability. The band gap of Na_2ZrO_3 was determined to be 4.3 eV, resulting in low electrical conductivity, a challenge addressed by coating carbon elements on the surface to enhance conductivity.

When Zr is employed as an electrolyte in SIBs, it contributes to increased stability and ionic conductivity. Seo et al. was particularly interested in NASICON-type Zr electrolytes and fluorophosphate ceramic electrolytes. An electrolyte synthesised using a glass ceramic, $(Na_2O + NaF)$-TiO_2-B_2O_3-P_2O_5-ZrF_4 (NTBPZ), exhibited a wide electrochemical window compared to other electrolytes. Additionally, a $Na_{1+x}Zr_2Si_xP_{3-x}O_{12}$ separator was found to perform well as a solid-state conductor. This innovative Zr-based separator achieved energy and coulombic efficiencies of up to 81.96% and 99.45%, respectively. The separator facilitated rapid transportation of Na ions and effectively prevented chemical crossover between electrodes.

18.4 Lithium–Sulphur Batteries

A promising electrochemical energy storage system for the next generation involves LSBs, employing lithium as the negative electrode and sulphur as the positive electrode [43, 44]. Despite numerous challenges associated with this battery technology, researchers are actively addressing these issues by incorporating Zr-based materials [45]. The commonly utilised electrode comprises sulphur-incorporated carbon nanotubes [46]. Innovatively, Yao et al. introduced ZrO_2 into this electrode using melting-diffusion technology. This modification led to a significant improvement in cyclic performance and charge/discharge ratings compared to traditional electrodes. The addition of ZrO_2 as a dopant also enhanced the mesopore properties of the electrodes, maintaining the structure of permselective channels crucial for lithium-ion intercalation and deintercalation [46]. These channels were effectively utilised to prevent polysulphide dissolution into the electrolyte, contributing to increased stability in retention capacity for both charge and discharge rates.

In another study, Anasori et al. explored the use of $Li_{1+x}Y_xZr_{2-x}(PO_4)3$ (LYZP) as a solid-state electrolyte in LSBs to mitigate the issue of polysulphide crossover. This investigation revealed an increase in cycling rates, as the LYZP membrane served both as an electrolyte to activate Li-ion flow and as a separator to insulate the electrodes. LYZP demonstrated excellent chemical compatibility with the various cell components. Furthermore, Yuan et al. incorporated ZrO_2 into gel polymer electrolytes, resulting in stable columbic efficiency and superior electrolyte performance. This modification addressed challenges associated with LSBs, contributing to the overall advancement of this electrochemical energy storage system.

18.5 Lithium–Air Batteries

In Lithium–air batteries, oxygen serves as the cathode, while lithium acts as the anode, offering higher energy density compared to LIBs [47]. These batteries, utilising advanced materials, present a viable option for energy sources dependent on oil while maintaining environmental sustainability [48–50]. Wang et al. discovered that employing pure zirconium (Zr) as the anode enabled lithium–air batteries to operate effectively even at elevated temperatures. The hexagonal close-packed (hcp) structure of Zr exhibited a broad range of structural and thermodynamic properties, making it an ideal candidate for the anode in Li–air batteries. The use of cubic-structured $Li_7La_3Zr_2O_{12}$ (c-LLZO) demonstrated exceptional performance as a solid-state electrolyte in lithium–air batteries. Its properties contribute to the overall efficiency and reliability of the battery system. In a separate study, a researcher utilised Ca-stabilised ZrO_2 (CSZ) as an electrolyte, revealing that oxygen could diffuse through the electrolyte as ions. The electrochemical reactions in the battery involving the air electrode and the fuel electrode are as follows:

$$O_2 + 4e^- = 2O_2 \tag{18.3}$$

$$2O_2^- = O_2 + 4e^- \tag{18.4}$$

Measurement of charging and discharging rates demonstrated the capability of the battery for reversible operations, facilitated by the uniform transport of the CSZ electrolyte. This finding enhances the overall versatility and applicability of lithium–air batteries.

18.6 Lithium–Metal Batteries

Lithium and its alloys serve as prominent electrode materials in lithium–metal batteries, owing to lithium's high specific capacity, exceptionally low negative electrochemical potential, and theoretically low density, thereby demonstrating superior battery performance [51–53]. However, a substantial challenge arises from the significant interfacial resistance between the electrode and electrolytes in Li–metal batteries [54]. To address this challenge, Zhang et al. utilised the atomic layer deposition (ALD) method to coat Al_2O_3 on $Li_7La_{2.75}Ca_{0.25}Zr_{1.75}Nb_{0.25}O_{12}$ (LLCZN). This innovative approach notably improved the stability of the electrolyte. The incorporation of Zr_{4+} and Nb_{5+} in the LLCZN composition not only increased Li-ion conductivity but also enhanced the overall stability of the battery.

Recognising the crucial role of separators in preventing the diffusion of Li ions, Fang et al. employed a layer-by-layer assembly process to fabricate polyhedral oligomeric silsesquioxane (POSS) and ZrO_2 multilayers on polyethylene separators. Prior to assembly, the separators underwent a carbon dioxide plasma treatment for surface activation. The electrostatic forces drove the assembly process of ZrO_2 and POSS, contributing to the creation of an effective barrier against Li-ion diffusion. These advancements underscore the continuous efforts to overcome challenges associated with interfacial resistance and separator performance in Li–metal batteries, paving the way for enhanced stability and overall efficiency in lithium-based energy storage systems [55].

18.7 Super Capacitors (SCs)

In the realm of next-generation electrochemical energy storage devices, SCs play a pivotal role in efficiently storing and rapidly delivering energy compared to traditional batteries [56–58]. Integration with fuel cells or batteries further enhances their performance, leading to higher power density [59]. One notable advancement involved the utilisation of Zr-based electrodes in SCs, aiming to achieve elevated energy density and cycling stability. In a separate study, Schipper et al. synthesised a series of graphene–ZrO_2–polyaniline ternary composites (GZP-I, II, and III). The observation revealed the growth of polyaniline on the graphene–ZrO_2 composite, significantly enhancing the SC's performance. This composite exhibited superior energy storage capacity attributing to the introduction of polyaniline nanofibres, which facilitated ion diffusion into ZrO_2 and promoted efficient electron transportation.

Choudhary et al. employed one-step chronopotentiometry technology to fabricate a 3D structure of ZrO_2 while simultaneously reducing graphene oxide and polypyrrole. This ternary composite exhibited enhanced cycling stability and specific capacitance. Moreover, a different approach by Chen et al. involved increasing the pore size of a membrane composed of ZrO_2 nanoparticles and graphene oxide, resulting in a remarkable 100% improvement in capacitance value. The material's stability was assessed through cyclic tests and morphological examinations highlighting the impact on the material's structure. In a distinct investigation, the incorporation of ZrO_2 into pure polypyrrole coatings on a carbon substrate led to a smoother coating, altering the porosity and structural morphology of the electrodes [60–63]. These developments underscore the diverse strategies researchers employ to enhance the efficiency, stability, and capacitance of SCs in the quest for advanced electrochemical energy storage solutions.

18.8 Conclusion

Technological advancements have propelled energy storage systems into the next generation, featuring LIBs, LSBs, SIBs, lithium–metal batteries, lithium–air batteries, and superconductors. The integration of Zr-based materials into these energy storage solutions has demonstrated exceptional chemical properties, high mechanical strength, substantial capacities, and enhanced cyclic performance. This chapter provides an overview of the incorporation of Zr-based materials in the mentioned batteries and SCs, highlighting their diverse applications and benefits. Among the cases discussed, Zr-doped Li[Li$_{0.2}$Mn$_{0.54}$Ni$_{0.13}$Co$_{0.13}$]O$_2$ exhibited improved cyclic performance and increased ion conductivity in batteries. In another instance, the introduction of Zr composites enhanced the energy density of superconductors.

However, challenges persist in the application process, indicating the need for further research. Some Zr-doped materials exhibited low capacities due to structural instability, emphasising the importance of maintaining the structure for improved cyclic performance and rate capability. Observations also indicated that Zr reduced electrochemical impedance and facilitated Li-ion diffusion, contributing to enhanced battery performance. Despite the low abundance of Zr in the Earth's crust (0.025%) and its dispersed distribution, there are challenges associated with its industrial use. The synthesis process requiring high temperatures poses additional hurdles in meeting industrial demands. Further analysis of the structure and behaviour of zirconia materials is essential to optimise their properties. The battery's performance, heavily dependent on ionic conductivity, was influenced by the monoclinic Li$_2$ZrO$_3$ structure, which improved Li-ion diffusion along a 3D path but presented stability challenges. On the other hand, Zr-doped LTO with a narrow structure showed potential for enhancing electrochemical properties. In conclusion, a comprehensive understanding of the relationship between the structure and electrochemical properties of Zr-based materials is crucial for improving the performance of Zr-based batteries and superconductors. Ongoing research efforts are needed to address challenges and unlock the full potential of Zr in advancing energy storage technologies.

REFERENCES

[1] Wang, F., Liu, Y., Zhao, Y., Wang, Y., Wang, Z., Zhang, W., & Ren, F. *Applied Sciences, 8*(1), (2017), 22.

[2] Ma, J., Wen, J., Li, Q., & Zhang, Q. *International Journal of Hydrogen Energy, 38*(34), (2013). 14896–14902.

[3] Zhu, J., Chroneos, A., Eppinger, J., & Schwingenschlögl, U. *Applied Materials Today, 5*, (2016). 19–24.

[4] Ma, J., Zhang, Y., Qin, C., Ren, F., & Wang, G. *International Journal of Hydrogen Energy, 45*(23), (2020). 13025–13034.

[5] Lee, J. H., Kim, J. W., Kang, H. Y., Kim, S. C., Han, S. S., Oh, K. H., ... & Joo, Y. C. *Journal of Materials Chemistry A, 3*(24), (2015). 12982–12991.

[6] Zhu, Y., Choi, S. H., Fan, X., Shin, J., Ma, Z., Zachariah, M. R., ... & Wang, C. *Advanced Energy Materials, 7*(7), (2017). 1601578.

[7] Huang, S., Wilson, B. E., Smyrl, W. H., Truhlar, D. G., & Stein, A. *Chemistry of Materials, 28*(3), (2016). 746–755.

[8] Hu, B., DeBruler, C., Rhodes, Z., & Liu, T. L. *Journal of the American Chemical Society, 139*(3), (2017). 1207–1214.

[9] Zheng, S., Li, X., Yan, B., Hu, Q., Xu, Y., Xiao, X., ... & Pang, H. *Advanced Energy Materials, 7*(18), (2017). 1602733.

[10] Lu, Y., Li, B., Zheng, S., Xu, Y., Xue, H., & Pang, H. *Advanced Functional Materials, 27*(44), (2017). 1703949.

[11] Manthiram, A., Chung, S. H., & Zu, C. *Advanced Materials, 27*(12), (2015). 1980–2006.

[12] Xu, S., & Negishi, E. I. *Accounts of Chemical Research, 49*(10), (2016). 2158–2168.

[13] Liang, J., Chen, R. P., Wang, X. Y., Liu, T. T., Wang, X. S., Huang, Y. B., & Cao, R. *Chemical Science, 8*(2), (2017). 1570–1575.

[14] Witoon, T., Chalorngtham, J., Dumrongbunditkul, P., Chareonpanich, M., & Limtrakul, J. *Chemical Engineering Journal, 293*, (2016). 327–336.

[15] Fu, C., Bai, S., Liu, Y., Tang, Y., Chen, L., Zhao, X., & Zhu, T. *Nature Communications, 6*(1), (2015). 8144.

[16] Chakravarty, D., Tiwary, C. S., Machado, L. D., Brunetto, G., Vinod, S., Yadav, R. M., ... & Ajayan, P. M. *Advanced Materials, 27*(31), (2015). 4534–4543.

[17] Alves, A. P. P., Koizumi, R., Samanta, A., Machado, L. D., Singh, A. K., Galvao, D. S., & Ajayan, P. M. *Nano Energy, 31*, (2017). 225–232.

[18] Wang, M., Hu, Y., Han, J., Guo, R., Xiong, H., & Yin, Y. *Journal of Materials Chemistry A, 3*(41), (2015). 20727–20735.

[19] Seo, I., Lee, C. R., & Kim, J. K. *Journal of Physics and Chemistry of Solids, 108*, (2017). 25–29.

[20] Yao, B., Zhang, J., Kou, T., Song, Y., Liu, T., & Li, Y. *Advanced Science*, *4*(7), (2017). 1700107.

[21] Anasori, M. Lukatskaya, R., and Gogotsi, Y. *Nature Reviews Materials*, *2*(2), (2017). 16098.

[22] Yuan, S., Bao, J. L., Wang, L., Xia, Y., Truhlar, D. G., & Wang, Y. *Advanced Energy Materials*, *6*(5), (2016). 1501733.

[23] Wang, L., Han, Y., Feng, X., Zhou, J., Qi, P., & Wang, B. *Coordination Chemistry Reviews*, *307*, (2016). 361–381.

[24] Xia, H., Xia, Q., Lin, B., Zhu, J., Seo, J. K., & Meng, Y. S. *Nano Energy*, *22*, (2016). 475–482.

[25] Liu, W., Chen, Z., Zhou, G., Sun, Y., Lee, H. R., Liu, C., ... & Cui, Y. *Advanced Materials*, *28*(18), (2016). 3578–3583.

[26] Liu, C., Neale, Z. G., & Cao, G. *Materials Today*, *19*(2), (2016). 109–123.

[27] Liu, W., Shi, Q., Qu, Q., Gao, T., Zhu, G., Shao, J., & Zheng, H. *Journal of Materials Chemistry A*, *5*(1), (2017). 145–154.

[28] Zhang, B., Chen, J., Sun, W., Shao, Y., Zhang, L., & Zhao, K. *Energies*, *15*(13), (2022). 4698.

[29] Fang, X., & Peng, H. *Small*, *11*(13), (2015). 1488–1511.

[30] Chi, S. S., Liu, Y., Song, W. L., Fan, L. Z., & Zhang, Q. *Advanced Functional Materials*, *27*(24), (2017). 1700348.

[31] He, Z., Wang, Z., Chen, H., Huang, Z., Li, X., Guo, H., & Wang, R. *Journal of Power Sources*, *299*, (2015). 334–341.

[32] Yu, L., Hu, H., Wu, H. B., & Lou, X. W. *Advanced Materials*, *29*(15), (2017). 1604563.

[33] Xu, S., Jacobs, R. M., Nguyen, H. M., Hao, S., Mahanthappa, M., Wolverton, C., & Morgan, D. *Journal of Materials Chemistry A*, *3*(33), (2015). 17248–17272.

[34] Schipper, F., Bouzaglo, H., Dixit, M., Erickson, E. M., Weigel, T., Talianker, M., & Aurbach, D. *Advanced Energy Materials*, *8*(4), (2018). 1701682.

[35] Yu, L., Wu, H. B., & Lou, X. W. D. *Accounts of Chemical Research*, *50*(2), (2017). 293–301.

[36] Yano, A., Ueda, A., Shikano, M., Sakaebe, H., & Ogumi, Z. *Journal of the Electrochemical Society*, *163*(2), (2015). A75.

[37] Wei, Q., Xiong, F., Tan, S., Huang, L., Lan, E. H., Dunn, B., & Mai, L. *Advanced Materials*, *29*(20), (2017). 1602300.

[38] Choudhary, N., Li, C., Moore, J., Nagaiah, N., Zhai, L., Jung, Y., & Thomas, J. *Advanced Materials*, *29*(21), (2017). 1605336.

[39] Zhou, L., Zhuang, Z., Zhao, H., Lin, M., Zhao, D., & Mai, L. *Advanced Materials*, *29*(20), (2017). 1602914.

[40] Sun, H., Mei, L., Liang, J., Zhao, Z., Lee, C., Fei, H., ... & Duan, X. *Science*, *356*(6338), (2017). 599–604.

[41] Cheng, L., Wu, C. H., Jarry, A., Chen, W., Ye, Y., Zhu, J., & Doeff, M. *ACS Applied Materials & Interfaces*, *7*(32), (2015). 17649–17655.

[42] Ai, W., Zhou, W., Du, Z., Sun, C., Yang, J., Chen, Y., & Yu, T. *Advanced Functional Materials*, *27*(19), (2017). 1603603.

[43] Tao, X., Liu, Y., Liu, W., Zhou, G., Zhao, J., Lin, D., & Cui, Y. *Nano Letters*, *17*(5), (2017). 2967–2972.

[44] Song, J., Gordin, M. L., Xu, T., Chen, S., Yu, Z., Sohn, H., & Wang, D. *Angewandte Chemie*, *127*(14), (2015). 4399–4403.

[45] Schipper, F., Dixit, M., Kovacheva, D., Talianker, M., Haik, O., Grinblat, J., & Aurbach, D. *Journal of Materials Chemistry A*, *4*(41), (2016). 16073–16084.

[46] Chung, S. H., Han, P., Singhal, R., Kalra, V., & Manthiram, A. *Advanced Energy Materials*, *5*(18), (2015). 1500738.

[47] Zhou, G., Paek, E., Hwang, G. S., & Manthiram, A. *Nature Communications*, *6*(1), (2015). 7760.

[48] Han, B., Zhang, W., Gao, D., Zhou, C., Xia, K., Gao, Q., & Wu, J. *Journal of Power Sources*, *449*, (2020). 227564.

[49] Häupler, B., Wild, A., & Schubert, U. S. *Advanced Energy Materials*, *5*(11), (2015). 1402034.

[50] Fu, K., Gong, Y., Dai, J., Gong, A., Han, X., Yao, Y., & Hu, L. *Proceedings of the National Academy of Sciences*, *113*(26), (2016). 7094–7099.

[51] Jalem, R., Morishita, Y., Okajima, T., Takeda, H., Kondo, Y., Nakayama, M., & Kasuga, T. *Journal of Materials Chemistry A*, *4*(37), (2016). 14371–14379.

[52] Xia, W., Xu, B., Duan, H., Guo, Y., Kang, H., Li, H., & Liu, H. *ACS Applied Materials & Interfaces*, *8*(8), (2016). 5335–5342.

[53] Liu, C., Rui, K., Shen, C., Badding, M. E., Zhang, G., & Wen, Z. *Journal of Power Sources*, *282*, (2015). 286–293.

 Introduction to Functional Nanomaterials

[54] Buannic, L., Orayech, B., Lopez Del Amo, J. M., Carrasco, J., Katcho, N. A., Aguesse, F., & Llordés, A. *Chemistry of Materials*, *29*(4), (2017). 1769–1778.

[55] Zhang, J., Zhao, N., Zhang, M., Li, Y., Chu, P. K., Guo, X., & Li, H. *Nano Energy*, *28*, (2016). 447–454.

[56] Han, F., Zhu, Y., He, X., Mo, Y., & Wang, C. *Advanced Energy Materials*, *6*(8), (2016). 1501590.

[57] Li, W., Li, Z., Yang, F., Fang, X., & Tang, B. *ACS Applied Materials & Interfaces*, *9*(40), (2017). 35030–35039.

[58] Kato, T., Iwasaki, S., Ishii, Y., Motoyama, M., West, W. C., Yamamoto, Y., & Iriyama, Y. *Journal of Power Sources*, *303*, (2016). 65–72.

[59] Afyon, S., Krumeich, F., & Rupp, J. L. *Journal of Materials Chemistry A*, *3*(36), (2015). 18636–18648.

[60] Van Den Broek, J., Afyon, S., & Rupp, J. L. *Advanced Energy Materials*, *6*(19), (2016). 1600736.

[61] Zhang, X., Liu, T., Zhang, S., Huang, X., Xu, B., Lin, Y., & Shen, Y. *Journal of the American Chemical Society*, *139*(39), (2017). 13779–13785.

[62] Hwang, J. Y., Myung, S. T., & Sun, Y. K. *Chemical Society Reviews*, *46*(12), (2017). 3529–3614.

[63] Wang, J., Xiao, X., Lu, Y., Wang, Y., Chen, C., & Pang, H. *ChemElectroChem*, *6*(7), (2019). 1949–1968.

19

Nanocomposite Materials for Biomedical and Energy Storage Devices

M. Devaiah
Geethanjali College of Engineering and Technology, Telangana, India

Sharan Kumar
Sharnbasva University, Kalaburagi, Karnataka, India

Mahesh R. Shukla
MKSSS Cummins College of Engineering for Women, Nagpur, India

Saravana Bavan
Dayananda Sagar University, Bangalore, South Karnataka, India

P. Vignesh
Chennai Institute of Technology, Chennai, India

19.1 Introduction

Nanocomposites, an innovative class of materials, have emerged as a transformative force in contemporary materials science, attracting researchers and engineers [1]. At their core, nanocomposites are distinguished by their composition – a matrix material integrated with nanoscale reinforcements. This union provides these materials with exceptional properties that overcome the limitations of traditional composites, opening up avenues for unprecedented advancements in various applications. The fundamental nature of nanocomposites, their fabrication methods, diverse types, and the immense potential they hold, particularly in the realms of biomedical applications and energy storage, are explored in the coming sections [2, 3].

Nanocomposites owe their extraordinary attributes to the synergistic effects arising from the incorporation of nanoparticles, nanotubes, or nanosheets within the matrix material [4]. This intentional inclusion at the nanoscale enhanced mechanical, thermal, and electrical properties, thus elevating the overall performance of the composite material [5]. The utilisation of various fabrication methods allows researchers to precisely engineer nanocomposites tailored to specific applications, ranging from medical advancements to sustainable energy solutions.

19.1.1 Fabrication Methods

The fabrication of nanocomposites is a meticulous process that involves integrating nanoscale constituents into matrices through various techniques. In the realm of polymer matrix nanocomposites, for instance, the melt blending or solution casting methods facilitate the homogenous dispersion of nanoparticles, leading to improved polymer properties. Metal matrix nanocomposites, on the other hand, may be synthesised through powder metallurgy or electrodeposition, harnessing the advantageous properties of metals for specific applications. Ceramic matrix nanocomposites, with their enhanced toughness, are realised through techniques like chemical vapour infiltration or spark plasma sintering.

DOI: 10.1201/9781003495437-19

19.2 Nanocomposites and Their Potential Applications

The diversity of nanocomposites extends across multiple material classes, each offering unique advantages for specific applications.

19.2.1 Carbon-Based Nanocomposites

Carbon-based nanocomposites represent a cutting-edge class of materials that leverage the extraordinary properties of carbon nanomaterials, such as graphene, carbon nanotubes (CNTs), and carbon nanofibres, within diverse matrices. This synergy between carbon nanomaterials and various matrix materials gives rise to a range of exceptional properties, making carbon-based nanocomposites highly attractive for a myriad of applications [6].

19.2.1.1 Graphene-Based Nanocomposites

Graphene, a single layer of carbon atoms arranged in a hexagonal lattice, is a key player in carbon-based nanocomposites. Its remarkable electrical conductivity, high surface area, and mechanical strength contribute to enhanced properties in various matrices.

Applications include energy storage devices, where graphene-based nanocomposites significantly improve the performance of batteries and supercapacitors, offering higher capacity, faster charge–discharge rates, and longer cycle life [7].

19.2.1.2 Carbon Nanotube (CNT) Nanocomposites

CNTs, cylindrical structures of carbon atoms, exhibit extraordinary tensile strength and electrical conductivity. When incorporated into matrices, they impart these properties to the composite material. CNT nanocomposites find applications in aerospace, automotive, and structural materials, where their lightweight nature and mechanical strength are highly advantageous.

19.2.1.3 Carbon Nanofibre Nanocomposites

Carbon nanofibres, similar to CNTs but with a different structural arrangement, are characterised by high thermal conductivity and mechanical strength. These nanocomposites are utilised in applications such as reinforcement in polymers, enhancing the overall structural integrity and thermal properties of the composite material.

19.2.2 Applications of Carbon-Based Nanocomposites

19.2.2.1 Energy Storage Devices

Carbon-based nanocomposites play a pivotal role in improving the performance of batteries and supercapacitors. The high conductivity of carbon materials enhances charge transport, leading to enhanced energy storage capacity and faster charging [6].

19.2.2.2 Conductive Polymers

Integration of carbon nanomaterials into polymers results in conductive composites. These materials are used in flexible electronics, sensors, and actuators, where electrical conductivity is crucial [8].

19.2.2.3 Structural Materials

CNTs and carbon nanofibres reinforce structural materials, providing lightweight yet robust components in applications ranging from automotive to construction.

19.2.2.4 Biomedical Devices

Carbon-based nanocomposites are explored in biomedical applications, such as drug delivery systems and imaging agents. Their biocompatibility and tuneable properties make them promising candidates for improving therapeutic and diagnostic capabilities [9].

19.3 Metal Oxide Nanocomposites

Metal oxide nanocomposites represent a groundbreaking class of materials that has shown widespread attention due to their novel combination of properties derived from the inclusion of metal oxide nanoparticles into various matrices. Among the prominent metal oxides explored in nanocomposites are titanium dioxide (TiO_2) and zinc oxide (ZnO) [10]. The deliberate incorporation of these nanoparticles into matrices results in materials with enhanced functionalities, making them versatile candidates for biomedical applications [11].

The distinctive physicochemical properties of metal oxide nanocomposites, such as their adjustable surface characteristics, reactivity, and biocompatibility, make them particularly compelling for use in various fields. This section explores the metal oxide nanocomposites, focusing on their integration into biomedical applications. Beyond their initial roles in drug delivery systems and medical imaging, these nanocomposites have proven to be indispensable in addressing complex challenges within the biomedical field, showcasing their potential to revolutionise diagnostics, therapeutics, and regenerative medicine.

19.4 Biomedical Applications of Metal Oxide Nanocomposites

19.4.1 Biocompatibility and Cytotoxicity

Metal oxide nanocomposites often possess inherent biocompatibility, a critical feature for biomedical applications. Their interaction with biological systems, including cells and tissues, has been extensively studied to ensure minimal cytotoxic effects, making them suitable candidates for medical use [12].

19.4.2 Antimicrobial Properties

Zinc oxide nanoparticles, in particular, exhibit potent antimicrobial properties. This antimicrobial activity is leveraged in various biomedical applications, such as wound healing and surface coatings for medical devices, to mitigate the risk of infections [13, 14].

19.4.3 Biosensors and Diagnostics

Metal oxide nanocomposites play a pivotal role in the development of biosensors for diagnostic purposes. Their unique electronic properties and surface reactivity enable the sensitive detection of biomolecules, contributing to the early diagnosis of diseases.

19.4.4 Therapeutic Platforms

Beyond drug delivery, metal oxide nanocomposites serve as platforms for innovative therapeutic strategies. The controlled release of therapeutic agents can be tailored to specific physiological conditions, optimising treatment efficacy while minimising side effects.

19.4.5 Tissue Engineering and Regenerative Medicine

Metal oxide nanocomposites contribute to advancements in tissue engineering by providing scaffolds with tailored properties. Their ability to mimic the extracellular matrix and support cellular growth makes them valuable in regenerative medicine for repairing or replacing damaged tissues.

19.4.6 Targeted Drug Delivery Systems

The surface characteristics of metal oxide nanocomposites can be modified to enable targeted drug delivery. Functionalisation with specific ligands allows for site-specific drug release, reducing systemic side effects and improving the overall therapeutic outcome.

19.5 Polymer–Clay Nanocomposites

Polymer–clay nanocomposites, resulting from the amalgamation of nanoscale clay particles into polymer matrices, have emerged as a compelling class of materials, drawing considerable attention across scientific and industrial domains. Meticulously fabricated through techniques like melt intercalation or solution mixing, these nanocomposites exhibit a well-structured composition with enhanced properties. Notably, the inclusion of clay nanoparticles elevates the mechanical strength, thermal stability, and barrier properties of polymers. Montmorillonite and kaolinite, two prevalent clay types, play distinctive roles in reinforcing the layered structure and improving mechanical and thermal characteristics, respectively [15]. The versatile applications of polymer–clay nanocomposites span various industries, including packaging materials, automotive components, construction materials, textiles, and aerospace engineering. Their reduced flammability, improved dimensional stability, and lightweight nature make them invaluable in addressing multifaceted challenges and advancing material science across diverse sectors, showcasing their potential to revolutionise the landscape of advanced materials.

19.5.1 Fabrication

The fabrication of polymer–clay nanocomposites typically involves the dispersion of clay nanoparticles within a polymer matrix using methods such as melt intercalation or solution mixing. This results in a well-dispersed structure, where the clay layers are separated within the polymer, imparting a range of improved properties.

19.6 Potential Applications of Polymer–Clay Nanocomposites

19.6.1 Supercapacitors

Polymer–clay nanocomposites, particularly those incorporating conductive polymers and clay with high surface area, have been explored for use in supercapacitors. The enhanced electrical conductivity and surface area contribute to improved charge storage capacity and faster charge–discharge rates, making them promising for energy storage applications.

19.6.2 Battery Electrodes

The incorporation of clay nanoparticles into polymer matrices has been studied for battery electrode materials. The nanocomposites offer improved mechanical strength and thermal stability, addressing challenges associated with the expansion and contraction of electrode materials during charge–discharge cycles.

19.6.3 Conductive Polymer Composites

Polymer–clay nanocomposites have been utilised to enhance the conductivity of polymers, making them suitable for use in energy storage devices. Conductive polymers, such as polyaniline, when combined with clay, form nanocomposites with improved electrical properties, contributing to the development of efficient energy storage materials.

19.6.4 Drug Delivery Systems

Polymer–clay nanocomposites serve as promising candidates for drug delivery systems. The large surface area and high aspect ratio of clay nanoparticles enable the efficient loading and controlled release of therapeutic agents. The enhanced barrier properties of the nanocomposites can also prevent premature drug release.

19.6.5 Biocompatible Scaffolds for Tissue Engineering

In the biomedical field, polymer–clay nanocomposites find applications in tissue engineering. The nanocomposites can be used to create biocompatible scaffolds with tailored mechanical properties, supporting cell adhesion, proliferation, and tissue regeneration.

19.6.6 Bioimaging Contrast Agents

The incorporation of clay nanoparticles into polymers enhances the contrast properties of imaging agents. Polymer–clay nanocomposites are explored for their potential use in bioimaging applications, providing improved visibility and accuracy in medical diagnostics.

19.6.7 Wound Healing

Polymer–clay nanocomposites, especially those containing antimicrobial agents, exhibit potential in wound healing applications. The controlled release of therapeutic agents from the nanocomposites can aid in preventing infections and promoting faster healing.

19.6.8 Implant Coatings

Polymer–clay nanocomposites have been investigated for coating implants to improve their biocompatibility and longevity. The enhanced mechanical properties and the ability to control drug release make these nanocomposites suitable for applications in orthopaedics and other implant-related fields.

19.7 Summary

In summary, nanocomposites have emerged as versatile materials with transformative applications in both biomedical and energy storage domains. Carbon-based nanocomposites, leveraging materials like graphene and carbon nanotubes, show promise in enhancing energy storage devices such as batteries and supercapacitors. Their remarkable electrical conductivity and mechanical strength make them key players in advancing energy storage technologies. Metal oxide nanocomposites, incorporating nanoparticles like titanium dioxide and zinc oxide, exhibit widespread utilisation in biomedical applications. Their biocompatibility, imaging contrast properties, and antimicrobial features make them integral in drug delivery systems, medical imaging, and various therapeutic platforms. Furthermore, polymer–clay nanocomposites, formed by integrating nanoscale clay particles into polymer matrices, showcase enhanced mechanical and thermal properties. These nanocomposites find applications in diverse fields, including biomedical scaffolds for tissue engineering and energy storage devices such as supercapacitors. The continual exploration and refinement of these nanocomposites underscore their pivotal role in addressing complex challenges and driving innovation in critical areas of science and technology.

REFERENCES

[1] A. Sharma and M. Hafeez, "Nanocomposite materials for biomedical and energy storage applications," in Ashutosh Sharma *Nanocomposite Materials for Biomedical and Energy Storage Applications*, IntechOpen, 2022. doi: 10.5772/intechopen.95130

[2] Y.-W. Mai and Z.-Z. Yu, "Polymer nanocomposites," 2006, Accessed: Dec. 10, 2023. https://books. google.com/books?hl=en&lr=&id=kbtQAwAAQBAJ&oi=fnd&pg=PP1&dq=nanocomposites&ots=f5 6YBRlmjk&sig=Mw-OSPK8KwBqgVD-0sysdc2b9yc

[3] S. Komarneni, "Nanocomposites," *Journal of Materials Chemistry*, vol. *2*, no. 12, pp. 1219–1230, 1992.

[4] E. Omanović-Mikličanin, A. Badnjević, A. Kazlagić, and M. Hajlovac, "Nanocomposites: A brief review," *Health Technology*, vol. *10*, no. 1, pp. 51–59, Jan. 2020, doi: 10.1007/s12553-019-00380-x.

[5] T. Hassan et al., "Functional nanocomposites and their potential applications: A review," *Journal of Polymer Research*, vol. *28*, no. 2, p. 36, Feb. 2021, doi: 10.1007/s10965-021-02408-1.

[6] S. S. Siwal, Q. Zhang, N. Devi, and V. K. Thakur, "Carbon-based polymer nanocomposite for high-performance energy storage applications," *Polymers*, vol. *12*, no. 3, p. 505, 2020.

[7] L. Ji, P. Meduri, V. Agubra, X. Xiao, and M. Alcoutlabi, "Graphene-based nanocomposites for energy storage," *Advanced Energy Materials*, vol. *6*, no. 16, p. 1502159, Aug. 2016, doi: 10.1002/aenm.201502159.

[8] R. Rahmawati et al., "The synthesis of Fe3O4/MWCNT nanocomposites from local iron sands for electrochemical sensors," in *AIP Conference Proceedings*, AIP Publishing, 2018. Accessed: Dec. 10, 2023. [Online]. Available: https://pubs.aip.org/aip/acp/article-abstract/1958/1/020016/756534

[9] D. Maiti, X. Tong, X. Mou, and K. Yang, "Carbon-based nanomaterials for biomedical applications: A recent study," *Frontiers in Pharmacology*, vol. *9*, p. 1401, 2019.

[10] S. Jafari, B. Mahyad, H. Hashemzadeh, S. Janfaza, T. Gholikhani, and L. Tayebi, "Biomedical applications of TiO$_2$ nanostructures: Recent advances," *Indian Journal of Nephrology*, vol. *15*, pp. 3447–3470, May 2020, doi: 10.2147/IJN.S249441.

[11] F. I. Abou El Fadl, D. E. Hegazy, N. A. Maziad, and M. M. Ghobashy, "Effect of nano-metal oxides (TiO$_2$, MgO, CaO, and ZnO) on antibacterial property of (PEO/PEC-co-AAm) hydrogel synthesized by gamma irradiation," *International Journal of Biological Macromolecules*, vol. *250*, p. 126248, 2023.

[12] M. Lee, et al., "ZrO$_2$/ZnO/TiO$_2$ nanocomposite coatings on stainless steel for improved corrosion resistance, biocompatibility, and antimicrobial activity," *ACS Applied Materials & Interfaces*, vol. *14*, no. 11, pp. 13801–13811, Mar. 2022, doi: 10.1021/acsami.1c19498.

[13] K. Kusdianto, T. D. Sari, M. A. Laksono, S. Madhania, and S. Winardi, "Fabrication and application of ZnO-Ag nanocomposite materials prepared by gas-phase methods," in *IOP Conference Series: Materials Science and Engineering*, IOP Publishing, 2021, p. 012023. Accessed: Dec. 10, 2023. [Online]. Available: https://iopscience.iop.org/article/10.1088/1757-899X/1053/1/012023/meta

[14] S. J. Owonubi, N. M. Malima, and N. Revaprasadu, "Metal oxide–based nanocomposites as antimicrobial and biomedical agents," in *Antibiotic Materials in Healthcare*, Elsevier, 2020, pp. 287–323. Accessed: Nov. 30, 2023. [Online]. Available: https://www.sciencedirect.com/science/article/pii/B9780128200544000161

[15] Y. Lan, Y. Liu, J. Li, D. Chen, G. He, and I. P. Parkin, "Natural clay-based materials for energy storage and conversion applications," *Advanced Science*, vol. *8*, no. 11, p. 2004036, Jun. 2021, doi: 10.1002/advs.202004036.

20

Therapeutic Capabilities of Advance Nanomaterials to Enhance the Diabetes Mellitus Treatment

S. Jeyabharathi, B. Rathinambal, and M. Sivadharani
Cauvery College for Women (Autonomous), Trichy, India

20.1 Introduction

20.1.1 Diabetes Mellitus

A disorder of insulin and glucagon metabolism, which are secreted in pancreas, is diabetes mellitus (DM). In fact, it ranks as one of the ten most common diseases worldwide that pose a threat to human life and health, and it is also one of the health crises with the fastest rate of growth in the 21st century [1]. While insulin lowers blood sugar level, the hormone glucagon raises blood sugar level and keeps them from falling too low [2]. Glucagon is secreted from pancreatic α-cells in the islet of Langerhans. Insulin is secreted by β-cells in the islets of Langerhans of the pancreas [3]. Type 1 and type 2 DM are the two subtypes of the metabolic disorder. When beta cells are destroyed, the pancreas cannot produce insulin, resulting in type 1 DM [4–9]. When our pancreas does not make enough insulin to keep our blood sugar levels in control, we have type 2 diabetes. It lessens the insulin's reaction. Insulin resistance is the term used to describe this condition [10–14].

20.1.2 Nano-Based Medicine

With the site-specific and target-oriented delivery of medication, nanotechnology offers numerous benefits for the diagnosis and treatment of chronic human disease. Recently, there is a wide range of development in nano-based therapy and site specification of drug target [15–17]. Different administration routes are possible: Intravenous injection, oral administration, or inhalation [18] (Figure 20.1).

20.2 Oral Insulin Administration in Nanosystem

Oral insulin administration is the painless and strong adaptability to the mechanism of action. Insulin formulation intended for oral administration is frequently developed [19]. Polymeric nanoparticle greatly improves insulin absorption and helps move insulin through the intestine through intracellular and paracellular pathways. The formulation uses nanotechnology to combine insulin with different materials that have specific function, with the goal of increasing the bioavailability of oral insulin. Oral insulin administration typically mimics the physiological insulin pathway by being absorbed from the intestinal lumen, travelling via portal circulation to the liver, and then reaching peripheral circulation at relatively low levels [20]. Because of enzymatic breakdown, low stability at pH values, and low permeability of the gastrointestinal tract, oral insulin administration is associated with poor bioavailability. As a result, the drug delivery nanosystem should accomplish the following goals:

- Acid reflux resistance.
- Breakdown of digestive enzymes.
- Entry into small intestinal epithelial cells and mucus layer.

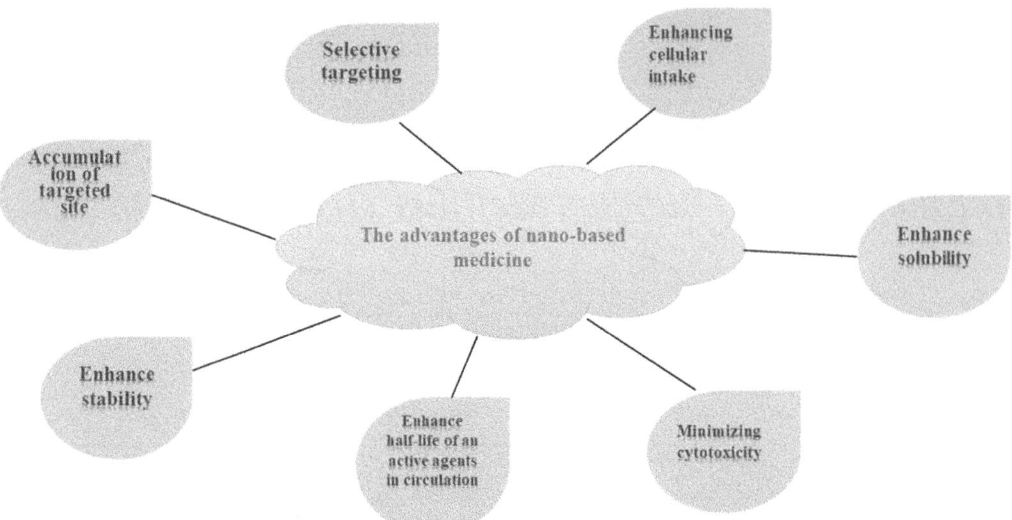

FIGURE 20.1   The advantages of nano-based medicine.

Oral insulin delivery nanosystems are constructed using carrier materials that load insulin. Different types of polymers are used to fabricate the drug that is administrated orally in nanosystem. The polymers are majorly divided into natural polymers and synthetic polymers [21]. Proteins, chitosan, sodium alginate, hyaluronic acid, starch, and bile acid are examples of common natural polymer carrier materials [22]. Polylactic acid (PLA), polylactic *co*-glycolic acid (PLGA), and polycaprolactone (PCL) are examples of synthetic polymers [23].

20.2.1 Chitosan

Chitosan, a derivative of chitin, has been broadly used as a building material for nanoparticles. Chitin is the second most abundant natural biopolymer. Its uses are minimal compared to chitosan. It shields pancreatic beta cells and stimulates insulin secretion, speeding up beta cell division and promoting insulin release and glycaemic regulation

20.2.1.1 Chitosan-Based Therapy

Antioxidant characteristics are exhibited by chitosan and its derivatives. A few of chitosan's special qualities are its non-toxicity, ease of grafting functional groups, muco-adhesion, absorption-enhancing effect, biocompatibility, and biodegradability [24, 25]. Reverse micelle formation, polyelectrolyte complexation ionic gelation, solvent evaporation technique, emulsion solvent diffusion, emulsion droplet coalescence, and complex coacervation are some of the techniques that can be used to synthesise chitosan nanoparticle. In order to shield insulin-loaded nanoparticle from premature release, chitosan and its derivatives have been employed as a coating material. Insulin can be transported either intracellularly or transcellularly to blood capillaries and then transported to the liver through portal vein by means of chitosan nanoparticles, which also extend absorption sites in the intestinal region.

20.3 Cellular Therapies

Nanotechnology has the potential to significantly reduce the immune response to newly produced insulin-producing cells in the field of cell-based therapy. When it comes to delivering medicines, nanoparticles have a lot of promise. Within the next ten years, nanomedicine will be a major factor in bettering diabetes management [26, 27].

20.3.1 3D Bioprinting of Cell

Bioprinting is the process for perfect specimen that grounds for visual analysis. Bioprinting falls under the category of 3D processes, which are used to create more sophisticated experimental tools for research and patient-specific treatment strategies. The basis of 3D printing was first introduced by Charles W. Hull in 1986 [28, 29]. 3D bioprinting can have a huge impact on a range of applications such as regenerative and personalised medicine, tissue engineering, drug discovery, and cosmetics. It aims to biofabricate the living functional tissue and organ for the transplantation process. This specimen combines living cells with bioactive agents that mimic organ and tissue cells and scaffolds with 3D interconnected structures [30, 31].

20.3.2 The Three Main Pillars of 3D Bioprinting

- Mini-tissue building blocks.
- Autonomous self-assembly.
- Biomimicry [32].

20.3.3 Characteristics of Material for Bioprinting

- Capability to print.
- Compatibility with biological system.
- Degradation.
- Mechanical and Structural properties.
- Material biomimicry [32].

20.3.4 Types of Bioprinting

The three types of 3D bioprinting technologies (1) extrusion-based, (2) light/laser-based, and (3) inkjet-based, also known as drop on demand [33, 34].

The most popular kind of printer for both biological and non-biological uses is the inkjet printer, sometimes referred to as a drop-on-demand printer. It is performed by either thermal or piezoelectric actuator method. By heating to the point where inflated bubble pushes the ink out of the narrow nozzle and onto the substrate is how ink droplets are created in thermal technology. It is an inexpensive method that is broadly used. Drops are produced by the transient pressure from a piezoelectric actuator in piezoelectric technology. Mammalian cells deposited piezoelectrically, such as human osteoblasts, fibroblasts, and bovine chondrocytes, are shown to have more than 90% viability [35]. The biomaterials used in extrusion-based bioprinting should be in the form of solutions, pastes, or dispersions that extrude onto a stationary substrate by correlating the motion of pneumatic pressure, plunger, or screw-based pressure in the form of a continual filament through a microscale nozzle orifice or a microneedle. Complete 3D patterns and constructs are eventually formed through layer-by-layer application [36]. Using laser as energy source, biomaterials are deposited onto a substrate in laser-based bioprinting [35].

20.3.5 Pancreatic Cells

To achieve normoglycaemia by substituting damaged or faulty pancreatic cells with new ones is the primary aim of diabetes bioprinting. Pancreatic cell is the embryonic source to restore the health from DM [37, 38]. The hardest part of bioprinting cells to mimic pancreatic function is deciding which type and source of cells to use. Induced pluripotent stem cells, which are obtained by cellular manipulation of somatic, have been studied for their ability to form islet cells, even though there is no standard protocol for obtaining pancreatic cells. Considering the donor-deficiency allogenic and xenogeneic sources, there is angst towards their inherent immunogenic response. Bioprinting is the way to reduce the rejection rate of transplant organ or tissue. Primary islets are considered as the preferred cells for bioprinting of pancreatic cells [39]. Encased in a semi-permeable membrane, the biofabricated pancreatic cells serve as a physical barrier against the host immune system. In order to provide immunological isolation, pancreatic islet encapsulation has made extensive use of natural polymers and their derivatives, including collagen, gelatin, gelma, and alginate [40]. To re-establish the normoglycaemia, implantation of large number of islets is required [41].

20.3.6 Bioprinting of Nanocomposites for Cardiovascular Applications

Patients with DM are more likely to develop heart failure-related conditions like coronary artery disease and hypertension [42]. Diabetes-induced endothelial dysfunction is a dependent pathology linked to myocardial dysfunction and compromised coronary perfusions in diabetic heart disease [43]. The limited ability of cardiac cells to kick-start makes their loss especially harmful in terms of functional reduction [44]. These restrictions connected to the available treatment options can be overcome with 3D bioprinting.

The quick fix for myocardial infraction is biofabrication of cardiac patch for bio-mimics the myocardium extracellular matrix that enables the unification with host tissue so they can be directly engraft in myocardial infraction [45]. In contrast to synthetic macromolecules like polyacrylamide and polyethylene glycol, natural hydrogels like collagen, fibrin, and Matrigel are frequently used for cardiac tissue regeneration. To obtain ascendency, the natural and synthetic hydrogels can be blended together [46, 47]. The scaffolded 3D biofabricated tissue is seeded with viable cells; then they are engrafted over the defective region of the heart. They easily coordinate with host cell and deterioration rate will be truncated [44].

20.4 Selenium (Se) Nanoparticles

There is an interdependence between DM and selenium level in serum. An eventual study from France concluded that old men with higher selenium level in serum had innocuous of developing type 2 DM, but there is no interdependence in old women. A study conducted among Chinese people shows that higher selenium level is directly proportional to the risk of type 2 DM. Hence there is an interrelationship between selenium and DM [48]. An experimental study of selenium nanoparticles was conducted on adult male albino rats of the Wistar strain. The selected rats were injected with Streptozotocin to destruct β-cells in islets of Langerhans to induce diabetes. Selenium nanoparticles were orally administrated to the diabetic rat (0.1mg/kg) for 28 days. The diabetic rat's blood was subjected to biochemical and histological investigations. Selenium has been reported to have outstanding antioxidant and anti-inflammatory actions, thus selenium-based nanoparticles can be used as hypoglycaemic agent and to treat wound infection caused by DM [49].

20.5 Inhalable Insulin Nanoparticles

The benefits of inhalable powder formulation for nasal delivery include low drug degradation, self-administration, a wide surface area for absorption, fast beginning of action, and systemic bioavailability [50]. Muco-adhesion plays a significant role in nasal delivery of drugs. Nanoparticles are used in both dry powder inhaler and nasal delivery of therapeutics. Particle size control must be done precisely for dry powder drug particles. Achieving high delivery efficacy and preserving colloidal stability during aerosolisation are the two main obstacles in the development of particle system for nasal drug delivery [51]. This method allows direct delivery of insulin molecules [52].

20.6 Conclusion

Patient-level management is necessary for diabetes, a condition that is spreading quickly throughout the world. Due to the drug carrier's minuscule size and ability to be eliminated from the body by excretory pathway, nanotechnology has a great deal of promise to improve the care of patients with diabetes. Future treatment and cures for diabetes mellitus appear to benefit from nanotechnology.

REFERENCES

[1] Guo, Y., Baldelli, A., Singh, A., et al. (2022). Production of high loading insulin nanoparticles suitable for oral delivery by spray drying and freeze drying techniques. *Scientific Reports*, *12*, 9949. https://doi.org/10.1038/s41598-022-13092-6

[2] Rix, I., Nexøe-Larsen, C., Bergmann, N. C., et al. Glucagon physiology. [Updated 2019 Jul 16]. In: Feingold KR, Anawalt B, Blackman MR, et al., editors. *Endotext* [Internet]. South Dartmouth (MA): MDText.com, Inc.; 2000. Available from: https://www.ncbi.nlm.nih.gov/books/NBK279127/

[3] Rahman, M. S., Hossain, K. S., Das, S., Kundu, S., Adegoke, E. O., Rahman, M. A., Hannan, M. A., Uddin, M. J., & Pang, M. G. (2021). Role of insulin in health and disease: An update. *International Journal of Molecular Sciences*, *22*(12), 6403. https://doi.org/10.3390/ijms22126403

[4] Souto, E. B., Souto, S. B., Campos, J. R., Severino, P., Pashirova, T. N., Zakharova, L. Y., Silva, A. M., Durazzo, A., Lucarini, M., Izzo, A. A., et al. (2019). Nanoparticle delivery systems in the treatment of diabetes complications. *Molecules*, *24*, 4209. https://doi.org/10.3390/molecules24234209

[5] Shaw, J. E., Sicree, R. A., & Zimmet, P. Z. (2010). Global estimates of the prevalence of diabetes for 2010 and 2030. *Diabetes Research and Clinical Practice*, *87*, 4–14. https://doi.org/10.1016/j.diabres.2009.10.007

[6] Wang, X., Sng, M. K., Foo, S., et al. (2015). Early controlled release of peroxisome proliferator-activated receptor beta/delta agonist GW501516 improves diabetic wound healing through redox modulation of wound microenvironment. *Journal of Controlled Release*, *197*, 138–147. https://doi.org/10.1016/j.jconrel.2014.11.001

[7] Ahmed, N. (2005). Advanced glycation endproducts—Role in pathology of diabetic complications. *Diabetes Research and Clinical Practice*, *67*, 3–21. https://doi.org/10.1016/j.diabres.2004.09.004

[8] Zhang, T., Tang, J. Z., Fei, X., Li, Y., Song, Y., Qian, Z., & Peng, Q. (2020). Can nanoparticles and nano-protein interactions bring a bright future for insulin delivery? *Acta Pharmaceutica Sinica B*, 1–20.

[9] Zhang, T., Tang, J. Z., Fei, X., Li, Y., Song, Y., Qian, Z., & Peng, Q. (2021). Can nanoparticles and nano-protein interactions bring a bright future for insulin delivery? *Acta Pharmaceutica Sinica B*, *11*(3), 651–667, ISSN 2211-3835, https://doi.org/10.1016/j.apsb.2020.08.016

[10] Goyal, R., Singhal, M., & Jialal, I. Type 2 Diabetes. [Updated 2023 Jun 23]. In: *StatPearls* [Internet]. Treasure Island (FL): StatPearls Publishing; 2023 Jan. Available from: https://www.ncbi.nlm.nih.gov/books/NBK513253/

[11] Sunkari, V. G., Lind, F., Botusan, I. R., et al. (2015). Hyperbaric oxygen therapy activates hypoxia-inducible factor 1 (HIF-1), which contributes to improved wound healing in diabetic mice. *Wound Repair and Regeneration*, *23*, 98–103. https://doi.org/10.1111/wrr.12253

[12] Kim, S. K., Lee, K. J., Hahm, J. R., et al. (2012). Clinical significance of the presence of autonomic and vestibular dysfunction in diabetic patients with peripheral neuropathy. *Diabetes and Metabolism Journal*, *36*, 64–69. https://doi.org/10.4093/dmj.2012.36.1.64

[13] Gregory, J. M., & Moore, D. J. (2013). Type 1 diabetes mellitus. *Pediatrics in Review*. https://doi.org/10.1016/B978-1-4377-1604-7.00561-3

[14] Debele, T. A., & Yoonjee P. (2022). Application of nanoparticles: Diagnosis, therapeutics, and delivery of insulin/anti-diabetic drugs to enhance the therapeutic efficacy of diabetes mellitus. *Lifestyles*, *12*(12), 2078. https://doi.org/10.3390/life12122078

[15] Patra, J. K., Das, G., Fraceto, L. F., et al. (2018). Nano based drug delivery systems: Recent developments and future prospects. *Journal of Nanbiotechnology*, *16*, 71. https://doi.org/10.1186/s12951-018-0392-8

[16] DiSanto, R. M., Subramanian, V., & Gu, Z. (2015 Jul-Aug). Recent advances in nanotechnology for diabetes treatment. *Wiley Interdisciplinary Reviews. Nanomedicine and Nanobiotechnology*, *7*(4), 548–564. https://doi.org/10.1002/wnan.1329

[17] Arcos, D. (2023 May 19). Nanomaterials in biomedicine 2022. *International Journal of Molecular Sciences*, *24*(10), 9026. https://doi.org/10.3390/ijms24109026

[18] Forest, V., & Pourchez, J. (2022). Nano-delivery to the lung –By inhalation or other routes and why nano when micro is largely sufficient? *Advanced Drug Delivery Reviews*, *183*, 114173, ISSN 0169-409X, https://doi.org/10.1016/j.addr.2022.114173

[19] Zhao, H., Ye, H., Zhou, J., Tang, G., Hou, Z., & Bai, H. (2020). Montmorillonite-enveloped zeolitic imidazolate framework as a nourishing oral nano-platform for gastrointestinal drug delivery. *ACS Applied Materials & Interfaces*, *12*(44), 49431–49441. https://doi.org/10.1021/acsami.0c15494

[20] Seyam, S., Nordin, N. A., & Alfatama, M. (2020). Recent progress of chitosan and chitosan derivatives-based nanoparticles: Pharmaceutical perspectives of oral insulin delivery. *Pharmaceuticals (Basel, Switzerland), 13*(10), 307. https://doi.org/10.3390/ph13100307

[21] Ahmad, J., Singhal, M., Amin, S., Rizwanullah, M., Akhter, S., Kamal, M. A., Haider, N., Midoux, P., & Pichon, C. (2017). Bile salt stabilized vesicles (bilosomes): A novel nano-pharmaceutical design for oral delivery of proteins and peptides. *Current Pharmaceutical Design, 23*(11), 1575–1588. https://doi.org/1 0.2174/1381612823666170124111142

[22] Kecman, S., Škrbić, R., Badnjevic Cengic, A., Mooranian, A., Al-Salami, H., Mikov, M., & Golocorbin-Kon, S. (2020). Potentials of human bile acids and their salts in pharmaceutical nano delivery and formulations adjuvants. *Technology and Health Care: Official Journal of the European Society for Engineering and Medicine, 28*(3), 325–335. https://doi.org/10.3233/THC-191845

[23] Wang, M., Wang, C., Ren, S., Pan, J., Wang, Y., Shen, Y., Zeng, Z., Cui, H., & Zhao, X. (2022). Versatile oral insulin delivery nanosystems: From materials to nanostructures. *International Journal of Molecular Sciences, 23*(6), 3362. https://doi.org/10.3390/ijms23063362

[24] Bravo-Anaya, L. M., Fernández-Solís, K. G., Rosselgong, J., Nano-Rodríguez, J. L. E., Carvajal, F., & Rinaudo, M. (2019). Chitosan-DNA polyelectrolyte complex: Influence of chitosan characteristics and mechanism of complex formation. *International Journal of Biological Macromolecules, 126*, 1037–1049. https://doi.org/10.1016/j.ijbiomac.2019.01.008

[25] He, Zhiyu, Santos, Jose Luis, Tian, Houkuan, Huang, Huahua, Hu, Yizong, Liu, Lixin, Leong, Kam W., Chen, Yongming, & Mao, Hai-Quan. (2017). Scalable fabrication of size-controlled chitosan nanoparticles for oral delivery of insulin. *Biomaterials, 130*, 28–41, ISSN 0142-9612, https://doi.org/10.1016/j.biomaterials.2017.03.028

[26] Lemmerman, L. R., Das, D., Higuita-Castro, N., Mirmira, R. G., & Gallego-Perez, D. (2020). Nanomedicine-based strategies for diabetes: Diagnostics, monitoring, and treatment. *Trends in Endocrinology and Metabolism: TEM, 31*(6), 448–458. https://doi.org/10.1016/j.tem.2020.02.001

[27] Jalilian, E., Elkin, K., & Shin, S. R. (2020). Novel cell-based and tissue engineering approaches for induction of angiogenesis as an alternative therapy for diabetic retinopathy. *International Journal of Molecular Sciences, 21*, 3496. https://doi.org/10.3390/ijms21103496

[28] Kim, J., Kang, K., Drogemuller, C. J., et al. (2019). Bioprinting an artificial pancreas for type 1 diabetes. *Current Diabetes Reports, 19*, 53. https://doi.org/10.1007/s11892-019-1166-x

[29] Birla, R. K., Borschel, G. H., Dennis, R. G., & Brown, D. L. (2005). Myocardial engineering in vivo: Formation and characterization of contractile, vascularized three-dimensional cardiac tissue. *Tissue Engineering, 11*(5–6), 803–813. https://doi.org/10.1089/ten.2005.11.803

[30] Yilmaz, Bengi, Al Rashid, Ans, Mou, Younss Ait, Evis, Zafer, & Koç, Muammer. (2021). Bioprinting: A review of processes, materials and applications. *Bioprinting, 23*, e00148, ISSN2405-8866, https://doi.org/10.1016/j.bprint.2021.e00148 (https://www.sciencedirect.com/science/article/pii/S2405886621 00021X).

[31] Zhang, S., Chen, X., Shan, M., Hao, Z., Zhang, X., Meng, L., Zhai, Z., Zhang, L., Liu, X., & Wang, X. (2023). Convergence of 3D bioprinting and nanotechnology in tissue engineering scaffolds. *Biomimetics (Basel, Switzerland), 8*(1), 94. https://doi.org/10.3390/biomimetics8010094

[32] Murphy, S. V., & Atala, A. (2014). 3D bioprinting of tissues and organs. *Nature Biotechnology, 32*, 773–785. https://doi.org/10.1038/nbt.2958

[33] Tan, C. T., Liang, K., Ngo, Z. H., Dube, C. T., & Lim, C. Y. (2020). Application of 3D bioprinting technologies to the management and treatment of diabetic foot ulcers. *Biomedicine, 8*(10), 441. https://doi.org/10.3390/biomedicines8100441

[34] Al-Atawi, S. (2023). Three-dimensional bioprinting in ophthalmic care. *International Journal of Ophthalmology, 16*(10), 1702–1711. https://doi.org/10.18240/ijo.2023.10.21

[35] Li, J., Chen, M., Fan, X., et al. (2016). Recent advances in bioprinting techniques: Approaches, applications and future prospects. *Journal of Translational Medicine, 14*, 271. https://doi.org/10.1186/s12967-016-1028-0

[36] Chien, K. B., Makridakis, E., & Shah, R. N. (2013). Three-dimensional printing of soy protein scaffolds for tissue regeneration. *Tissue Engineering. Part C, Methods, 19*(6), 417–426. https://doi.org/10.1089/ten.TEC.2012.0383

[37] Kumar, S. A., Delgado, M., Mendez, V. E., & Joddar, B. (2019). Applications of stem cells and bioprinting for potential treatment of diabetes. *World Journal of Stem Cells, 11*(1), 13–32. https://doi.org/10.4252/wjsc.v11.i1.13

[38] Klak, M., Wszoła, M., Berman, A., Filip, A., Kosowska, A., Olkowska-Truchanowicz, J., Rachalewski, M., Tymicki, G., Bryniarski, T., Kołodziejska, M., et al. (2023). Bioprinted 3D bionic scaffolds with pancreatic islets as a new therapy for type 1 diabetes—Analysis of the results of preclinical studies on a mouse model. *Journal of Functional Biomaterials, 14*(7), 371. https://doi.org/10.3390/jfb14070371

[39] Ghosh, A., Sanyal, A., & Mallick, A. (2023). Recent advances in the development of bioartificial pancreas using 3D bioprinting for the treatment of type 1 diabetes: A review. Preprint, 2023100071. https://doi.org/10.20944/preprints202310.0071.v1

[40] Parvaneh, S., Kemény, L., Ghaffarinia, A., Yarani, R., & Veréb, Z. (2023). Three-dimensional bioprinting of functional β-islet-like constructs. *International Journal of Bioprinting, 9*(2), 665. https://doi.org/10.18063/ijb.v9i2.665

[41] Soetedjo, Andreas Alvin Purnomo, Lee, Jia Min, Lau, Hwee Hui, Goh, Guo Liang, An, Jia, Koh, Yexin, Yeong, Wai Yee, & Teo, Adrian Kee Keong. (2021). Tissue engineering and 3D printing of bioartificial pancreas for regenerative medicine in diabetes. *Trends in Endocrinology and Metabolism, 32*(8), 609–622, ISSN 1043-2760, https://doi.org/10.1016/j.tem.2021.05.007

[42] Kannel, W. B., Hjortland, M., & Castelli, W. P. (1974). Role of diabetes in congestive heart failure: The Framingham study. *The American Journal of Cardiology, 34*(1), 29–34. https://doi.org/10.1016/0002-9149(74)90089-7

[43] Lew, J. K., Pearson, J. T., Schwenke, D. O., & Katare, R. (2017). Exercise mediated protection of diabetic heart through modulation of microRNA mediated molecular pathways. *Cardiovascular Diabetology, 16*(1), 10. https://doi.org/10.1186/s12933-016-0484-4

[44] Tasnim, N., De la Vega, L., Anil Kumar, S., et al. (2018). 3D bioprinting stem cell derived tissues. *Cellular and Molecular Bioengineering, 11*, 219–240. https://doi.org/10.1007/s12195-018-0530-2

[45] Loukelis, K., Helal, Z. A., Mikos, A. G., & Chatzinikolaidou, M. (2023). Nanocomposite bioprinting for tissue engineering applications. *Gels (Basel, Switzerland), 9*(2), 103. https://doi.org/10.3390/gels9020103

[46] Li, Z., & Guan, J. (2011). Hydrogels for cardiac tissue engineering. *Polymers, 3*, 740–761. https://doi.org/10.3390/polym3020740

[47] Vanaei, S., Parizi, M. S., Vanaei, S., Salemizadehparizi, F., & Vanaei, H. R. (2021). An overview on materials and techniques in 3d bioprinting toward biomedical application. *Engineered Regeneration, 2*, 1–18, ISSN 2666-1381, https://doi.org/10.1016/j.engreg.2020.12.001

[48] Guan, B., Yan, R., Li, R., & Zhang, X. (2018). Selenium as a pleiotropic agent for medical discovery and drug delivery. *International Journal of Nanomedicine, 14*(13), 7473–7490.

[49] Al-Quraishy, S., Dkhil, M. A., & Abdel Moneim, A. E. (2015). Anti-hyperglycemic activity of selenium nanoparticles in streptozotocin-induced diabetic rats. *International Journal of Nanomedicine, 10*, 6741–6756. https://doi.org/10.2147/IJN.S91377

[50] Muralidharan, Priya, Malapit, Monica, Mallory, Evan, Hayes, Don, & Mansour, Heidi M. (2015). Inhalable nanoparticulate powders for respiratory delivery. *Nanomedicine: Nanotechnology, Biology and Medicine, 11*(5), 1189–1199, ISSN 1549-9634, https://doi.org/10.1016/j.nano.2015.01.007

[51] Matthews, A. A., Ee, P. L. R., & Ge, R. (2020). Developing inhaled protein therapeutics for lung diseases. *Molecular Biomedicine, 1*, 11. https://doi.org/10.1186/s43556-020-00014-z

[52] Gupta, R. (2017). Diabetes treatment by nanotechnology. *Journal of Biotechnology and Biomaterials, 7*, 268. https://doi.org/10.4172/2155-952X.1000268

21

Machine Learning Applications in Designing Advanced Energy Storage Materials: A Comprehensive Exploration

S. K. Jameer Basha and E. Ajith Jubilson
VIT-AP University, Amaravati, Guntur, Andhra Pradesh, India

21.1 Introduction to the Confluence of Machine Learning and Materials Science

Materials science and artificial intelligence (AI) converge in a transformative confluence that reshapes the landscape of advanced energy storage solutions. This section initiates the exploration by navigating the intricate intersection of AI and materials science, unravelling the synergies poised to revolutionise energy storage technologies [1].

21.1.1 Navigating the Intersection of AI and Materials Science

At the heart of this confluence lies the seamless integration of AI methodologies with the complexities of materials science. Navigating this intersection involves leveraging AI's computational prowess to decipher the intricate behaviours and properties of materials at the atomic and molecular levels. The application of machine learning (ML) algorithms facilitates the extraction of meaningful patterns from vast datasets, offering a novel lens through which materials can be understood and engineered [2].

21.2 Unveiling the Potential for Advanced Energy Storage Solutions

The integration of ML brings forth a paradigm shift in envisioning advanced energy storage solutions. By harnessing the predictive capabilities of AI, researchers gain insights into the behaviours of materials under diverse conditions. This unveiling of potential goes beyond traditional empirical approaches, allowing for the identification of materials with tailored properties for enhanced energy storage [3]. The synergy holds promise for overcoming conventional limitations and unlocking unprecedented efficiency in energy storage systems.

21.3 Objectives of Synergistic Exploration

As we embark on this synergistic exploration, the objectives become clear. The primary goal is to decipher how ML can augment the capabilities of materials science, propelling advancements in energy storage. By understanding the fundamentals of AI algorithms in a materials context, the exploration aims to unlock the full potential of predictive modelling for identifying novel materials. Additionally, the exploration seeks to optimise energy storage parameters using AI, fine-tuning the intricate variables that influence performance [4]. This confluence does not merely aim at isolated advancements; it aspires

DOI: 10.1201/9781003495437-21

to catalyse a holistic transformation in how we perceive and engineer materials for sustainable energy storage. The objectives extend to unravelling the ethical considerations, challenges, and opportunities that emerge at the nexus of AI and materials science. By delineating these objectives, the exploration lays the foundation for a comprehensive understanding of the confluence's impact on energy storage technologies.

21.4 Fundamentals of Machine Learning Algorithms in Materials Design

ML algorithms have emerged as pivotal tools in revolutionising materials design, ushering in a new era of efficiency and innovation. This section explores the fundamental aspects of these algorithms, their versatile applications across materials science, and the crucial customisation required to tackle the unique challenges posed by energy storage materials.

21.4.1 Unpacking Machine Learning Algorithms in Materials Context

ML algorithms serve as sophisticated interpreters within the context of materials science. Unpacking these algorithms involves a meticulous exploration of their underlying principles, spanning classical methodologies to advanced deep learning models. These algorithms act as key instruments for decoding intricate patterns and correlations at the atomic and molecular levels, providing invaluable insights into material behaviours [5]. A comprehensive understanding of how these algorithms navigate the material landscape is essential for unlocking their full potential in materials science.

21.4.2 Applications across Materials Science

ML algorithms find multifaceted applications within materials science, transforming traditional approaches to materials research. This section underscores their pivotal role in predicting material behaviours under diverse conditions, expediting the discovery of novel compounds, and revolutionising materials design. In the context of energy storage, ML becomes a linchpin, contributing significantly to the identification of materials with optimal electrochemical properties, thermal stability, and mechanical strength. The applications extend to the precise design of materials tailored for batteries, capacitors, and other critical components integral to advanced energy storage systems.

21.4.3 Customising Algorithms for Energy Storage Challenges

The adaptability of ML algorithms is paramount when confronted with the distinctive challenges presented by energy storage materials. Moving beyond the selection of generic algorithmic architectures, customisation becomes imperative. Fine-tuning these algorithms to consider specific parameters, such as charge–discharge cycles, degradation patterns, and thermal influences, ensures alignment with the unique requirements of energy storage [6]. This level of precision not only enhances the efficacy of the algorithms but also holds the promise of tailoring materials with exacting specifications for sustainable and efficient energy storage solutions.

21.4.4 Navigating the Interplay between Algorithms and Material Properties

The interplay between ML algorithms and material properties is a dynamic process requiring a nuanced understanding of both domains. Algorithms, ranging from classical regression models to intricate neural networks, serve as invaluable tools for deciphering intricate relationships governing material behaviours. Whether predicting the stability of a novel compound or optimising the electrochemical performance of a material for energy storage, the navigation through this interplay involves iterative learning [7]. Continuous refinement of algorithms based on real-world feedback refines their predictive capabilities, marking a cyclical process of improvement.

21.4.5 Challenges and Future Trajectories in Algorithmic Materials Design

Despite the transformative potential, algorithmic materials design encounters challenges. The interpretability of complex algorithms remains a concern, especially when dealing with intricate material behaviours. Striking a balance between model complexity and interpretability is an ongoing pursuit. Moreover, the need for robust datasets for training algorithms presents a challenge in domains where data availability is limited. Looking to the future, the trajectory of algorithmic materials design holds exciting possibilities. The integration of advanced computational techniques, such as quantum ML, promises unparalleled precision in simulating complex material behaviours [8]. Meta-learning approaches, enabling algorithms to learn from a spectrum of tasks, enhance adaptability and efficiency. The collaborative synergy of robotics and ML in material experimentation heralds an era of high-throughput research, accelerating the pace of material discovery and optimisation. This exploration into the fundamentals of ML algorithms in materials design exemplifies the transformative synergy within the field of materials science. Unpacking these algorithms, understanding their applications, and customising them for specific challenges pave the way for unprecedented advancements in materials research and development.

21.5 Predictive Modelling for Novel Material Identification and Energy Storage

Predictive modelling stands as a pivotal tool in the domain of materials science, particularly in the identification of novel materials tailored for advanced energy storage applications. This section intricately explores the processes involved in harnessing predictive modelling, ranging from its fundamental application in materials discovery to the construction of specialised models for identifying materials optimised for energy storage [9]. The discussion extends to the critical considerations of data sources and features, which form the foundation for effective predictions.

21.5.1 Harnessing Predictive Modelling in Materials Discovery

Predictive modelling plays a central role in expediting the materials discovery process. By leveraging historical data encompassing a spectrum of material properties, behaviours, and performance metrics, predictive models navigate the vast expanse of chemical space to pinpoint promising candidates. This involves the development of sophisticated mathematical models that encapsulate the intricate relationships governing various material attributes. Serving as virtual laboratories, these models simulate and predict material behaviours under diverse conditions. The harnessing of predictive modelling offers a systematic, data-driven approach to identifying compounds with precisely tailored characteristics, accelerating the pace of materials discovery.

21.5.2 Constructing Models for Identifying Novel Energy Storage Materials

The creation of predictive models assumes greater importance in the field of energy storage. These models are meticulously designed to unravel the complex interplay of factors influencing the performance of materials within energy storage devices. Whether applied to batteries, capacitors, or other energy storage systems, the construction process involves integrating a multitude of parameters, including electrochemical properties, thermal stability, and structural characteristics [10]. The challenge lies in crafting models that not only accurately predict material behaviours but also provide insights into the underlying mechanisms dictating energy storage efficiency. The construction of these models demands a profound understanding of the unique challenges posed by energy storage materials and a keen insight into the specific properties necessary for optimal performance.

21.5.3 Data Sources and Features for Effective Predictions

The efficacy of predictive modelling is contingent upon the quality and relevance of data sources and features incorporated into the models. This segment underscores the pivotal role of comprehensive datasets derived from experiments, simulations, and extensive material databases. Diverse data sources contribute to a holistic comprehension of material behaviours, enabling models to generalise across various scenarios. Equally crucial is the meticulous selection of features representing specific material attributes. These features could encompass structural details, electronic properties, or thermodynamic characteristics, contingent upon the material under consideration and the targeted application. The thoughtful curation of data sources and features ensures that predictive models are robust, reliable, and capable of providing accurate forecasts.

21.5.4 Integration of Machine Learning Techniques in Predictive Modelling

The construction of predictive models often involves the integration of ML techniques. Supervised learning, where models are trained on labelled datasets, proves effective in predicting material properties based on known examples. Unsupervised learning techniques, such as clustering, aid in the identification of patterns within datasets, offering insights into novel material groupings. Additionally, the application of reinforcement learning enhances the adaptability of predictive models, allowing them to learn from real-world feedback and refine predictions over time [11]. The integration of these ML techniques amplifies the predictive capabilities of models, enabling more accurate and nuanced predictions for novel energy storage materials.

21.5.5 Challenges and Future Directions in Predictive Modelling for Energy Storage Materials

While predictive modelling has significantly advanced materials discovery, challenges persist. The limited availability of high-quality labelled data for training models poses a hurdle, especially for emerging materials. The need for interpretability in complex models remains a concern, impacting the trustworthiness of predictions. Future directions involve addressing these challenges through collaborative efforts, developing standardised datasets, and refining interpretability tools.

21.6 Optimising Energy Storage: AI's Role in Enhancing Battery Performance

21.6.1 Fine-Tuning Parameters for Improved Energy Storage

AI's contribution to optimising energy storage is prominently exemplified through the fine-tuning of critical parameters within battery systems. ML algorithms, equipped with vast datasets encompassing battery materials, configurations, and operational conditions, engage in sophisticated analyses. By discerning optimal combinations of electrode materials, electrolytes, and structural configurations, AI facilitates substantial enhancements in energy storage capacity, charging rates, and overall battery performance [12]. This fine-tuning process is pivotal for advancing battery technologies across diverse applications, ranging from electric vehicles to grid-scale energy storage systems.

21.6.2 Predicting Degradation Patterns in Energy Storage Materials

AI algorithms play a pivotal role in predicting degradation patterns within energy storage materials. Continuous monitoring and analysis of operational data enable ML models to forecast potential degradation mechanisms, such as electrode wear, electrolyte breakdown, or thermal stresses. Predictive maintenance strategies can then be deployed to address these degradation issues, extending the lifespan of energy storage systems and optimising their overall efficiency. The ability to predict degradation patterns contributes significantly to the development of durable and long-lasting energy storage solutions.

21.6.3 Insights into Efficient Energy Utilisation

AI's analytical capabilities provide invaluable insights into efficient energy utilisation within storage systems. ML models delve into the complex relationships between charge–discharge cycles, environmental conditions, and energy efficiency. By identifying optimal operating conditions and recommending adaptive energy management strategies, AI ensures that energy storage systems operate at peak performance while minimising waste. This optimisation is fundamental for achieving sustainable and cost-effective energy storage solutions across various applications.

21.7 Nanotechnology Innovations: Integrating Machine Learning for Energy Storage

21.7.1 Nanomaterials Revolution in Energy Storage

The confluence of nanotechnology and ML heralds a revolutionary era in energy storage. Nanomaterials, designed and optimised through computational approaches, have redefined the landscape of energy storage devices. The integration of ML techniques expedites the discovery of nanomaterials with exceptional properties, such as high conductivity, enhanced surface area, and improved electrochemical performance. This nanomaterials revolution transcends conventional energy storage limitations, paving the way for compact and efficient energy storage solutions.

21.7.2 Machine Learning-Driven Nanomaterial Design

ML takes centre stage in the design of nanomaterials for energy storage applications. By leveraging vast datasets on nanomaterial properties and behaviours, ML models navigate the intricate design space, predicting the performance of hypothetical nanomaterial structures. This predictive capability accelerates the identification of nanomaterials with tailored properties, enabling the design of energy storage devices with superior performance characteristics. ML-driven nanomaterial design represents a paradigm shift, offering a systematic and data-driven approach to nanotechnology innovations in energy storage [13].

21.7.3 Synergies for Advanced Energy Storage Solutions

The integration of ML with nanotechnology establishes synergies that transcend individual technological capabilities. ML algorithms, when applied to nanomaterial design, enhance the precision and efficiency of material discovery processes. This synergy leads to the development of advanced energy storage solutions that leverage the unique properties of nanomaterials. From high-performance batteries to efficient capacitors, the collaboration between ML and nanotechnology unfolds new possibilities for energy storage technologies, driving innovation in the pursuit of sustainable and high-performance energy storage systems.

21.8 Computational Approaches for Materials Discovery in Energy Storage Systems

21.8.1 Computational Methods Shaping Materials Science

In the method of materials science, computational methods play a pivotal role in reshaping the traditional paradigms of materials discovery. Advanced computational approaches, including quantum mechanical simulations, density functional theory, and molecular dynamics simulations, provide insights into the behaviour of materials at the atomic and molecular levels. These methods enable researchers to explore vast material spaces and identify candidates with desirable properties for energy storage applications. The synergy between computational methods and ML algorithms enhances the efficiency and accuracy of materials discovery, paving the way for accelerated advancements in energy storage systems.

21.8.2 Machine Learning's Impact on Accelerated Discovery

ML accelerates materials discovery by harnessing the power of data-driven insights. In the context of energy storage, ML models analyse vast datasets encompassing material properties, synthesis conditions, and performance metrics. By identifying patterns and correlations within this data, ML facilitates the prediction of novel materials with tailored properties. The integration of ML with computational methods enables researchers to navigate complex materials landscapes more efficiently, expediting the discovery of materials for advanced energy storage systems [14].

21.8.3 Bridging Computational and Experimental Approaches

The confluence of computational and experimental approaches represents a critical juncture in materials discovery. While computational methods provide predictive capabilities, experimental validation is essential to confirm the real-world performance of newly discovered materials. ML acts as a bridge between these approaches, guiding experimental efforts by predicting the properties of materials that are most likely to exhibit the desired characteristics. This synergy ensures a more seamless transition from theoretical predictions to practical applications in the development of energy storage materials.

21.9 Smart Energy Storage Systems: Real-Time Optimisation through Machine Learning

21.9.1 Evolution of Intelligent Battery Management Systems

The evolution of intelligent battery management systems (BMS) epitomises the transformative impact of ML on energy storage technologies. Traditional BMS focused on basic monitoring and control functions, but the integration of ML introduces a new era of intelligence. ML algorithms, embedded within BMS, continuously analyse real-time data from batteries, adapting to changing conditions and optimising performance. This evolution enhances the efficiency, safety, and longevity of energy storage systems, making them more adaptable to dynamic operational scenarios.

21.9.2 Real-Time Monitoring and Adaptive Control Strategies

ML empowers real-time monitoring and adaptive control strategies in smart energy storage systems. Through constant analysis of operational data, ML models detect patterns indicative of optimal and suboptimal conditions. These models enable adaptive control strategies that adjust parameters such as charging rates, discharging profiles, and thermal management in response to dynamic environmental factors. Real-time monitoring and adaptive control contribute to the overall resilience and responsiveness of energy storage systems, ensuring optimal performance under varying conditions [15].

21.9.3 Advancements in Charging and Discharging Algorithms

The application of ML extends to the development of advanced charging and discharging algorithms. Traditional algorithms were rule based and lacked adaptability to diverse usage patterns. ML algorithms, however, learn from historical usage data and user behaviour to optimise charging and discharging cycles. This personalised approach improves the overall efficiency of energy storage systems, tailoring their operation to user-specific requirements. Advancements in charging and discharging algorithms represent a critical facet of smart energy storage systems, aligning energy delivery with user preferences and grid dynamics.

21.10 Challenges and Opportunities in the Fusion of Machine Learning and Materials Science

21.10.1 Tackling Computational and Data Challenges

The fusion of ML and materials science presents inherent challenges related to the computational complexity and quality of data. Tackling these challenges requires advancements in algorithms, computing infrastructure, and data quality assurance. Researchers delve into innovative approaches to streamline computations, enhance algorithmic efficiency, and address data quality concerns, ensuring the robustness of ML models in materials science applications.

21.10.2 Seizing Opportunities for Innovation

Amid challenges lie significant opportunities for innovation. Researchers explore novel avenues to harness the power of ML for materials discovery, design, and optimisation. Opportunities include leveraging advanced algorithms for more accurate predictions, integrating multi-modal data for comprehensive insights, and expanding collaborative efforts to pool diverse expertise. Seizing these opportunities contributes to the transformative impact of ML on materials science, unlocking new frontiers in advanced material functionalities.

21.10.3 Future Prospects at the Intersection

The intersection of ML and materials science holds promising future prospects. Ongoing research focuses on overcoming existing challenges and capitalising on emerging opportunities. The integration of ML models with quantum computing, advanced imaging techniques, and high-throughput experimentation heralds a future where materials science evolves into a more dynamic and efficient discipline. The continual advancement of ML methodologies ensures sustained growth and innovation at the intersection of these two fields.

21.11 Case Studies: Machine Learning Applications in Sustainable Energy Ecosystems

21.11.1 Smart Grids Orchestrated by Machine Learning

ML applications in smart grids revolutionise energy distribution and consumption. Case studies showcase how ML algorithms optimise grid operations, predict energy demand, and enhance overall efficiency. By orchestrating smart grids, ML contributes to a more adaptive and resilient energy ecosystem, ensuring optimal utilisation of resources and minimising wastage.

21.11.2 Integrating Renewable Energy with Advanced Algorithms

ML plays a pivotal role in the integration of renewable energy sources into the grid. Case studies highlight how advanced algorithms forecast renewable energy production, manage intermittency challenges, and enable seamless integration into existing energy systems. The synergy between ML and renewable energy fosters a more sustainable and reliable energy ecosystem, reducing dependence on non-renewable sources.

21.11.3 Evaluating Environmental Impact through Data-Driven Insights

Case studies delve into the use of ML to assess the environmental impact of energy systems. From life cycle assessments to real-time monitoring of emissions, ML provides data-driven insights into the environmental footprint of energy technologies. These studies contribute to informed decision-making, guiding the development of energy solutions that prioritise environmental sustainability.

21.12 Ethical and Environmental Considerations in Machine Learning-Enhanced Materials Science

21.12.1 Responsible AI Practices in Materials Science

Ethical considerations in ML-enhanced materials science encompass responsible AI practices. Case studies highlight how researchers adhere to ethical guidelines in data collection, algorithm development, and model deployment. Transparency, fairness, and accountability form the pillars of responsible AI practices, ensuring that advancements in materials science align with ethical standards.

21.12.2 Assessing Environmental Impact of Advanced Materials

Ethical considerations extend to assessing the environmental impact of materials designed with ML. Case studies explore methodologies for evaluating the life cycle of materials, considering factors such as resource extraction, manufacturing processes, and end-of-life disposal. This holistic approach ensures that materials science advancements align with environmental sustainability goals.

21.12.3 Establishing Ethical Frameworks for Sustainable Engineering

Researchers actively contribute to establishing ethical frameworks that guide ML applications in sustainable engineering. Case studies showcase collaborative efforts across disciplines to formulate guidelines that prioritise safety, fairness, and accountability. These ethical frameworks evolve with technological advancements, ensuring that ML in materials science aligns with evolving ethical norms.

21.13 Future Trajectories: Integrating Machine Learning for Materials Science Advancements

21.13.1 Exploring Emerging Technologies in Materials Science and AI

The exploration of future trajectories involves delving into emerging technologies that further integrate ML with materials science. Case studies showcase how researchers explore quantum ML, advanced imaging techniques, and innovative data-driven approaches. These explorations anticipate breakthroughs that will shape the future of materials science.

21.13.2 Fostering Collaborative Research and Interdisciplinary Approaches

Case studies emphasise the importance of collaborative research and interdisciplinary approaches. Future trajectories involve fostering partnerships between materials scientists, data scientists, and engineers. By breaking down traditional silos and encouraging cross-disciplinary collaboration, researchers aim to accelerate advancements at the intersection of ML and materials science.

21.13.3 Steering towards a Sustainable Future with Advanced Engineering

The future trajectory of integrating ML into materials science aims to steer towards a sustainable future. Case studies highlight how advancements contribute to sustainable engineering practices. Whether through the development of eco-friendly materials, energy-efficient technologies, or waste reduction strategies, ML propels materials science towards a future where innovation aligns seamlessly with environmental and societal needs.

21.14 Conclusion: A Holistic Perspective on Achievements and Ongoing Advances in Materials Science

21.14.1 Recapitulation of the Synergy between Machine Learning and Materials Science

The conclusion provides a comprehensive recapitulation of the synergy between machine learning and materials science. It reflects on the transformative impact of machine learning in reshaping materials discovery, design, and optimisation. The recapitulation underscores the dynamic interplay between algorithms and materials, yielding unprecedented possibilities for scientific innovation.

21.14.2 Contributions to Energy Storage and Environmental Preservation

A central focus of the conclusion is highlighting the contributions of machine learning to energy storage and environmental preservation. Case studies showcased throughout the exploration demonstrate how machine learning accelerates the development of advanced materials for energy storage, contributes to smart and sustainable energy ecosystems, and assesses environmental impacts.

21.14.3　Continued Pathways for Advancements in the Field

The conclusion outlines continued pathways for advancements in the field. It emphasises the need for ongoing research, collaboration, and interdisciplinary approaches to address challenges and unlock new potentials. The dynamic nature of the intersection between machine learning and materials science anticipates a future marked by continuous breakthroughs, driving sustainable innovation in engineering practices.

In conclusion, the exploration of machine learning and materials science unveils a landscape of innovation, challenges, and ethical considerations. From fundamental principles to real-world applications, the synergy between these two domains propels advancements that shape the future of materials science and engineering.

REFERENCES

[1] Ogunjobi, O.A., Eyo-Udo, N.L., Egbokhaebho, B.A., Daraojimba, C., Ikwue, U. and Banso, A.A., 2023. Analyzing historical trade dynamics and contemporary impacts of emerging materials technologies on international exchange and us strategy. *Engineering Science & Technology Journal, 4*(3), pp. 101–119.

[2] Joseph Arockiam, A., Rajesh, S. and Karthikeyan, S., Development of fish scale particle reinforced PLA filaments for 3D printing applications. *Journal of Applied Polymer Science, 141*(12), p. e55132.

[3] Goodenough, J.B., Abruna, H.D. and Buchanan, M.V., 2007. Basic research needs for electrical energy storage. Report of the basic energy sciences workshop on electrical energy storage, April 2–4, 2007. DOESC (USDOE Office of Science (SC)).

[4] Zheng, L., Zhang, S., Huang, H., Liu, R., Cai, M., Bian, Y., Chang, L. and Du, H., 2023. Artificial intelligence-driven rechargeable batteries in multiple fields of development and application towards energy storage. *Journal of Energy Storage, 73*, p. 108926.

[5] Arockiam, A.J., Rajesh, S., Karthikeyan, S., Thiagamani, S.M.K., Padmanabhan, R.G., Hashem, M., Fouad, H. and Ansari, A., 2023. Mechanical and thermal characterization of additive manufactured fish scale powder reinforced PLA biocomposites. *Materials Research Express, 10*(7), p. 075504.

[6] Krishen, K., 1992. Fifth Annual Workshop on Space Operations Applications and Research (SOAR'91). NASA CP-3127. In Fifth Annual Workshop on Space Operations Applications and Research (SOAR 1991) (Vol. 3127).

[7] Vignesh, P., Ramanathan, S., Ramesh, K. and Arockiam, A.J., 2022. Finite element modelling to predict hardness of Mg alloy reinforced with Ti/hydroxyapatite hybrid composites–An axisymmetric approach. *Materials Today: Proceedings, 68*, pp. 1830–1834.

[8] Badini, S., Regondi, S. and Pugliese, R., 2023. Unleashing the power of artificial intelligence in materials design. *Materials, 16*(17), p. 5927.

[9] Pilania, G., 2021. Machine learning in materials science: From explainable predictions to autonomous design. *Computational Materials Science, 193*, p. 110360.

[10] Arockiam, A.J., Subramanian, K., Padmanabhan, R.G., Selvaraj, R., Bagal, D.K. and Rajesh, S., 2022. A review on PLA with different fillers used as a filament in 3D printing. *Materials Today: Proceedings, 50*, pp. 2057–2064.

[11] Lv, C., Zhou, X., Zhong, L., Yan, C., Srinivasan, M., Seh, Z.W., Liu, C., Pan, H., Li, S., Wen, Y. and Yan, Q., 2022. Machine learning: an advanced platform for materials development and state prediction in lithium-ion batteries. *Advanced Materials, 34*(25), p. 2101474.

[12] Lombardo, T., Duquesnoy, M., El-Bouysidy, H., Årén, F., Gallo-Bueno, A., Jørgensen, P.B., Bhowmik, A., Demortière, A., Ayerbe, E., Alcaide, F. and Reynaud, M., 2021. Artificial intelligence applied to battery research: hype or reality?*Chemical Reviews, 122* (12), pp. 10899–10969.

[13] Chandrashekhar, K.G., Laxmaiah, G., Ram Kumar, P and Ramesh, B., 2023. Load-bearing characteristics of a hybrid Si_3N_4-epoxy composite. *Biomass Conversion and Biorefinery*, pp. 1–9.

[14] Lu, Z., 2021. Computational discovery of energy materials in the era of big data and machine learning: A critical review. *Materials Reports: Energy, 1*(3), p. 100047.

[15] Akram, U., Nadarajah, M., Shah, R. and Milano, F., 2020. A review on rapid responsive energy storage technologies for frequency regulation in modern power systems. *Renewable and Sustainable Energy Reviews, 120*, p. 109626.

22

Sustainable Solutions on the Horizon: A Critical Review on Bioplastics as Key Materials in Modern Food Science

K. Nikitha, B. Thanuja, and E. Agalya
Cauvery College for Women (Autonomous), Tiruchirappalli, India

22.1 Introduction

Bioplastic materials serve as a viable alternative to conventional plastics, constituting just about 1% of the total global plastic production but are projected to experience a substantial growth rate of approximately 30% by 2025. Unlike traditional plastics derived from chemical processes, bioplastics originate from renewable biomass sources and encompass various types such as elastomers, thermosets, thermoplastics, and synthetic fibres. While these materials have historically relied on petrochemicals, there is a growing demand for their production from renewable sources [1].

Although all plastics theoretically degrade, their breakdown process is notably slow, rendering them largely non-biodegradable. The degradation of bioplastics depends on factors like their chemical makeup, environmental conditions (such as moisture, oxygen, temperature, and pH), and microbial activity from various organisms like actinobacteria, bacteria, and fungi. Incorporating compounds like ester, amide, or hydrolysable carbonate enhances their biodegradability, making composting the most environmentally friendly disposal method [2].

In terms of environmental impact, bioplastics generate lower greenhouse gas emissions than conventional plastics throughout their lifecycle, contributing to a more sustainable society. They possess similar properties to traditional plastics while offering the added benefit of a reduced carbon footprint. Strategies to reinforce bioplastics explore eco-friendly alternatives like lignocellulosic fibres and lignin instead of non-biodegradable materials such as glass and carbon fibres [3].

Innovative techniques like thermal treatment, dehydrothermal treatment, and ultrasound applications promise to enhance bioplastic's mechanical properties and functionalities. Recent research has also introduced a one-step water-based process to convert vegetable waste into biodegradable bioplastic films with comparable properties to other bioplastics [4].

Polyhydroxyalkanoates (PHAs) emerge as potential substitutes for conventional plastics due to their biocompatibility, biodegradability, and diverse thermal and mechanical characteristics. However, their industrial production remains less cost-effective, prompting exploration into cost-efficient carbon sources like macroalgae, peanut oil, crude glycerol, and whey [5].

Ongoing research focuses on developing advanced polymers like poly(ethylene 2, 5-furan dicarboxylate) (PEF) derived from biomass and exploring green microalgae cells as raw materials for cell plastic production [1, 6, 7].

DOI: 10.1201/9781003495437-22

Essentially, the goal of this chapter is to provide a thorough and informative explanation of the function that bioplastics play as sustainable materials in contemporary food science, with the potential to transform procedures and lessen the environmental impact that comes with the widespread use of regular plastics.

22.2 Bioplastics in Modern Food Packaging

Food production regulations have prompted a major transformation in the food industry, with packaging receiving a great deal of attention and developing into a separate sector. Packaging made of biopolymers has become the industry standard because of its vital role in promoting sustainability in the food supply chain, from the manufacturing plants to the homes of the customers. Even while biodegradable and compostable polymers hold great potential for providing superior food storage, there are still certain barriers to their widespread use. There is a tendency among large food wholesalers, though, who are thinking more and more about making the switch to bioplastics. In the industry's search for more environmentally friendly packaging options, this change represents a major advancement [8].

Indeed, the diverse nature of foods demands tailored packaging solutions to uphold their freshness and quality. Fruits and vegetables, characterised by high respiration rates, necessitate packaging that meticulously manages carbon dioxide and oxygen levels. This packaging also needs to regulate light exposure, provide structural integrity, and effectively block odours. In contrast, raw meat requires packaging that allows for high oxygen permeability to preserve its colour. This requirement is often fulfilled through techniques such as vacuum sealing or the use of active packaging, incorporating layers that absorb oxygen, as highlighted in the research by Baldino et al. [9]. Such specific packaging considerations cater to the unique needs of different food types, ensuring their optimal preservation and presentation. Dairy products have different packaging needs, requiring low oxygen permeability to prevent oxidation and microbial growth. Polysaccharides such as pectins, primarily sourced from fruits and vegetables, serve as effective barriers against oxidation and odours. For example, research on edible coatings made from polysaccharides showcased reduced mould growth on cheese surfaces compared to uncoated cheese, underscoring their effectiveness in preserving cheese quality [10]. The Applications of Biopolymers in Food Industries has been presented in Table 22.1.

TABLE 22.1

Applications of Biopolymers in Food Industries [11]

Packaging	Biopolymer
Starch based	
Milk chocolates	Corn-starch trays
Organic tomatoes	Corn-based packaging
Cellulose based	
Kiwi	Bio-based trays wrapped
Sweets	Metallised cellulose film
Organic pasta	Cellulose-based packaging
Potato chips	Metallised cellulose film
PLA (Polylactic acid)	
Yoghurt	PLA cups
Fresh salads	PLA bowls
Beverages	PLA cups
Bread	Paper bags with PLA

TABLE 22.2

Classification of Bioplastics [5]

	Bio Based	**Petroleum Based**
Biodegradable	Bioplastic examples: Starch, cellulose, polylactic acid, polyhydroxyalkanoates	Bioplastic examples: Polycaprolactone, polybutylene adipate, terephthalate, polybutylene succinate
Non-biodegradable	Bioplastic examples: Bio-polypropylene, bio-polyethylene	Bioplastic examples: PVC (Polyvinyl chloride), polyethylene, polypropylene

22.3 Categorisation of Bioplastics

Plastics can be biodegradable or non-biodegradable, and they can be produced using resources derived from either bio- or fossil fuels. Biodegradable plastics are created from fossil fuels or a combination of fossil fuels and renewable components, whereas bioplastics are entirely made from renewable resources. The classification of bioplastic is shown in Table 22.2. The following three categories of bioplastics exist:

- Bio-based and biodegradable.
- Fossil-based and biodegradable.
- Petroleum-based known as plastics and non-biodegradable.

While starch and cellulose are not inherently plastic, they can be made into plastics through fermentation or polymer technology through a variety of processes such as mixing, extrusion, injection moulding, and casting [5]. The prefix 'bio' denotes that bio-ethylene and bio-propylene are derived from renewable resources and have no characteristics with polymers derived from petrochemicals [12].

22.4 Sources of Bioplastic

Bioplastics have become a viable way to lessen the harm that conventional plastics do to the environment. Their many sources – mostly renewable biomass like sugarcane, maize starch, and materials containing cellulose – showcase the compositional diversity of the materials. The synthesis of diverse bioplastics, such as polylactic acid (PLA), PHA, and starch-based polymers, is fuelled by this diversity, signalling a major shift towards sustainable material alternatives [8]. Recent developments highlight a critical shift in the production of bioplastics, emphasising the use of agricultural waste and other rejected materials [13]. Reusing food waste, such as vegetable and fruit scraps, has gained popularity as a possible source for producing biodegradable polymers. This method not only addresses the issue of waste management but also turns organic waste that is thrown away into useful materials for the manufacturing of bioplastics [14].

Furthermore, microbes have become important participants in the search for precursors to bioplastics. Significant interest has been aroused by their capacity to synthesise PHA from a variety of carbon sources, including waste streams. This microbial-based method has great potential to transform the production of sustainable materials. Using these microbes to their full potential increases the possibility of turning waste materials into useful bioplastics. Using waste streams and agricultural by-products to make bioplastics is in line with the circular economy's tenets of material repurposing, decreasing reliance on finite resources, and minimising environmental damage [15]. The development of bioplastics from diverse biomass sources, such as waste streams and microbial synthesis, offers an engrossing story for a sustainable future. This paradigm shift in the manufacture of materials marks a step forward in minimising the ecological footprint and opens the door for creative and eco-friendly substitutes for traditional plastics [16] (Figure 22.1).

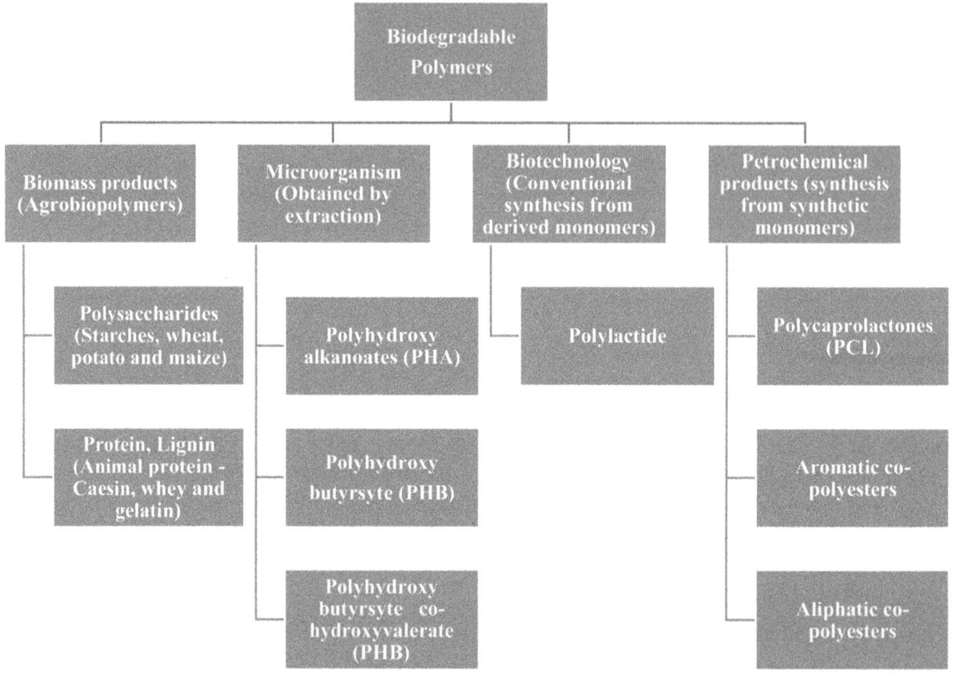

FIGURE 22.1   Sources of bioplastics [17].

22.5 Nanomaterial-Based Food Packaging

Food packaging plays a crucial role in the food industry by preserving the nutritional value and organoleptic qualities of food. Food packaging shields food's delicate bioactive components from abrasive physical and environmental factors in addition to preventing spoiling. Many studies are being conducted on nanotechnology to enhance food packaging [18]. It has been discovered that functional nanoparticles increase the strength, durability, and barrier qualities of packaging materials, all of which contribute to the extension of food products' shelf lives [19]. There are two main ways to create nano-based packaging materials for extending the shelf life and preserving the quality of food products: Either by adding nanoparticles to conventional food packaging materials like films and containers or by creating nanocomposite multilayer packaging materials and organic, inorganic, and combined nanocoatings by rubbing, spraying, or immersion [20]. Better packaging with enhanced mechanical and physical qualities, active packaging with antibacterial, antioxidant, and UV absorption capabilities, and smart packaging with monitored and regulated food conditions are all made possible by nanotechnology [21]. According to Statistics MRC, at a compound annual growth rate of 11%, the global market for smart packaging is expected to reach $32.7 billion by 2022. Several predictions have indicated that intelligent packaging will grow at the highest rate of any smart packaging because of its distinctive, interactive, and cost-effective qualities [22].

22.6 Ecological Considerations of Bioplastic

22.6.1 Ecological Footprint and Sustainability

It is possible to accurately assess the sustainability of the entire family of bioplastics if all the material processes, such as sourcing, production, use, and disposal, are taken into account. To be more specific, each of these stages has a weight assigned to its economic and environmental qualities. For instance, the production of bio-composites for use in buildings directly reduces the ecological footprint of the entire construction engineering sector.

Food and agricultural processing waste, as well as renewable material sources that do not compete with food production chains, are the sources of bio-based sustainable packaging [13]. We can apply a biofuel classification scheme, dividing first-, second-, and third-generation feedstocks, to categorise the sources of materials used. Edible biomass such as sugarcane, whey, and maize are used as first-generation feedstock. The second generation consists of biomasses that are inedible and come from lignocellulosic sources, such as urban trash, byproducts of animal processing, forests, and agriculture. Algal biomass is one of the most unusual sources; it is classified as a third-generation feedstock [23].

22.6.2 Disposal Techniques and Environmental Effects

Bioplastics are mostly available for private consumption because their market prices are still higher than those of conventional plastics when packaging applications are taken into account.

This leads us to conclude that household consumption accounts for most bioplastic disposal routes. Composting is therefore the primary method of disposal; however, due to errors in the process or even separation by screening in the disposal plants, a constant portion of the total mass reaches the residual waste and is ultimately sent to incinerators [24].

A study on the potential interactions between bioplastic bags and fermentation disposal plants in Germany was published by Grundmann and Wonschik [25]. This behaviour has been tested using hydrolysis analysis and anaerobic fermentation. The results demonstrate how thermophilic characteristics are truly required for fermentation processes to occur, with higher degradation degrees typically falling between 20% and 30%.

The following phases have been taken into consideration in the analysis of an extended life cycle assessment for bio-polythene systems [26]:

- Production of the primary materials (bio-polyethylene).
- Transportation of the new product to the processing facility.
- Production of the film products.
- Transportation of the film products.
- Films disposal.
- Utilisation of films.
- Credits allocated for the use of secondary materials and secondary energy from recycling and disposal processes.
- Accounting (credit) for the CO_2 bound in the bio-PE.

Compared to fossil-based plastics, bio-PE responds better to the impacts of climate change and the consumption of fossil resources, but it is deficient in aspects such as acidification, eutrophication, and human toxicity [17].

22.7 Benefits and Drawbacks of Bioplastic

The primary environmental contaminant that is utilised regularly is plastic [27, 28]. Therefore, instead of using products based on petrochemicals, we should use bioplastics to reduce environmental pollution [24]. Numerous environmental problems can be resolved in this way [29]. The special qualities of bioplastics are their energy efficiency, compostability, biodegradability, and environmental friendliness [30].

Biodegradable plastics exhibit great potential in the future. The following are a few benefits of bioplastics:

- Energy efficiency.
- Decreased carbon footprint.

- Eco-safety.
- Partly based on natural feedstock [31].

However, using bioplastics could lead to issues. The following are a few drawbacks of bioplastics:

- Expensive.
- Thermal instability.
- Brittleness.
- A variety of recycling issues [32].

22.8 Market for Bioplastics

Numerous academics have created resources to help with plastic selection decision-making. The Spectrum of Plastic is one of them. Bio-based bioplastics show up higher on this spectrum because they come from renewable resources, degrade naturally, and can be composted. Given that bioplastic packing films and containers have a shorter lifespan and are more likely to wind up in landfills, they are a better choice. In the realm of biomedicine, biodegradable bioplastics find numerous uses, including the production of bone plates and screws, scaffolds for tissue engineering, and drug delivery vehicles [33]. In America, McDonald's has already begun to produce biodegradable containers for their fast-food delivery and service. Numerous businesses, including Dow Cargill, Nike, Danone, Bayer, and DuPont, are manufacturing biodegradable packaging materials [34, 35].

The worldwide bioplastics market is expanding at a rate of approximately 15% to 20% annually. By 2024, bioplastics are projected to account for 15% to 20% of the global plastics market, with an expected market share of 30% to 35%. The bioplastics market is anticipated to have grown from $1 billion in US dollars in 2007 to approximately $20 billion by 2024. The bioplastics sector is seeing a surge in investments, which is positive for the environment [11].

22.9 Upcoming Developments in Bioplastics

The environmental crisis caused by traditional plastic pollution has spurred increasing research and development in the field of bioplastics. These materials offer a promising alternative, as they are derived from renewable resources like biomass and often possess biodegradability, reducing their environmental impact. This review explores the latest advancements in bioplastics, highlighting key developments and prospects.

22.9.1 New Biopolymer Sources and Production Techniques

Recent research focuses on exploring new sources for biopolymer production beyond traditional crops like corn and sugarcane. Studies show promising results with utilising agricultural waste like rice husk and straw, food waste, and even marine algae for bioplastic production [36, 37]. Additionally, advancements in fermentation techniques using bacteria and fungi are enabling the production of biopolymers like PHAs with improved properties and reduced production costs [38, 39].

22.9.2 Enhanced Bioplastic Properties

One major challenge with bioplastics has been their limited performance compared to traditional plastics. However, current research is focusing on developing bioplastics with improved mechanical strength, thermal stability, and barrier properties. This is being achieved through various strategies, including blending different biopolymers, adding reinforcing agents, and employing nanotechnology [40].

22.9.3 Addressing Biodegradability Concerns

The biodegradability of bioplastics can vary widely, prompting recent research efforts to focus on creating bioplastics that degrade entirely under practical composting conditions. The goal is to minimise the potential for microplastic formation. This endeavour involves investigating innovative biopolymers such as PLA that exhibit enhanced biodegradability. Additionally, scientists are exploring additives designed to enhance the biodegradation process, as highlighted in the work of [41, 42]. These advancements aim to ensure that bioplastics break down completely, addressing concerns about their environmental impact and paving the way for more eco-friendly materials in various applications.

22.9.4 Circular Economy Approaches

The bioplastics industry is swiftly adopting circular economy principles, focusing on designing bioplastics for easier recycling and composting. This includes aligning bioplastics with existing waste management systems and exploring closed-loop applications. It involves not just creating recyclable materials but also reintroducing bioplastics into production cycles to minimise waste and maximise resource efficiency. Collaboration among stakeholders is crucial for establishing efficient collection and recycling methods tailored for bioplastics. This concerted effort highlights the industry's commitment to sustainable practices and revolutionising the life cycle management of these materials [43, 44].

22.9.5 Regulatory Landscape and Consumer Acceptance

The regulatory landscape for bioplastics is evolving rapidly, with governments implementing policies to encourage their adoption and address potential environmental concerns. Additionally, consumer awareness and acceptance of bioplastics are increasing, fuelled by growing environmental consciousness. This positive societal shift is crucial for driving the widespread adoption of bioplastics [45–47].

The field of bioplastics is rapidly evolving, with significant advancements in new materials, production techniques, and improved properties. As research and development continue, bioplastics are poised to play a crucial role in addressing the global plastic pollution crisis and creating a more sustainable future.

22.10 Conclusion

Bioplastics are a shining example of contemporary food science's efforts to find sustainable solutions. This analysis emphasises how they came from a variety of renewable sources, including agricultural waste, sugarcane, and maize starch. The creative move towards turning waste streams – vegetable scraps and fruit peels – into biodegradable polymers transforms our understanding of materials and promotes a circular economy. This revolution is further propelled by the microbial synthesis of bioplastics, which use microorganisms to transform trash into useful materials. This thorough analysis redefines waste as a useful resource and emphasises how bioplastics might reduce plastic pollution and environmental deterioration.

Bioplastics are an important component of food science's sustainable development. These materials' inventive approaches, wide range of sources, and dedication to environmental responsibility open the door to a future in which they are indispensable in promoting a society that is resource-aware and environmentally conscientious.

Acknowledgement

The authors thank Cauvery College for Women (Autonomous), Tiruchirappalli, for the financial support under the scheme of DST-CURIE Core grant for Women PG colleges [Reference No. DST/CURIE-PG/2022/8 (G)].

REFERENCES

[1] Nakanishi, A., Iritani, K., & Sakihama, Y. (2020). Developing neo-bioplastics for the realization of carbon sustainable society. *Journal of Nanotechnology and Nanomaterials*, *1*(2), 72–85.

[2] Emadian, S. M., Onay, T. T., & Demirel, B. (2017). Biodegradation of bioplastics in natural environments. *Waste Management*, *59*, 526–536.

[3] Yang, J., Ching, Y. C., & Chuah, C. H. (2019). Applications of lignocellulosic fibers and lignin in bioplastics: A review. *Polymers (Basel)*, *11*(5), 1–26.

[4] Perotto, G., et al. (2018). Bioplastics from vegetable waste: Via an ecofriendly water-based process. *Green Chemistry*, *20*(4), 894–902.

[5] Khatami, K., Perez-Zabaleta, M., Owusu-Agyeman, I., & Cetecioglu, Z. (2021). Waste to bioplastics: How close are we to sustainable polyhydroxyalkanoates production? *Waste Management*, *119*, 374–388.

[6] Hwang, K. R., Jeon, W., Lee, S. Y., Kim, M. S., & Park, Y. K. (2020). Sustainable bioplastics: Recent progress in the production of bio-building blocks for the bio-based next-generation polymer PEF. *Chemical Engineering Journal*, *390*, 124636.

[7] Iben Nasser, I., Algieri, C., Garofalo, A., Drioli, E., Ahmed, C., & Donato, L. (2016). Hybrid imprinted membranes for selective recognition of quercetin. *Separation and Purification Technology*, *163*, 331–340.

[8] Yadav, B., Pandey, A., Kumar, L. R., & Tyagi, R. D. (2020). Bioconversion of waste (water)/residues to bioplastics—A circular bioeconomy approach. *Bioresource Technology*, *298*(2019), 122584.

[9] Baldino, A., Smith, J., & Garcia, M. (2018). Advanced packaging technologies for extending the shelf life of perishable foods. *Journal of Food Science and Technology*, *45*(3), 211–225.

[10] Cerqueira, M. A., Vicente, A. A., & Pastrana, L. M. (2021). Advances in food packaging: A review. *Food Research International*, *143*, Article 110253.

[11] Shah, M., Rajhans, S., Pandya, H. A., & Mankad, A. U. (2021). Bioplastic for future: A review then and now. *World Journal of Advanced Research and Reviews*, *09*(02), 056–067.

[12] Asrofi, M., Sapuan, S. M., Ilyas, R. A., & Ramesh, M. (2020). Characteristic of composite bioplastics from tapioca starch and sugarcane bagasse fiber: Effect of time duration of ultrasonication (bath-type). *Materials Today: Proceedings 33*, 118–123.

[13] Reichert, C. L., et al. (2020). Bio-based packaging: Materials, modifications, industrial applications and sustainability. *Polymers*.

[14] Hamidon, N., et al. (2018). Potential of production bioplastic from potato starch. *Sustainable Environmental Technologies*, *1*, 115–123.

[15] Soykeabkaew, N., Tawichai, N., Thanomsilp, C., & Suwantong, O. (2017). Nanocellulose reinforced green composite materials. *Walailak Journal Science & Technology*, *14*(5), 353–368.

[16] Rahman, M. H., & Bhoi, P. R. (2021). An overview of non-biodegradable bioplastics. *Journal of Cleaner Production*, *294*, 126218.

[17] Coppola, G., Gaudio, M. T., Lopresto, C. G., Calabro, V., Curcio, S., & Chakraborty, S. (2021). Bioplastic from renewable biomass: A facile solution for a greener environment. *Earth Systems and Environment*, *5*(2), 231–251.

[18] Ashfaq, A., Khan, S. A., & Ahmad, F. (2022). Enhancing food packaging through nanotechnology: A review of functional nanoparticles. *Food Packaging Technology*, *8* (2), 115–130.

[19] Gregor-Svetec, D. (2018). Intelligent packaging. In *Nanomaterials For Food Packaging: Materials, Processing Technologies, and Safety Issues* (pp. 203–247). Elsevier.

[20] Shukla, P., Chaurasia, P., Younis, K., Qadri, O. S., Faridi, S. A., & Srivastava, G. (2019). Nanotechnology in sustainable agriculture: Studies from seed priming to postharvest management. *Nanotechnology for Environmental Engineering*, *4*(1), 17–38. Springer Science and Business Media Deutschland GmbH.

[21] Tsagkaris, A. S., Tzegkas, S. G., & Danezis, G. P. (2018). Nanomaterials in food packaging: State of the art and analysis. *Journal of Food Science and Technology*, *55*(8), 2862–2870. Springer.

[22] Primožič, M., Knez, Ž, & Leitgeb, M. (2021). (Bio)nanotechnology in food science—food packaging. *Nanomaterials*, *11*(2), 1–31. MDPI AG.

[23] Naik, S. N., Goud, V. V., Rout, P. K., & Dalai, A. K. (2018). Production of first and second generation biofuels: A comprehensive review. *Renewable and Sustainable Energy Reviews*, *14*(2), 578–597.

[24] Ahamed, A., et al. (2021). Life cycle assessment of plastic grocery bags and their alternatives in cities with confined waste management structure: A Singapore case study. *Journal of Cleaner Production*, *278*, 123956.

[25] Grundmann, M., & Wonschik, S. (2011). Interactions between bioplastic bags and fermentation disposal plants: Hydrolysis analysis and anaerobic fermentation. *Journal of Environmental Science and Engineering A, 1*(2), 45–53.

[26] Detzel, A., & Kauertz (2015). Phases considered in the analysis of an extended life cycle assessment for bio-PE systems. *Journal of Environmental Assessment and Management, 25*(3), 112–126.

[27] Al Battashi, H., Al-Kindi, S., Gupta, V. K., & Sivakumar, N. (2020). Polyhydroxyalkanoate (PHA) production using volatile fatty acids derived from the anaerobic digestion of waste paper. *Journal of Polymers and the Environment, 1*(10), 13–17.

[28] Zeb, A., et al. (2019). Advanced nanomaterials in food packaging: A review of recent developments in nanotechnology. *Critical Reviews in Food Science and Nutrition, 59*(5), 683–699.

[29] Ateş, M., & Kuz, P. (2020). Starch-based bioplastic materials for packaging industry. *Journal of Sustainable Construction Materials and Technologies, 5*(1), 399–406.

[30] Beckstrom, B. D., Wilson, M. H., Crocker, M., & Quinn, J. C. (2020). Bioplastic feedstock production from microalgae with fuel co-products: A techno-economic and life cycle impact assessment. *Algal Research, 46*, 101769.

[31] Ebrahimian, F., Karimi, K., & Kumar, R. (2020). Sustainable biofuels and bioplastic production from the organic fraction of municipal solid waste. *Waste Management, 116*, 40–48.

[32] El-Malek, F. A., Khairy, H., Farag, A., & Omar, S. (2020). The sustainability of microbial bioplastics, production and applications. *International Journal of Biological Macromolecules, 157*, 319–328.

[33] Bhati, R. (2019). Biodegradable plastics production by cyanobacteria. In *Biotechnology Products in Everyday Life* (1st ed., p. 250). Springer.

[34] Alshehrei, F. (2019). Production of polyhydroxybutyrate (PHB) by bacteria isolated from soil of Saudi Arabia. *Journal Pure Applied Microbiology, 13*(2), 897.

[35] Younis, K., et al. (2022). Available application of nanotechnology in food packaging: Pros and Cons. *Journal of Agriculture and Food Research, 7*, 100270.

[36] Kumar, P., Singh, R. K., Singh, R., & Sarkar, A. (2023). Sustainable bioplastic production from rice husk: Current status and future prospects. *Renewable and Sustainable Energy Reviews, 181*, 113975.

[37] Saravanakumar, S. S., Kumar, S. R., Karthikeyan, C., Kumar, S. S., & Pugazhendhi, A. (2023). Sustainable bio-plastic production from marine algae: An eco-friendly approach. *Bioresource Technology Reports, 20*, 101091.

[38] Chen, G. Q. (2023). Recent advances in microbial production of polyhydroxyalkanoates (PHAs). *Current Opinion in Biotechnology, 82*, 169–177.

[39] Kasmuri, N., & Zait, M. S. A. (2018). Enhancement of bio-plastic using eggshells and chitosan on potato starch based. *International Journal of Engineering and Technology, 7*(3), 110–115.

[40] Shah, A. A., Hasan, F., Hameed, A., & Ahmed, S. (2022). Strategies for improving the mechanical and thermal properties of bioplastics. *International Journal of Biological Macromolecules, 213*, 92–122.

[41] Sudhakar, M., Yaakob, H., Kumar, G., & Prasath, M. (2023). Biodegradable polylactic acid—A comprehensive review. *International Journal of Biological Macromolecules, 222*, 102–120.

[42] Bhatia, S. K., et al. (2019). Bioconversion of plant biomass hydrolysate into bioplastic (polyhydroxyalkanoates) using Ralstoniaeutropha 5119. *Bioresource Technology, 271*, 306–315.

[43] Reddy, D. V., Singh, S., & Prakash, O. (2022). Circular economy models for bioplastics: Opportunities and challenges. *Bioresource Technology, 351*, 126847.

[44] Bilo, F., et al. (2018). A sustainable bioplastic obtained from rice straw. *Journal of Cleaner Production, 200*, 357–368.

[45] Krishnamurthy, R., Bhattacharya, R., & Bhattacharjee, S. (2023). Consumer perception and willingness to pay for bioplastics: A cross-country analysis. *Journal of Cleaner Production, 373*, 133629.

[46] Al Battashi, H., et al. (2019). Production of bioplastic (poly-3-hydroxybutyrate) using waste paper as a feedstock: Optimization of enzymatic hydrolysis and fermentation employing Burkholderiasacchari. *Journal of Cleaner Production, 214*, 236–247.

[47] Kumar, S., & Thakur, K. S. (2017). Bioplastics—Classification, production and their potential food application. *Journal of Hill Agriculture, 8*(2), 118–129.

23

Review on Performance Comparison of Contemporary Multi-level Inverter Structures for Energy Storage Applications

Titus Sigamani
K. Ramakrishnan College of Engineering, Tiruchirappalli, India

Ram Prakash Ponraj and Vijay Ravindran
Saranathan College of Engineering, Tiruchirappalli, India

Venkateswaran Madhu
Lendi Institute of Engineering and Technology, Vizianagaram, India

Paranthagan Balasubramanian and Satheesh Ragunathan
Saranathan College of Engineering, Tiruchirappalli, India

23.1 Introduction

Electric and hybrid-electric vehicle development presents the power electronics sector with numerous new opportunities and problems, especially in the core traction motor drive [1]. Induction motors provide traction in most of the current and future electric car designs. Heavy-duty trucks and military combat vehicles with enormous electric drives require modern power electronic inverters (>250 kW). Making these huge trucks as electric cuts down pollution and may increase the performance (acceleration and braking). Multilevel inverters (MLIs) are suitable due to their high volt-ampere (VA) ratings. To drive the motor, an alternator can be used with a back-to-back diode-clamped converter. Conventional 3-level inverters are mostly used in electrical drives and their applications. Even though the conventional 180° and 120° conduction modes of inverter operation using pulse width modulation (PWM) techniques are very popular, the MLIs are substituted by 3-levelinverters [2, 3]. Problems occur with 3-level high-frequency PWM inverters for automobile drives. High frequency switching might compound the situation by repeatedly applying the common mode voltage to the motor. Due to the higher switching loss, PWM-controlled inverters require more heat removal. As the common mode voltage is applied to the motor more frequently, high frequency switching can exacerbate the issue of high switching losses and therefore the PWM-operated inverters require more heat evacuation. MLIs solve these concerns by switching at a lower frequency than PWM-controlled inverters, allowing them to run at high efficiency [4, 5]. Various MLIs utilised in e-vehicle applications are discussed in this chapter. The cascaded H-bridge (CHB) inverter, capacitor-clamped (CC) inverter, and diode-clamped inverter are the fundamental designs which are explained by various authors in different forms [6]. The CHB inverter with series connection which consists of an isolated DC source gives the appropriate amplitude and frequency with non-sinusoidal output voltage. However, in CC and diode-clamped inverters, single input DC is used for different voltages of capacitor bank. The increase in cost and complexity is caused due to blocking voltages which increase the levels in diode-clamped and CC MLIs.

Converter module which has huge components not only expands the size of the module, but also increases the circuit complexity with huge number of gate drivers and different switching sequences [7]. This concept initiates researchers and industrialists to develop modern converter topologies with minimum component and control techniques [8]. Basic design parameters of MLIs are shown in Figure 23.1(a).

DOI: 10.1201/9781003495437-23

MLI topologies are broadly classified into symmetric and asymmetric converter topologies [9], shown in Figure 23.1(b). The symmetric converter topologies consist of separate DC sources with equal magnitudes, whereas asymmetric converter topology has an uneven magnitude which makes the symmetrical circuits to have modulation control. Subsequently, the asymmetric converter topologies have no modulation control but needs devices with different blocking voltage capability [10].

The simplified block diagram of an electric vehicle, shown in Figure 23.1(c), contains a DC–DC converter supplied by renewable energy sources. An inverter is required to convert DC voltage into AC voltage when AC motor is used. Both converters can work bi-directionally as the battery is charged while

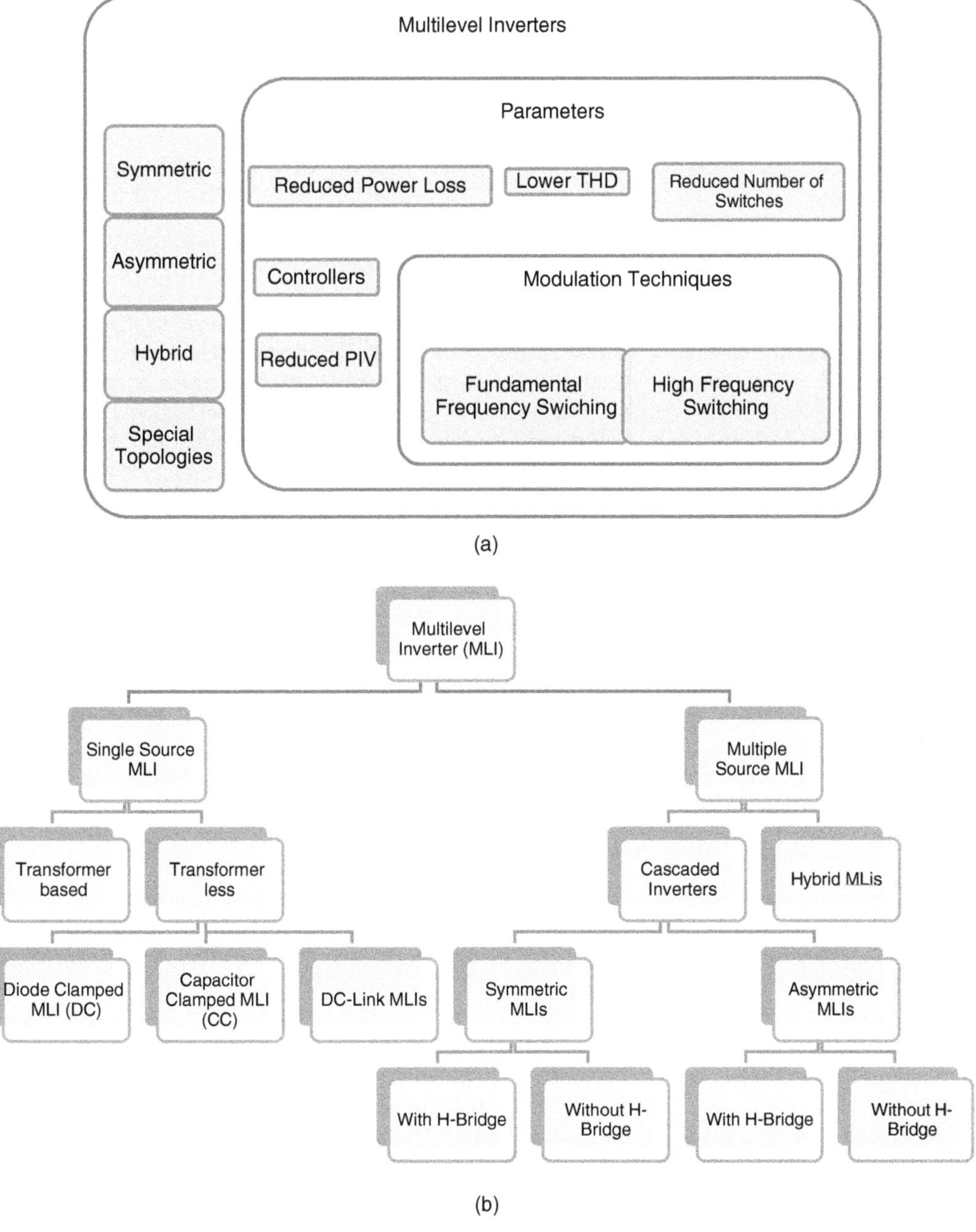

FIGURE 23.1 (a) Multilevel inverter design parameters. (b) Hierarchy of multilevel inverters. (c) Block diagram of an electric vehicle with energy storage.

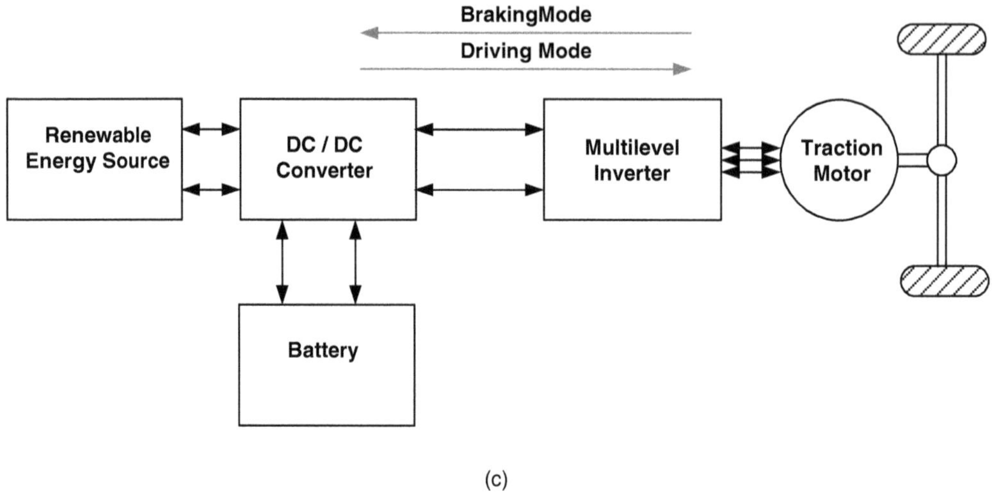

(c)

FIGURE 23.1   (Continued)

braking as well as in motoring modes. Usage of two-level inverters will cause high *dv/dt* value, electro-magnetic interference, square waveform, and high harmonics. MLI is the best alternative for conventional inverters because of its advantages like low electromagnetic interference, dv/dt, and near-sinusoidal output [11]. Even though there is a dearth of literature addressed towards reduction of switching component, there is still scope for alternate and reduced component count configurations. In many topologies, it is not possible for further reduction in the circuit component count, but the novel converter topology encourages that because of attractive design modules, reducing capacitor values, intelligent switching techniques, etc. The conventional modulation of MLIs is greatly challenged by new inverter topologies and introducing a new switching topology leads to complexity in control circuitry. The converter which consists of MLI is widely used in industrial sector, is unique, and does wonderful operations. The tremendous development of PWM technique makes the MLIs to produce challenging performance in electric vehicle applications. Hence MLI topologies with respect to their design parameters, switching techniques, and circuit perfor-mance of hybrid electric vehicle with contemporary power applications are elaborately discussed in this chapter.

23.2 MLIs of the Cascaded H-Bridge (CHB) Design with a Level Doubling Network (LDN)

The modified structures are introduced for reducing switch count and minimising the voltage stress across the switches. Figure 23.2(a)–(c) shows the unidirectional and bi-directional switches used in the MLIs. Figure 23.2(d) shows the generalised structure of MLIs with the corresponding output voltage waveform. Total harmonic distortion (THD) value, output voltage, and currents are compared with the existing single or multiple source topologies. These topologies have 'm' sub-cells containing 'p' number of switches in each sub-cell. Each sub-cell has either single source or 'k' number of sources. These sub-cell connec-tions are made in series or parallel to produce the expected levels of output voltage and also connected with an H-bridge for polarity change. As illustrated in Figure 23.3(a), Sumit K created a new symmetric CHB topology with LDN. To double the output voltage levels of a symmetric CHB inverter, this architec-ture uses an LDN half-bridge inverter. To provide asymmetric performance, this topology was designed. Prabaharan and Palanisamy [12] employed multicarrier PWM (MCPWM) to increase output quality in symmetric and asymmetric topologies while decreasing switch peak inverse voltage (PIV). Table 23.1 shows the detailed general analysis of the MLIs with LDN. CHB-based MLIs were proposed for heavy-duty vehicles and trucks because of the controllability of the high-voltage motors using low-voltage

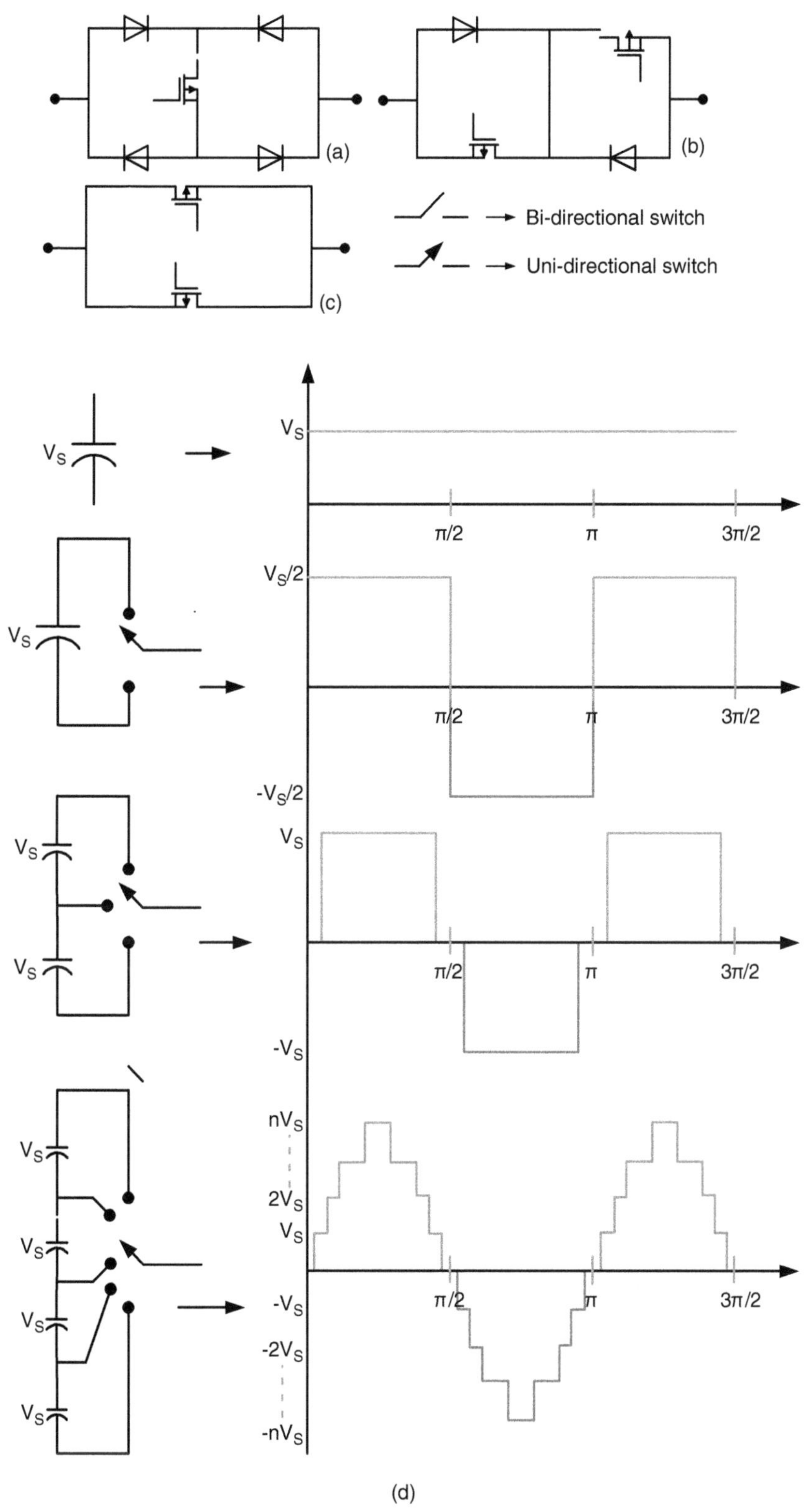

FIGURE 23.2 (a)–(c) Basic structures of bi-directional switches; (d) general MLI structure with output voltage.

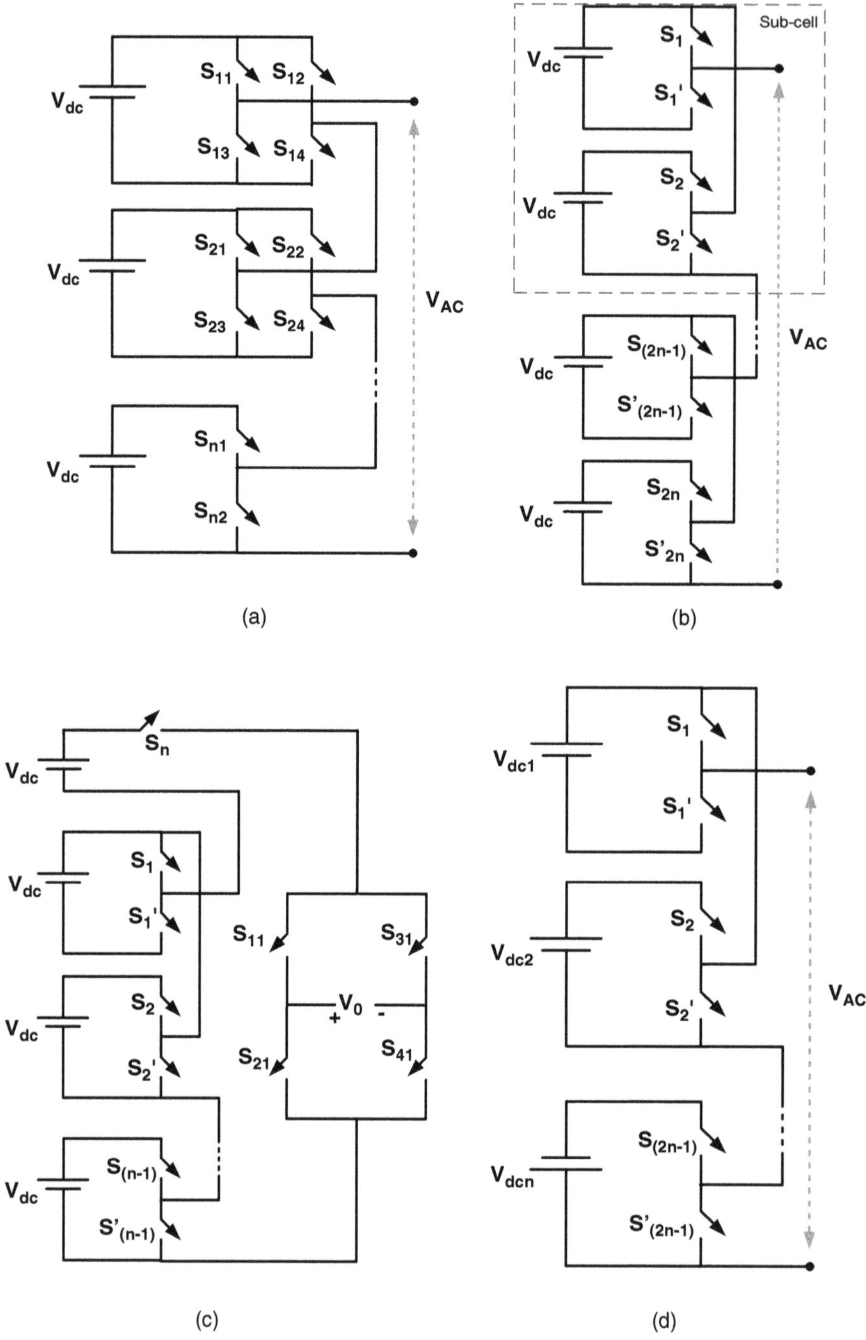

FIGURE 23.3 (a) CHB MLI with level doubling network; (b) two-leg CHB basic structure; (c) modified structure with H-bridge; (d) modified structure with asymmetric configuration.

TABLE 23.1

Properties and Parameters of CHB MLIs with LDN for 'm' Number of Basic Cells

Topology	Configuration	N_{LEVEL}	N_{SOURCE}	N_{SWITCH}	Switching Method
[12]	Symmetric without LDN	$2m+1$	M	$4m$	Multicarrier PWM (MCPWM)
	Symmetric with LDN	$4m+1$	$m+1$	$4m+2$	
	Asymmetric without LDN	$2 \times \left(\sum_{i=0}^{m-1} 2^i \right) + 1$	M	$4m$	
	Asymmetric with LDN	$2 \times \left[2 \times \left(\sum_{i=0}^{m-1} 2^i \right) + 1 \right] - 1$	$m+1$	$4m+2$	
[13]	Symmetric	$2m+1$	2m	4m	Phase-shifted multicarrier
[14]	Asymmetric	$2^{m+1}-1$	$2m$	$4m$	Fundamental frequency
		3^m	$2m$	$4m$	switching (FFS)
[15]	Symmetric	$2m+3$	$m+1$	$2m+5$	MCPWM
[16]	Asymmetric	$2^m - 1$	M	$2m$	MCPWM

Properties of CHB Inverter with LDN
- Voltage source connected in LDN is positive with respect to ground.
- Self-balancing capability.
- Reducing peak inverse voltage (PIV).
- Uniform loading of each cell in symmetric topology.

switches and sources. The CHB circuit is used to drive the traction motor supplied by a set of batteries or fuel cells also.

Waltrich et al. introduced a modified inverter topology with two legs as shown in Figure 23.3(b). This eliminates the H-bridge but it requires two independent sources for positive and negative output voltages, respectively [13]. Each sub-cell comprises four switches and two sources to produce a three-level output. Series connection of 'm' sub-cells will create the required levels in output voltage and this topology is presented by Babaei et al. especially for electric drive applications [14]. This system used modified phase-shifted PWM for harmonic elimination. Nagarajan introduced a modified topology with H-bridge [15] and it reduces the number of switches compared to the basic circuit as shown in Figure 23.3(c). This circuit is presented for electrical drive application and uses MCPWM – phase disposition (PD), phase opposition disposition (POD), alternative phase opposition disposition (APOD), etc. Mahrous Ahmed [16] presented the same topology with modified structure in asymmetric configuration as shown in Figure 23.3(d). This topology ignores the H-bridge for low dv/dt value and used MCPWM technique for harmonic elimination. Table 23.1 shows the properties and applications of MLI topologies with LDN.

23.3 Cascaded H-Bridge MLIs with Sub-cells

Kangarlu and Babaei et al. [17] introduced a modified sub-cell inverter topology as shown in Figure 23.4(a). They also proposed the optimal structures for symmetric and asymmetric configurations in terms of various parameters. This topology aims to reduce the number of H-bridges in CHB. It requires 'm' number of series-connected sub-cells which are equal to the number of H-bridges. Each sub-cell contains 'k' number of sources and switches. Babaei presented another topology which has a similar structure without direct connection between source and load as shown in Figure 23.4(b) [18]. These topologies eliminate the requirement of H-bridge for each voltage source and hence reduce the switching losses, blocking voltage, and *dv/dt*. Anish K analysed the topology shown in Figure 23.4(c) with MCPWM like PD, POD, APOD, and variable frequency (VF) to improve the output quality [19]. Siddharth Pachpor et al. [20] proposed the modified topology with switched capacitor circuit shown in Figure 23.4(d). This circuit not only reduces the switch count but also eliminates the requirement of bi-directional switches compared to Figure 23.5(a) and (b). In this structure, each sub-cell has three sources for symmetric and asymmetric

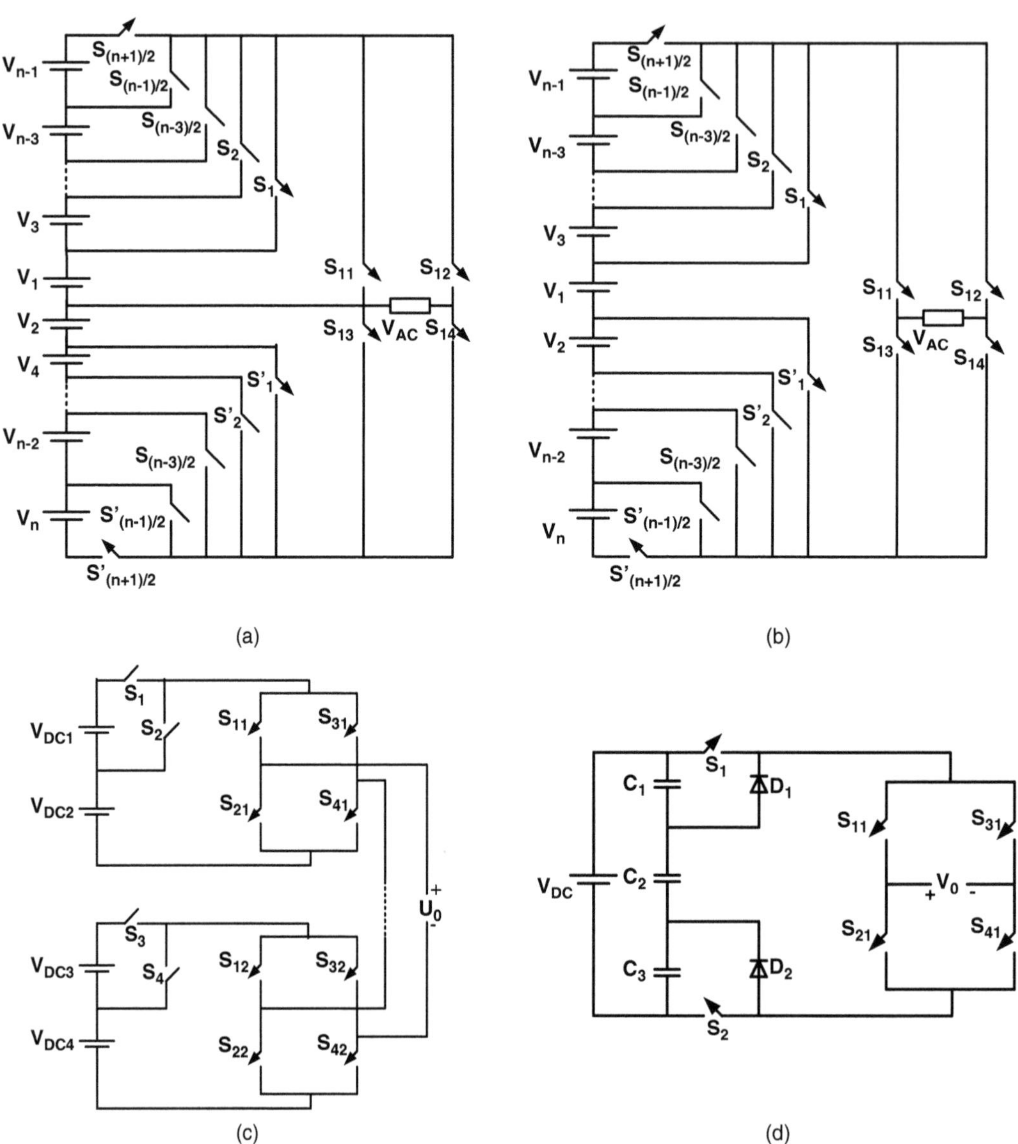

FIGURE 23.4 (a, b) General structures of MLI with sub-cells; (c) two sources per cell structure; (d) modified structure with three sources per cell.

modes explained with Inductive (RL) load. Karthikeyan et al. [21] described the same topology with LDN for photovoltaic applications. The topology has been analysed for RL load with modified nearest level modulation (NLM) PWM technique. This article also analysed the topology with DC–DC converter and voltage balancing circuit that was also presented by Jiang [22]. Table 23.2 gives the properties and applications of the topologies with sub-cells presented in Figure 23.4(a)–(d).

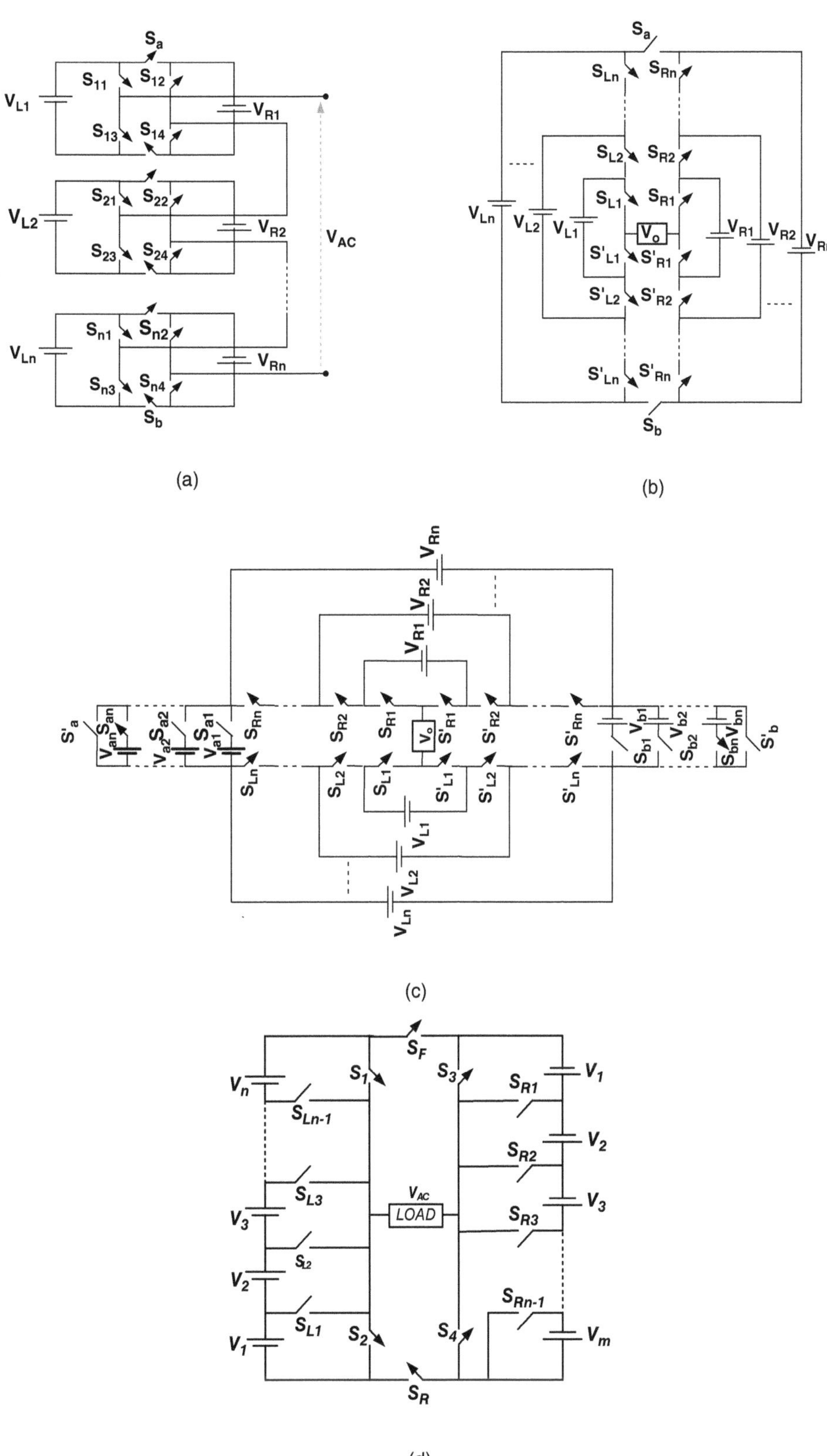

(a)

(b)

(c)

(d)

FIGURE 23.5 (a)–(d) Multilevel inverter structures with modified H-bridge.

TABLE 23.2

Properties and Parameters of CHB MLIs with Sub-cells

Topology	Configuration	N_{LEVEL}	N_{SOURCE}	N_{SWITCH}	Switching Method
[17]	S	$2mk+1$	$2mk+4$	$2m+4$ for $k=1$	FFS
	AS	$2(k+1)^m-1$		$m(k+2)+4$ for $k \geq 2$	
[18, 19]	S	$2mk+1$	$2m(k_n+1)$	$2m(k+1)$ for $k=1,2$	FFS [23], Sine PWM SPWM [24]
				$m(k+5)$ for $k \geq 3$	
	AS	$\prod_{i=1}^{m}(2k_i+1)$		$2m(k_n+1)$ for $k_n=1,2$	
				k_n+5 for $k_n \geq 3$	
[20]	S	$6m+1$	m	$6m$, $2m$ diodes, $3m$ capacitors	FFS
	AS	7^m			
[21, 22]	S	$6m+1$	m	$6m$, $2m$ diodes, $3m$ capacitors	Modified nearest voltage-level modulation
	AS	$12m+1$	$m+1$	6m+2, 2m diodes, 3m+1 capacitors	

Properties of Topologies Presented in Figure 23.4(a)–(d)
- Constructed using series connected sub-multilevel cells.
- Voltage rating of switches is equal in both symmetric and asymmetric topologies.
- Provides low harmonic distortion and high output root mean square (RMS) voltage.

23.4 MLIs with Modified H-Bridge

Figure 23.5(a) represents a new topology with modified H-bridge circuit, similar to previous topologies having sub-cells and H-bridges. The basic sub-cell has six switches in the H-bridge connected with two sources for five-level output voltage [25, 26]. This topology has modified versions as shown in Figure 23.5(b) and (c) by increasing the number of switches in the H-bridge [27] or by connecting voltage sources in series with the switches as shown in Figure 23.5(d) [28–30]. Table 23.3 gives the properties and applications of the topologies with modified H-bridge presented in Figure 23.5(a)–(d).

TABLE 23.3

Properties of Multilevel Inverter Structures with Modified H-Bridge

Topology	Configuration	N_{LEVEL}	N_{SOURCE}	N_{SWITCH}	Switching Method
[25]	S	$4m+1$	$2m$	$6m$	FFS
	AS	$6m+1$			
		$2^{m+2}-1$			
[26]	AS	$2^{2m+1}-1$	$2m$	$4n+2$	FFS
[27]	AS	$16n-1$	$4m$	$6n+2$	FFS
[28]	S	$n^2 + (n+1)^2$	$2m$	$2(n-1)+6$	Sine tracking algorithm
	AS	$n^2 + (n+1)^2 + 2n.x$		$2(n-1)+6+x$	
[29]	S	$2m+5$	$2m$	$2m+6$	NLM
[30]	S	$2m+5$	$2m$	$2m+6$	SPWM
	AS	$2m+9$			

Properties of Topologies Presented in Figure 23.5(a)–(d)
- Requirement of different voltage amplitudes is less.
- Reduced blocking voltages.
- Provides high efficiency.
- Lower THD.

23.5 MLIs with Series/Parallel Switching

Edwin Jose and Titus [31] discussed about the common topology with '*n*' isolated DC voltage sources integrated with H-bridge inverter for hybrid systems in Figure 23.6(a). Hinago et al. [32] stated that these topologies have combined voltage sources in series and parallel by switching appropriate devices to achieve the desired output voltage. The output of topologies in Figure 23.6(b) and (c) is not zero due to the numerous switch configurations in a sub-cell. Compared to other topologies, these modules have less voltage level [33–35]. Choi et al. proved that these topologies are also suitable for grid-connected photovoltaic (PV) systems [36]. Table 23.4 gives the properties and applications of topologies with modified H-bridge presented in Figure 23.6(a)–(d).

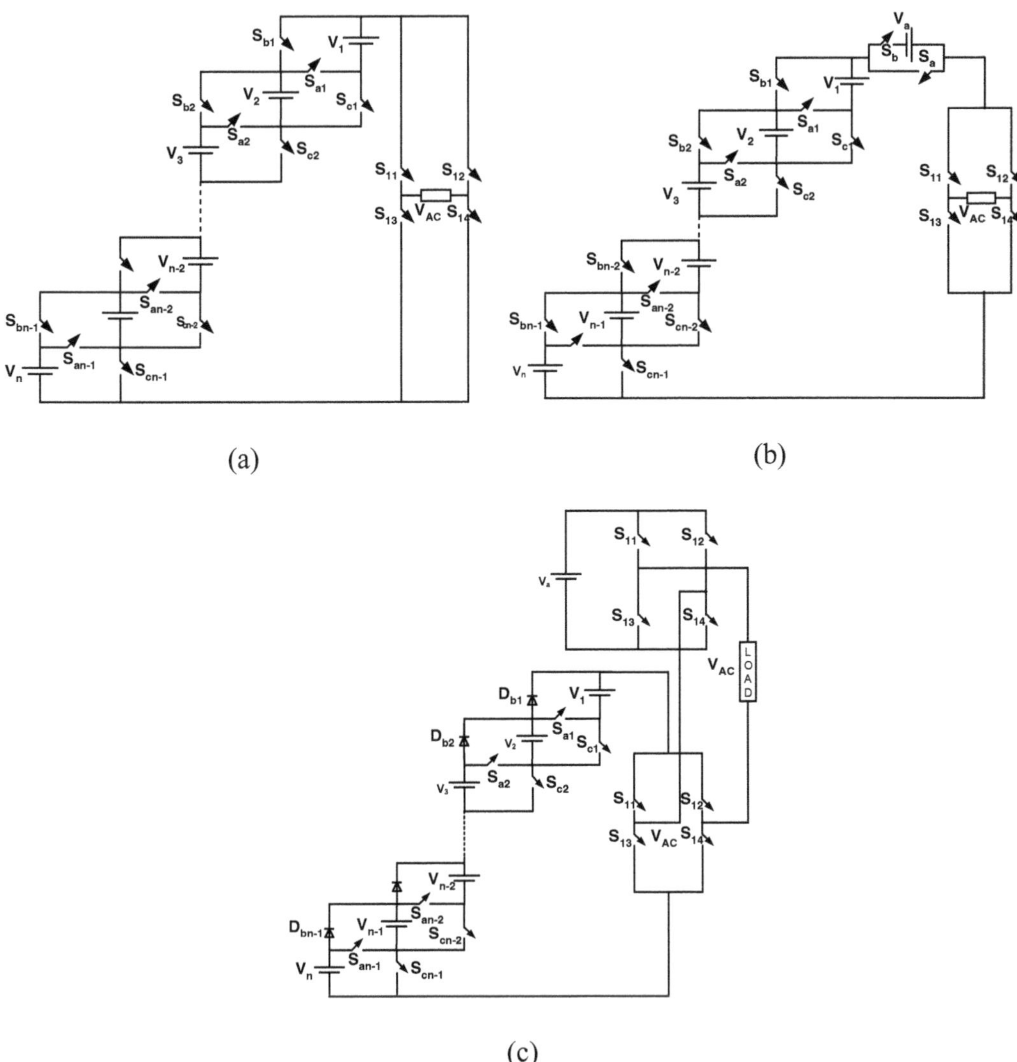

FIGURE 23.6   (a)–(c) Series/parallel switching multilevel inverter topologies.

TABLE 23.4

Characteristics of Multilevel Inverter Structures with Series versus Parallel Switching

Topology	Configuration	N_{LEVEL}	N_{SOURCE}	N_{SWITCH}	Switching Method
[31, 32]	S	$2m$	m	$3m+4$	Hybrid PWM
	AS	$4m+2$			
[33]	S	$2(m+1)$	m	$3m+6$	Hybrid PWM
	AS	$4(m+1)$			
[34]	S	$4m+7$	$2m+2$	$2m+8$, m diodes	Hybrid PWM
[35, 36]	S	$4m+3$	$2m$	$10m$, $2m$ diodes	FFS

Properties of Topologies Presented in Figure 23.6(a)–(d)
- Series/parallel voltage sources.
- Can be operated with or without zero output voltage level.
- Reduces THD value of output waveform.

TABLE 23.5

Properties of Contemporary Multilevel Inverter Structures

Topology	Configuration	N_{LEVEL}	N_{SOURCE}	N_{SWITCH}	Switching Method
[37]	SC	$6m+3$	$3m+1$	$5m+6$	FFS
	ASC	$12m-3$			
[38]	ASC	$3N_{\text{IGBT}}-17$	$\dfrac{1}{6}\left(N_{\text{LEVEL}}+5\right)$	$\dfrac{1}{3}\left(N_{\text{LEVEL}}+17\right)$	SHEPWM
[39–41]	ASC	$[k(k+1)+1]^m$	mk	$2m(k+1)$	FFS
[42, 43]	SC	$\displaystyle\prod_{i=1}^{m} k_m = k^m$	mk	mk	FFS
[44]	SC	$2m+1$	m	$m+5$	NLM
[45]	SC	$2m+3$	$2m$	$m+4$, m diodes	SHEPWM with fish swarm opt
[46]	SC	$2(m+2)+1$	$2m+2$, with separate source	$2m+6$	MCPWM
	ASC	$2^{(2m+4/2)}-1$			
		$2^{(2m+5/2)}-1$	$2m+1$, without separate source		
[47]	SC	$4m+1$	$4m$	$10m$	MCPWM
	ASC	$12m+1$			
	ASC	$16m+1$			
[48]	SC	$6m+5$	$3m+2$	$3m+6$, m diodes	NLM
	ASC	$12m-1$			

Properties of Topologies Presented in Figure 23.7(a)–(i)
- Same structure for symmetric and asymmetric configurations.
- Reduces total standing voltage (TSV).
- Has improved efficiency.

SC, symmetric configuration; ASC, asymmetric configuration; SHEPWM, selective harmonic elimination PWM.

TABLE 23.6

Parameter Comparison of Reduced Switch Multilevel Inverters

Topology	Configuration	N_{LEVEL}	N_{SOURCE}	N_{SWITCH}	Switching Method
[49]	SC	$2m+1$	m	$2m+4$	FFS
	ASC	$2^{m+1}-1$			
[50]	SC	$2m+1$			FFS
	ASC	$2^{m+1}-1$			
	ASC	$4m-1$			
[51]	SC	$2m+1$			SHEPWM with genetic algorithm (GA)
[52]	ASC	$2^{m+1}-1$			FFS
	SC	$2m+1$	m	$2m+2$	
[53]	SC	$2m^2-1$	m	$m+4$, m diodes	Degree switching
[54]	SC	$2m+1$	m	$m+4$, m diodes	SHEPWM
[55]	SC	$2m+1$	m	$m+4$, $(m-1)$ diodes	IPD PWM
[56]	SC	$2m+1$	m	$m+4$, $(m-1)$ diodes	SHEPWM
[57]	SC	$2m+1$	m	$m+4$, $(m-1)$ diodes	SPWM
[58]	SC	$2m+5$	$m+2$	$2m+4$	SPWM
[59]	ASC	$2^{m+1}+1$	m	$5m$	SVPWM
[60]	SC	$2mk+1$	m	$2m(k+1)$	SPWM
[61]	SC	$2mk+1$	m	$2m(k+1)$	Modified MCPWM
[23]	ASC	$16m+1$	$4m$	$12m$	NLM
[24]	ASC	$16m+1$	$4m$	$10m$	FFS
[62]	SC	$2m+1$	1	$3m+4$, 'm' capacitors	SPWM
[63]	SC	$2m+1$	1	$m+5$, $(3m-7)/2$ diodes, $(m-2)$ capacitors	SPWM
[64]	SC	$2m+1$	m	$m+4$, 'm' diodes	SHEPWM using GA

23.6 Contemporary Multilevel Inverter Structures

The new MLI structure shares voltage stress across power switches, reduces component count, and has distinct DC voltage sources. The overall framework of modular topology is shown in Figure 23.7(a) which contains more than one sub-cell module; each contains a source, switches, and level count increasing (LCI) module to produce an offset DC voltage and inter-linking H-bridge (ILHB) module. Babaei et al. [37] used this structure for hybrid power conversion systems, since it has separate DC offset source which is also called as developed cascaded MLI and can be used for dynamic voltage restorer (DVR). An asymmetric topology presented in Figure 23.7(b) used several DC sources of $3V_{\text{DC}}$ to produce required levels of output voltage [38]. Figure 23.7(c) and (d) shows the MLI with and without polarity-changing H-bridge circuit [39–42], the topology uses bi-directional switches which increase the complexity over other inverters but it offers lower blocking voltage and low-voltage operation of H-bridge [43]. Figure 23.7(e) shows an inverter proposed by Panda et al. with H-bridge. It has 2 sources and a single switch in a sub-cell [44]. It is connected with a sub-cell containing a single source and a switch. The main sub-cell can produce five levels of output voltage and the additional cell can produce 3 levels. As proposed by Vinothkumar et al., this topology presented in Figure 23.7(f) uses bypass diodes to bypass the sub-cell when it is not required in the particular level of output voltage [45]. Absolute DC-link voltages will be blocked by H-bridge inverter switches; therefore, devices with high blocking capability will be chosen. This MLI uses half the number of active current switches and carrier waves as a CHB inverter. The more active devices in the conduction channel, the larger the voltage drop and thus lower the output voltage.

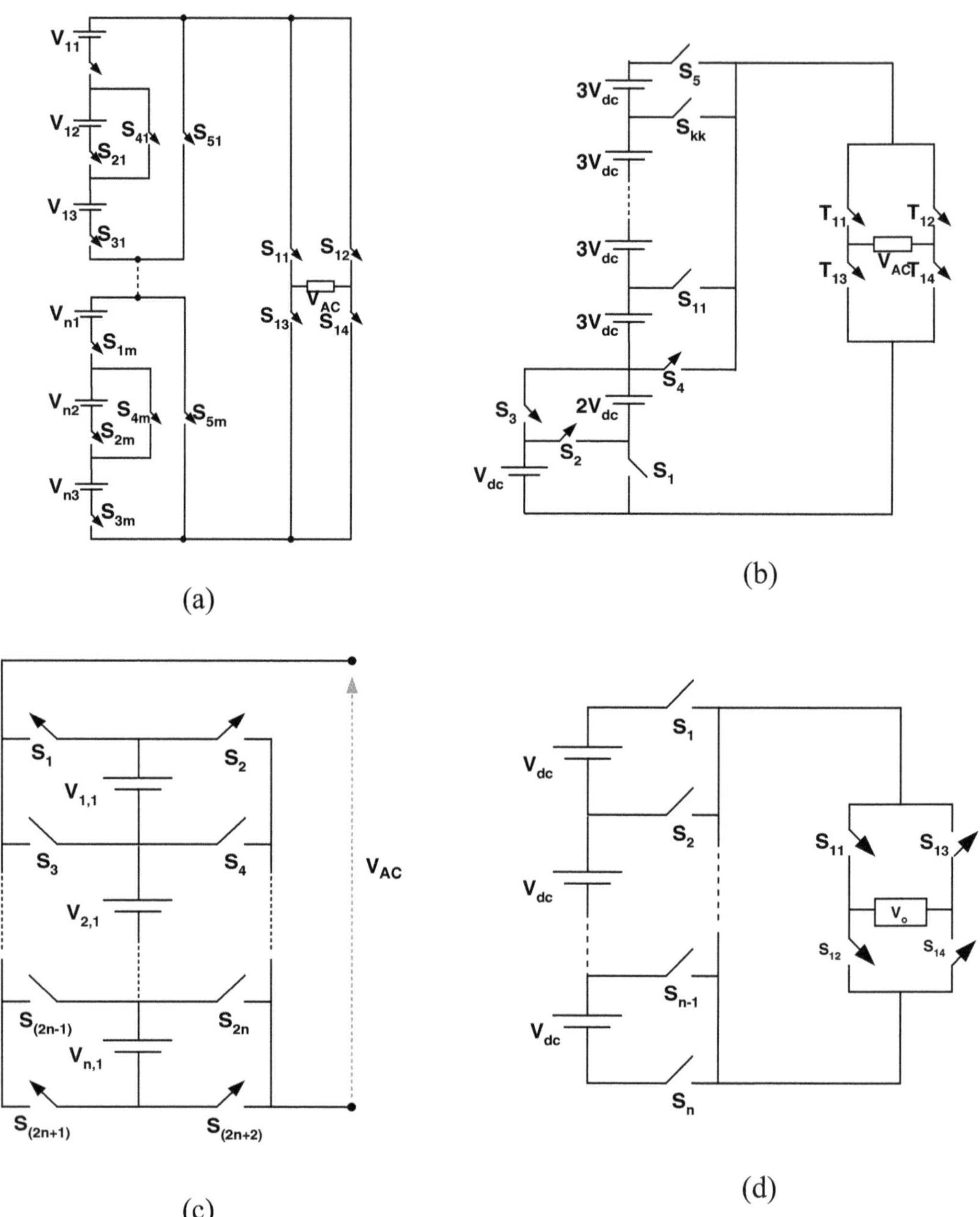

FIGURE 23.7 (a)–(i) Contemporary structures.

Kishore Thakre et al. presented a modified MLI structure with cascaded connection of basic cells [46]. The output level can be enhanced by adding a separate DC source as indicated in Figure 23.7(g). Figure 23.7(h) shows a modified MLI structure with cascade connection of basic units [47], with more output levels and fewer ON state switches than other topologies were discussed in this section, [48] as shown in Figure 23.7(i).

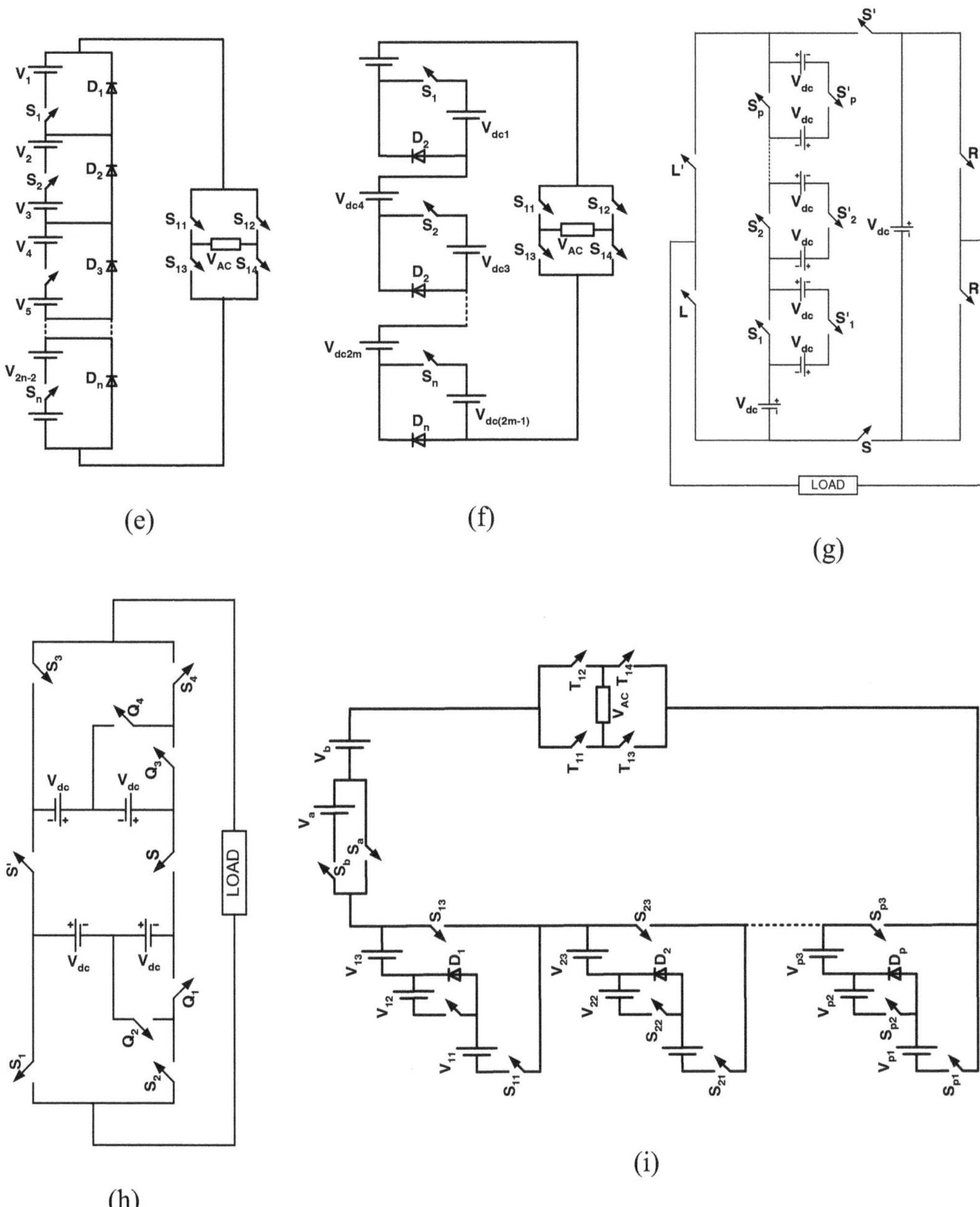

(e) (f)

(g)

(h) (i)

FIGURE 23.7 (Continued)

23.7 Cascaded MLI with Reduced Switch

If the topologies are configured in asymmetric modes, the voltage sources are unequal which require less voltage sources and switches with an output voltage 'nV_{DC}.' Figure 23.8(a)–(e) shows the modern structures with reduced switches and THD. Topologies 'a'–'e' share the same structure, whereas 23.8(b) uses bi-directional switches. Topologies 23.8(c) and (d) use diodes in place of bi-directional switches and observed significant reduction in complexity of the circuit. These circuits use H-bridge to change

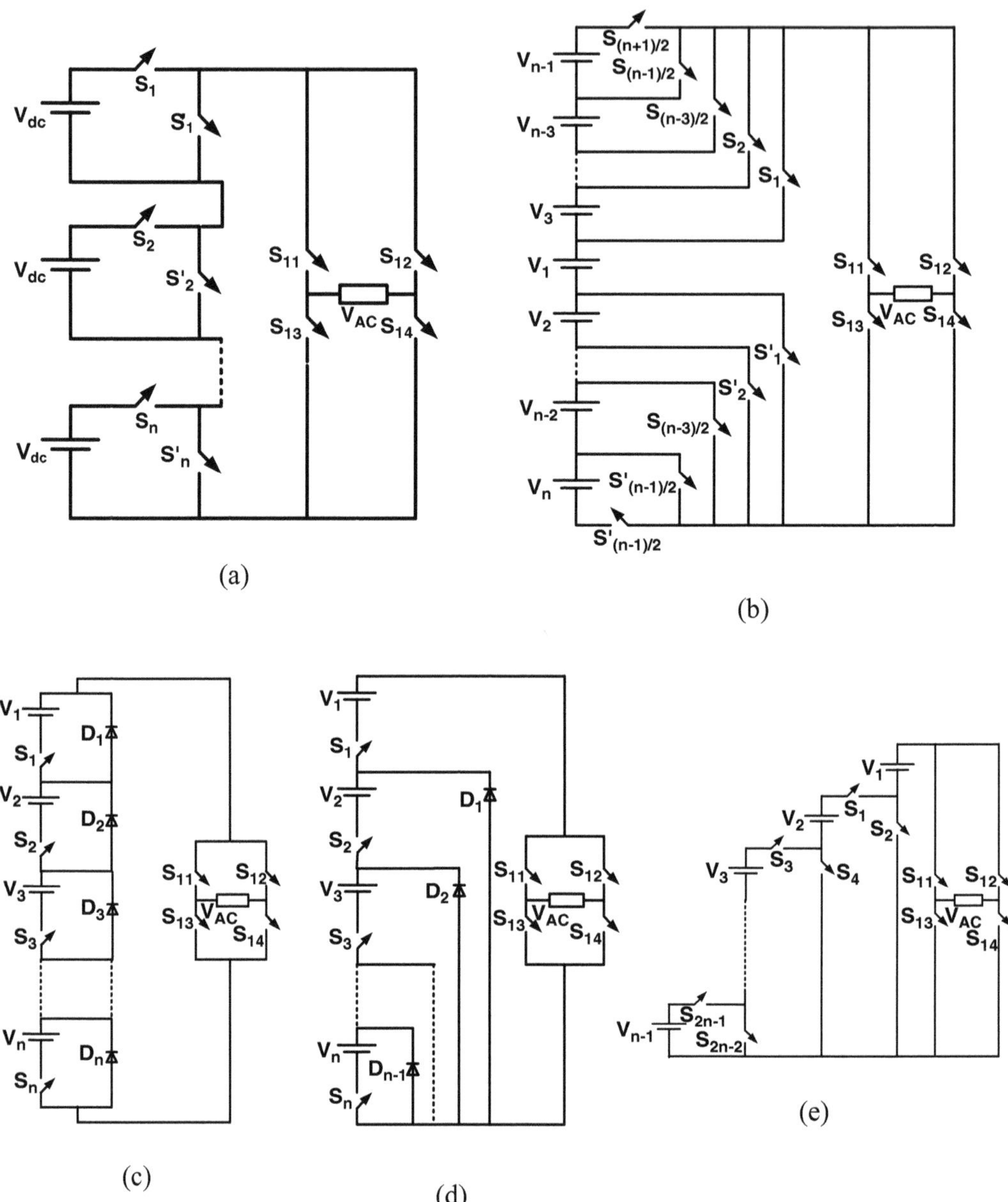

FIGURE 23.8   (a)–(i) Modern structures of reduced switch multilevel inverter with similar structure.

the output voltage polarity. It also gives the overall component comparison for conventional and modern inverters. Topologies in Figure 23.8(f)–(h) have the self-voltage balancing capability as they use capacitors and single voltage source compared to other topologies. These comparisons show the reduction in switch count and significant improvement in number of levels for modern topologies over the conventional topologies. Topology in Figure 23.8(i) has the least 'ON' state switching devices compared to other topologies. Topologies in Figure 23.8(c), (d), and (i) have no path for reactive current as other topologies have that advantage despite higher number of switches.

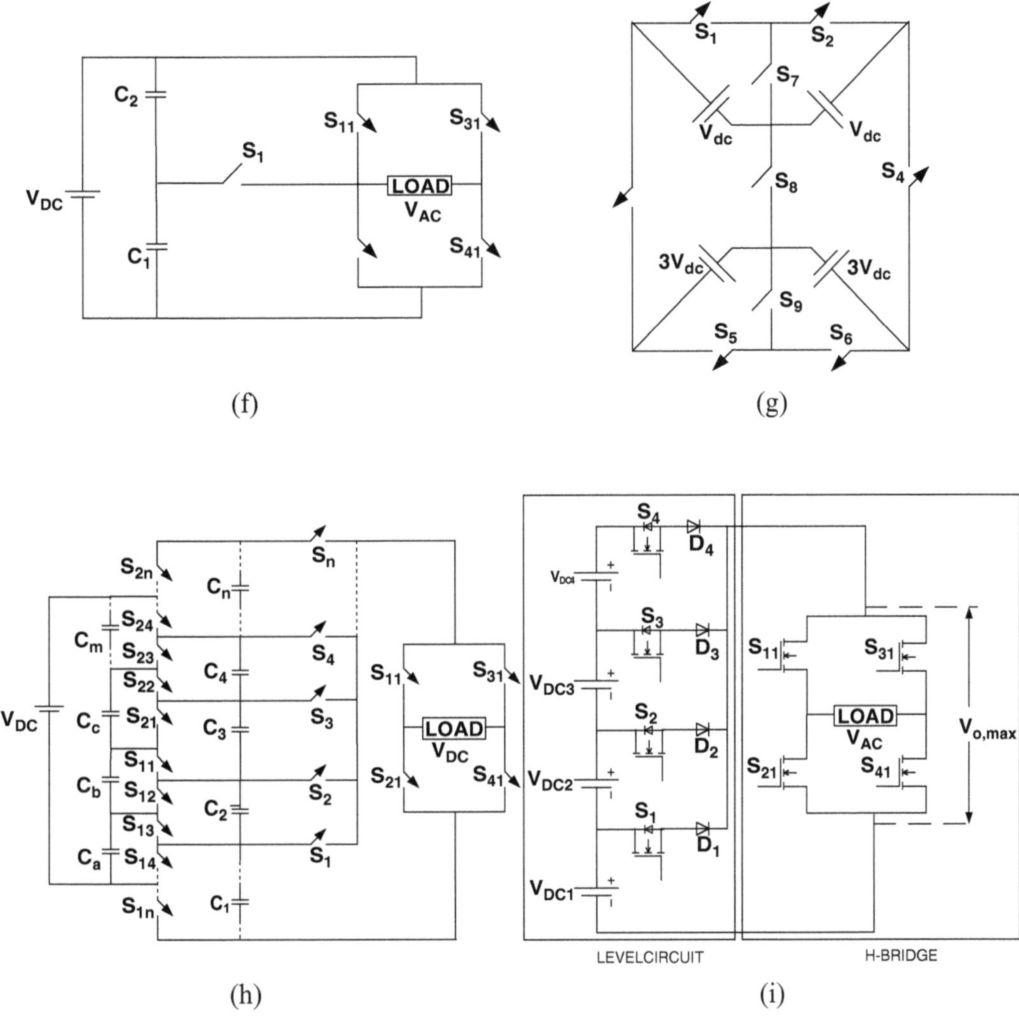

FIGURE 23.8 (Continued)

23.8 Results and Waveforms

The number of levels in a symmetric MLI is proportional to the number of DC voltages or switches. Figure 23.9(a) and (b) demonstrate the output voltage of a three-source symmetric MLI whereas Figure 23.10(a) and (b) describe the output of four voltage sources.

Asymmetric MLI produces nonlinear output voltage with a number of voltage sources and switches used. Figure 23.11 shows an 11-level output voltage for an asymmetric inverter. The number of levels is increased or decreased in asymmetric inverter by selecting the appropriate switching sequence. Figure 23.12 shows a13-level output voltage of an asymmetric inverter. Harmonic analysis of output voltage and current describes the quality of output power of an inverter. Low value of THD means high quality of power and vice versa.

The THD spectra of a symmetric and an asymmetric MLI for the same number of input sources are shown in Figures 23.13 and 23.14, respectively. Since asymmetric inverter has higher number of levels compared to symmetric inverters, the THD value of asymmetric inverter is lower than that of the symmetric inverter for the same operating conditions.

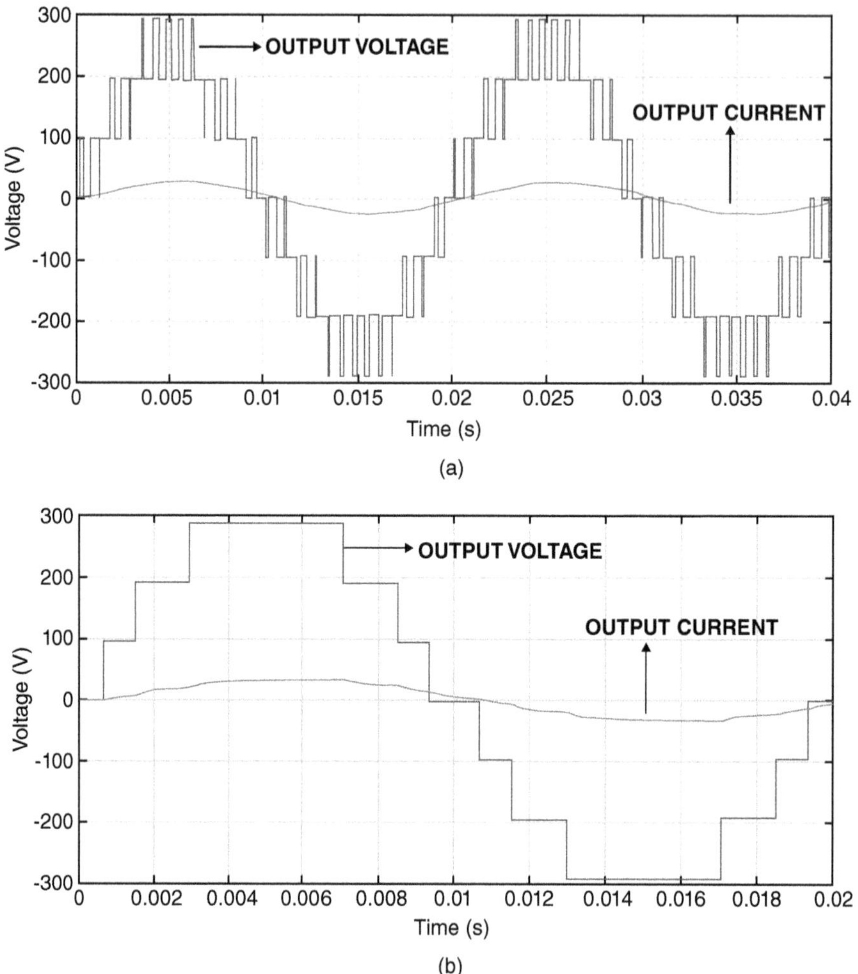

FIGURE 23.9   Seven-level output voltage of a symmetric multilevel inverter at different PWM methods.

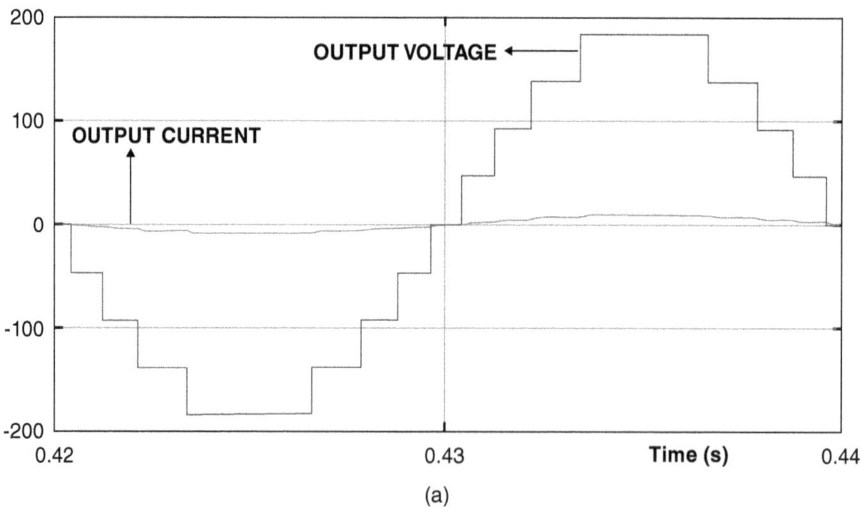

FIGURE 23.10   Nine-level output voltage of a symmetric multilevel inverter at different PWM methods.

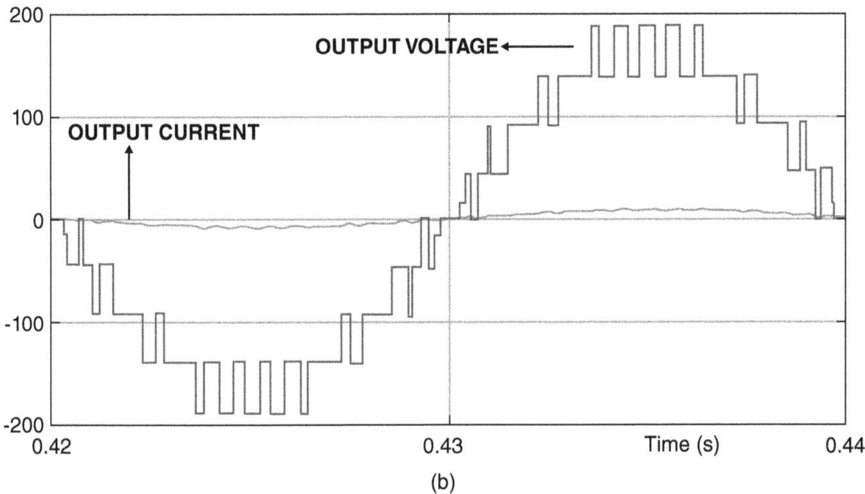

FIGURE 23.10 (Continued)

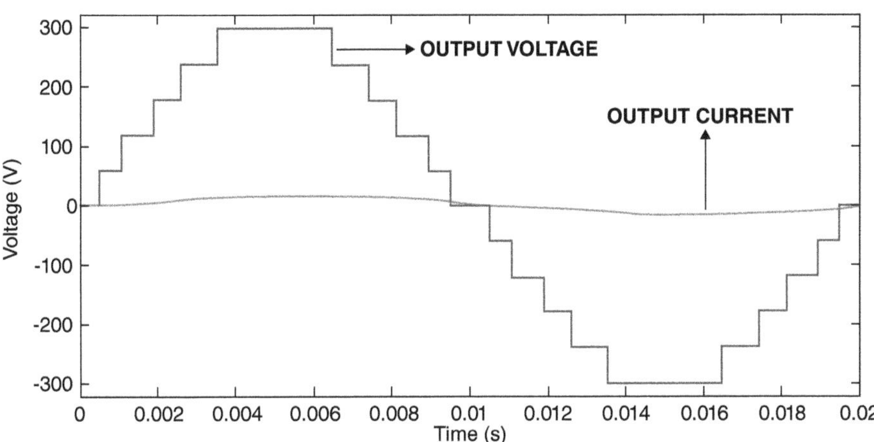

FIGURE 23.11 Eleven-level output voltage of asymmetric multilevel inverter.

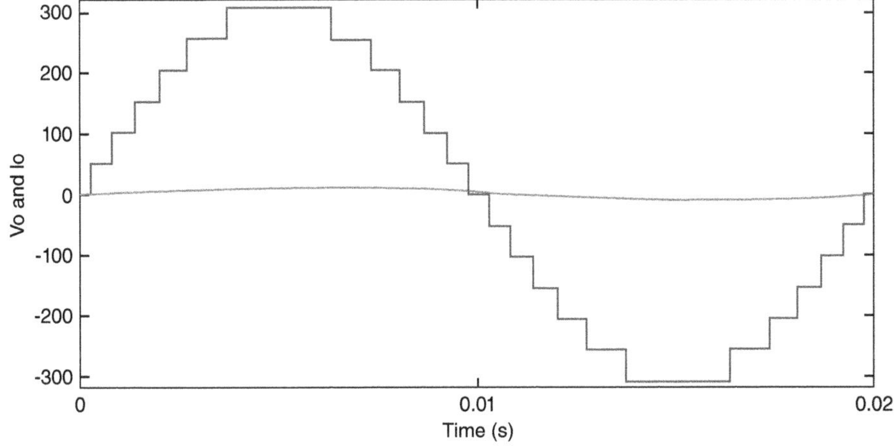

FIGURE 23.12 Thirteen-level output voltage of asymmetric multilevel inverter.

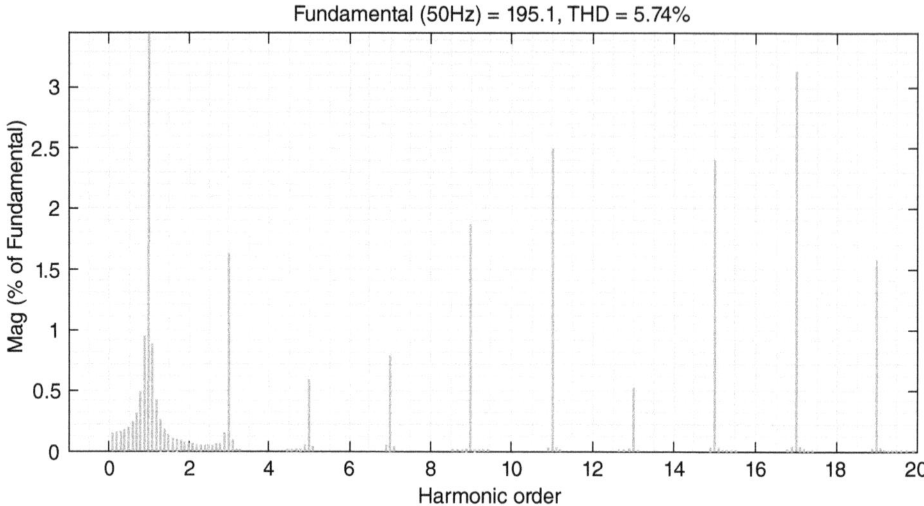

FIGURE 23.13   Harmonic spectrum of a symmetric multilevel inverter.

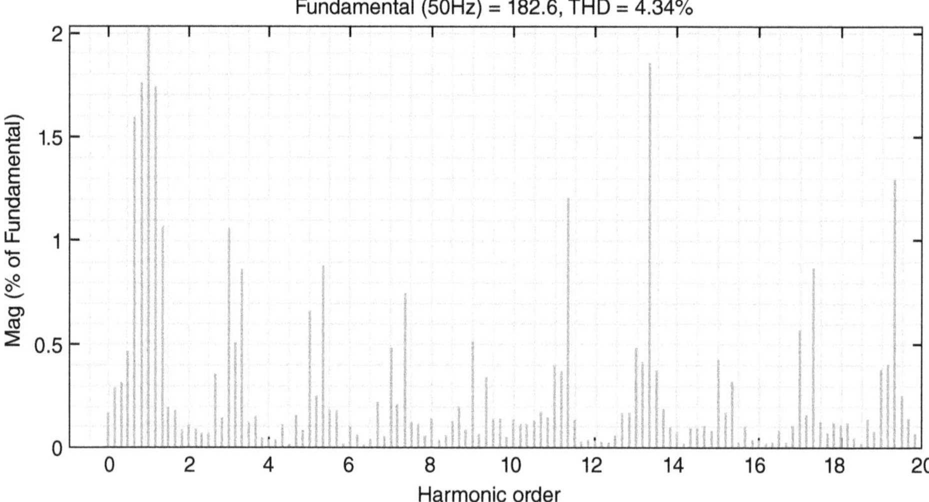

FIGURE 23.14   Harmonic spectrum of an asymmetric multilevel inverter.

23.9 Conclusion

This chapter clearly discussed and presented about the different multilevel converter structures in perspective of switching components and control techniques used for energy storage and e-vehicle applications. Based on the analysis, it is concluded that contemporary inverters have the following advantages:

- Reduction in the number of switches leads to less complexity in driver circuits.
- Minimum switching losses and low *dv/dt*.
- Low electromagnetic interference.
- Near-sinusoidal output voltage.
- Low harmonic distortion.
- Higher efficiency.

- High power quality not only in high switching frequency but also in lower switching frequency.
- Reduction of passive filter requirements.
- Ability to integrate unequal voltage sources with a common load.
- Ability to integrate low-voltage sources with high and medium power applications.

Conventional MLI is facing difficulty due to complicate design in control circuit, and increase in switch count proportionally leads to more driver circuits. In this chapter, cascaded MLIs with multiple sources, widely used in electric vehicles and other modern power applications, have been discussed. Low filter requirement in MLIs makes these circuits useful for battery operation in electric vehicles. The contemporary MLI structures are mainly focused on optimising the switching components whereas modulation techniques were minimally engrossed in literature. Hence, it is suggested that researchers can focus and extend their studies on investigation of various modulation techniques to optimise the various MLI configurations with respect to number of switches, dv/dt value, and number of levels, for e-vehicle.

REFERENCES

[1] Tolbert, L.M., Peng, F.Z. and Habetler, T.G., 2018. Multilevel inverters for electric vehicle applications. *IEEE Power Electronics in Transportation, 1*, pp. 30–35.

[2] Vijayalakshmi, S., Raj, L.H.T., Palaniyappan, S. and Rajkumar, A., 2021. A review on multilevel H-bridge cascaded inductor less hybrid inverter for electric vehicles with PWM control. *Materials Today: Proceedings, 45*, pp. 1644–1650.

[3] Qashqai, P., Sheikholeslami, A., Vahedi, H. and Al-Haddad, K., 2015, October. A review on multilevel converter topologies for electric transportation applications. In *2015 IEEE Vehicle Power and Propulsion Conference (VPPC)* (pp. 1–6). IEEE.

[4] Chakravarthi, B.C.V., Naveen, P., Pragaspathy, S. and Raju, V.N., 2021, March. Performance of Induction Motor with hybrid Multi level inverter for Electric vehicles. In *2021 International Conference on Artificial Intelligence and Smart Systems (ICAIS)* (pp. 1474–1478). IEEE.

[5] Kiran, B.M. and Ram, B.S., 2016 February. Fault detection and mitigation in cascaded MLI fed EV. In *2016 2nd International Conference on Advances in Electrical, Electronics, Information, Communication and Bio-Informatics (AEEICB)* (pp. 491–498). IEEE.

[6] Rodriguez, J., Lai, J.S. and Peng, F.Z., 2002. Multilevel inverters: a survey of topologies, controls, and applications. *IEEE Transactions on Industrial Electronics, 49*(4), pp. 724–738.

[7] Ozpineci, B., Tolbert, L.M. and Du, Z., 2004. Optimum fuel cell utilization with multilevel inverters. *IEEE 35th Annual Power Electronics Specialists Conference PESC'04* (pp. 1572–1578).

[8] Ramkumar, S., Kamaraj, V., Thamizharasan, S. and Jeevananthan, S., 2012. A new series parallel switched multilevel dc-link inverter topology. *International Journal of Electrical Power and Energy Systems, 36*(1), pp. 93–99.

[9] Peng, F.Z., Lai, J.S., McKeever, J.W. and Van Coevering, J., 1996. A multilevel voltage-source inverter with separate DC sources for static VAR generation. *IEEE Transactions on Industry Applications, 32*(5), pp. 1130–1138.

[10] Nabae, Akira, Takahashi, I. and Akagi, H., 1981. A new neutral-point clamped PWM inverter. *IEEE Transactions on Industry Applications, IA-17*(5), pp. 518–523.

[11] Poorfakhraei, A., Narimani, M. and Emadi, A., 2021. A review of multilevel inverter topologies in electric vehicles: Current status and future trends. *IEEE Open Journal of Power Electronics, 2*, pp. 155–170.

[12] Prabaharan, N. and Palanisamy, K., 2017. Analysis of cascaded H-bridge multilevel inverter configuration with double level circuit. *IET Power Electronics, 10*(9), pp. 1023–1033.

[13] Waltrich, Gierri and Barbi, Ivo, 2009. Three-phase cascaded multilevel inverter using power cells with two inverter legs in series. In *2009 IEEE Energy Conversion Congress and Exposition*, (pp. 3085–3092). IEEE.

[14] Babaei, Ebrahim, Dehqan, Ali and Sabahi, Mehran, 2013. Improvement of the performance of the cascaded multilevel inverters using power cells with two series legs. *Journal of Power Electronics, 13*(2), 223–231.

[15] Nagarajan, R. and Saravanan, M., 2014. Performance analysis of a novel reduced switch cascaded multilevel inverter. *Journal of Power Electronics, 14*(1), pp. 48–60.

[16] Ahmed, M., Sheir, A. and Orabi, M., 2018. Asymmetric cascaded half-bridge multilevel inverter without polarity changer. *Alexandria Engineering Journal*, 57(4), pp. 2415–2426.

[17] Kangarlu, M. F. and Ebrahim, B., 2013. A generalized cascaded multilevel inverter using series connection of sub multilevel inverters. *IEEE Transactions on Power Electronics*, 28(2), pp. 625–636.

[18] Babaei, E., et al., 2012. Cascaded multilevel inverter using sub-multilevel cells. *Journal of Electric Power System Research*, 96, pp. 101–110.

[19] Anish, K. and Ramprakash, P., 2014. Performance evaluation of sub-multilevel cell cascaded multilevel inverter using different PWM techniques. *International Journal of Applied Engineering Research*, 9 (8), pp. 961–975.

[20] Pachpor, S., Dutta, S., Sathik, M.J. and Babel, M., 2020. A new compact multilevel inverter design with less power electronics component. In Janusz Kacprzyk (ed.) *Intelligent Communication, Control and Devices* (pp. 661–669). Springer, Singapore.

[21] Karthikeyan, D., Vijayakumar, K. and Sathik, J., 2019. Generalized cascaded symmetric and level doubling multilevel converter topology with reduced THD for photovoltaic applications. *Electronics*, 8(2), p. 161.

[22] Jiang, W., Chincholkar, S.H. and Chan, C.Y., 2017. Investigation of a voltage-mode controller for a dc-dc multilevel boost converter. *IEEE Transactions on Circuits and Systems II: Express Briefs*, 65(7), pp. 908–912.

[23] Samadaei, E., Sheikholeslami, A., Gholamian, S.A. and Adabi, J., 2017. A square T-type (ST-type) module for asymmetrical multilevel inverters. *IEEE Transactions on power Electronics*, 33(2), pp. 987–996.

[24] Siddique, M.D., Mekhilef, S., Shah, N.M., Sarwar, A. and Memon, M.A., 2020. A new single-phase cascaded multilevel inverter topology with reduced number of switches and voltage stress. *International Transactions on Electrical Energy Systems*, 30(2), p. e12191.

[25] Babaei, E., Alilu, S. and Laali, S., 2013. A new general topology for cascaded multilevel inverters with reduced number of components based on developed H-bridge. *IEEE Transactions on Industrial Electronics*, 61(8), pp. 3932–3939.

[26] Babaei, E., Laali, S. and Alilu, S., 2014. Cascaded multilevel inverter with series connection of novel H-bridge basic units. *IEEE Transactions on Industrial Electronics*, 61(12), pp. 6664–6671.

[27] Babaei, Ebrahim and Laali, Sara, 2016. A multilevel inverter with reduced power switches. *Arabian Journal for Science and Engineering*, 41(9), pp. 3605–3617.

[28] Thiruvengadam, Annamalai, 2019. An enhanced H-bridge multilevel inverter with reduced THD, conduction, and switching losses using sinusoidal tracking algorithm. *Energies*, 12(1), p. 81.

[29] Ponraj, R.P. and Sigamani, T., 2021. A novel design and performance improvement of symmetric multilevel inverter with reduced switches using genetic algorithm. *Soft Computing*, 25, pp. 4597–4607.

[30] Hamidi, M.N., Ishak, D., Zainuri, M.A.A.M. and Ooi, C.A., 2020. Multilevel inverter with improved basic unit structure for symmetric and asymmetric source configuration. *IET Power Electronics*, 13 (7), pp. 1445–1455.

[31] Edwin Jose, S. and Titus, S., 2016 July. A level dependent source concoction multilevel inverter topology with a reduced number of power switches. *Journal of Power Electronics*, 16 (4), pp. 1316–1323.

[32] Hinago, Youhei and Koizumi, Hirotaka, 2009. A single-phase multilevel inverter using switched series/ parallel dc voltage sources. *IEEE Transactions on Industrial Electronics*, 57 (8), pp. 2643–2650.

[33] Bassi, Hussain Mohammad and Salam, Zainal, 2019. A new hybrid multilevel inverter topology with reduced switch count and dc voltage sources. *Energies*, 12 (6), p. 977.

[34] Titus, S., 2017. A stipulation based sources insertion multilevel inverter (SBSIMLI) for waning the component count and separate dc sources. *Journal of Electrical Engineering & Technology*, 12 (4), pp. 1519–1528.

[35] Zamiri, E., Vosoughi, N., Hosseini, S.H., Barzegarkhoo, R. and Sabahi, M., 2016. A new cascaded switched-capacitor multilevel inverter based on improved series–parallel conversion with less number of components. *IEEE Transactions on Industrial Electronics*, 63 (6), pp. 3582–3594.

[36] Choi, Won-Kyun and Kang, Feel-soon, 2009. H-bridge based multilevel inverter using PWM switching function. In *INTELEC 2009-31st International Telecommunications Energy Conference* (pp. 1–5). IEEE.

[37] Babaei, E., Laali, S. and Bayat, Z., 2014. A single-phase cascaded multilevel inverter based on a new basic unit with reduced number of power switches. *IEEE Transactions on Industrial Electronics*, 62(2), pp. 922–929.

[38] Siddique, M.D., Mekhilef, S., Shah, N.M., Sarwar, A., Memon, M.A., Seyedmahmoudian, M., Horan, B., Stojcevski, A., Ogura, K., Rawa, M. and Bassi, H., 2018 December. Asymmetrical multilevel inverter topology with reduced number of components. In *2018 IEEE International Conference on Power Electronics, Drives and Energy Systems (PEDES)* (pp. 1–5). IEEE.

[39] Laali, Sara, Babaei, Ebrahim and Bagher, Mohammad, 2014. A new basic unit for cascaded multilevel inverters with the capability of reducing the number of switches. *Journal of Power Electronics, 14* (4), pp. 671–677.

[40] Laali, S., Babaei, E. and Saadatizadeh, Z., 2016. Optimum structures of a cascaded multilevel inverter for minimum number of power electronic devices. *Iranian Journal of Electrical and Electronic Engineering, 12* (3), pp. 206–221.

[41] Babaei, E., Hosseini, S.H., Gharehpetian, G.B., Haque, M.T. and Sabahi, M., 2007. Reduction of dc voltage sources and switches in asymmetrical multilevel converters using a novel topology. *Electric Power Systems Research, 77* (8), pp. 1073–1085.

[42] Ebrahimi, Javad, Babaei, Ebrahim and Gharehpetian, Gevorg B., 2011. A new multilevel converter topology with reduced number of power electronic components. *IEEE Transactions on Industrial Electronics, 59*(2), pp. 655–667.

[43] Vinothkumar, N., Kumar Chinnaiyan, V., Pradish, M. and Prabhakar Karthikeyan, S., 2018. A recommended cascaded structure for multilevel inverter with minimum power electronic components. *Electric Power Components and Systems, 46* (13), pp. 1432–1447.

[44] Almakhles, D.J., Ali, J.S.M., Padmanaban, S., Bhaskar, M.S., Subramaniam, U. and Sakthivel, R., 2020. An original hybrid multilevel dc-ac converter using single-double source unit for medium voltage applications: hardware implementation and investigation. *IEEE Access, 8*, pp. 71291–71301.

[45] Panda, Kaibalya Prasad, Lee, Sze Sing and Panda, Gayadhar, 2019. Reduced switch cascaded multilevel inverter with new selective harmonic elimination control for standalone renewable energy system. *IEEE Transactions on Industry Applications, 55* (6), pp. 7561–7574.

[46] Thakre, K., Mohanty, K.B., Chatterjee, A. and Kommukuri, V.S., 2019. A modified circuit for symmetric and asymmetric multilevel inverter with reduced components count. *International Transactions on Electrical Energy Systems, 29* (6), p. e12011.

[47] Mohamed Ali Jagabar Sathik, Rasoul Shalchi Alishah, N. Sandeep, Hosseini Seyed Hossein, Babaei, Ebrahim, Vijayakumar, Krishnasamy and Yaragatti, Udaykumar R., 2018. A new generalized multilevel converter topology based on cascaded connection of basic units. *IEEE Journal of Emerging and Selected Topics in Power Electronics, 7* (4), pp. 2498–2512.

[48] Karasani, R.R., Borghate, V.B., Meshram, P.M. and Suryawanshi, H.M., 2016. A modified switched-diode topology for cascaded multilevel inverters. *Journal of Power Electronics, 16* (5), pp. 1706–1715.

[49] Babaei, Ebrahim and Hosseini, Seyed Hossein, 2009. New cascaded multilevel inverter topology with minimum number of switches. *Energy Conversion and Management, 50* (11), pp. 2761–2767.

[50] Bektas, Enes and Karaca, Hulusi, 2019. GA based selective harmonic elimination for multilevel inverter with reduced number of switches: An experimental study. *Elektronika ir Elektrotechnika, 25* (3), pp. 10–17.

[51] Mallikarjun, Chithaj, Shanbog, Niteesh S. and Modi, Sangeeta, 2019. Comparative analysis of conventional and modified H-bridge inverter configuration. arXiv preprint arXiv:1908.10032.

[52] Babaei, Ebrahim, Kangarlu, Mohammad Farhadi and Mazgar, Farshid Najaty, 2012. Symmetric and asymmetric multilevel inverter topologies with reduced switching devices. *Electric Power Systems Research, 86*, pp. 122–130.

[53] Tarmizi, Tarmizi, Taib, Soib and Desa M.K., 2019. Design and verification of improved cascaded multilevel inverter topology with asymmetric DC sources. *Journal of Power Electronics, 19* (5), pp. 1074–1086.

[54] Nandhini, G.M., Ram Prakas, P. and Amarnath, M., 2014. Performance evaluation of modified cascaded multilevel inverter. *JApSc, 14*(15), pp. 1750–1756.

[55] Garapati, Durga Prasad, Jegathesan, V. and Veerasamy, Moorthy, 2018. Minimization of power loss in newfangled cascaded H-bridge multilevel inverter using in-phase disposition PWM and wavelet transform based fault diagnosis. *Ain Shams Engineering Journal, 9* (4), pp. 1381–1396.

[56] Pradhan, Ajoya Kumar, Kar, Sanjeeb Kumar, Mohanty, Mahendra Kumar and Behra, Navneet, 2019. Design and simulation of cascaded and hybrid multilevel inverter with reduced number of semiconductor switches. *International Journal of Ambient Energy, 42* (8) pp.950–960.

[57] Rajalakshmi, Sambasivam and Rangarajan, Parthasarathy, 2019. Investigation of modified multilevel inverter topology for PV system. *Microprocessors and Microsystems, 71*, p. 102870.

[58] Ebadpour, Mohsen, Sharifian, Mohammad Bagher Bannae and Hosseini, Seyed Hossein, 2011. A new structure of multilevel inverter with reduced number of switches for electric vehicle applications. *Energy and Power Engineering, 3* (02), p. 198.

[59] Tuluhong, A., Wang, W., Li, Y. and Xu, L., 2018. A novel hybrid T-type three-level inverter based on SVPWM for PV application. *Journal of Electrical and Computer Engineering*, 1–12.

[60] Chokkalingam, Bharatiraja, Bhaskar, Mahajan Sagar, Padmanaban, Sanjeevikumar, Ramachandaramurthy, Vigna K. and Iqbal, Atif, 2019. Investigations of multi-carrier pulse width modulation schemes for diode free neutral point clamped multilevel inverters. *Journal of Power Electronics, 19*(3), pp. 702–713.

[61] Amir, A., Amir, A., Selvaraj, J. and Abd Rahim, N., 2020. Grid-connected photovoltaic system employing a single-phase T-type cascaded H-bridge inverter. *Solar Energy, 199*, pp. 645–656.

[62] Matam, Manjunatha Budagavi, Venkata, Ashok Kumar Devarasetty and Mallapu, Vijaya Kumar, 2018. Analysis and implementation of impedance source based switched capacitor multi-level inverter. *Engineering Science and Technology, An International Journal, 21*(5), pp. 869–885.

[63] Ponraj, R.P., Sigamani, T. and Subramanian, V., 2021. A developed H-bridge cascaded multilevel inverter with reduced switch count. *Journal of Electrical Engineering & Technology, 16*, pp. 1445–1455.

[64] Ponraj, R.P., Sigamani, T. and Ravindran, V., 2023. Analysis of multi-input multilevel boost inverter circuit with optimal firing angles using dSPACE. *Journal of Electrical Engineering & Technology, 18*(2), pp. 1135–1145.

24

Detection of Energy Storage Failure in Mechanical Industries Using Machine Learning

M. Swapna

Stanley College of Engineering Technology for Women, Hyderabad, India

24.1 Introduction

Being able to analyse foresight process failure, identify product expansion, and then test it is an important part of testing process development. Defect detection can provide insight into projects post-test, while developers can reduce costs by leveraging defect prediction before testing. Identifying programming errors early in the development process can be difficult. However, analysing the relationship between the characteristics of test results and the failure rate aids in diagnosis. The problems caused by defects cannot be resolved by conventional methods like testing or simulation due to high cost and time overhead. Managing software development is a challenging task, with most projects exceeding their allocated budgets. The cost of discovering and resolving the issue increases significantly from the requirements gathering stage to the project's final deployment. Improved software development management can increase revenue and customer satisfaction, making it critical to integrate software quality monitoring principles as early as possible in the project life cycle.

This study aims to replace conventional maintenance methods with contemporary machine learning (ML) approaches to detect and prevent machine losses and faults. Software defect prediction (SDP) can reduce the time and effort required for the code review process. Additionally, the software system would benefit from enhanced stability and reliability.

24.2 Domain

Computer algorithms that learn from data are the focus of the artificial intelligence area known as 'machine learning.' The main components of ML, which aims to replicate human learning on computers, are observing an event and drawing generalisations from the observations. The two primary divisions of ML are supervised learning and unsupervised learning. Given that it divides instances into two or more categories, supervised learning is also known as categorisation. Prior SDP studies using a variety of ML classifiers, also known as classifier algorithms, have been presented for supervised learning. These families include classification rules, probabilistic classifiers, decision trees, and neural networks.

The ML algorithms belonging to the 'deep learning' (DL) category employ multiple layers to systematically eliminate unnecessary elements from the input data. For example, in image processing, lower layers can detect borders, while upper layers can recognise important human objects such as numbers, symbols, or faces. Using input data, weights, and biases, DL neural networks, also known as neural networks, attempt to replicate the functions of the human brain. Together, these elements properly identify, categorise, and characterise entities in the data.

24.3 Literature Survey (Table 24.1)

TABLE 24.1

Survey from Last 5 Years

S.No	Year	Author	Method
1	2022	R. Mahajan, S. K. Gupta, and R. K. Bedi [13]	Bayesian Regularization algorithm
2	2022	Umesh and G. N. Srinivasan [8]	Rejuvenation model
3	2021	Y. Abdi and S. Parsa [23]	Particle swarm optimisation
4	2019	M. R. Mesbahi and A. M. Rahmani [10]	Label propagation
5	2019	B. Liu and L. Lucia [5]	Support Vector Machine, K-Nearest Neighbour, decision tree
6	2019	Xiang Gong and Wei Qiao [22]	Wideband synchronous sampling algorithm
7	2018	Kim-Anh Nguyen, Phuc Do, and Antoine Gral [19]	Multi-component system
8	2020	Srikanth Namuduri, Barath Narayanan, Venkata Salini Priyamvada Davuluru, Lamar Burton, and Shekhar Bhansali [18]	Logistic Regression, K-Nearest Neighbour
9	2015	Ashu Jain M. and Avadhnam Madhav Kumar [4]	ANN
10	2017	Rui Zhao, Dongzhe Wang, et al. [15]	Feature-based gated recurrent unit (FBGRU)

24.4 Methodology

LSTM stands for 'long short-term memory,'" and **it** is a type of **recurrent neural network** (RNN). In RNN, the output **of** the previous step is **used** as **the** input for the next **step**. Hochreiter and Schmidhuber are credited with the development of LSTMs. It solves the problem of relying on the long-term RNN, where the RNN cannot predict the message stored in long-term memory but succeeds when utilising huge current data. RNN has poor performance as the gap length increases. LSTM is widely used to store data for a very long time. Its processing, prediction, and classification procedures are constructed using temporal data, Sentiment analysis, character creation, and handwriting recognition are just a few examples of sequence modelling where LSTM has been successful. To elucidate the dynamic characteristics of computer systems, considering substantial temporal dependencies in predicting failures, an LSTM-based forecasting network is employed in this study. To solve this issue, memory cells – long-lasting storage components – are utilised in LSTMs. The input gate, memory gate, and an output gate together control memory. These gates determine the data that must be entered into is removed from the memory cell and output is taken [14]. The input table chooses which data to include in the memory cell. The memory port determines what data is removed from the memory cell. The output port also controls the output of the memory cell. LSTM networks can detect long-term dependencies by choosing which elements to store and which to reject as it moves through the network. Nested LSTMs called deep LSTM networks may extract even more complex patterns from sequential data in Figure 24.1.

A similar architecture for reference is provided in Figure 24.1.

24.4.1 Logistic Regression (LR)

LR is one of the best-known ML algorithms and a subset of supervised learning. A set of distinct factors is employed to forecast the category-specific variable. LR anticipates the output of a dependent categorical variable. As a result, the outcome should be a distinct or categorical value. It furnishes probability values within the range of 0–1, instead of delivering a definitive response of 0 or 1. It could be either accurate or incorrect, 0 or 1, or affirmative or negative. There are some traits that LR and linear regression have in common.

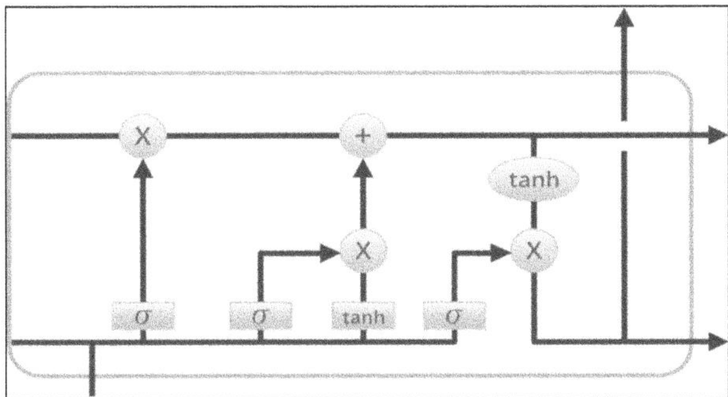

FIGURE 24.1 Architecture of LSTM.

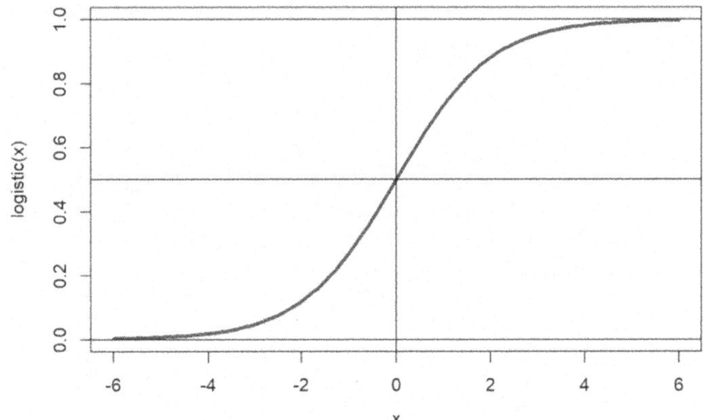

FIGURE 24.2 Logistic regression curve.

In contrast to LR, which is employed for resolving classification challenges [16, 17], linear regression is applied to address regression issues, depending on the weight of the data, whether the cells are malignant, etc. LR holds significance in ML as it can assign probabilities and classify new data across continuous and discrete datasets.

The dependent variable must be categorical in nature, whereas the independent variable should not be multi-collinear. It is easy to identify the most useful factors for the classification and may be used to categorise the observations utilising several data sources (Figure 24.2).

24.5 Dataset

Machine failure dataset comprises of machine values with different speed, heat, and temperature values as features. The pre-processed machine failure dataset is shown in this dataset. In this step, characteristics of the dataset are entered [20]. The computation involves determining the variance in time series data, resulting in the creation of a new dataset [21] with values different from the previous values. Utilising this dataset, predictions for future values are made by considering the variance as a feature and the subsequent value as the corresponding label (Figures 24.3–24.5).

Figure 24.5 shows the machine failure dataset, which is between 0 and 1. In this project, datasets include a feature with memory and a label with 0 and 1.

	col1	col2	col3	col4	col5
0	1511933834320	67	30	52	89
1	1511933854257	65	53	71	99
2	1511933874226	56	55	83	128
3	1511933894135	42	63	99	154
4	1511933914098	60	53	100	160
5	1511933934093	51	77	89	165
6	1511933954011	54	68	116	161
7	1511933973979	46	66	129	177
8	1511933993959	68	55	148	187
9	1511934013900	51	60	147	168

FIGURE 24.3 Machine features given in an excel sheet.

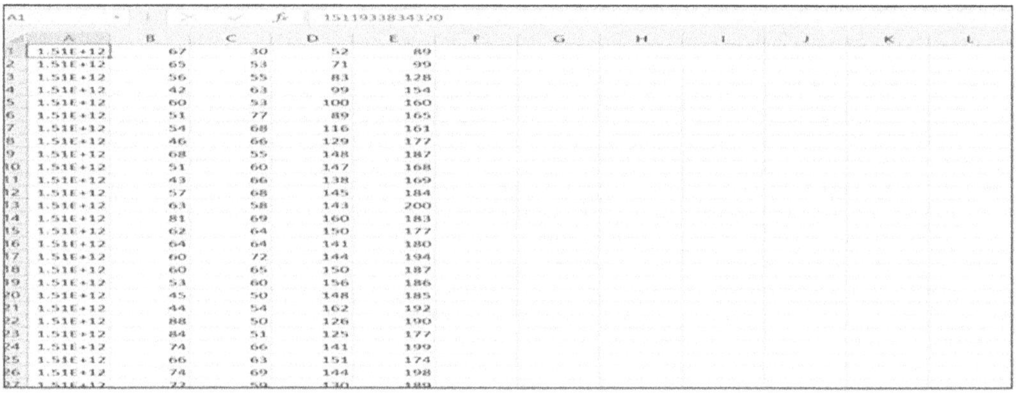

A1 f_x 1511933834320

	A	B	C	D	E	F	G	H	I	J	K	L
1	1.51E+12	67	30	52	89							
2	1.51E+12	65	53	71	99							
3	1.51E+12	56	55	83	128							
4	1.51E+12	42	63	99	154							
5	1.51E+12	60	53	100	160							
6	1.51E+12	51	77	89	165							
7	1.51E+12	54	68	116	161							
8	1.51E+12	46	66	129	177							
9	1.51E+12	68	55	148	187							
10	1.51E+12	51	60	147	168							
11	1.51E+12	43	66	138	169							
12	1.51E+12	57	68	145	184							
13	1.51E+12	63	58	143	200							
14	1.51E+12	81	69	160	183							
15	1.51E+12	62	64	150	177							
16	1.51E+12	64	64	141	180							
17	1.51E+12	60	72	144	194							
18	1.51E+12	60	65	150	187							
19	1.51E+12	53	60	156	186							
20	1.51E+12	45	50	148	185							
21	1.51E+12	44	54	162	192							
22	1.51E+12	88	50	126	190							
23	1.51E+12	84	51	125	177							
24	1.51E+12	74	66	141	199							
25	1.51E+12	66	63	151	174							
26	1.51E+12	74	69	144	198							
27	1.51E+12	72	58	130	189							

FIGURE 24.4 Pre-processed machine dataset.

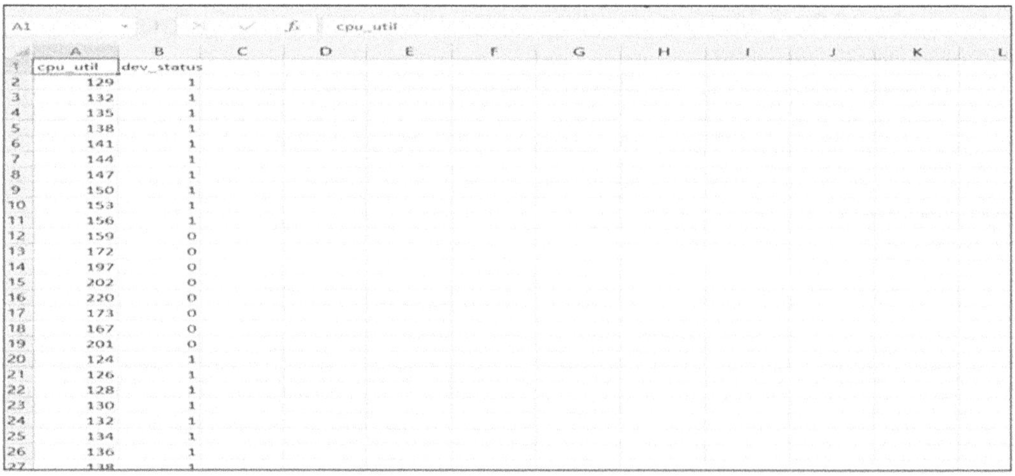

A1 f_x cpu_util

	A	B	C	D	E	F	G	H	I	J	K	L
1	cpu_util	dev_status										
2	129	1										
3	132	1										
4	135	1										
5	138	1										
6	141	1										
7	144	1										
8	147	1										
9	150	1										
10	153	1										
11	156	1										
12	159	0										
13	172	0										
14	197	0										
15	202	0										
16	220	0										
17	173	0										
18	167	0										
19	201	0										
20	124	1										
21	126	1										
22	128	1										
23	130	1										
24	132	1										
25	134	1										
26	136	1										
27	138	1										

FIGURE 24.5 Machine failure dataset.

24.6　Result

After generating labels from the machine failure dataset using the LSTM algorithm, the algorithm trains the dataset and compares accuracy with wanted and predicted values, which is depicted in the graph in Figure 24.6.

This output displays machine values that were supplied as input to determine if the values are typical or indicative of a machine failure. The input dataset is employed to feed information into the LSTM model. The LSTM's output, in turn, serves as input for the trained LR model, generating an alert message for predicted failure values, as depicted in Figure 24.7.

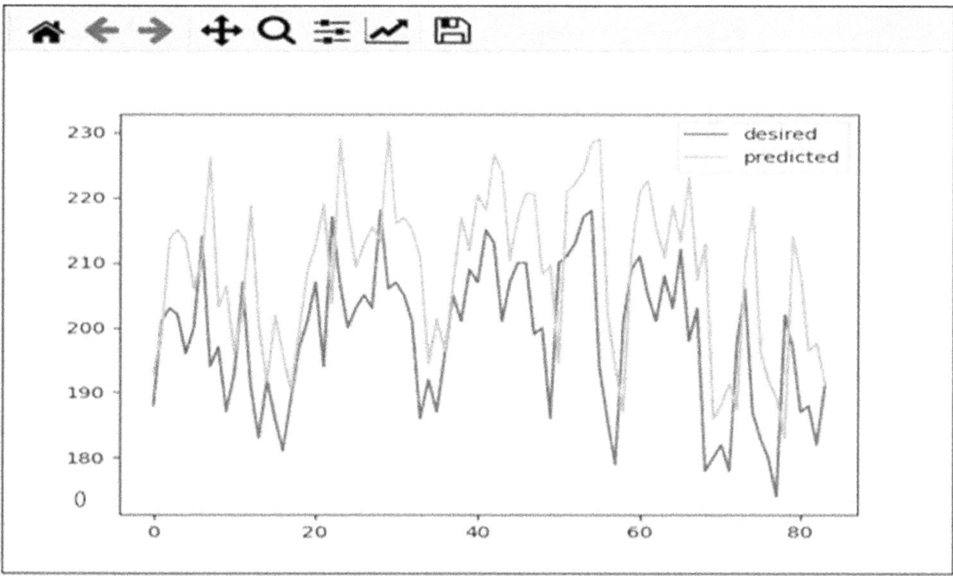

FIGURE 24.6　Comparison graph for desired and prediction.

```
present Machine Value is: [142]
expected next Machine value could be  [153.04355362]
Machine status is Normal
present Machine Value is: [144]
expected next Machine value could be  [155.04198149]
Machine status is Normal
present Machine Value is: [146]
expected next Machine value could be  [157.04112408]
Machine status is Normal
present Machine Value is: [148]
expected next Machine value could be  [159.04065657]
Aleart Message For Machine Failure
present Machine Value is: [150]
expected next Machine value could be  [161.04040122]
Aleart Message For Machine Failure
present Machine Value is: [152]
expected next Machine value could be  [163.04026228]
Aleart Message For Machine Failure
present Machine Value is: [154]
```

FIGURE 24.7　Predicted result for failure or normal conditions.

24.7 Conclusion and Future Work

The objective of the project is to investigate the capabilities of machine learning models in identifying software faults and suggest methods to improve their application by tackling current challenges. Various machine learning classifiers were evaluated to identify the ones that are most effective in detecting defects. The study suggests a technique for short-term prediction utilising the LSTM network that can improve the accuracy of fault detection when compared to the existing SVM and naive Bayes models. The proposed method predicts machine failure and displays a warning signal by comparing the predicted value to values from previous datasets. The study contributes to the body of research on machine learning techniques for software quality and addresses the challenges associated with detecting faults or vulnerabilities in software. The project utilised both deep learning and machine learning models and methods for fault identifications. These models were selected based on their popularity and existing research in other software quality domains. Future research may explore more advanced machine learning models as the field of machine and deep learning continues to evolve rapidly. The methods and techniques explored in this study can potentially benefit various aspects of software engineering, particularly software quality, which is a crucial aspect of software development. Software faults should be detected and fixed as soon as possible, and they share some characteristics with rare and hard-to-detect software vulnerabilities.

REFERENCES

[1] A. Makanju, A. N. Zincir-Heywood, and E. E. Milios, "Clustering event logs using iterative partitioning," in *Proceedings of the 15th ACM SIGKDD International Conference on Knowledge Discovery and Data Mining*. ACM, 2009, pp. 1255–1264.

[2] Graves, Alex and N. Jaitly, "Towards end-to-end speech recognition with recurrent neural networks," in *Proceedings of the 31th International Conference on Machine Learning*, ICML 2014, Beijing, China, 21–26 June 2014, 2014, pp. 1764–1772.

[3] Graves, Alex, "Generating sequences with recurrent neural networks," arXiv preprint arXiv:1308.0850, 2013.

[4] Ashu Jain, and Avadhnam Madhav Kumar, "Hybrid neural network models for hydrologic time series forecasting," *Applied Soft Computing*, vol. *7*, pp. 585–592, 2007.

[5] Liu, Bing, Shiva Nejati, and Lionel C. Briand. "Improving fault localization for Simulink models using search-based testing and prediction models." *2017 IEEE 24th International Conference on Software Analysis, Evolution and Reengineering (SANER)*. IEEE, 2017.

[6] Jin, Cong and Shu-Wei Jin, "Prediction approach of software fault proneness based on hybrid artificial neural network and quantum particle swarm optimization," *Applied Soft Computing*, vol. *35*, pp. 717–725, 2015.

[7] Alsmadi, Izzat and Hassan Najadat, "Evaluating the change of software fault behavior with dataset attributes based on categorical correlation," *Advances in Engineering Software*, vol. *42*, pp. 535–546, 2011.

[8] Umesh, I. M. and G. N. Srinivasan, "Dynamic software aging detection-based fault tolerant software rejuvenation model for virtualized environment," in *Proceedings of the International Conference on Data Engineering and Communication Technology*, 2017, pp. 779–787.

[9] Nguyen, Kim-Anh, Phuc Do, and Antoine Grall, "Multi-level predictive maintenance for multi-component systems", *Reliability Engineering & System Safety*, vol. *144*, pp. 83–94, 2015.

[10] M. R. Mesbahi, A. M. Rahmani, and M. Hosseinzadeh, "Reliability and high availability in cloud computing environments: a reference roadmap," *Human-Centric Computing and Information Sciences*, vol. *8*, p. 20, 2018.

[11] N. Srivastava, E. Mansimov, and R. Salakhutdinov, "Unsupervised learning of video representations using lstms," arXiv preprint arXiv:1502.04681, 2015.

[12] P. J. García Nieto, E. García-Gonzalo, F. Sánchez Lasheras, and F. J. de Cos Juez, "Hybrid PSO–SVM-based method for forecasting of the remaining useful life for aircraft engines and evaluation of its reliability," *Reliability Engineering & System Safety*, vol. *138*, pp. 219–231, 2015.

[13] R. Mahajan, S. K. Gupta, and R. K. Bedi, "Design of software fault prediction model using BR technique," *Procedia Computer Science*, vol. *46*, pp. 849–858, 2015.

[14] R. Vaarandi et al., "A data clustering algorithm for mining patterns from event logs," in *Proceedings of the 2003 IEEE Workshop on IP Operations and Management (IPOM)*, 2003, pp. 119–126.

[15] Zhao, Rui, et al., "Machine health monitoring using local feature-based gated recurrent unit networks", *IEEE Transactions on Industrial Electronics*, vol. *65*(2), pp. 1539–1548, 2017.

[16] S. Chatterjee, and A. Roy, "Web software fault prediction under fuzzy environment using MODULO-M multivariate overlapping fuzzy clustering algorithm and newly proposed revised prediction algorithm," *Applied Soft Computing*, vol. *22*, pp. 372–396, 2014.

[17] A. Souri, M. Hoseyninezhad Hussien, and M. Norouzi, "A systematic review of IoT communication strategies for an efficient smart environment," *Transactions on Emerging Telecommunications Technologies*, vol. *33*(3), p. e3736, 2022.

[18] Srikanth Namuduri, Barath Narayanan, Venkata Salini Priyamvada Davuluru, Lamar Burton and Shekhar Bhansali, "Deep learning methods for sensor based predictive maintenance and future perspectives for electrochemical sensors", https://iopscience.iop.org/journal/1945-7111

[19] T. Kimura, A. Watanabe, T. Toyono, and K. Ishibashi, "Proactive failure detection learning generation patterns of large scale network logs," in *Network and Service Management (CNSM), 2015 11th International Conference* on, Nov 2015, pp. 8–14.

[20] T. Li, F. Liang, S. Ma, and W. Peng, "An integrated framework on mining logs files for computing system management," in *Proceedings of the Eleventh ACM SIGKDD International Conference on Knowledge Discovery in Data Mining*. ACM, 2005, pp. 776–781.

[21] V. Balasubramanian, F. Zaman, M. Aloqaily, I. Al Ridhawi, Y. Jararweh, and H. B. Salameh, "A mobility management architecture for seamless delivery of 5G-IoT services," in *ICC 2019–2019 IEEE International Conference on Communications (ICC)*, 2019.

[22] Xiang Gong, IEEE Member, Wei Qiao, and IEEE Senior Member, "Current-based mechanical fault detection for direct-drive wind turbines via synchronous sampling and impulse detection," *IEEE Transactions on Industrial Electronics*, vol. *62*, no. 3, March 2015.

[23] Y. Abdi, S. Parsa, and Y. Seyfari, "A hybrid one-class rule learning approach based on swarm intelligence for software fault prediction," *Innovations in Systems and Software Engineering*, vol. *11*, pp. 289–301, 2015.

25

Research Tendency Route to Extremely Efficient and Stable Dye-Sensitised Solar Cell Applications

S. Afnan and R. Gayathri
Cauvery College for Women (Autonomous) (affiliated to Bharathidasan University), Trichy, India

25.1 Introduction

The increasing demand for fuel has hastened the gradual elimination of fossil fuels. It is estimated that the entire globe's fossil fuel reserves will last just 40 decades for fluid, a total of 60 years for gas from natural sources, and a century for lignite [1]. Global worries regarding these sorts of problems are presently motivating the technological ambition of renewable and sustainable energy sources [2]. Photovoltaic technology, among the various renewable energy technologies, has great potential for the immediate transformation of sunlight into powerful electrical production. Fortunately, because of excessive manufacturing and environmental expenses, current silicon-based photovoltaic cells are limited to the commercial photovoltaic manufacturing [3]. Still, for many reasons, organic solar cells have been developed in recent decades that combine the compact size and mobility of molecules of organic matter with an affordable component expense [4]. The dye-sensitised solar cell (DSSC) is one of the most desirable alternatives to solar energy cells made of silicon when it comes to performance and cost-effectiveness in manufacture [5]. The benefits of DSSCs opened up opportunities for a great deal of attraction, as seen by an enormous spike in the number of journals over the recent years [6]. Even though O'Regan and Grätzel started doing substantial research on DSSCs in 1991 [7], shown in Figure 25.1 solar current trends in development. The research has evolved at a quick speed, and significant effort has been made to increase the device effectiveness from more than to the level judged essential for commercial application [8].

25.2 Postulate of DSSC

Solar radiation can be directly converted into electric current using a DSSC, a type of photovoltaic semiconductor device [9]. In order to convert photons into electricity, the DSSC hypothesis includes four essential steps [10].

1. A photon first absorbs the light that comes into it (electron), which cause particles in the dye to transfer from its state of neutrality (S+/S) to the excited state of action (S+/S*). In the overwhelming majority of the dye, absorption occurs in the 700 nm range, which corresponds to the photons' electrical power of approximately 1.72 eV [11–13].

2. Nowadays, a nanoporous titanium dioxide (TiO_2) electrode's charge that sits underneath the highly excited state of the dye and absorbs very little ultraviolet (UV) radiation from the UV wavelength has been exposed to nanosecond-long excited electron infusions into its conduction bandwidth. The pigment oxidises as a result [14, 15].

DOI: 10.1201/9781003495437-25

3. These injected electrons flow towards the back interaction (transparent conducting oxide [TCO]) through TiO_2 nanoparticles [16, 17]. In Figure 25.2, the electrons reach the counter electrode through the external circuit.

4. The electrons at the opposing electrode decrease I_3^- to I^-; hence, dye regeneration of the dye's ground state occurs as a result of electron acceptance from the I^- ion electron a neutral, and I^- undergoes oxidation to I_3^- (oxidised state) [18, 19].

5. The oxidised intermediate (I_3^-) spreads towards the neutral electrode and decreases to I ion a second time.

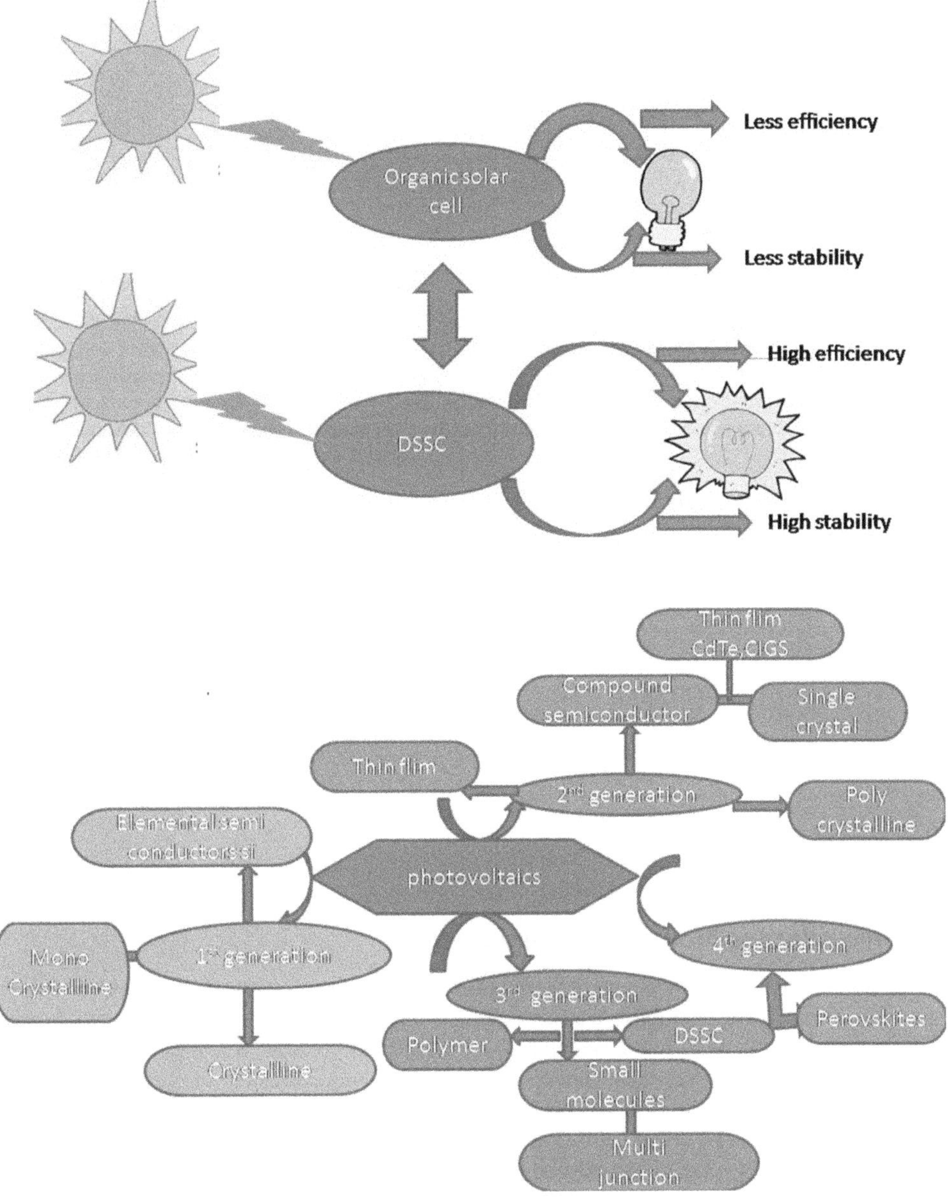

FIGURE 25.1 Category of a solar cell and current trends of development.

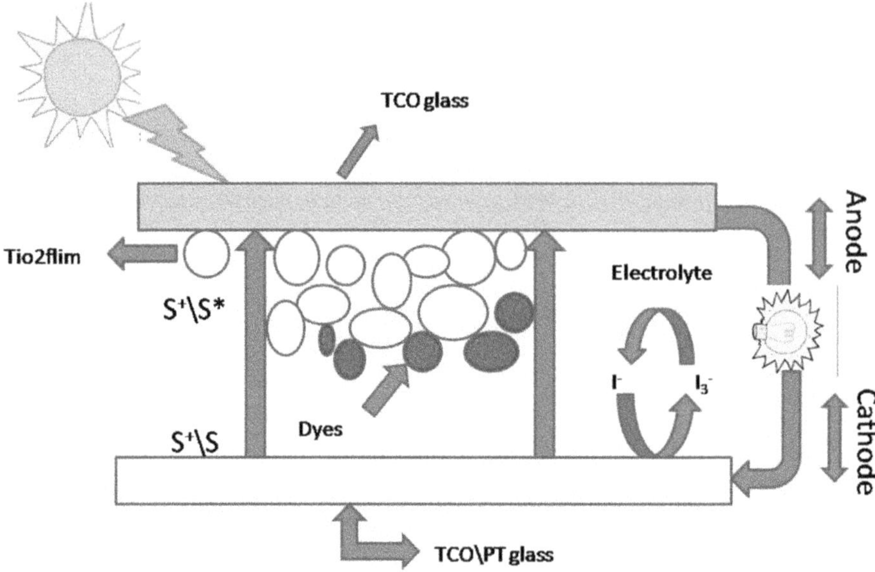

FIGURE 25.2 Schematic representation of the components and the basic operating principle of DSSCs and the schematic illustration of two similar types of visible light sensitisation of TiO₂. (i) Dye sensitisation (indirect): (1) excitation of the dye by visible light absorption, (2) electron transfer from the excited state of the dye to TiO₂, (3) recombination, (4) electron transfer to the acceptors, and (5) regeneration of the sensitiser by an electron donor; (ii) ligand to metal charge transfer: (1) visible light induced, (2) recombination, (3) electron transfer to the acceptor, and (4) regeneration of adsorption by an electron donor and the sensitiser, electron donor, and electron acceptor, respectively (s_0: groundstate and s^* and s_1: excited state of the sensitiser and adsorbate).

$$S^+ \backslash S + h\nu \Rightarrow S^+ \backslash S$$

$$S^+ \backslash S^* \Rightarrow S^+ \backslash S + e^- \left(TiO_2\right)$$

$$S^+ \backslash S^* + e^- \Rightarrow S^+ \backslash S$$

$$I_3^- + 2e^- \Rightarrow 3I^-$$

25.3 Semiconducting Materials of Photoanode

In most cases, photographic anode materials consist of a metallic oxide transistor with an enormous surface area for sufficient dye adsorption, a structure that is highly permeable for efficient mass distribution through propagation, and an elastic bandgap [20–24]. The factor's cumulative improvements contribute to greater charge-collection effectiveness. Due to their strong photo corrosion stability (transparent to the majority of the solar spectrum) and good electrical characteristics, wide band gap metal oxide semiconductors (E_g > 3eV) such as TiO₂, ZnO, SnO₂, and Nb₂O₅ have been widely employed as photoanode semiconductors [25–28]. Photo corrosion produced by the oxidation of gaps (formed by band gap stimulation) with reactive electrolytes may impair semiconductor performance. For the performance of DSSCs using different photoanode elements at the moment, the most effective photoanode material in DSSCs is TiO₂ nanoparticle, which has the highest observed efficiency (12.3%) [29, 30]. TiO₂ is also an inexpensive,

TABLE 25.1

Highest Efficiency Achieved Based on the Different Semiconductor in Photoanode Materials in Dye-Sensitised Solar Cells

Photoanode Materials	Band Gap	Efficiency
TiO_2	3.23	12.3
ZnO	3.3	5.60
Nb_2O_5	3.49	5.00
Zn_2SnO_4	3.7	3.70
$SrTiO_3$	3.2	1.80
WO_3	2.6–3.0	0.75
SnO_2	3.6	0.30

widely available, non-toxic, and adaptable substance. It has been utilised in medical items as well as household uses [31]. ZnO, on the other hand, has an equivalent conduction rate.

The spectrum edge and operational function of TiO_2, but with more carrier flexibility than TiO_2, were regarded as encouraging [32] as shown in Table 25.1. A DSSC photoanode. However, the acidic volatility of ZnO and the generation of dye clumps degrade its performance [33]. Other semiconductor components, such as Zn_2SnO_4, WO_3, and $SrTiO_3$, have been used to make DSSCs substrates. However, the effectiveness of these DSSCs remains inferior to TiO_2-based DSSCs [34–36]. TiO_2 is frequently discovered in three solid forms: the element rutile with an energy band E_g of 3.05 eV, the substance with an E_g of 3.23 eV, and limestone with an E_g of 3.26 eV. The TiO_2 photoanode has an additional dye-loading area on its surface and a greater electron transport coefficient than the element rutile nanoparticle photoanode [37]. However, because limestone is difficult to generate, it has been excluded from the DSSC application [38]. As a result, the anatase nanoparticle of TiO_2 is frequently utilised as the main element of the photoanode technology in DSSCs.

25.4 Mechanism of the Dye-Sensitisation

The dye-sensitisation method is used to increase the catalytic wavelength spectrum for photocatalysts; it has been used to breakdown chlorophenol, a substance called insecticides, and paraffin [39, 40].

A substance called benzyl alcohol is also present. Moving metal-based dyes are the best chemicals for achieving excellent electron transport to the semiconductor, but they are costly and not harmful to the environment; thus, dyes made from organic materials can be employed in the sensitisation process [41, 42] as shown in Figure 25.3. A proton travels from its highest occupied molecular orbital (HOMO) to its lowest unoccupied molecular orbital (LUMO), and then to the TiO_2 conduction band (CB); after a dye is released on the catalyst, the surface gets activated by apparent illumination.

The dye in an approach on the opposite together may absorb illumination, and the electron in the LUMO combines with oxygen that is present to generate the anion of the superoxide charge in nature [43]. Dyes must have specific features in order to be categorised as photosensitisers. Some of this powerful UV) being absorbed, even in the near infrared (NIR) region, visibility (unless self-sensitised degradation is required), anchoring groups (SO_3H, –COOH, –H_2PO_3, etc.) which encourage robust dye molecule's structure connecting onto the TiO_2 surface, and the photosensitiser state of excitement bringing a greater energy level than the TiO_2CB edge are some of these properties. As a result, the flow of electrons through the activated pigment and TiO2 will improve thermodynamics. and TiO_2 CB and Rhodamine B, frequently referred to as RhB, are cost effective [44] and contain an identical binding block (–COOH), which renders compounds outstanding detectors. They had been selected as stimulants for their low harmful effects, chemical makeup, and relatively straightforward learning.

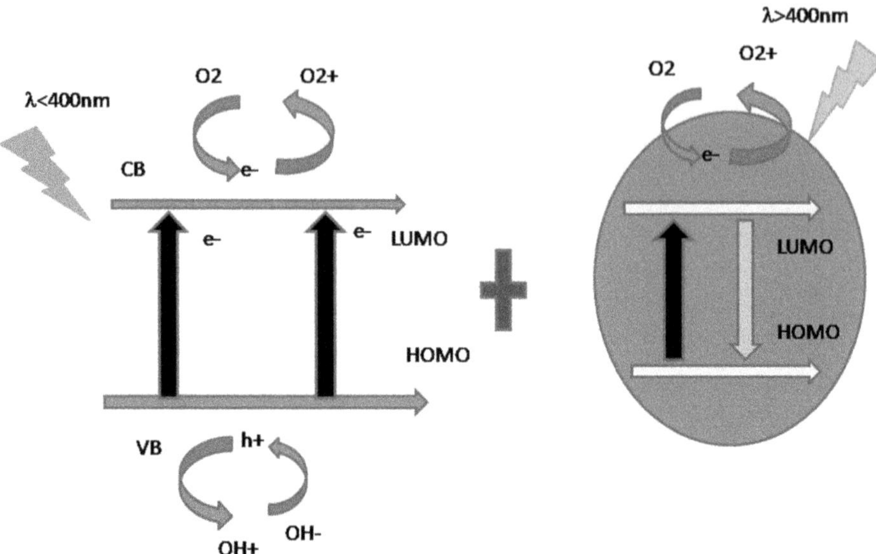

FIGURE 25.3 Mechanism involving adsorbed dye adsorbed onto the catalyst surface and dye in the bulk based on HOMO and LUMO in DSSC which HOMO the TiO$_2$CB, after a dye released on the catalyst surface gets activated by apparent illumination and the electron in the LUMO combines with oxygen that is present to generate the anion of the superoxide charge in nature.

25.5 Peculiar Framework of DSSC

All DSSCs depend on an electrolyte to function. It works as an electric charge carrier, receiving electrons from the cathode's surface and delivering them to the molecules of the dye. When addressing the cell productivity, the iodide/triiodide (I$^-$/I$_3^-$) electron couple in an organic framework is often a solution to the most frequently utilised electrolyte [45]. However, there are unwanted essential characteristics of a liquid electrolyte that have an important effect on a device's long-term viability and operating stability. For example, not only would harmful organic solvent leaks pollute the environment but dispersion of unstable iodide particles will raise total resistance inside through decreasing energy-focused attention. In order to avoid these drawbacks, researchers have worked to create unusual electrolytes such as low-temperature ionising liquids, quasi-solid state, and solid-state electrolytes.

A solid-state electrode is considered to be an improved choice to a liquid electrolyte since it has greater mechanical properties and easier manufacturing techniques. To prevent evaporation and leakage of the fluid electrolyte, it alternates the fluid electrolyte with solid p-type semiconductors or organic hole transporting substances. In this context, dye-sensitised photovoltaic panels are in concept comparable to usual Si-based p–n junction solar cells due to the n-type semiconductor and p-type sensitisers. In order to control the charge-transfer process, an acceptable hole transport substance must have a band gap topology suitable with the dye's maximum unoccupied molecular orbital (HOMO) level and TiO$_2$'s CB structure. The key challenge in manufacturing such a solid-state electrolyte technology is making close interaction at the p–n junction surfaces.

The materials CuSCN and CuI and the first p-type semiconductors that served as hole transporters were introduced by Tennakone et al. [46] and O'Regan and Schwartz [47]. They took their initial step, because unfilled gaps in the TiO2 intercrystallite, which could lead to an overall efficiency of less than 1%. During the coating technique, p-CuI crystalline semiconductors develop quickly. Kumara et al. [48] discovered a powerful crystal growth inhibitor where the growth of a tiny crystal with a size greater than those of TiO$_2$ pores will cause problems with the refilling technique, a considerable number of pores remain unfilled. Heng et al. [49] published their findings that triethylaminehydrothiocyannate (THT) was used to boost cell productivity to 3.75%. Recently, an additional working electrode TiO$_2$/CuI/Cu has been created.

The majority of studies have found that the conversion rate of DSSCs utilising solid electrolytes is inadequate, often less than 5%. This happens due to weak interaction at the TiO_2/electrolyte junction and the relatively low electrical conductivity of such substances everywhere [50]. To overcome the aforementioned limitations, a different method has been created and evaluated in which both liquid and solid electrolytes are blended to generate gel electrolyte. Wang et al. [51] showed that the combination of PVDF polymer can confine 3.methoxypropionitrile (MPN)-based liquid electrolyte into gel without compromising with the transfer of charges of the triiodide/iodide pair inside the polymer network. A rate of conversion outcome of more than 6% was achieved. Furthermore, the cell with such gel electrolyte sustained 94% of its initial efficacy after 1000 hours of heating at 80 degree Celsius and subsequently, the electrostatic liquid was semi-solidified by incorporating quartz nanoparticles, creating an ionic liquid-based quasi-solid-state electrolyte with a 7% capacity [52].

25.6 Semantic Influence of Nano in DSSC

Electrons in nanoparticles of TiO_2 encounter several DE-trapping incidents, throughout the charge transportation manipulate, dye-infused particles can be confined by states of trapping or spontaneously released and returned to propagation [53]. As a result of mechanical flaws at the surfaces, determine situations may arise and boundary lines of titanium dioxide (TiO_2) nanoparticles has the efficiency of 12.3 compare to other nanostructures [54], shown in (Table 25.2) and they can also occur within TiO_2 Nanocrystallites [55]. The defects create a condition of trapped electrons, reducing particle transit and raising the distance travelled in advance of combination [56]. Kopidakis et al. [57] determined the relationship with hardness parameter and the whole trapping densities for estimating the positioning of transport limiting clutches.

The information's best fit demonstrated an immediate association between the amount of internal surface area and the number of captured electron devices in the film, demonstrating that the trap is mainly situated at the outermost layers of the particles of TiO_2, compared to at the mass of the particles as a whole or at the interparticle restrictions of the grain [58]. Ansari-Rad et al. discovered that the increasing with smaller fragments was caused by the trap refilling electrical energy effects rather than an architectural consequence, and that the nanoparticles' dimension had no effect on the transmission of large quantities or the exterior.

25.7 Future Prospectus

A more affordable substitute for sophisticated semiconductor solar cells is the DSSC. This low-cost method is very effective, steady throughout a long period of time, and relatively easy to manufacture. Furthermore, crystalline silicon cells are more immune to visible light as well as light that enters at an angle that is shallow than DSSCs. In light of those features, DSSCs can function as reliable sources of electricity in areas with limited illumination. In labs, DSSCs are being studied and created. For use in reality, a lot more research needs to be done for enhancing their capacity for electricity conversion. Since the light-to-electrical transformation in DSSCs depends mainly on the ability to absorb light of sensitive

TABLE 25.2

Comparison Performance of Nanostructure with Different Nanomaterials

ID Nanostructure	Efficiency
TiO_2 nanowires	9.33
TiO_2 nanorod	9.00
TiO_2 nanotube arrays	9.1
TiO_2 nanoparticles	2.3

dyes, dyes, as an essential component of DSSCs, have attracted a lot of attention in research. To create panchromatic dyes, an enormous amount of study has been done for generations. Despite co-sensitised technological advances, this novel design presents an additional method for DSSCs to detect more light. Alongside increasing efficiency, decreasing the cost of materials is an essential problem that needs to be handled in subsequent research. The attempt had been made by Wang and Kerr to fabricate flexible DSSCs (productiveness 2.66%) from tissue. Using the recently developed nickel coating process on a paper platform can offer low-cost roll-to-roll large-scale production options, considering the fact that Ni-coated paper substrate is vulnerable to breaking throughout calcinations that In DSSCs, the expensive metal platinum is employed as a catalyst on the cathode; around 10–100 mg/cm^2 of palladium is needed to function [59]. Several noble metal-free materials have been developed in order to reduce the cost of the materials and achieve an acceptable transparency in appearance. However, colors with significant absorption are generally not perceived as widely. In an innovative approach presented by Hardin et al. photons with high energy are absorbed by highly photoluminescent chromophores that are free of titania and send their energy to the sensitisation dye using conduction synchronous transmission. The technique will be studied in order to assess the possibility that the chromophores being investigated are related to DSSC (Wang et al. [60, 61]. Recently, it was shown that COS (cobalt, sulphide) nanoparticles produced electrochemically on flexible ITO/PEN film function similarly to Pt when used as a triiodide reductions catalysis in DSSCs. Unexpectedly, cobalt is several hundred times less expensive than platinum. Employing conventional diagram nanoplatelets, Kavan et al. [62] produced semi-transparent (485%) thin films on top of F-doped SnO$_2$ (FTO). This graphene cathode operates comparably to the Pt cathode in ionic liquid-based DSSCs. Additionally transparency in optics is an essential property that makes it possible for increasing the use of certain industries to openings, gable sections, and double cells. Tandem pn-DSSC not only enhances the effectiveness of each component in DSSCs but also presents a further option for further study. Because they use spectrum of similar dyes, dual cells that integrate p-type DSSCs with standard n-type DSSCs can collect light from the environment with greater efficiency than single-junction DSSCs. Recently, a series of donor–acceptor dyes with variable length oligothiophene bridges were created by Nattestad et al. [63] in order to regulate the geographical separation of the carriers of charges generated by photosynthetic. It has been established that these dyes may reach an absorption photonic to particle conversion rate of up to 96% and significantly slow down the process of charge recombination in p-type DSSC. Such p-type DSSCs perform faster in simultaneous cells than in their individual parts. If it is possible to successfully adjust dye spectral equivalents and semiconductor energy band topologies, this could happen soon.

25.8 Conclusion

Dye se-sensitized solar cells (DSSCs) are a particular renewable energy source that might be competing with conventional solar cells. Additionally, the dye-sensitized solar cell is nearing readiness for industrial development, and commercialisation predictions for this kind of technology are similar to those for other photovoltaic techniques. In order to improve DSSC science and technology, it must be productive in boosting the device's reliability and efficiency while reducing material as well as manufacturing costs in order to become more economically viable. Techniques for quantitatively choosing optimal electronic components, dyes, and electrolytes – as well as how they are combined – must be researched for the purpose of achieving these goals. A strong model might be used to speed up the research process while reducing significant research time as well as expenses. Real strategies, in addition to reasoning design and simulation, are necessary for improving the capture of light and minimise electron losses. Furthermore, the solid-state and simultaneous DSSCs represent two preferred semiconductor designs due to their outstanding durability and outstanding productivity, correspondingly. These two attractive features of perovskite photovoltaic cells, which were created from DSSCs, would make them an intriguing new category of meso-superstructure solar cells (MSSCs).

REFERENCES

[1] Li, B., Wang, L., Kang, B., Wang, P., & Qiu, Y. (2006). Review of recent progress in solid-state dye-sensitized solar cells. *Solar Energy Materials and Solar Cells*, *90*(5), 549–573.

[2] http://www.energy.eu

[3] O'Regan, B., & Grätzel, M. (1991). A low-cost, high-efficiency solar cell based on dye-sensitized colloidal TiO$_2$ films. *Nature*, *353*(6346), 737–740.

[4] Hoppe, H., & Sariciftci, N. S. (2004). Organic solar cells: An overview. *Journal of Materials Research*, *19*(7), 1924–1945.

[5] Günes, S., Neugebauer, H., & Sariciftci, N. S. (2007). Conjugated polymer-based organic solar cells. *Chemical Reviews*, *107*(4), 1324–1338.

[6] Gong, J., Sumathy, K., Qiao, Q., & Zhou, Z. (2017). Review on dye-sensitized solar cells (DSSCs): Advanced techniques and research trends. *Renewable and Sustainable Energy Reviews*, *68*, 234–246.

[7] Mathew, S., Yella, A., Gao, P., Humphry-Baker, R., Curchod, B. F., Ashari-Astani, N., ... & Grätzel, M. (2014). Dye-sensitized solar cells with 13% efficiency achieved through the molecular engineering of porphyrin sensitizers. *Nature Chemistry*, *6*(3), 242–247.

[8] Gao, F., Wang, Y., Shi, D., Zhang, J., Wang, M., Jing, X., ... & Gratzel, M. (2008). Enhance the optical absorptivity of nanocrystalline TiO$_2$ film with high molar extinction coefficient ruthenium sensitizers for high performance dye-sensitized solar cells. *Journal of the American Chemical Society*, *130*(32), 10720–10728.

[9] Sima, C., Grigoriu, C., & Antohe, S. (2010). Comparison of the dye-sensitized solar cells performances based on transparent conductive ITO and FTO. *Thin Solid Films*, *519*(2), 595–597.

[10] Cai, H., Tang, Q., He, B., Li, R., & Yu, L. (2014). Bifacial dye-sensitized solar cells with enhanced rear efficiency and power output. *Nanoscale*, *6*(24), 15127–15133.

[11] Liu, J., Tang, Q., He, B., & Yu, L. (2015). Cost-effective, transparent iron selenide nanoporous alloy counter electrode for bifacial dye-sensitized solar cell. *Journal of Power Sources*, *282*, 79–86.

[12] Hagfeldt, A., & Grätzel, M. (2000). Molecular photovoltaics. *Accounts of Chemical Research*, *33*(5), 269–277.

[13] O'Regan, B., Lenzmann, F., Muis, R., & Wienke, J. (2002). A solid-state dye-sensitized solar cell fabricated with pressure-treated P25–TiO$_2$ and CuSCN: Analysis of pore filling and IV characteristics. *Chemistry of Materials*, *14*(12), 5023–5029.

[14] Taguchi, T., Zhang, X. T., Sutanto, I., Tokuhiro, K. I., Rao, T. N., Watanabe, H., ... & Fujishima, A. (2003). Improving the performance of solid-state dye-sensitized solar cell using MgO-coated TiO$_2$ nanoporous film. *Chemical Communications*, *1*(19), 2480–2481.

[15] Murakoshi, K., Kogure, R., Wada, Y., & Yanagida, S. (1998). Fabrication of solid-state dye-sensitized TiO$_2$ solar cells combined with polypyrrole. *Solar Energy Materials and Solar Cells*, *55*(1–2), 113–125.

[16] Hagen, J., Schaffrath, W., Otschik, P., Fink, R., Bacher, A., Schmidt, H. W., & Haarer, D. (1997). Novel hybrid solar cells consisting of inorganic nanoparticles and an organic hole transport material. *Synthetic Metals*, *89*(3), 215–220.

[17] Arango, A. C., Johnson, L. R., Bliznyuk, V. N., Schlesinger, Z., Carter, S. A., & Hörhold, H. H. (2000). Efficient titanium oxide/conjugated polymer photovoltaics for solar energy conversion. *Advanced Materials*, *12*(22), 1689–1692.

[18] Calandra, P., Calogero, G., Sinopoli, A., & Gucciardi, P. G. (2010). Metal nanoparticles and carbon-based nanostructures as advanced materials for cathode application in dye-sensitized solar cells. *International Journal of Photoenergy*, *2010*, 1–3.

[19] Lan, Z., Wu, J., Wang, D., Hao, S., Lin, J., & Huang, Y. (2007). Quasi-solid-state dye-sensitized solar cells based on a sol–gel organic–inorganic composite electrolyte containing an organic iodide salt. *Solar Energy*, *81*(1), 117–122.

[20] Sawatsuk, T., Chindaduang, A., Sae-Kung, C., Pratontep, S., & Tumcharern, G. (2009). Dye-sensitized solar cells based on TiO$_2$–MWCNTs composite electrodes: Performance improvement and their mechanisms. *Diamond and Related Materials*, *18*(2–3), 524–527.

[21] Jang, S. R., Vittal, R., & Kim, K. J. (2004). Incorporation of functionalized single-wall carbon nanotubes in dye-sensitized TiO$_2$ solar cells. *Langmuir*, *20*(22), 9807–9810.

[22] Lee, T. Y., Alegaonkar, P. S., & Yoo, J. B. (2007). Fabrication of dye sensitized solar cell using TiO_2 coated carbon nanotubes. *Thin Solid Films, 515*(12), 5131–5135.

[23] Kim, S. L., Jang, S. R., Vittal, R., Lee, J., & Kim, K. J. (2006). Rutile TiO_2-modified multi-wall carbon nanotubes in TiO_2 film electrodes for dye-sensitized solar cells. *Journal of Applied Electrochemistry, 36*, 1433–1439.

[24] Usui, H., Matsui, H., Tanabe, N., & Yanagida, S. (2004). Improved dye-sensitized solar cells using ionic nanocomposite gel electrolytes. *Journal of Photochemistry and Photobiology A: Chemistry, 164*(1–3), 97–101.

[25] Saito, M., & Fujihara, S. (2008). Large photocurrent generation in dye-sensitized ZnO solar cells. *Energy & Environmental Science, 1*(2), 280–283.

[26] Onwona-Agyeman, B., Kaneko, S., Kumara, A., Okuya, M., Murakami, K., Konno, A., & Tennakone, K. (2005). Sensitization of nanocrystalline SnO_2 films with indoline dyes. *Japanese Journal of Applied Physics, 44*(5L), L731.

[27] Guo, P., & Aegerter, M. A. (1999). RU (II) sensitized Nb_2O_5 solar cell made by the sol-gel process. *Thin Solid Films, 351*(1–2), 290–294.

[28] Lenzmann, F. O., & Kroon, J. M. (2007). Recent advances in dye-sensitized solar cells. *Advances in OptoElectronics, 2007*, 4–6.

[29] Yella, A., Lee, H. W., Tsao, H. N., Yi, C., Chandiran, A. K., Nazeeruddin, M. K., ... & Grätzel, M. (2011). Porphyrin-sensitized solar cells with cobalt (II/III)–based redox electrolyte exceed 12 percent efficiency. *Science, 334*(6056), 629–634.

[30] Jose, R., Thavasi, V., & Ramakrishna, S. (2009). Metal oxides for dye-sensitized solar cells. *Journal of the American Ceramic Society, 92*(2), 289–301.

[31] Grätzel, M. (2003). Dye-sensitized solar cells. *Journal of Photochemistry and Photobiology C: Photochemistry Reviews, 4*(2), 145–153.

[32] Minoura, H., & Yoshida, T. (2008). Electrodeposition of ZnO/dye hybrid thin films for dye-sensitized solar cells. *Electrochemistry, 76*(2), 109–117.

[33] Tan, B., Toman, E., Li, Y., & Wu, Y. (2007). Zinc stannate (Zn_2SnO_4) dye-sensitized solar cells. *Journal of the American Chemical Society, 129*(14), 4162–4163.

[34] Zheng, H., Tachibana, Y., & Kalantar-Zadeh, K. (2010). Dye-sensitized solar cells based on WO_3. *Langmuir, 26*(24), 19148–19152.

[35] Burnside, S., Moser, J. E., Brooks, K., Grätzel, M., & Cahen, D. (1999). Nanocrystalline mesoporous strontium titanate as photoelectrode material for photosensitized solar devices: Increasing photovoltage through flatband potential engineering. *The Journal of Physical Chemistry B, 103*(43), 9328–9332.

[36] Dinh, N. N., Bernard, M. C., Hugot-Le Goff, A., Stergiopoulos, T., & Falaras, P. (2006). Photoelectro-chemical solar cells based on SnO_2 nanocrystalline films. *ComptesRendus Chimie, 9*(5–6), 676–683.

[37] Park, N. G., Schlichthörl, G., Van de Lagemaat, J., Cheong, H. M., Mascarenhas, A., & Frank, A. J. (1999). Dye-sensitized TiO_2 solar cells: structural and photoelectrochemical characterization of nano-crystalline electrodes formed from the hydrolysis of $TiCl_4$. *The Journal of Physical Chemistry B, 103*(17), 3308–3314.

[38] Park, N. G., Van de Lagemaat, J., & Frank, A. A. (2000). Comparison of dye-sensitized rutile-and anatase-based TiO_2 solar cells. *The Journal of Physical Chemistry B, 104*(38), 8989–8994.

[39] Yun, E. T., Yoo, H. Y., Kim, W., Kim, H. E., Kang, G., Lee, H., ... & Lee, J. (2017). Visible-light-induced activation of periodate that mimics dye-sensitization of TiO_2: Simultaneous decolorization of dyes and production of oxidizing radicals. *Applied Catalysis B: Environmental, 203*, 475–484.

[40] Chowdhury, P., Athapaththu, S., Elkamel, A., & Ray, A. K. (2017). Visible-solar-light-driven photo-reduction and removal of cadmium ion with Eosin Y-sensitized TiO_2 in aqueous solution of triethanol-amine. *Separation and Purification Technology, 174*, 109–115.

[41] Safaralizadeh, E., Darzi, S. J., Mahjoub, A. R., & Abazari, R. (2017). Visible light-induced degradation of phenolic compounds by Sudan black dye sensitized TiO_2 nanoparticles as an advanced photocatalytic material. *Research on Chemical Intermediates, 43*, 1197–1209.

[42] Rani, M., & Tripathi, S. K. (2016). Electron transfer properties of organic dye sensitized ZnO and ZnO/TiO_2 photoanode for dye sensitized solar cells. *Renewable and Sustainable Energy Reviews, 61*, 97–107.

[43] Vignesh, K., Suganthi, A., Rajarajan, M., & Sakthivadivel, R. (2012). Visible light assisted photodecol-orization of eosin-Y in aqueous solution using hesperidin modified TiO_2 nanoparticles. *Applied Surface Science, 258*(10), 4592–4600.

[44] He, J., Zhao, J., Shen, T., Hidaka, H., & Serpone, N. (1997). Photosensitization of colloidal titania particles by electron injection from an excited organic dye–antennae function. *The Journal of Physical Chemistry B, 101*(44), 9027–9034.

[45] Penny, M., Farrell, T., & Will, G. (2008). A mathematical model for the anodic half cell of a dye-sensitised solar cell. *Solar Energy Materials and Solar Cells, 92*(1), 24–37.

[46] Tennakone, K., Kumara, G. R. R. A., Kumarasinghe, A. R., Wijayantha, K. G. U., & Sirimanne, P. M. (1995). A dye-sensitized nano-porous solid-state photovoltaic cell. *Semiconductor Science and Technology, 10*(12), 1689.

[47] O'Regan, B., & Schwartz, D. T. (1995). Efficient photo-hole injection from adsorbed cyanine dyes into electrodeposited copper (I) thiocyanate thin films. *Chemistry of Materials, 7*(7), 1349–1354.

[48] Kumara, G. R. A., Kaneko, S., Okuya, M., & Tennakone, K. (2002). Fabrication of dye-sensitized solar cells using triethylamine hydrothiocyanate as a CuI crystal growth inhibitor. *Langmuir, 18*(26), 10493–10495.

[49] Heng, L., Wang, X., Yang, N., Zhai, J., Wan, M., & Jiang, L. (2010). p–n-junction-based flexible dye-sensitized solar cells. *Advanced Functional Materials, 20*(2), 266–271.

[50] Kron, G., Egerter, T., Werner, J. H., & Rau, U. (2003). Electronic transport in dye-sensitized nanoporous TiO_2 solar cells comparison of electrolyte and solid-state devices. *The Journal of Physical Chemistry B, 107*(15), 3556–3564.

[51] Wang, P., Zakeeruddin, S. M., Moser, J. E., Nazeeruddin, M. K., Sekiguchi, T., & Grätzel, M. (2003). A stable quasi-solid-state dye-sensitized solar cell with an amphiphilic ruthenium sensitizer and polymer gel electrolyte. *Nature Materials, 2*(6), 402–407.

[52] Wang, P., Zakeeruddin, S. M., Comte, P., Exnar, I., & Grätzel, M. (2003). Gelation of ionic liquid-based electrolytes with silica nanoparticles for quasi-solid-state dye-sensitized solar cells. *Journal of the American Chemical Society, 125*(5), 1166–1167.

[53] Usami, A., & Ozaki, H. (2001). Computer simulations of charge transport in dye-sensitized nanocrystalline photovoltaic cells. *The Journal of Physical Chemistry B, 105*, 4577–4583.

[54] Nelson, J., & Chandler, R. E. (2004). Random walk models of charge transfer and transport in dye sensitized systems. *Coordination Chemistry Reviews, 248*, 1181–1194.

[55] Wang, Y., Wu, D., Fu, L.-M., Ai, X.-C., Xu, D., & Zhang, J.-P. (2014). Density of state determination of two types of intra-gap traps in dye-sensitized solar cells and its influence on device performance. *Physical Chemistry Chemical Physics, 16*, 11626–11632.

[56] Hsiao, P.-T., Tung, Y.-L., & Teng, H. (2010). Electron transport patterns in TiO_2 nanocrystalline films of dye-sensitized solar cells. *The Journal of Physical Chemistry C, 114*, 6762–6769.

[57] Kopidakis, N., Neale, N. R., Zhu, K., van de Lagemaat, J., & Frank, A. J. (2005). Spatial location of transport-limiting traps in TiO_2 nanoparticle films in dye-sensitized solar cells. *Applied Physics Letters, 87*, 202106.

[58] Ansari-Rad, M., Abdi, Y., & Arzi, E. (2012). Monte Carlo random walk simulation of electron transport in dye-sensitized nanocrystalline solar cells: Influence of morphology and trap distribution. *The Journal of Physical Chemistry C, 116*, 3212–3218.

[59] Hardin, B. E., Hoke, E. T., Armstrong, P. B., Yum, J. H., Comte, P., Torres, T., ... & McGehee, M. D. (2009). Increased light harvesting in dye-sensitized solar cells with energy relay dyes. *Nature Photonics, 3*(7), 406–411.

[60] Wang, B., & Kerr, L. L. (2011). Dye sensitized solar cells on paper substrates. *Solar Energy Materials and Solar Cells, 95*(8), 2531–2535.

[61] Wang, M., Anghel, A. M., Marsan, B., Cevey Ha, N. L., Pootrakulchote, N., Zakeeruddin, S. M., & Grätzel, M. (2009). CoS supersedes Pt as efficient electrocatalyst for triiodide reduction in dye-sensitized solar cells. *Journal of the American Chemical Society, 131*(44), 15976–15977.

[62] Kavan, L., Yum, J. H., & Grätzel, M. (2011). Optically transparent cathode for dye-sensitized solar cells based on graphene nanoplatelets. *ACS Nano, 5*(1), 165–172.

[63] Nattestad, A., Mozer, A. J., Fischer, M. K., Cheng, Y. B., Mishra, A., Bäuerle, P., & Bach, U. (2010). Highly efficient s for dye-sensitized tandem solar cells. *Nature Materials, 9*(1), 31–35.

26

An Outlook of Organic Photosensitiser for Highly Efficient Dye-Sensitised Solar Cell Applications

J. Priyadharshini and R. Gayathri
Cauvery College for Women(Autonomous) (affiliated to Bharathidasan University), Trichy, Tamil Nadu, India

26.1 Introduction

The Sun supplies solar energy to the entire earth perpetually, without limitation [1]. The benefits of solar radiation include its environmental friendliness, renewable nature, and sustainability. Moreover, various types of power can be produced using solar energy [2]. The dye-sensitised solar cell (DSSC), created by Professor Michael Grätzel in 1991, is the most promising low-cost method of absorbing solar energy among all the other forms of solar cells [3]. DSSCs, which are less costly than conventional silicon solar cells, are a part of the thin film family of solar cells [4]. The manufacture of DSSCs is simple, affordable, and efficient [5]. Even in lower solar radiation, whether at dawn and twilight or in overcast weather, DSSCs perform better. Because they can exploit dispersed light so well, DSSCs are a great option for interior applications like sunroofs and windows. The advantages of DSSCs have led to a notable increase in research interest, as evidenced by the astounding increase in publications over the past 10 years [6]. The five primary parts of the DSSCs are an electrolyte, a semiconductor, a transparent conducting electrode, a photosensitiser, and a counter electrode. Among all the components, the photosensitiser is seen to be crucial since it is in charge of light gathering, electron transfer, and injection. As a result, scientists are working hard to create a wide variety of useful organic dyes that can be identified by their unique beginning materials, high material costs, molecular architecture's versatility, and improved light-harvesting efficiency (LHE) [7] (Figure 26.1).

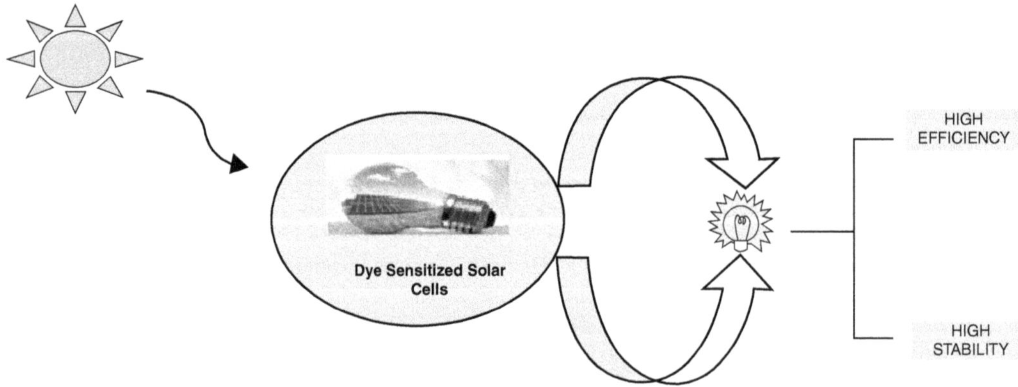

FIGURE 26.1 Schematic representation of dye-sensitised solar cells.

DOI: 10.1201/9781003495437-26

26.2 Dye-Sensitised Solar Cells

The DSSC generation is recognised as the third generation of solar cells [8, 9]. Though not as high as crystalline solar cells, these solar cells' efficiency is nevertheless higher than thin-film solar cells. In 1991, O'Regan and Grätzel produced the initial DSSC [10]. The substrate, the transparent conducting layer, the electrolyte, dyes, the TiO_2 nanoparticles, the photo electrode, and the counter electrode are some of the layers that make up a standard DSSC [11]. The two tasks are separated here, in contrast to traditional systems where the semiconductor performs both the role of charge carrier conveyance and light absorption. A broadband semiconductor absorbs light by attaching a sensitiser to its surface. Charge separation at the contact happens as a result of photo-induced electron injection from the dye into the solid's conduction band [12]. From the charge collector to the conduction band of the semiconductor, carriers are transferred. By combining sensitisers with a broad absorption band with oxide coatings, a considerable amount of sunlight can be captured [13].

26.2.1 Requirements for a Sensitiser's Effective Operation

Among the special qualities, a sensitiser ought to possess are:

- Considerable absorption in the wavelengths of light.
- Outstanding durability in the oxidised, excited, and ground states.
- Suitable pH level.
- High efficiency in the processes of charge injection and regeneration.

Molecular electron pumping is the role of a sensitiser in DSSC. It helps with the following processes: picking up an electron from the redox pair, pumping an electron in the conduction band of the TiO_2 layer, absorbing ultraviolet radiation, and repeating the procedure [14, 15].

26.2.2 Working Mechanism of DSSC

There are four fundamental phases that comprise the working concept of DSSC:

- Light absorption
- Electron injection
- Carrier transit
- Current collection

The process of turning photons into current involves the following phases:

- The incident photon is first absorbed by a photosensitiser. The outcome is the promotion of the dye's electrons from their ground state to their excited state.
- An injection of excited electrons with a microsecond lifespan is made into the conduction band of the nanoporous TiO_2 electrode, which is situated beneath the excited state of the dye. The minimal amount of UV sun rays that the TiO_2 absorbs causes the colour to get oxidised.
- The injected electrons diffuse towards the back contact as they pass between the TiO_2 nanoparticles. The counterpart electrode receives electrons from the external circuit.
- The dye is oxidised to its ground state by the acceptance of electrons from the redox mediator, which lowers the electron concentration at the counter electrode and results in dye regeneration.
- As it moves in the direction of the counter electrode, the oxidised mediator diffuses and turns into an ion (Figure 26.2).

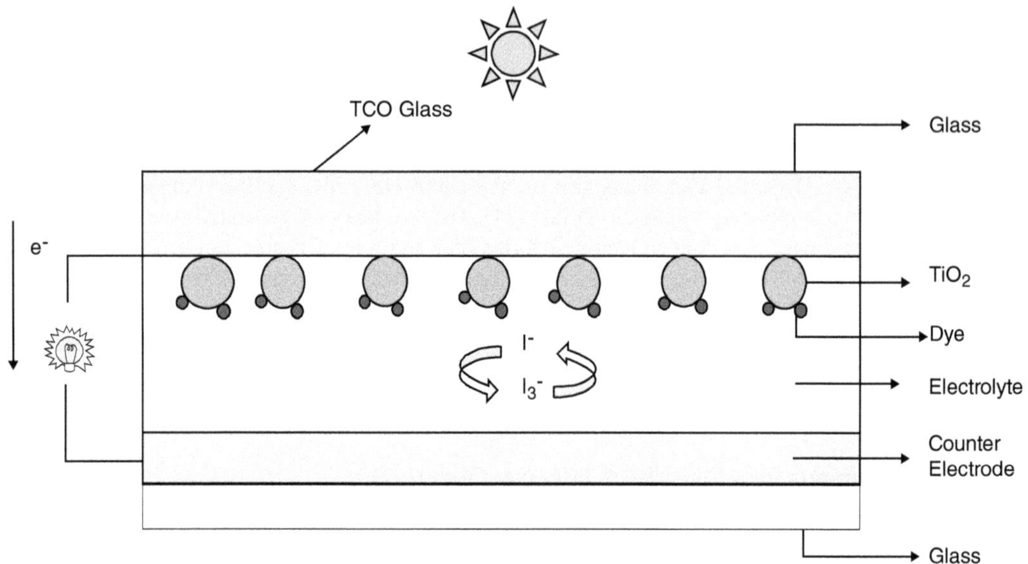

FIGURE 26.2   Setup of a dye-sensitised solar cell.

26.3 Dyes Involved in DSSC Applications

Organic (sometimes known as natural) and inorganic dyes are the two categories of dyes [16]. As of now, inorganic dyes like Ruthenium (Ru) are thought to be the most crucial for creating extremely effective dye-sensitive solar cells. They are costly and challenging to clean, though [17]. Natural dyes are seen to be the best options for replacing costly and uncommon inorganic sensitisers. On the other hand, the main advantages of using natural colouring agents as sensitisers are their affordable cost of manufacture, ease of use, short compensation periods, versatility, ease of access to raw materials, and minimal risks. Additionally, organic dyes are environmentally friendly and work exceptionally well in diffuse and multicoloured lighting conditions [18–20].

Numerous pigments that are readily extracted and used in DSSC can be found in naturally occurring fruits, flowers, leaves, microbes, and other objects [21]. These organic pigments have a number of advantages when used in DSSC, including being relatively abundant, simple to make, having high absorption coefficients in the visible spectrum, and being environmentally friendly [19, 22]. The wavelengths that are reflected or transmitted by the plant tissue are altered by the interaction between sunlight and the electrical structure of these plant pigments [23]. This process leads to plant pigmentation, where each pigment is identified by its specific wavelength of maximal absorption (λ_{max}) and the hue that is perceived by mankind [24]. When compared to artificial dyes, natural materials can be easily extracted for use as natural dye pigments. These pigments include flavonoid, carotenoid, and chlorophyll [25, 26].

26.3.1 Flavonoids

Plant pigments called flavonoids are commonly utilised. More than 5000 organically produced antioxidants have been identified from plant species based on their chemical structure, and they are categorised into seven categories: flavonoids, flavones, flavanones, isoflavones, catechins, anthocyanins, and chalcones [27]. Because flavonoid molecules have free electrons, lowest unoccupied molecular orbital (LUMO) may be electronically excited with less energy, allowing pigment molecules to be driven by visible light. Fruits that entice animals to eat them and disperse fruit and flower seeds that draw pollinating insects are known to contain flavonoids [28].

26.3.2 Carotenoids

Plant chromoplasts and chloroplasts, as well as some other physiological species including yeast and bacteria, contain organic pigments known as carotenoids [29]. Carotenoids are essential components of plants and algae because they protect chlorophyll from sun damage and absorb sunlight that is necessary for synthesis. Carotenoids give many shrubs and trees their distinctive red, yellow, and orange colours. They also offer these plants a spectrum of smells. More than 600 different forms of carotenoids have been identified. These include oxygen-containing xanthophylls and pure hydrocarbons called carotenes [30, 31].

26.3.3 Chlorophyll

The majority of plants, microorganisms, and organisms contain the green pigment known as chlorophyll (Chl) in their outermost layer of leaves. Chlorophyll is present in six dye hues, the most popular of which is Chla [32]. Chlorophyll is a chelate, a large metal ion that binds to important molecules, including hydrogen, carbon, and other elements like oxygen and nitrogen. During photosynthesis, chlorophyll uses absorbed energy to convert carbon dioxide into carbohydrates and water into oxygen. Solar radiation is converted into a form that plants can use through this mechanism. Countless confinement, microscopic hydrocarbons, a magnesium-centred chlorine ring, and other components make up the molecular structure of chlorides, depending on their kind [28, 33].

The closed structure of pigments, which is regulated by sunlight and alters the wavelengths used in interactions, causes plants to become coloured [34]. The pigment commonly absorbs light in a number of ways in the visible range. Using the same sensitising dye as intended in DSSC, the same parameters were measured: open-circuit voltage (V_{oc}), short-circuit current (J_{sc}), and energy conversion efficiency (η). Variations in natural substance dyes, like photosensitisers for DSSC, have an impact on various portions of a plant [35, 36].

The dye must meet specific criteria in order to function as a superior dye sensitiser, including:

- Having a strong dye bond with semiconductor materials.
- Having a broad absorption spectrum.
- Having the capacity to inject electrons into semiconductor materials [37].

26.4 Photovoltaic Features

26.4.1 Light-Harvesting Efficiency

In order to convert solar energy into electricity, two essential processes carried out by DSSCs are the conversion of photons into carriers and the passage of photogenerated carriers through the device [38]. Photosensitisers typically have an effect on DSSC efficiency. The LHE of DSSCs is by far its most significant feature. The dye's LHE must be as high as feasible to maximise the photocurrent response. An indicator of the efficacy of DSSC, the LHE, quantifies a dye's ability to interact with light [39]. For the charge transfer process of the DSSC, LHE values are essential. The amount of dye injected simultaneously with the molar absorption coefficient affects the LHE [40]. The oscillator strength will increase as LHE does [41]. A few factors that could impact a material's LHE are its bandgap, absorption properties, and ability to form and split electron–hole pairs in response to light [42].

26.4.2 Open Circuit Photo Voltage (V_{oc})

V_{oc} is considered a critical quantity that depends on several factors, such as light sources, charge carrier recombination, and material energy content. A DSSC sensitiser functions optimally when both the open circuit voltage (V_{oc}) and the short-circuit current density (J_{sc}) reach their optimum values [43]. V_{OC} discloses the number and mobility of charge carriers over the contact. Increased V_{oc} leads to higher J_{sc} levels [44]. A high V_{oc} value indicates reduced loss due to charge recombination, hence, increasing cell efficiency [45].

26.4.3 Electron Injection Ability (ΔG_{inj})

Separate energy levels correspond to the dye's excited-state LUMO level and the semiconductor TiO_2. E_{CB} drives the process of infusing dye-derived electrons into the semiconductor TiO_2 [46]. The negative sign of ΔG_{inj} indicates that the process is spontaneous. The electron injection ability (ΔG_{inj}) measures the free energy changes that occur when an electron is injected from the dye-excited state to the CB. The high ΔG_{inject} has led to high electron injection efficiency and an increase in the photo conversion efficiency (PCE) of the DSSC. Increasing ΔG_{inject} will help the DSSC's J_{sc} [47].

26.4.4 Power Conversion Efficiency

The PCE of a solar cell is the most often used metric to determine its capacity to convert sunlight into electrical power. Various photovoltaic parameters impact PCE (η), which is one of the ultimate performance measures of DSSC devices. The following formula can be used to predict it:

$$\eta = FF V_{oc} J_{sc} / P_{inc}$$

where V_{oc} is the open-circuit voltage, J_{sc} is the short-circuit current density, P_{inc} is the incident photon to current efficiency, and FF is the fill factor.

26.5 Development Obstacles in the Enhancement of DSSCs

- Poor contact between the electrodes.
- Diminished electrolyte properties due to UV radiation-induced light absorption.
- Low extinction coefficient requires a large surface area.
- The photo electrode and counter electrode, the two primary components of the DSSC, have poor encapsulation, which leads to electrolyte leakage and lower cell conversion efficiency [48–50].

26.6 Advantages of DSSCs

- **Reduced Weight**

  Solar cells and panels can be made lightweight by using plastic substrates instead of crystal or semiconductor materials. Because aesthetics are so important, DSSCs can be used in situations where other solar cells cannot [51].
- **Colourful and Transparent**

  There is a large selection of coloured and translucent cells due to the usage of organic dye. DSSC's transparency and range of colours make them suitable for window and sunroof decoration [52].
- **Ecologically Friendly and Recyclable**

  There are no hazardous substances included in the DSSCs component materials. Compared to solar cell panels, recycling has the benefit that the materials can be easily separated and used for other purposes [53].
- **Adaptable and Compact Architecture**

  By using aggregation of tiny optical conversion material particles, flexible thin-film solar cells can be produced.

26.7 Prospectus of the Future

As a key component of DSSCs, dyes have drawn a lot of attention from researchers since the light absorption of sensitised dyes is essential to DSSCs ability to convert light into electricity. The DSSC has demonstrated potential as an affordable substitute for costly silicon solar cells. This low-cost technique is very efficient, stable over the long term, and simple to produce [54]. Moreover, crystalline silicon cells are less sensitive to visible light and light that enters at a shallow angle than DSSCs. Because of these characteristics, DSSCs can function as a dependable power source in dimly lit areas. DSSCs are currently undergoing research and development in labs. For practical applications, much more work needs to be done to improve its energy conversion efficiency [55]. The ultimate goal of DSSC research is to use semiconductor materials with organic dye to produce electrical energy. Methods to reduce material costs while maintaining performance are an important question that has to be answered in further studies. When compared to earlier research reported in the literature, current high-efficiency values have already been attained. Nonetheless, there is still opportunity for advancement in DSSC technology in the future, especially in the area of conversion value stability research. Ultimately, the potential to create DSSCs in a solid form would allow for a vast area of study and advancement in DSSC technology [56–58].

26.8 Conclusion

In order for DSSC technology to advance and become more competitive, it must be able to reduce material and production costs while simultaneously improving device stability and efficiency. Techniques for statistically choosing appropriate semiconductors, dyes, and electrolytes – as well as how they are combined – must be developed in order to achieve these goals. Real solutions are required to maximise light capture and decrease electron losses, in addition to logic design and simulation. Thoroughly characterising the chemical changes caused by varying light and temperature, as well as the resulting effects on device efficiency, is essential for stability testing.

REFERENCES

[1] Atli, A., & Yildiz, A. (2022). Opaque Pt counter electrodes for dye-sensitized solar cells. *International Journal of Energy Research, 46*(5), 6543–6552.

[2] Yildiz, A., Chouki, T., Atli, A., Harb, M., Verbruggen, S. W., Ninakanti, R., & Emin, S. (2021). Efficient iron phosphide catalyst as a counter electrode in dye-sensitized solar cells. *ACS Applied Energy Materials, 4*(10), 10618–10626.

[3] Okoye, I. F. (2020). *Basic Applications in Energy and Power.* Ahmadu Bello University Publisher and Press Limited:Zaria.

[4] Alizadeh, A., Roudgar-Amoli, M., Shariatinia, Z., Abedini, E., Asghar, S., & Imani, S. (2023). Recent developments of perovskites oxides and spinel materials as platinum-free counter electrodes for dye-sensitized solar cells: A comprehensive review. *Renewable and Sustainable Energy Reviews, 187,* 113770.

[5] Muñoz-García, A. B., Benesperi, I., Boschloo, G., Concepcion, J. J., Delcamp, J. H., Gibson, E. A., … & Freitag, M. (2021). Dye-sensitized solar cells strike back. *Chemical Society Reviews, 50*(22), 12450–12550.

[6] Rahman, S., Haleem, A., Siddiq, M., Hussain, M. K., Qamar, S., Hameed, S., & Waris, M. (2023). Research on dye sensitized solar cells: Recent advancement toward the various constituents of dye sensitized solar cells for efficiency enhancement and future prospects. *RSC Advances, 13*(28), 19508–19529.

[7] El Mzioui, S., Bouzzine, S. M., Sidir, İ., Bouachrine, M., Bennani, M. N., Bourass, M., & Hamidi, M. (2019). Theoretical investigation on π-spacer effect of the D–π–A organic dyes for dye-sensitized solar cell applications: A DFT and TD-BH and H study. *Journal of Molecular Modeling, 25,* 1–12.

[8] Okoye, I. F., Nwokoye, A. O. C. and Ahmad, G. (2021). Power voltage characteristics of fabricated DSSC incorporating multiple organic dyes as photosensitizer. *Energy and Power Engineering, 13,* 221–235.

[9] Gerrit, B. (2019). Improving the performance of dye-sensitized solar cells. *Journal of Physical Chemistry and Chemical Physics, 11,* 278–293.

[10] Ananthakumar, S., Balaji, D., Kumar, J. R. and Babu, S. M. (2019). Role of co-sensitization in dye-sensitized and quantum dot-sensitized solar cells. *SN Applied Sciences, 1,* Article No. 186.

[11] Sokolsky, M., & Cirák, J. (2010). Dye-sensitized solar cells: Materials and processes. *Acta Electro technica et Informatica, 10,* 78–81.

[12] Kabirad, F., Bhuiyan, M. M., Manira, M. S., Rahaman, M. S., Khana, M. A., & Ikegamic, T. I. (2019). Development of dye-sensitized solar cell based on combination of natural dyes extracted from Malabar spinach and red spinach. *Results in Physics, 14*, Article ID: 102474.

[13] Ugwu, L. O., Ozuomba, J. O., Ekwo, P. I., & Ekpunobi, A. J. (2015). The optical properties of anthocyanin-doped nanocrystalline-TiO_2 and the photovoltaic efficiency on DSSC. *Der Chemica Sinica, 6*, 42–48.

[14] Mariotti, N., Bonomo, M., Fagiolari, L., Barbero, N., Gerbaldi, C., Bella, F., & Barolo, C. (2020). Recent advances in eco-friendly and cost-effective materials towards sustainable dye-sensitized solar cells. *Green Chemistry, 22*(21), 7168–7218.

[15] Mehmood, U., Rahman, S., Harrabi, K., et al. (2014). Recent advances in dye sensitized solar cells. *Advances in Materials Science and Engineering, 2014.*

[16] Reagan, B. O., & Gratzel, M. (1991). A low-cost, high efficiency solar cell based on dye-sensitized colloidal TiO_2 films. *Nature, 353*, 737–739.

[17] Bradha, M., Balakrishnan, N., Suvitha, A., Arumanayagam, T., Rekha, M., Vivek, P., ...& Steephen, A. (2021). Experimental, computational analysis of Butein and Lanceoletin for natural dye-sensitized solar cells and stabilizing efficiency by IoT. *Environment, Development and Sustainability, 24*, 1–16.

[18] Mwalukuku, V. M., Liotier, J., Riquelme, A. J., Kervella, Y., Huaulmé, Q., Haurez, A., ...& Demadrille, R. (2023). Strategies to improve the photochromic properties and photovoltaic performances of naphthopyran dyes in dye-sensitized solar cells. *Advanced Energy Materials, 13*(8), 2203651.

[19] Gomez-Ortiz, N. M., Vazquez-Maldonado, I. A., PerezEspadas, A. R., et al. (2010). Dye-sensitized solar cells with natural dyes extracted from achiote seeds. *Solar Energy Materials and Solar Cells, 94*, 40–44.

[20] Al-Ghamdi, A. A., Gupta, R. K., Kahol, P. K., et al. (2014). Improved solar efficiency by introducing grapheme oxide in purple cabbage dye sensitized TiO_2 based solar cell. *Solid State Communications, 183*, 56–59.

[21] Chang, H., & Lo, Y. J. (2010). Pomegranate leaves and mulberry fruit as natural sensitizers for dye-sensitized solar cells. *Solar Energy, 84*, 1833–1837.

[22] Im, J. H., Lee, C. R., Lee, J. W., et al. (2011). 6.5% efficient perovskite quantum-dot-sensitized solar cell. *Nanoscale, 3*, 4088–4093.

[23] Davies, K. M. (2004). *Plant Pigments and Their Manipulation (14th)*. Blackwell publishing Ltd.: USA.

[24] Kumara, N. T. R. N., Ekanayake, P., Lim, A., et al. (2013). Layered co-sensitization for enhancement of conversion efficiency of natural dye sensitized solar cells. *Journal of Alloys and Compounds, 581*, 186–191.

[25] Yella, A., Baker, R. H., Curchod, B. F. E., et al. (2013). Molecular engineering of a fluorene donor for dye-sensitized solar cells. *Chemistry of Materials, 25*, 2733–2739.

[26] Wang, X. F., Xiang, J., Wang, P., & Koyama, Y. (May 2019). Dye-sensitized solar cells using chlorophyll a derivate as the sensitizer and carotenoids having different conjugation lengths as redox spacers. *Optik, 185*, 620–625.

[27] Bashar, H., Bhuiyan, M. M. H., Hossain, M. R., Kabir, F., Rahaman, M. S., Manir, M. S., & Ikegami, T. (2019). Study on combination of natural red and green dyes to improve the power conversion efficiency of dye sensitized solar cells. *Optik, 185*, 620–625.

[28] Kabir, F., Bhuiyan, M. M. H., Manir, M. S., Rahaman, M. S., Khan, M. A., & Ikegami, T. (2019). Development of dye-sensitized solar cells based on a combination of natural dyes extracted from Malabar spinach and red spinach. *Results in Physics, 14*, 102474.

[29] Ammar, A. M., Mohamed, H. S., Yousef, M. M., Abdel-Hafez, G. M., Hassanien, A. S., & Khalil, A. S. (2019). Dye-sensitized solar cells (DSSCs) based on extracted natural dyes. *Journal of Nanomaterials, 1*, 2019.

[30] Kocak, Y., Atli, A., Atilgan, A., & Yildiz, A. (2019). Extraction method dependent performance of bio-based dye-sensitized solar cells (DSSCs). *Materials Research Express, 6*, 095512.

[31] Atli, A., Atilgan, A., Altinkaya, C., Ozel, K., & Yildiz, A. St. (2019). Lucie cherry, yellow jasmine, and madder berries as novel natural sensitizers for dye-sensitized solar cells. *International Journal of Energy Research, 43*, 3914–3922.

[32] Castillo-Robles, J. A., Rocha-Rangel, E., Ramírez-de-León, J. A., Caballero-Rico, F. C., & Armendáriz-Mireles, E. N. (2021). Advances on dye-sensitized solar cells (DSSCs) nanostructures and natural colorants: A review. *Journal of Composites Science, 5*(11), 288.

[33] Maurya, I. C., Singh, S., Srivastava, P., Maiti, B., & Bahadur, L. (2019). Natural dye extract from Cassia fistula and its application in dye-sensitized solar cells: Experimental and density functional theory studies. *Optical Materials, 90*, 273–280.

[34] Scheer, H. I. (2003). The pigments. In *Light-Harvesting Antennas in Photosynthesis*, Green, B. R. and Parson, W. W. (eds.). Kluwer Academic Publishers: Dordrecht. 513–530. *Chemical Physics Letters* 2005; 408:409–414.

[35] Shalini, S., Kumar, T. S., Prasanna, S., & Balasundaraprabhu, R. (2020). Investigations on the effect of co-doping in enhancing the performance of nanostructured TiO$_2$ based DSSC sensitized using extracts of Hibiscus Sabdariffa calyx. *Optik, 212*, 164672.

[36] Shalini, S., Balasundaraprabhu, R., Kumar, T. S., Sivakumaran, K., & Kannan, M. D. (2018). Synergistic effect of sodium and yeast in improving the efficiency of DSSC sensitized with extract from petals of Kigelia Africana. *Optical Materials, 79*, 210–219.

[37] Adachi, M., Sakamoto, M., Jiu, J., Ogata, Y., & Isoda, D. (2006). Determination of parameters of electron transport in dye-sensitized solar cells using electrochemical impedance spectroscopy. *Journal of Physical Chemistry B, 110*, 13872–13880.

[38] Narayan, M., & Raturi, A. (2011). Investigation of some common Fijian flower dyes as photosensitizers for dye sensitized solar cells abstract. *Applied Solar Energy, 47*, 112–117.

[39] Meenakshi, R. (2021). FT-IR and FT-RAMAN analysis and light-harvesting efficiency (LHE) enhancement for DSSC applications of hydrazide derivatives. *Journal of the Iranian Chemical Society, 18*(5), 1179–1198.

[40] Al-Qurashi, O. S., & Wazzan, N. (2021). Prediction of power conversion efficiencies of diphenylthienylamine-based dyes adsorbed on the titanium dioxide nanotube. *ACS Omega, 6*(13), 8967–8975.

[41] Lazrak, M., Toufik, H., Bouzzine, S. M., Ennehary, S., & Lamchouri, F. (2023). Theoretical analysis on D-π-A triphenylamine-based dyes for dye-sensitized solar cells: Effect of π-bridges on the optoelectronic, and photovoltaic properties. *Journal of Molecular Modeling, 29*(8), 1–12.

[42] Hagfeldt, A., Boschloo, G., Sun, L., Kloo, L., & Pettersson, H. (2010). Dye-sensitized solar cells. *Chemical Reviews, 110*(11), 6595–6663.

[43] Mustafa, F. M., Abdel Khalek, A. A., Mahboob, A. A., & Abdel-Latif, M. K. (2023). Designing efficient metal-free dye-sensitized solar cells: A detailed computational study. *Molecules, 28*(17), 6177.

[44] Zhong, Y., Tada, A., Izawa, S., Hashimoto, K., & Tajima, K. (2014). Enhancement of V_{OC} without loss of J_{SC} in organic solar cells by modification of donor/acceptor interfaces. *Advanced Energy Materials, 4*(5), 1301332.

[45] Bi, P., Zhang, S., Chen, Z., Xu, Y., Cui, Y., Zhang, T., …& Hou, J. (2021). Reduced non-radiative charge recombination enables organic photovoltaic cell approaching 19% efficiency. *Joule, 5*(9), 2408–2419.

[46] Fatima, K., Pandith, A. H., Manzoor, T., & Qureashi, A. (2023). DFT studies on a metal oxide@ graphene-decorated D–π1–π2–A novel multi-junction light-harvesting system for efficient dye-sensitized solar cell applications. *ACS Omega, 8*(9), 8865–8875.

[47] Wang, H., Liu, Q., Liu, D., Su, R., Liu, J., & Li, Y. (2018). Computational prediction of electronic and photovoltaic properties of anthracene-based organic dyes for dye-sensitized solar cells. *International Journal of Photoenergy, 2018*, 119–155.

[48] Lee, Y., & Kang, M. (2010). The optical properties of nanoporous structured titanium dioxide and the photovoltaic efficiency on DSSC. *Materials Chemistry and Physics, 122*, 284–289.

[49] Matthews, D., Infelta, P., & Grätzel, M. (1996). Calculation of the photocurrent-potential characteristics for regenerative sensitized semi-conductor electrodes. *Solar Energy Materials and Solar Cells, 44*, 119–155.

[50] Ekanayake, A. W. M. V., Kumara, G. R. A., Rajapaksa, R. M. G., & Pallegedara, A. (2018). Increasing the efficiency of a dye-sensitized solid-state solar cell by iodine elimination process in hole conductor material. *International Conference on Sustainable Built Environment, Kandy*, 13–15 December 2018, 282–287.

[51] Kimpa, I. M. (2016). Fabrication and analysis of dye sensitized solar cells using natural dyes extracted from pawpaw leaf and flame tree flower and the synthetic ruthenium dye. Master's Thesis, Usmanu Danfodiyo University, Sokoto.

[52] Hoseinnezhad, M. and Gharanjig, K. (2018). Synthesis and application of an organic dye in nanostructure solar cells device. *World Academy of Science, Engineering and Technology: International Journal of Chemical and Molecular Engineering, 12*, 438–441.

[53] Ahmed, M. H. M., Muraza, O., Galadima, A., Yoshioka, M., Yamani, Zain H., & Yokoi, T. (2019). Choreographing boron-aluminum acidity and hierarchical porosity in BEA zeolite by in-situ hydrothermal synthesis for a highly selective methanol to propylene catalyst. *Microporous and Mesoporous Materials, 273*, 249–255.

[54] Kavan, L., Yum, J. H., & Grätzel, M. (2011). Optically transparent cathode for dye-sensitized solar cells based on graphene nanoplatelets. *ACS Nano*, *5*(1), 165–172.

[55] Hardin, B. E., Hoke, E. T., Armstrong, P. B., Yum, J. H., Comte, P., Torres, T., ... & McGehee, M. D. (2009). Increased light harvesting in dye-sensitized solar cells with energy relay dyes. *Nature Photonics*, *3*(7), 406–411.

[56] He, J., Zhao, J., Shen, T., Hidaka, H., & Serpone, N. (1997). Photosensitization of colloidal titania particles by electron injection from an excited organic dye– Antennae function. *The Journal of Physical Chemistry B*, *101*(44), 9027–9034.

[57] Penny, M., Farrell, T., & Will, G. (2008). A mathematical model for the anodic half cell of a dye-sensitised solar cell. *Solar Energy Materials and Solar Cells*, *92*(1), 24–37.

[58] Tennakone, K., Kumara, G. R. R. A., Kumarasinghe, A. R., Wijayantha, K. G. U., & Sirimanne, P. M. (1995). A dye-sensitized nano-porous solid-state photovoltaic cell. *Semiconductor Science and Technology*, *10*(12), 1689.

27

Study on Behaviour of Sustainable Green Building Materials in Conoidal Shell Structure

P. Manikandan
Federal TVT Institute, Ethiopia, Africa

E. Veeramanipriya
Vivekanandha College of Engineering for Women (Autonomous), Thiruchengode, India

27.1 Introduction

Concrete is one of the building materials most frequently used in modern architecture, which resulted in a significant loss of natural resources [1]. However, as urbanisation picked up speed, old buildings built in the previous century faced the real possibility of being demolished. It is a substantial build-up of building waste that might have contributed to the situation known as the 'urban encirclement of waste' [2].

The use of recycled concrete in engineering has become a primary path within sustainable architectural and infrastructure development with the rise of the sustainable development ideology. In order to reduce the construction industry's dependency on virgin resources, recycled concrete predominantly obtained its constituent ingredients from the reconstruction of abandoned concrete and other construction waste. Recycled concrete additionally actively reduced carbon emissions in the building industry, which helped to mitigate unfavourable environmental effects [3].

One of the main sources of garbage produced by the building and demolition industries is concrete waste. A number of shredding procedures turn concrete trash into recycled material, which is said to be an eco-friendly and workable way to get rid of the massive number of concrete debris. The production of high-quality recycled aggregate, which can replace natural aggregate in concrete partially or completely, has been made possible by extensive research on the effects of various parameters, including the quality and replacement ratio of recycled aggregate on concrete properties and technological advancement [4–5].

Multi-recycling concrete has gained popularity in the last ten or so years as a means of achieving truly sustainable development. The idea behind concrete recycling repeatedly is this: In order to produce first-generation recycled aggregate concrete (RAC), natural aggregates are substituted with recycled coarse and fine aggregates that are obtained from crushing concrete waste, primarily natural aggregate concrete. After being crushed, the concrete yields recycled coarse and fine aggregates that can be utilised to create a second-generation recycled aggregate (RA).

The most common method is grinding up construction debris to create recycled aggregates, which may replace natural aggregate (NA) entirely or in part when making recycled concrete [6]. Aggregates have historically made up between 60% and 75% of the volume of concrete. However, over-exploitation and continued use of NA led to worries about resource scarcity and drove up prices for NA. Recycled aggregates were used in place of NA, which lessened environmental impact and effectively lowered building expenses.

Many studies have examined the mechanical [7–10], durability [11–14], and microstructural [15, 16] characteristics of periodically recycled aggregate concrete in both its fresh and hardened forms. However, research on the repeated recycling of concrete has also primarily focused on the effects of recycled coarse aggregate, with fewer studies examining the impacts of recycled fine aggregate. It has been noted that studies on concrete recycling are primarily focused on coarse aggregate [17]. In order to achieve zero

DOI: 10.1201/9781003495437-27

waste and multi-recycling, recycled concrete powder (RCP), the best recycled material that can be recovered from concrete waste must be used. RCP is inexorably formed during the concrete recycling process.

Das et al. (1991) analysed the conoidal shell structures by applying the finite difference approach, taking bending into consideration. Even with a microcomputer, which is common in design firms, the solution can be achieved with less CPU time and memory usage. According to comparative theoretical conclusions of a particular problem, a safe preliminary design of such structures can be achieved by using this method. Additionally, three conoidal shell specimens were the subject of experimental examinations, and the results show that these shells exhibit arching action as reinforcement levels rise [18].

Ghosh et al. (1993) previously presented the bending analysis of conoidal shells using curved quadratic isoparametric elements. The same formulation serves as the foundation for the current investigation into the impact of cut-outs in conoidal shells. The behaviour of the shell is examined and described in detail using five different types of cut-out instances, comparing it to a solid truncated conoidal shell under identical boundary conditions and loading intensities. The study also compares the performance of longitudinal and transverse cut-outs of the same size [19].

Another work reports on a general finite-element formulation using an eight-noded isoparametric-curved quadratic element that may be applied to the analysis of doubly curved laminated composite shell structures by Chakavitra Dipankar et al. (1994). Research has been done on the bending behaviour of composite shells in the shape of paraboloids of revolution. This work extends the use of finite-element simulation to analyse conoidal shells. For two isotropic conoidal shell issues and one laminated composite shell problem with eight distinct ply lay-ups and layers, numerical solutions are achieved. The two isotropic issues that the authors solved compare favourably to results that have been published in the literature. To determine the relative performance of the eight distinct types of the third problem for the uniformly distributed loading, the numerical results of the deflection and stress resultants are discussed. In order to examine the relative performance of two laminated composite materials, the authors also suggest using a comparative performance matrix, which will be very beneficial for designers [20].

The dynamic behaviours of conoidal shells that have been strengthened and subjected to temporary loads are studied by Nayak and associates (2006). The authors have modified their previous finite-element code to accurately perform the dynamic response analysis of stiffened conoidal shells. In order to choose the right stiffener number, orientation, and type as well as stiffener depth to shell thickness ratio, the current parametric study of clamped stiffened conoidal shells subjected to uniformly distributed transient load (Load case I) reveals the characteristic behaviour and intrinsic features of displacement response. Furthermore, it can be seen from the dynamic response analysis of stiffened conoidal shells subjected to the three distinct transient load cases examined here that, because the sine load is applied and withdrawn gradually, the step loads (Load cases I and II) are more severe than the sine load (Load case III) [21].

A ruled and aesthetically pleasing shape, the conoidal arrangement has long been favoured by civil and structural engineers (Ashwani Kota, 2013). The exploration of the many behavioural characteristics of composite conoids has been made possible by the introduction of laminated composite as the structural material. However, the main challenge in analysing these shells is the diversity in curvature. In light of the aforementioned, a finite-element analysis is performed utilising Sanders' strain displacement relationships and an eight-noded isoparametric element with five degrees of freedom per node. A wide range of composite conoidal shell problems with cross-ply laminates are solved by varying aspect ratio and degree of truncation for different stacking sequence and clamped boundary condition under uniformly distributed pressure. Benchmark problems are solved to validate the current approach. Results are shown for both full and truncated conoid, and a parametric study is used to draw some conclusions [22].

A roof is a crucial component of every building project because it protects the structural elements of any concrete structure. Concrete shells and steel frames are typically provided in long-span constructions. A solid component situated between two closely spaced covered surfaces is known as a shell structure. The shell structure's thickness "t" is determined by the separation of the two boundary surfaces. These structures have a single linear axis of formation. A pier that is either a section of a sphere or a hyperboloid is referred to as a double curvature shell, while a portion of a cylinder is known as a single curvature shell. When intermediary walls or columns must be absent from a building's interior, shells are frequently utilised.

These roofs can be flat, but they are usually curved, taking the shape of a parabolic structure, a dome, or a cylindrical form. Shells are pure compression shapes devoid of tensile forces. When compared to more conventional roofs, they use less materials overall due to their structural efficiency. In this study, conoidal shells are created using wire mesh, chicken mesh, and cement mortar. In shell structures, these meshes are regarded as reinforcement. The edge beams cast by the 6 mm diameter reinforcements balance these shells. In this research, the structural behaviour of a conoidal shell under various loading circumstances is studied using wire mesh and chicken mesh.

27.2 Material Properties

27.2.1 Fine Aggregate (Sand)

The sand is collected from Palar River bed nearby SRM University. The sand has been sieved in 4.75 mm sieve before it is used.

27.2.2 Cement

The cement used for this study is Ordinary Portland Cement and is conforming to Indian Standard IS: 8112 of Grade 53. The properties of the cement are given in Table 27.1.

27.2.3 Water

Available potable water in university premises which was free from all impurities is used for the entire work whenever it is required.

27.2.4 Steel

Steel for reinforcement Fe 415 (Tor steel) was used.

27.2.5 Aggregates

Aggregates must adhere to IS383-1970 specifications. Natural aggregate shall be referred to as much as possible. The essential components of a concrete are the aggregates given in Table 27.2. They give concrete body, lessen shrinkage, and have an impact on the economy. One of the most critical criteria for generating workable concrete is adequate gradation of particles. Good grading implies that a sample includes the necessary proportions of aggregates with a smaller number of voids. To fill in the voids in the well-graded aggregate samples with the fewest possible voids, the smallest amount of paste is needed. Less water and cement will be needed with minimum paste, which will also result in greater economy, strength, less shrinkage, and durability.

TABLE 27.1

Testing Results for Cement

Materials	Tests Conducted	Results
Cement	Fineness test	10%
	Standard consistency	31%
	Initial setting time (min)	31.30
	Final setting time (min)	450
	Specific gravity test	3.12

TABLE 27.2

Details of Specific Gravity and Sieve Analysis

Materials	Property	Results
Fine aggregate	Fineness modulus test	3.12
	Specific gravity test	2.5
	Water absorption	1.2%
Coarse aggregate (crushed stone)	Specific gravity	2.7
	Water absorption	0.6%

27.3 Experimental Setup

27.3.1 Specification Conoidal Shell

The Figure 27.1 shows the Casted conoidal shell with edge beam. The specifications are given below,

Length of conoidal shell= 910 mm
Breadth of conoidal shell = 910 mm
Rise from front = 175 mm
Rise from back = 0 mm.

27.4 Results and Discussions

27.4.1 Testing of Conoidal Shells

Conoidal shells undergo a 28-day curing and casting process. To transfer the stress from the frame to the shell, a steel frame is carefully welded. Five node points of needles are provided to transport and transfer the load via the steel frame, and the frame has nine needle points to endure the conoidal shells' over slope surface. A jack with a 25 T capacity is used to apply load to the steel frame, and the ultimate load is checked. Here, it is necessary to measure the conoidal shell's ultimate load and deflection. The testing setup should be displayed beneath as in Figure 27.2.

FIGURE 27.1 Casted conoidal shell with edge beam.

FIGURE 27.2 Testing setup for conoidal shell.

The deflection of the beam at various node points for chicken mesh (25 mm thickness) is shown in Figure 27.3. The deflection values for the beam and the conoidal shell (25 mm thickness) using chicken mesh at various points are tabulated in Table 27.3. From the results of experimental investigations on effects of load in conoidal shell, the ultimate load of conoidal shell using wire mesh value is higher than the conoidal shell with chicken mesh values.

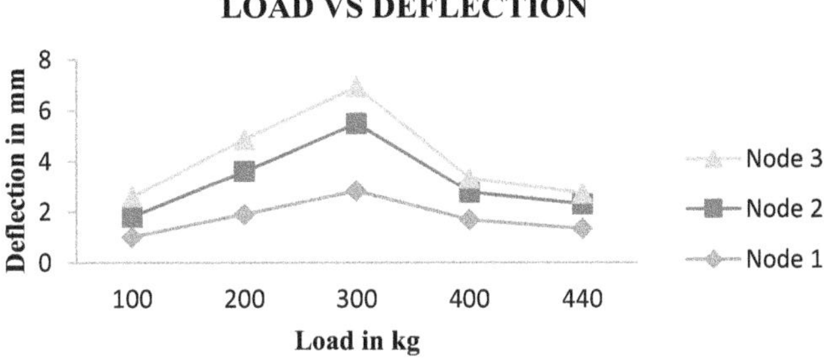

FIGURE 27.3 Deflection for 25 mm thick conoidal shell (chicken mesh).

TABLE 27.3

Deflection for 25 mm Thick Conoidal Shell (Chicken Mesh)

Load in kg	Deflection at Shell in mm			Deflection at Beam in mm		
	Node 1	Node 2	Node 3	Beam 1	Beam 2	Beam 3
100	1.01	0.80	0.79	0.1	0.09	0.1
200	1.9	1.7	1.26	0.26	0.22	0.22
300	2.82	2.66	1.45	0.40	0.36	0.41
400	1.68	1.08	0.56	0.23	0.19	0.22
440	1.33	0.97	0.43	0.21	0.18	0.23

The deflection of the beam at various node points for wire mesh (25 mm thickness) is shown in Figure 27.4. The deflection values for the beam and the conoidal shell (25 mm thickness) using wire mesh at various points are tabulated in Table 27.4. Then, 50 mm thick shell strength values are high compared to 25 mm thick shells. The strength of the shells increases according to the thickness of shells and reinforcements. The strength of the shell is increased up to 50–80%. The initial crack occurs at the conoidal shell with chicken mesh is 440 kg for 25 mm thick shells and 550 kg for 50 mm thick shells.

The deflection of the beam at the three sides of the conoidal shells using chicken mesh (50 mm thickness) is shown in Figure 27.5. The deflection values for the beam and the conoidal shell (50 mm thickness) using chicken mesh at various points are tabulated in Table 27.5. The deflection of the conoidal shell at various node points using wire mesh (50 mm thickness) is shown in Figure 27.6. The deflection values for the beam and the conoidal shell (50 mm thickness) using wire mesh at various points are tabulated in Table 27.6. The strength and ultimate load of the shell (wire mesh) are increasing compared to the shell with chicken mesh values up to 30–60%. The self-weight of the conoidal shell is high as compared to 50 mm thick shells. The value of ultimate loading is varied according to reinforcements of the shell. The deflection at the beams is similar to each side of the conoidal shell.

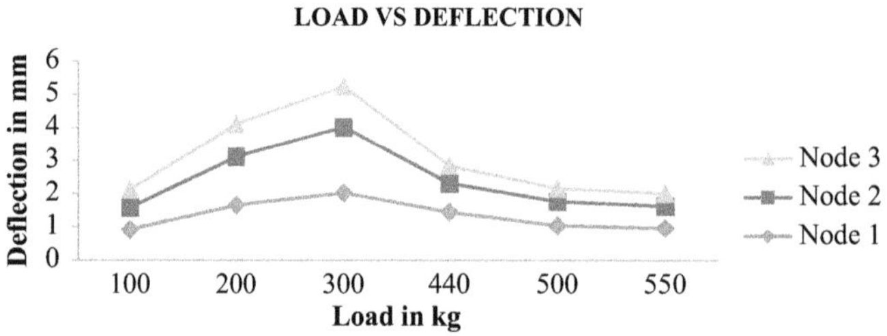

FIGURE 27.4   Deflection for 25 mm thick conoidal shell (wire mesh).

TABLE 27.4

Deflection for 25 mm Thick Conoidal Shell (Wire Mesh)

	Deflection at Shell in mm			Deflection at Beam in mm		
Load in kg	Node 1	Node 2	Node 3	Beam 1	Beam 2	Beam 3
100	0.91	0.67	0.54	0.07	0.09	0.07
200	1.65	1.46	1.00	0.16	0.17	0.17
300	2.02	1.99	1.23	0.33	0.32	0.36
440	1.44	0.87	0.53	0.29	0.25	0.29
500	1.03	0.73	0.40	0.26	0.20	0.24
550	0.96	0.66	0.39	0.20	0.18	0.15

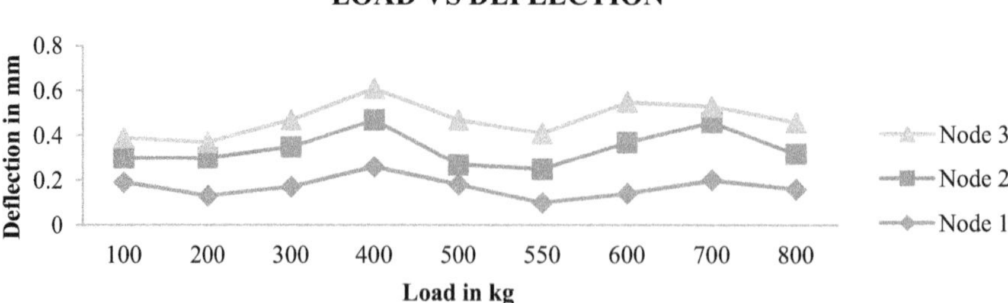

FIGURE 27.5   Deflection for 50 mm thick conoidal shell (chicken mesh).

TABLE 27.5

Deflection for 50 mm Thick Conoidal Shell (Chicken Mesh)

Load in kg	Deflection at Shell in mm			Deflection at Beam in mm		
	Node 1	Node 2	Node 3	Beam 1	Beam 2	Beam 3
100	0.19	0.11	0.09	0.19	0.17	0.14
200	0.13	0.17	0.07	0.21	0.18	0.18
300	0.17	0.18	0.12	0.20	0.11	0.17
400	0.26	0.21	0.14	0.15	0.15	0.09
500	0.18	0.09	0.20	0.10	0.14	0.17
550	0.10	0.15	0.16	0.18	0.10	0.19
600	0.14	0.23	0.18	0.22	0.19	0.21
700	0.20	0.26	0.07	0.13	0.13	0.15
800	0.16	0.16	0.14	0.17	0.16	0.14

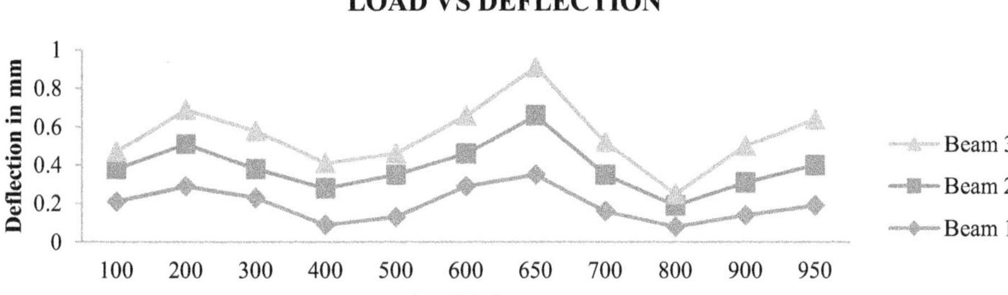

FIGURE 27.6 Deflection for 50 mm thick conoidal shell (wire mesh).

TABLE 27.6

Deflection for 50 mm Thick Conoidal Shell (Wire Mesh)

Load in kg	Deflection at Shell in mm			Deflection at Beam in mm		
	Node 1	Node 2	Node 3	Beam 1	Beam 2	Beam 3
100	0.10	0.09	0.09	0.21	0.17	0.09
200	0.06	0.08	0.07	0.29	0.22	0.18
300	0.05	0.08	0.03	0.23	0.15	0.20
400	0.20	0.60	0.09	0.09	0.19	0.13
500	0.13	0.17	0.07	0.13	0.22	0.11
600	0.12	0.15	0.11	0.29	0.17	0.20
650	0.06	0.10	0.06	0.35	0.31	0.25
700	0.01	0.09	0.04	0.16	0.19	0.17
800	0.09	0.19	0.13	0.08	0.11	0.06
900	0.12	0.17	0.05	0.14	0.17	0.19
950	0.10	0.15	0.09	0.19	0.21	0.24

27.4.2 X-Ray Diffraction (XRD) Analysis

All of the spectra were normalised in order to make comparisons between the crystalline phases easier and more accurate. The main peak of the mixes of conoidal shell (chicken mesh) is illustrated in Figure 27.7 which is located at the scattering angle (2θ) of $24.6°$. For conoidal shell (wire mesh), it is followed by $24.2°$ and $58°$ and is shown in Figure 27.8.

Using wire mesh and chicken mesh in the recycled concrete did not result in the creation of any new crystalline phases. The figure indicates that the cumulative intensity of the peaks in the conoidal shell using wire mesh is greater than that of the conoidal shell using chicken mesh. Additionally, it is evident that when the conoidal shell using wire mesh is exposed to up to 0.2%, peak intensities grow. This observation is consistent with the findings of the scanning electron microscopy (SEM) investigation and may be explained by the conoidal shell (wire mesh) uniform pore filling and aggregation behaviour at low and high doses, respectively.

27.4.3 SEM Analysis

SEM in conjunction with energy-dispersive spectroscopy (SEM-EDS) is used to do a thorough analysis of the surface morphology and of a variety of RCAs that are produced through serial recycling of concrete. Extra evidence for this identification process came from similarities among the morphological characteristics that are derived from SEM data reported in the literature.

Upon closer examination of Figure 27.9(a) and (b), it can be observed that the chicken mesh RCA is populated by some unevenly spread portlandite crystals and clusters of flaky that resemble the foil. An increase in the replacement rate of wire mesh RA in concrete in Figure 27.10(a) and (b) resulted in an improvement in its frost resistance; however, an excessively high replacement rate could cause a

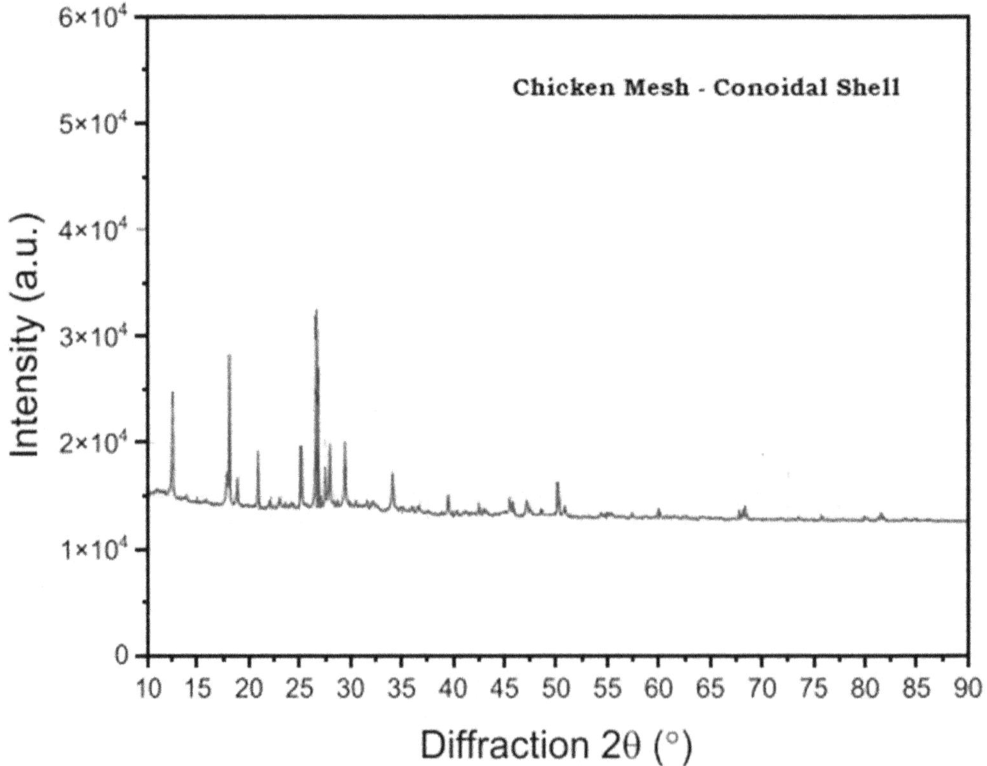

FIGURE 27.7 XRD pattern for chicken mesh conoidal shell.

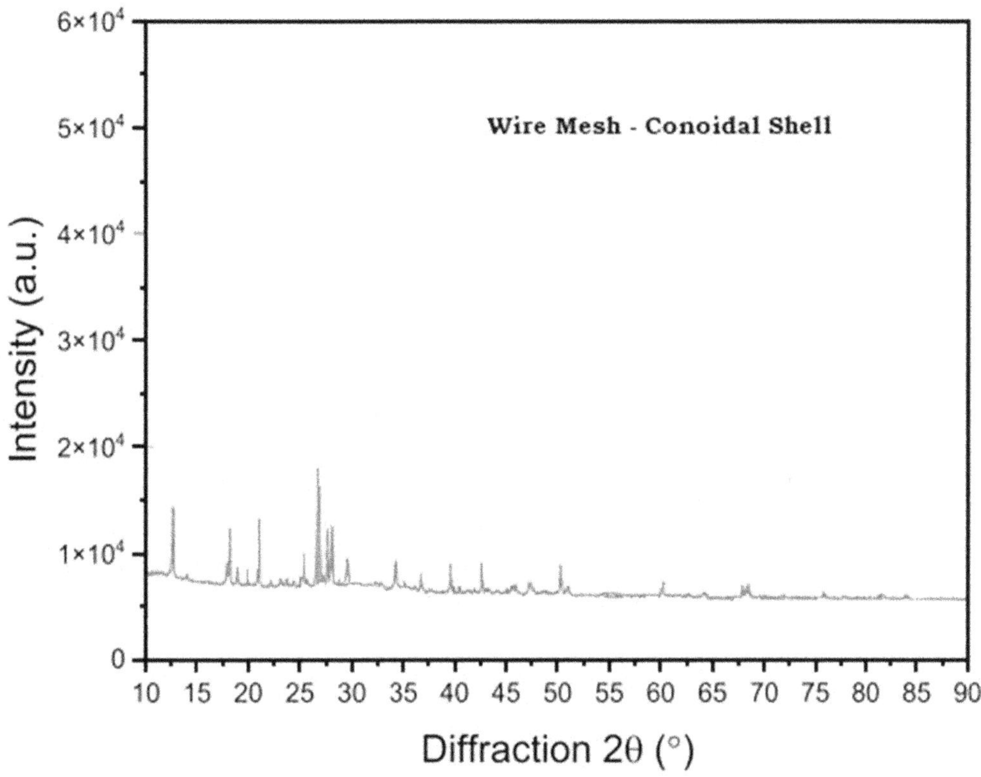

FIGURE 27.8 XRD pattern for wire mesh conoidal shell.

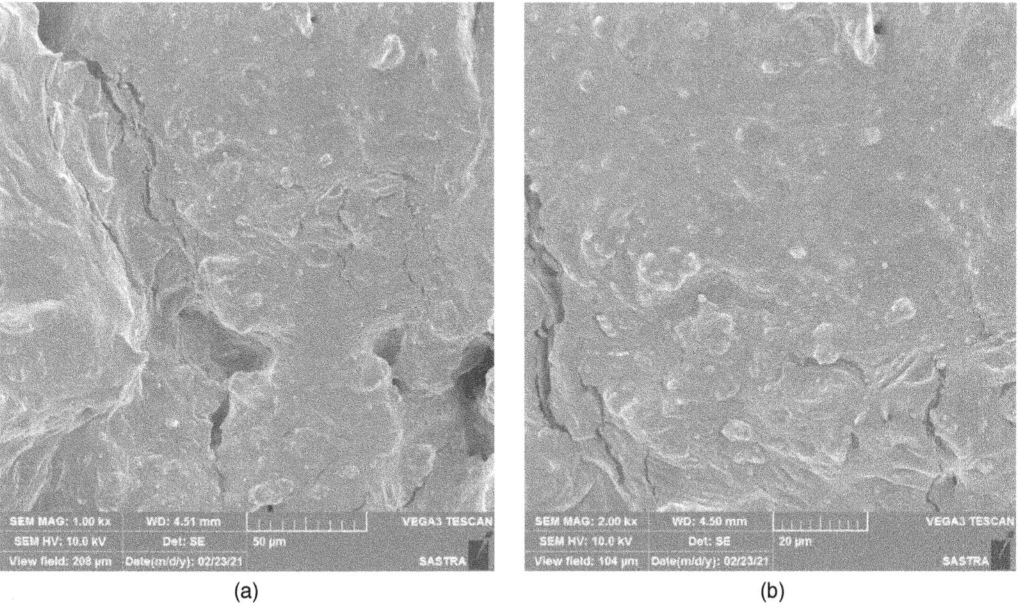

FIGURE 27.9 (a) SEM image of the conoidal shell (chicken mesh) at 1000 X and (b) SEM image of the conoidal shell (chicken mesh) at 2000 X.

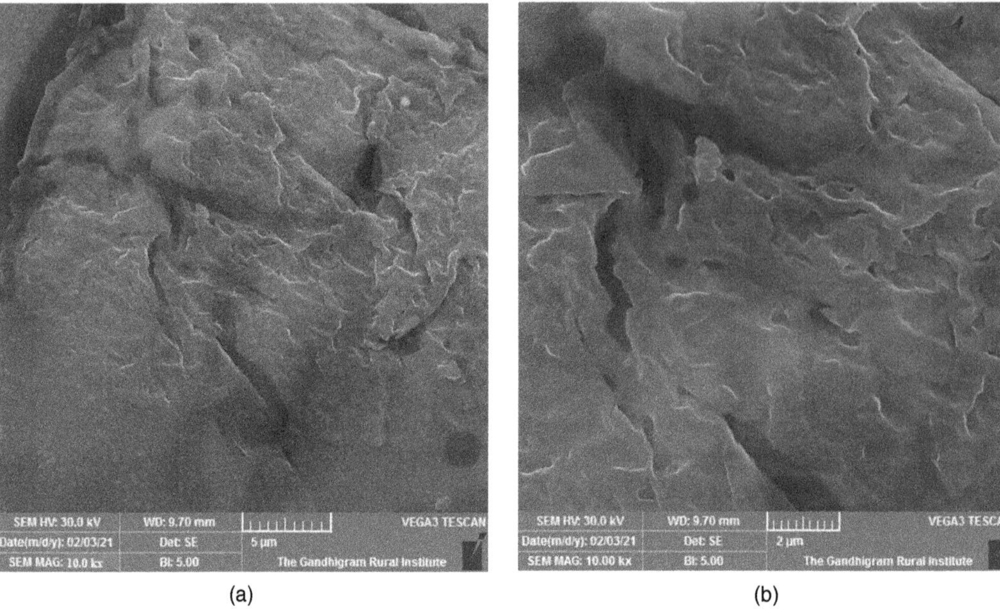

(a)   (b)

FIGURE 27.10   (a) SEM image of the conoidal shell (wire mesh) at 1000 X and (b) SEM image of the conoidal shell (wire mesh) at 10 KX.

significant decrease in the concrete's initial strength. Consequently, frost resistance and good mechanical performance could be achieved at the same time when a wire mesh RCA replacement rate of between 30% and 60% was maintained.

27.5 Conclusion

The outcomes of experimental studies on the effects of load in conoidal shells led to the following findings. Conoidal shells with wire mesh values have a larger ultimate load than those with chicken mesh values. Strength numbers for 50 mm thick shells are higher than those for 25 mm thick shells. The thickness of the reinforcements and shells both increase the strength of the shells. Up to 50–80% more strength is added to the shell. The first fracture at the conoidal shell with chicken mesh weighs 440 kg for a shell that is 25 mm thick and 550 kg for a shell that is 50 mm thick. When shells with chicken mesh values of up to 30–60% are compared, the ultimate load and strength of the shell increase.

REFERENCES

[1] Cachim P.B., 2009, Mechanical properties of brick aggregate concrete, *Constr. Build. Mater. 23* (3) 1292–1297.

[2] Meng T., Wei H., Dai D., 2022, Effect of brick aggregate on failure process of mixed recycled aggregate concrete via X-CT, *Constr. Build. Mater. 327*, 126934.

[3] Rosa L., Becattini V., Gabrielli P., 2022, Carbon dioxide mineralization in recycled concrete aggregates can contribute immediately to carbon-neutrality, *Resour. Conserv. Recycl. 184*, 106436.

[4] Kim J., 2021, Construction and demolition waste management in Korea: Recycled aggregate and its application, *Clean Techn. Environ. Policy.* 2223–2234, https://doi.org/10.1007/s10098-021-02177-x

[5] Silva R.V., Jiménez J.R., Agrela F., de Brito J., 2019, Real-scale applications of recycled aggregate concrete, in: *New Trends Eco-Efficient Recycle Concrete*, Elsevier, pp. 573–589. https://doi.org/10.1016/B978-0-08-102480-5.00021-X

[6] Kumar B.M.V., Ananthan. H., Balaji K.V.A., 2017, Experimental studies on cement stabilized masonry blocks prepared from brick powder, fine recycled concrete aggregate and pozzolanic materials, *J. Build. Eng. 10*, 80–88.

[7] Huda S.B., Alam M.S., 2014, Mechanical behavior of three generations of 100% repeated recycled coarse aggregate concrete, *Constr. Build. Mater. 65*, 574–582, https://doi.org/10.1016/j.conbuildmat.2014.05.010

[8] Kim J., Yang S., Kim N., 2023, Effect of plasticizer dosage on properties of multiple recycled aggregate concrete, *J. Mater. Cycles Waste Manag. 25* (3), 1457–1469, https://doi.org/10.1007/s10163-023-01624-9

[9] Abreu V., Evangelista L., de Brito J., 2018, The effect of multi-recycling on the mechanical performance of coarse recycled aggregates concrete, *Constr. Build. Mater. 188*, 480–489, https://doi.org/10.1016/j.conbuildmat.2018.07.178

[10] Salesa A., Pérez-Benedicto J.A., Esteban L.M., Vicente-Vas R., Orna-Carmona M., 2017, Physico-mechanical properties of multi-recycled self-compacting concrete prepared with precast concrete rejects, *Constr. Build. Mater. 153*, 364–373, https://doi.org/10.1016/J.CONBUILDMAT.2017.07.087

[11] Zhu P., Hao Y., Liu H., Wei D, Liu L., Gu S., 2019, Durability evaluation of three generations of 100% repeatedly recycled coarse aggregate concrete, *Constr. Build. Mater. 210*, 442–450, https://doi.org/10.1016/j.conbuildmat.2019.03.203

[12] Liu H., Hua M., Zhu P., Chen C., Wang X., Qian Z., Dong Y., 2021, Effect of freeze-thaw cycles on carbonation behavior of three generations of repeatedly recycled aggregate concrete, *Appl. Sci. 11*, 2643, https://doi.org/10.3390/app11062643

[13] Liu H., Zhu X., Zhu P., Chen C., Wang X., Yang W., Zong M., 2022, Carbonation treatment to repair the damage of repeatedly recycled coarse aggregate from recycled concrete suffering from coupling action of high stress and freeze-thaw cycles, *Constr. Build. Mater. 34*, 128688, https://doi.org/10.1016/j.conbuildmat.2022.128688

[14] Silva S., Evangelista L., de Brito J., 2021, Durability and shrinkage performance of concrete made with coarse multi-recycled concrete aggregates, *Constr. Build. Mater. 272*, 121645, https://doi.org/10.1016/J.CONBUILDMAT.2020.121645

[15] Thomas C., de Brito J., Cimentada A., Sainz-Aja J.A., 2020, Macro- and micro- properties of multi-recycled aggregate concrete, *J. Clean. Prod. 245*, 118843, https://doi.org/10.1016/j.jclepro.2019.118843

[16] Thomas C., de Brito J., Gil J., Sainz-Aja J.A., Cimentada A., 2018, Multiple recycled aggregate properties analysed by X-ray microtomography, *Constr. Build. Mater. 166*, 171–180, https://doi.org/10.1016/j.conbuildmat.2018.01.130

[17] Villagran-Zaccardi Y.A., Marsh A.T.M., Sosa M.E., Zega C.J., De Belie N., Bernal S.A., 2022, Complete re-utilization of waste concretes–Valorisation pathways and research needs, *Resour. Conserv. Recycl. 177*, 105955, https://doi.org/10.1016/J.RESCONREC.2021.105955

[18] Das A.K., Bandyopadhyay J.N., 1993, Theoretical and experimental studies on conoidal shells, *Comput. Struct. 49*(3), 531–536.

[19] Ghosh A., Chakravorty D., 2017, Failure analysis of civil engineering composite shell roofs, *Science Direct 173*, 1642–1649.

[20] Chakravorty D., Bandyopadhyay J.N., 1994, Effects of release of boundary constraints on the natural frequencies of clamped, thin, cylindrical shells, *Computers and Structures*, 52(3):489–493.

[21] Nayak A.N., Bandyopadhyay J.N., 2002, Free vibration analysis and design aids of stiffened conoidal shells, *J. Eng. Mech.*, 128(4), 419–427.

[22] Kota A., 2013, Effect of aspect ratio and degree of truncation on bending behaviour of laminated composite conoidal shell roof, *Int. J. Innov. Res. Sci. Eng. Technol.*, 2(6).

28

Strategic and Systematic Approach of Nanomaterials as well as Simulation Studies to Enhance Cancer Therapy

M. Ayisha Zeenath and R. Gayathri
Cauvery College for Women (Autonomous) (affiliated to Bharathidasan University), Trichy, India

28.1 Introduction

Every nation on earth ranks cancer as the greatest cause of mortality. Nearly 10 million individuals died from cancer in 2020, and, by 2040, that number is predicted to increase to 16.3 million [1, 2]. To identify cures for cancer, researchers worldwide working extremely hard. Additionally, the emergence of multidrug-resistant cells contributes to the failure of numerous cancer treatments. Consequently, not every patient can benefit from the same course of treatment. Because cancer may develop a resistance to treatments that once effectively healed, recurrence is frequently a possibility [3, 4]. First-line cancer treatments include chemotherapy, radiation, immunotherapy, and surgical intervention. Over the past several decades, they have remained the most efficient ways to treat cancer [5]. Despite this, they have significant drawbacks, such as limited tumour selectivity, off-target toxicity, systemic toxicity, multi-drug resistance, and significant adverse effects on human health [6, 7]. Nanomaterials have recently drawn much interest from researchers focusing on cancer therapy because of their distinct chemical and physical properties [8]. Nanoparticles are minuscule objects according to some definitions, with a size of 1–100 nm, or even up to 1 micron. Compared to traditional cancer therapies, nanoparticle delivery technologies in cancer therapies offer higher absorption of chemicals used in medicine and diagnosis within the human body at lower risk [9, 10]. Nanomedicine's long-term objective is to transform the medical field by more effectively combating fatal diseases [11, 12]. Additionally, the components of cells actively interact with nanoparticles of the cell due to their small size, which is analogous to that of the biological system, and has a potential future in biological applications, both in vivo and in vitro. The integration of biological with nanotechnology has been hailed as a groundbreaking technological advancement with a plethora of uses, such as biosensing, drug delivery systems (DDSs), specialised therapeutic processes, and diagnostic instruments [13, 14]. Liposomes [15], dendrimers [16], carbon nanotubes (CNTs) [17], inorganic [18, 19], and polymer-based NPs [20] are among the many diverse forms of NPs, each with its own set of features. The use of nanoparticles as transporters for therapeutic compounds has been the subject of numerous reports. The following sub-categories of these methods include photodynamic, photothermal, magnetic, and neutron-capturing devices [21]. Although experimental attempts to quantify, arrange, optimise the dynamics and structure of NPs are becoming increasingly sophisticated, theoretical and practical limitations will always exist. For instance, many DDSs exhibit promise in vitro but fail in vivo, which presents a challenge for DDS development. This is primarily due to the inadequate mechanistic understanding offered by experimentation and failure experiments [22]. Theoretical techniques, both computational and analytical, can enable initial variable evaluating to predict favourable circumstances for additional research [23]. In addition to supporting the logical creation of innovative compositions with higher efficacies, computer simulation can be used to supplement tests. In this review, it has been shown that strategic and systematic approaches to nanomaterials combined with simulation studies have

DOI: 10.1201/9781003495437-28

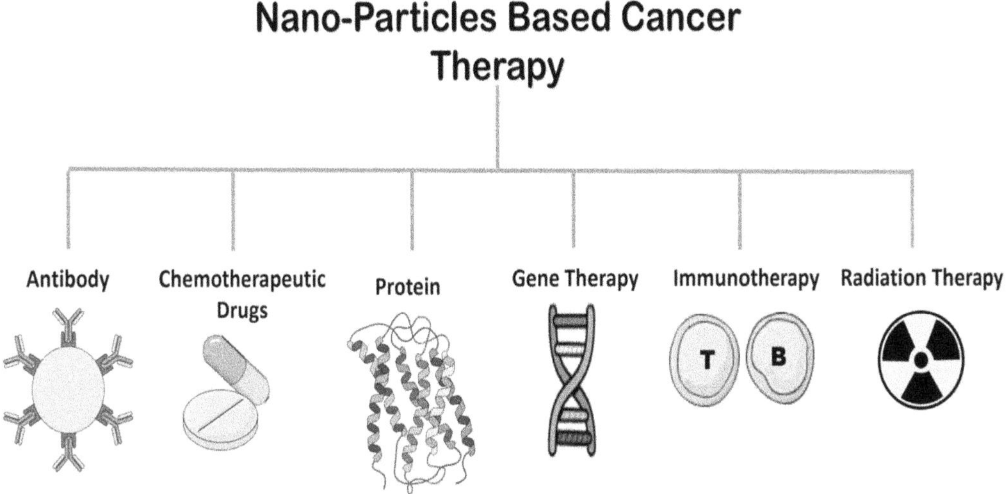

FIGURE 28.1 Applications of nanoparticles in cancer therapy.

enormous potential to revolutionise cancer treatment. Researchers want to develop more individualised, personalised, and targeted therapy approaches for improved patient outcomes in the fight against cancer by utilising computational modelling to make use of special features of nanomaterials. Figure 28.1 shows the application of nanoparticles in cancer therapy.

28.2 Nanomaterials in Cancer Therapy

Nanotechnology in cancer therapy may go beyond the administration of drug delivery and entail the creation of innovative treatments that can eradicate tumours with minimal damage to the surrounding healthy tissues and organs in addition to the early identification and elimination of cancerous cells. Binding to specific target molecules, ligands, nucleic acids, peptides, or antibodies can also confer extra functionality upon these comparatively small particles. Here, we have listed multifunctional nanomaterials that have been utilised to successfully target and destroy cancer cells, as well as some theories about how they can work as a cancer treatment (Figure 28.2).

28.2.1 Noble Metal Nanoparticles

Noble metal NPs in particular are adaptable substances with many different medicinal purposes including medication and gene transport, thermal ablation, radiation enhancement, and highly sensitive diagnostic assays. Noble metal NPs have also been suggested as harmless carriers for applications involving the delivery of drugs and genes. Theranostics, which explores the unique properties of nanoparticle-based systems for better therapeutic moiety penetration and tracking within the body, can provide simultaneous diagnostic and therapeutic services, making them more effective than conventional therapies [24]. Currently, gold, silver, platinum, copper, lead, and other noble metal elements are the principal components of the noble metal nanoparticles that are frequently utilised in the treatment of tumours [25]. Figure 28.3 shows the types of noble-metal-based nanoparticles and their mode of action.

28.2.1.1 Gold Nanoparticles

Gold nanoparticles (GNPs) have garnered a lot of interest from researchers lately due to their exceptional features. The electrochemical approach [26], the two-phase method [27], and the seed growth method

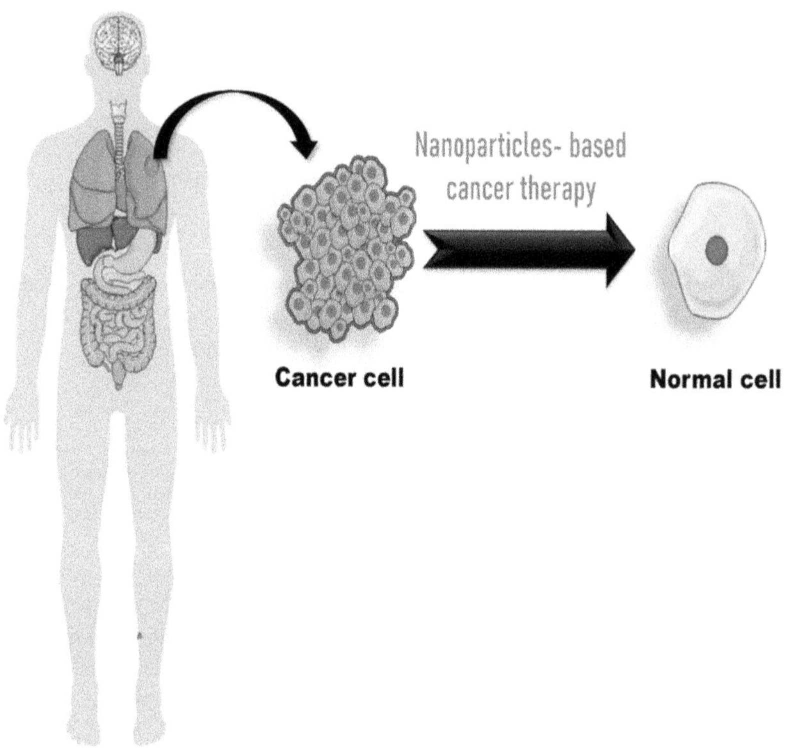

FIGURE 28.2 Nanoparticles-based cancer therapy.

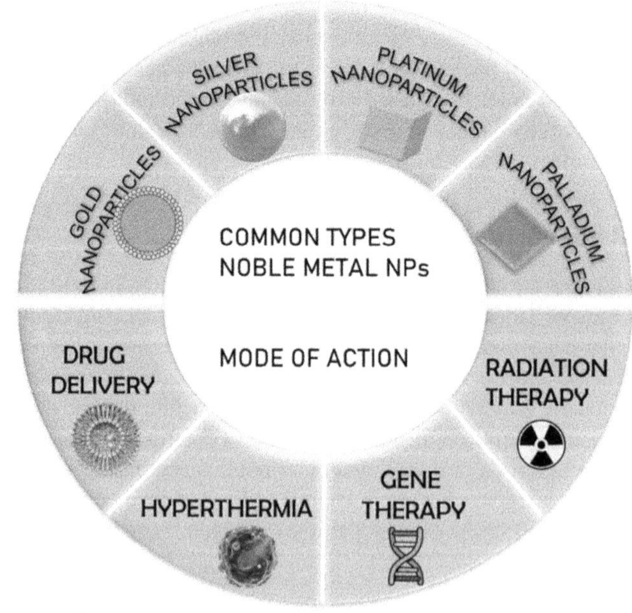

FIGURE 28.3 Types and applications of noble metal nanoparticles.

[28] are now effective techniques frequently utilised for the production of GNPs. Applications for GNPs in biomedicine are common due to their colloidal stability, high tissue permeability, small particle size, and good monodispersity [29]. In addition to causing hyperthermia and delivering anticancer drugs, they can be utilised to cure tumours via preventing angiogenesis. Conversely, GNPs possess a localized surface plasmon resonance (LSPR) effect that effectively absorbs light energy in the near-infrared (NIR) and converts it to thermal energy. To kill cancer cells, hyperthermia is applied to the tumour location to cause cell apoptosis and protein denaturation [30]. Common gold nanomaterial structures include nanowires, nanorods, nanospheres, and nanocubes [31]. Figure 28.4 shows the various morphologies of GNPs. GNPs are widely used in biomedical disciplines because of their small particle size, good monodispersity, high tissue permeability, and colloidal stability [29]. By preventing angiogenesis, causing hyperthermia, and delivering anticancer medications, they can be utilised to treat tumours. To kill cancer cells, hyperthermia is applied to the tumour location. This causes cell apoptosis and protein denaturation [30]. Gold nanorods or nanospheres are employed both functionalised and as the matrix modifications are applied to their surface in the design and production of anticancer medicines that can be used for photothermal treatment [32]. Additionally, GNPs can be used as DDSs to load a variety of pharmaceuticals, enhance their bioavailability, and enable controlled release of the medication. This can be done directly or after surface modification and drug combination through covalent or noncovalent bonding [33]. Sun et al. [34] used positron emission tomography (PET) for optical imaging while using polyethylene glycol (PEG) and 64Cu to modify the surface of gold nanorods as the carrier. This DDS demonstrated strong targeting capabilities, allowing for the realisation of individualised treatment while avoiding side effects. Future research will continue to focus on achieving economic security, high efficacy, and low toxicity of GNPs.

28.2.1.2 Silver Nanoparticles

Silver nanoparticles (AgNPs) have a slightly lower stability than GNPs and will gradually oxidise in oxygen-containing fluids. AgNPs have demonstrated strong antibacterial activity, outstanding LSPR and surface-enhanced Raman scattering (SERS) effects capabilities, and catalytic performance, making them the most studied inorganic nanomaterials. Low-temperature nonequilibrium contact is one of the most creative and sustainable ways to make silver nanoparticles [35, 36]. Silver nanowires, nanocubes, and

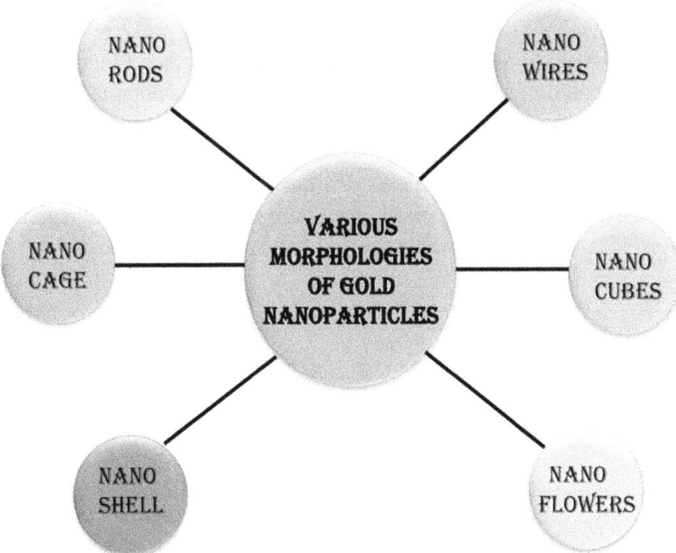

FIGURE 28.4 Various morphologies of gold nanoparticles.

nanospheres are currently prevalent silver nanostructures. By altering the size, shape, and structure of silver nanoparticles, ideal optical characteristics can be attained [37]. The amount of research on using silver as a bactericidal agent to treat tumours has increased dramatically during the 1990s. It is crucial to keep in mind that the oxidative stress of silver nanoparticles mediates the toxicity of AgNPs rather than coming directly from free Ag+ [38]. According to Gurunathan et al. [39], silver nanoparticles can cause oxidative stress by releasing reactive oxygen species(ROS) in cells, which causes cancer cells to become toxic, undergo apoptosis, and die. This discovery opens up a new avenue to cure tumours by utilising silver NPs' anti-bacterial properties.

28.2.1.3 Platinum Nanoparticles

Medical experts have paid close attention to the usage of compound metals in the medicine of tumours ever since researchers first discovered cisplatin's anticancer action in 1969 [40]. Gold, silver, platinum, palladium, rhodium, and other noble metal compounds have been created based on cisplatin research and can be used to treat and diagnose cancer. Organo-platinum compounds are biological alkylating agents used in anticancer medications [41], which can cause tumour cells to undergo apoptosis and necrosis by altering the target DNA's structure and impairing cell division. According to studies, platinum metal nanoparticles can boost the effectiveness of treatment against tumours by modulating several communication pathways and stimulating the immune system. Because of their superior catalytic performance, imaging capability, outstanding photothermal conversion ability, and radio-sensitisation ability, platinum nanoparticles have been used extensively in medical purposes [42, 43]. Platinum nanoparticles are a good candidate for cancer therapy because of their adaptability and great selectivity. However, acute kidney damage (AKI) is a common side effect of treatment for platinum compounds [44, 45].

28.2.1.4 Palladium Nanoparticles

Palladium nanomaterials frequently replace GNPs in the therapy of tumours due to their superior photo-thermal stability, superior catalytic activity, and significant cost advantages [46]. Huang et al.'s ultrathin hexagonal palladium nanosheets [47] demonstrated remarkable stability in a two-dimensional form when illuminated by NIR light. Palladium nanoparticles could also significantly enhance photothermal conversion capability and biocompatibility when compared to gold/silver nanostructures. Studies have demonstrated that palladium nanoparticles' biocompatibility, aqueous dispersion, and physiological stability can all be markedly enhanced by functionalising their surfaces with polymers [46]. The performance of palladium nanoparticles depends on the synthesis process, and surface functionalisation can considerably increase the catalytic activity and utilisation rate. Palladium nanoparticles are currently prepared using a variety of techniques, most notably the chemical reduction approach, the biological reduction method, and others. The chemical reduction method is frequently coupled with other preparation techniques because it is an effective and established conventional technique [48]. Palladium nanoparticles have enormous advantage for research and use in the realm of curing cancer due to their well-developed preparation processes, significant cost advantages, and good optical characteristics.

28.3 Carbon-Based Nanoparticles

Because of their unique physicochemical characteristics and safety, carbon-based nanomaterials are extensively employed in biological applications, including for the cure of cancer. This improves their capacity to supply drugs to the cell without having any unfavourable side effects. Some of the most promising materials for targeted medication delivery are those that can be surface functionalised with chemical and biological molecules. Carbon nanomaterials have significant NIR absorption, which is advantageous for photothermal tumour ablation. These materials also can emit heat in a radiofrequency field, which can be used to heat cancer cells to death. They can, therefore, be used in photodynamic and photothermal therapies. Activated carbon, fullerenes, CNTs, carbon nanofibers, carbon black, and other carbon-based nanomaterials can all be categorised [4]. Figure 28.5 shows carbon-based nanoparticles

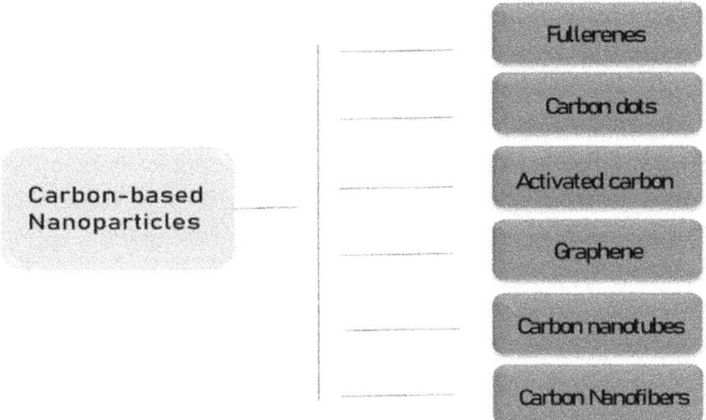

FIGURE 28.5 Carbon-based nanoparticles for cancer therapy.

for cancer therapy. The remarkable biocompatibility, biodegradability, and versatile applications of these nanomaterials in biosensing, bioimaging, diagnostics, and treatments have further increased their promise for biomedical applications [49].

28.4 Conjugated Polymer Nanoparticles

Conjugated polymer nanoparticles (CPNs) have become a novel family of potentially effective cancer theragnostic agents because of their favourable optical properties, such as their photothermal and photodynamic qualities, high absorption coefficients, and photoluminescence quantum yields ranging from ultraviolet to near-infrared [50]. CPNs have been used to produce real-time diagnostics in both in vitro and in vivo fluorescence imaging due to their superior light-harvesting and light-amplifying characteristics [51–53]. Quantum dots (QDs), silica nanoparticles, and organic fluorescent dyes are examples of common fluorescence imaging agents. Photo-bleaching and cellular toxicity are two common problems with fluorescent dyes. QDs are cytotoxic because they leach hazardous metals from their core of nanocrystals while having better brightness and photostability. These details serve as inspiration for the development of novel materials fluorescent for fluorescence photography imaging. One effective tactic is the creation of fluorescent-conjugated polymer materials, which has proven to be effective platforms for sensitive sensing due to the signal amplification caused by a system response [54–57]. Additionally, for prospective uses in combination cancer therapy, such conjugated polymers could act as multifunctional drug carriers [58].

28.5 Simulation Approach

In the near future, computational tools for drug development will frequently use molecular dynamics (MD) and related techniques. As a result, it is possible to estimate thermodynamics and kinetics linked to drug-target identification and binding with increasing precision as more sophisticated algorithms and hardware architecture are used. The theoretical foundation of MD and enhanced sampling techniques, emphasising free-energy perturbation, meta dynamics, guided MD, and other approaches frequently employed to investigate drug-target binding [59]. The proper analysis of the nanoparticle's potential interactions with biological surfaces should be a part of the computational design of nanoparticles. MD and free energy calculations are two approaches that, at least computationally, provide a better way to simulate those interactions. Other techniques, like quantitative structure–activity relationship (QSAR), may be more efficient at eliminating functional groups from the vast array, allowing for a closer look at MD. Free energy calculation, particularly umbrella sampling and the adaptive biassing force (ABF) method, is one of the most researched methods for understanding nanoparticle adsorption onto membranes. Comer et al. recently predicted the adsorption affinity of 30 small aromatic compounds with various functional groups on CNTs that are both bare and hydroxylated utilising the ABF approach. With this strategy, it is shown that in silico methods can be successfully used for the creation of novel nano-materials. As it shall be noted before, theoretical methods like QSAR, docking, and then MD simulations have been used to a considerable extent to estimate medication dosage and product durability. To prevent losing the pharmacological activity of drug-loaded liposomes, leaking is recommended as a crucial component that should be assessed during storage. It was discovered that 67 FDA-approved medications satisfied the requirements to be loaded in such liposomes. Different drug positions in nanospheres, including those for curcumin, paclitaxel, and vitamin D3, were determined using the docking method. According to the binding affinities study, the quantity of hydrogen bonds, the drug's hydrophobicity, the number of aromatic contacts, etc. were all connected. Since it becomes extremely important for drug administration and release, modelling nanoparticles under various pH settings is a common area of study. For instance, solid tumours in cancer are demonstrated to have modified pH gradients inside their individual cell divisions and an acidic extracellular environment. Therefore, the status of nanoparticles at a certain pH must be adequately represented in molecular models [60–68]. Computational techniques can be used to expedite and improve the target and drug discovery process. Furthermore, there exist novel and efficacious computational techniques for DDS simulation, which are now undergoing further development. Nevertheless, laboratory experiments will never fully replace computational searches.

28.6 Conclusion

This chapter discussed the strategic and systematic approach of nanomaterials in combination with simulation studies that have the potential to significantly advance cancer therapy. The nanomaterials include a noble metal, carbon-based, conjugated polymer that holds the potential to enhance cancer therapy. Nanoparticle-based therapy is covered in length as one of the most significant and well-researched steps, and we have highlighted the crucial elements that affect how successfully it interacts with tumour cells. Preclinical and clinical evidence currently available supports the idea that NPs can offer a way to deliver medications to certain cancer targets at a sustained rate. These targeted NPs will offer the enhanced cancer treatment choices that are desperately needed after being optimised. However, contemporary researchers are highlighting the challenges, and a number of endeavours are being made to further simulation studies in the real-world uses of nanotherapeutic platforms for the treatment of cancer.

REFERENCES

[1] Sung, H., Ferlay, J., Siegel, R. L., Laversanne, M., Soerjomataram, I., Jemal, A., & Bray, F. (2021). Global cancer statistics 2020: GLOBOCAN estimates of incidence and mortality worldwide for 36 cancers in 185 countries. *CA: A Cancer Journal for Clinicians, 71*(3), 209–249.

[2] Ferlay, J., Ervik, M., Lam, F., Colombet, M., Mery, L., Piñeros, M., ... & Bray, F. (2018). Global cancer observatory: Cancer today. *Lyon, France: International Agency for Research on Cancer, 3*(20), 2019.

[3] Bukhari, S. Z., Zeth, K., Iftikhar, M., Rehman, M., Munir, M. U., Khan, W. S., & Ihsan, A. (2021). Supramolecular lipid nanoparticles as delivery carriers for non-invasive cancer theranostics. *Current Research in Pharmacology and Drug Discovery, 2*, 100067.

[4] Kashyap, B. K., Singh, V. V., Solanki, M. K., Kumar, A., Ruokolainen, J., & Kesari, K. K. (2023). Smart nanomaterials in cancer theranostics: Challenges and opportunities. *ACS Omega, 8*(16), 14290–14320.

[5] Hanna, N. H., & Einhorn, L. H. (2014). Testicular cancer—Discoveries and updates. *New England Journal of Medicine, 371*(21), 2005–2016.

[6] Birgisson, H., Påhlman, L., Gunnarsson, U., & Glimelius, B. (2007). Late adverse effects of radiation therapy for rectal cancer–A systematic overview. *Acta Oncologica, 46*(4), 504–516.

[7] Rios Velazquez, E., Parmar, C., Liu, Y., Coroller, T. P., Cruz, G., Stringfield, O., ... & Aerts, H. J. (2017). Somatic mutations drive distinct imaging phenotypes in lung cancer somatic mutations and radiomic phenotypes. *Cancer Research, 77*(14), 3922–3930.

[8] Palazzolo, S., Hadla, M., Spena, C. R., Bayda, S., Kumar, V., Lo Re, F., ... & Rizzolio, F. (2019). Proof-of-concept multistage biomimetic liposomal DNA origami nanosystem for the remote loading of doxorubicin. *ACS Medicinal Chemistry Letters, 10*(4), 517–521.

[9] Praetorius, N. P., & Mandal, T. K. (2007). Engineered nanoparticles in cancer therapy. *Recent Patents on Drug Delivery & Formulation, 1*(1), 37–51.

[10] McNeil, S. E. (2005). Nanotechnology for the biologist. *Journal of Leukocyte Biology, 78*(3), 585–594.

[11] Giljohann, D. A., & Mirkin, C. A. (2009). Drivers of biodiagnostic development. *Nature, 462*(7272), 461–464.

[12] Bhattacharyya, S., Kudgus, R. A., Bhattacharya, R., & Mukherjee, P. (2011). Inorganic nanoparticles in cancer therapy. *Pharmaceutical Research, 28*, 237–259.

[13] Naveed, N. (2017). Nanomedicine-the solution to modern medicine's unsolved problems. *Journal of Science and Technology, 08*, 5140–5143.

[14] Guo, J., Rahme, K., He, Y., Li, L. L., Holmes, J. D., & O'Driscoll, C. M. (2017). Gold nanoparticles enlighten the future of cancer theranostics. *International Journal of Nanomedicine, 12*, 6131.

[15] Pattni, B. S., Chupin, V. V., & Torchilin, V. P. (2015). New developments in liposomal drug delivery. *Chemical Reviews, 115*(19), 10938–10966.

[16] Gillies, E. R., & Frechet, J. M. (2005). Dendrimers and dendritic polymers in drug delivery. *Drug Discovery Today, 10*(1), 35–43.

[17] Bianco, A., Kostarelos, K., & Prato, M. (2005). Applications of carbon nanotubes in drug delivery. *Current Opinion in Chemical Biology, 9*(6), 674–679.

[18] Ghosh, P., Han, G., De, M., Kim, C. K., & Rotello, V. M. (2008). Gold nanoparticles in delivery applications. *Advanced Drug Delivery Reviews, 60*(11), 1307–1315.

[19] Kim, C. S., Tonga, G. Y., Solfiell, D., & Rotello, V. M. (2013). Inorganic nanosystems for therapeutic delivery: Status and prospects. *Advanced Drug Delivery Reviews, 65*(1), 93–99.

[20] Soppimath, K. S., Aminabhavi, T. M., Kulkarni, A. R., & Rudzinski, W. E. (2001). Biodegradable polymeric nanoparticles as drug delivery devices. *Journal of Controlled Release, 70*(1–2), 1–20.

[21] Li, J., Fan, C., Pei, H., Shi, J., & Huang, Q. (2013). Smart drug delivery nanocarriers with self-assembled DNA nanostructures. *Advanced Materials, 25*(32), 4386–4396.

[22] Zhang, Y., Chan, H. F., & Leong, K. W. (2013). Advanced materials and processing for drug delivery: The past and the future. *Advanced Drug Delivery Reviews, 65*(1), 104–120.

[23] Huynh, L., Neale, C., Pomès, R., & Allen, C. (2012). Computational approaches to the rational design of nanoemulsions, polymeric micelles, and dendrimers for drug delivery. *Nanomedicine: Nanotechnology, Biology and Medicine, 8*(1), 20–36.

[24] Conde, J., Doria, G., & Baptista, P. (2012). Noble metal nanoparticles applications in cancer. *Journal of Drug Delivery, 2012*.

[25] Zhao, R., Xiang, J., Wang, B., Chen, L., & Tan, S. (2022). Recent advances in the development of noble metal NPs for cancer therapy. *Bioinorganic Chemistry and Applications, 2022*, 2444516.

[26] Ye, W., Yan, J., Ye, Q., & Zhou, F. (2010). Template-free and direct electrochemical deposition of hierarchical dendritic gold microstructures: Growth and their multiple applications. *The Journal of Physical Chemistry C, 114*(37), 15617–15624.

[27] Brust, M., Walker, M., Bethell, D., Schiffrin, D. J., & Whyman, R. (1994). Synthesis of thiol-derivatised gold nanoparticles in a two-phase liquid–liquid system. *Journal of the Chemical Society, Chemical Communications*, (7), 801–802.

[28] Jana, N. R., Gearheart, L., & Murphy, C. J. (2001). Seed-mediated growth approach for shape-controlled synthesis of spheroidal and rod-like gold nanoparticles using a surfactant template. *Advanced Materials, 13*(18), 1389–1393.

[29] Gu, Y. J., Cheng, J., Man, C. W. Y., Wong, W. T., & Cheng, S. H. (2012). Gold-doxorubicin nanoconjugates for overcoming multidrug resistance. *Nanomedicine: Nanotechnology, Biology and Medicine, 8*(2), 204–211.

[30] Wang, L., Yuan, Y., Lin, S., Huang, J., Dai, J., Jiang, Q., … & Shuai, X. (2016). Photothermo-chemotherapy of cancer employing drug leakage-free gold nanoshells. *Biomaterials, 78*, 40–49.

[31] Chen, H., Kou, X., Yang, Z., Ni, W., & Wang, J. (2008). Shape-and size-dependent refractive index sensitivity of gold nanoparticles. *Langmuir, 24*(10), 5233–5237.

[32] Gao, F., Sun, M., Xu, L., Liu, L., Kuang, H., & Xu, C. (2017). Biocompatible cup-shaped nanocrystal with ultrahigh photothermal efficiency as tumor therapeutic agent. *Advanced Functional Materials, 27*(24), 1700605.

[33] Kim, S. T., Chompoosor, A., Yeh, Y. C., Agasti, S. S., Solfiell, D. J., & Rotello, V. M. (2012). Dendronized gold nanoparticles for siRNA delivery. *Small (Weinheim an der Bergstrasse, Germany), 8*(21), 3253.

[34] Sun, X., Huang, X., Yan, X., Wang, Y., Guo, J., Jacobson, O., … & Chen, X. (2014). Chelator-free 64Cu-integrated gold nanomaterials for positron emission tomography imaging guided photothermal cancer therapy. *ACS Nano*, *8*(8), 8438–8446.

[35] Wang, L., Zhang, T., Li, P., Huang, W., Tang, J., Wang, P., … & Chen, C. (2015). Use of synchrotron radiation-analytical techniques to reveal chemical origin of silver-nanoparticle cytotoxicity. *ACS Nano*, *9*(6), 6532–6547.

[36] Saito, G., & Akiyama, T. (2016). Nanomaterial synthesis using plasma generation in liquid. *Journal of Nanomaterials*, *16*(1), 299–299.

[37] Bian, K., Zhang, X., Liu, K., Yin, T., Liu, H., Niu, K., … & Gao, D. (2018). Peptide-directed hierarchical mineralized silver nanocages for anti-tumor photothermal therapy. *ACS Sustainable Chemistry & Engineering*, *6*(6), 7574–7588.

[38] Soenen, S. J., Parak, W. J., Rejman, J., & Manshian, B. (2015). (Intra) cellular stability of inorganic nanoparticles: Effects on cytotoxicity, particle functionality, and biomedical applications. *Chemical Reviews*, *115*(5), 2109–2135.

[39] Gurunathan, S., Han, J. W., Eppakayala, V., Jeyaraj, M., & Kim, J. H. (2013). Cytotoxicity of biologically synthesized silver nanoparticles in MDA-MB-231 human breast cancer cells. *BioMed Research International*, *2013*, 535796.

[40] Rosenberg, B., Vancamp, L., Trosko, J. E., & Mansour, V. H. (1969). Platinum compounds: A new class of potent antitumour agents. *Nature*, *222*(5191), 385–386.

[41] Douple, E. B. (1984). Cis-diamminedichloroplatinum (II): Effects of a representative metal coordination complex on mammalian cells. *Pharmacology & Therapeutics*, *25*(3), 297–326.

[42] Kim, N. R., & Kim, Y. J. (2019). Oxaliplatin regulates myeloid-derived suppressor cell-mediated immunosuppression via downregulation of nuclear factor-κB signaling. *Cancer Medicine*, *8*(1), 276–288.

[43] Wong, D. Y. Q., Yeo, C. H. F., & Ang, W. H. (2014). Immuno-chemotherapeutic platinum (IV) prodrugs of cisplatin as multimodal anticancer agents. *Angewandte Chemie International Edition*, *53*(26), 6752–6756.

[44] Fan, J., Yin, J. J., Ning, B., Wu, X., Hu, Y., Ferrari, M., … & Nie, G. (2011). Direct evidence for catalase and peroxidase activities of ferritin–platinum nanoparticles. *Biomaterials*, *32*(6), 1611–1618.

[45] Foxwell, D. A., Pradhan, S., Zouwail, S., Rainer, T. H., & Phillips, A. O. (2020). Epidemiology of emergency department acute kidney injury. *Nephrology*, *25*(6), 457–466.

[46] Bharathiraja, S., Bui, N. Q., Manivasagan, P., Moorthy, M. S., Mondal, S., Seo, H., … & Oh, J. (2018). Multimodal tumor-homing chitosan oligosaccharide-coated biocompatible palladium nanoparticles for photo-based imaging and therapy. *Scientific Reports*, *8*(1), 1–16.

[47] Huang, X., Tang, S., Mu, X., Dai, Y., Chen, G., Zhou, Z., … & Zheng, N. (2011). Freestanding palladium nanosheets with plasmonic and catalytic properties. *Nature Nanotechnology*, *6*(1), 28–32.

[48] Alvarez, G. F., Mamlouk, M., Senthil Kumar, S. M., & Scott, K. (2011). Preparation and characterisation of carbon-supported palladium nanoparticles for oxygen reduction in low temperature PEM fuel cells. *Journal of Applied Electrochemistry*, *41*, 925–937.

[49] Hosnedlova, B., Kepinska, M., Fernandez, C., Peng, Q., Ruttkay-Nedecky, B., Milnerowicz, H., & Kizek, R. (2019). Carbon nanomaterials for targeted cancer therapy drugs: A critical review. *The Chemical Record*, *19*(2–3), 502–522.

[50] Koralli, P., Tsikalakis, S., Goulielmaki, M., Arelaki, S., Müller, J., Nega, A. D., … & Chochos, C. L. (2021). Rational design of aqueous conjugated polymer nanoparticles as potential theranostic agents of breast cancer. *Materials Chemistry Frontiers*, *5*(13), 4950–4962.

[51] Petros, R. A., & DeSimone, J. M. (2010). Strategies in the design of nanoparticles for therapeutic applications. *Nature Reviews Drug Discovery*, *9*(8), 615–627.

[52] Qian, C. G., Chen, Y. L., Feng, P. J., Xiao, X. Z., Dong, M., Yu, J. C., … & Gu, Z. (2017). Conjugated polymer nanomaterials for theranostics. *Acta Pharmacologica Sinica*, *38*(6), 764–781.

[53] Hua, Q., Qiang, Z., Chu, M., Shi, D., & Ren, J. (2019). Polymeric drug delivery system with actively targeted cell penetration and nuclear targeting for cancer therapy. *ACS Applied Bio Materials*, *2*(4), 1724–1731.

[54] Rohatgi, C. V., Harada, T., Need, E. F., Krasowska, M., Beattie, D. A., Dickenson, G. D., … & Kee, T. W. (2018). Low-bandgap conjugated polymer dots for near-infrared fluorescence imaging. *ACS Applied Nano Materials*, *1*(9), 4801–4808.

[55] Li, D., Gao, D., Qi, J., Chai, R., Zhan, Y., & Xing, C. (2018). Conjugated polymer/graphene oxide complexes for photothermal activation of DNA unzipping and binding to protein. *ACS Applied Bio Materials*, *1*(1), 146–152.

[56] Urbano, L., Clifton, L., Ku, H. K., Kendall-Troughton, H., Vandera, K. K. A., Matarese, B. F., ... & Harvey, R. D. (2018). Influence of the surfactant structure on photoluminescent π-conjugated polymer nanoparticles: Interfacial properties and protein binding. *Langmuir*, *34*(21), 6125–6137.

[57] Feng, X., Lv, F., Liu, L., Tang, H., Xing, C., Yang, Q., & Wang, S. (2010). Conjugated polymer nanoparticles for drug delivery and imaging. *ACS Applied Materials & Interfaces*, *2*(8), 2429–2435.

[58] Xu, L., Cheng, L., Wang, C., Peng, R., & Liu, Z. (2014). Conjugated polymers for photothermal therapy of cancer. *Polymer Chemistry*, *5*(5), 1573–1580.

[59] De Vivo, M., Masetti, M., Bottegoni, G., & Cavalli, A. (2016). Role of molecular dynamics and related methods in drug discovery. *Journal of Medicinal Chemistry*, *59*(9), 4035–4061.

[60] Duarte, Y., Márquez-Miranda, V., Miossec, M. J., & González-Nilo, F. (2019). Integration of target discovery, drug discovery and drug delivery: A review on computational strategies. *Wiley Interdisciplinary Reviews: Nanomedicine and Nanobiotechnology*, *11*(4), e1554.

[61] Zheng, M., Liu, X., Xu, Y., Li, H., Luo, C., & Jiang, H. (2013). Computational methods for drug design and discovery: Focus on China. *Trends in Pharmacological Sciences*, *34*(10), 549–559.

[62] Van't Veer, L. J., & Bernards, R. (2008). Enabling personalized cancer medicine through analysis of gene-expression patterns. *Nature*, *452*(7187), 564–570.

[63] Sliwoski, G., Kothiwale, S., Meiler, J., & Lowe, E. W. (2014). Computational methods in drug discovery. *Pharmacological Reviews*, *66*(1), 334–395.

[64] Shen, Y., Tang, H., Radosz, M., Van Kirk, E., & Murdoch, W. J. (2008). pH-responsive nanoparticles for cancer drug delivery. *Drug Delivery Systems*, 183–216.

[65] Reddy, A. S., Pati, S. P., Kumar, P. P., Pradeep, H. N., & Sastry, G. N. (2007). Virtual screening in drug discovery-a computational perspective. *Current Protein and Peptide Science*, *8*(4), 329–351.

[66] Gao, Q., Yang, L., & Zhu, Y. (2010). Pharmacophore based drug design approach as a practical process in drug discovery. *Current Computer-Aided Drug Design*, *6*(1), 37–49.

[67] Martinho, N., Florindo, H., Silva, L., Brocchini, S., Zloh, M., & Barata, T. (2014). Molecular modeling to study dendrimers for biomedical applications. *Molecules*, *19*(12), 20424–20467.

[68] Kapetanovic, I. (2008). Computer-aided drug discovery and development (CADDD): In silico-chemico-biological approach. *Chemico-Biological Interactions*, *171*(2), 165–176.

29

Role of Nanomaterials in Food Packaging Applications

Ravi Prakash Balasundaram, Praveen Gunasekaran, and Arulmari Ramaraj
Agricultural Engineering College and Research Institute, Tamil Nadu Agricultural University, Tiruchirappalli, India

Gitanjali Jothiprakash
Agricultural Engineering College and Research Institute, Tamil Nadu Agricultural University, Coimbatore, India

Deepa Jaganathan
Hindustan College of Engineering and Technology, Coimbatore, India

29.1 Introduction

Food packaging plays crucial functions in the handling, shipping, and protection of food goods via complex international supply chains prior to consumption. It also conveys information about the food products it contains and protects. But the ability of traditional packaging materials – such as glass, metal, plastic, and paperboard – to completely stop food waste resulting from oxidation, spoiling, moisture loss, and microbial growth is limited. It is imperative to improve packaging functionality because over 1.3 billion tonnes of food are lost each year worldwide. Furthermore, industry, government, and consumers all have a strong desire for cutting-edge, highly efficient, and environmentally friendly packaging solutions due to the negative environmental effects of conventional packaging materials, including greenhouse gas emissions during production and the generation of waste after use [1] (Figure 29.1).

Nanotechnology emerges as a promising avenue for developing next-generation packaging that not only improves preservation and significantly extends shelf life but also incorporates intelligent features such as sensors, lightweight attributes, and recyclable or biodegradable properties. Nanomaterials and nano-enabled technologies, involving the manipulation and production of materials, particles, fibres, and integrated systems at the nanoscale (typically between 1 and 100 nanometres), hold the key to these advancements [3]. At these dimensions, materials exhibit greatly enhanced mechanical, catalytic, thermal, electrical, optical, and other properties compared to their larger forms. In the realm of food packaging, nanomaterials offer the potential to provide superior gas barriers, improved mechanical performance, smart functionalities and sensors, antibacterial and antioxidant activities, as well as controlled release capabilities [4].

29.2 Types of Nanomaterials for Food Packaging

A wide range of engineered nanomaterials is currently being explored and developed for incorporation into food packaging systems. These materials can be broadly classified into several categories.

DOI: 10.1201/9781003495437-29

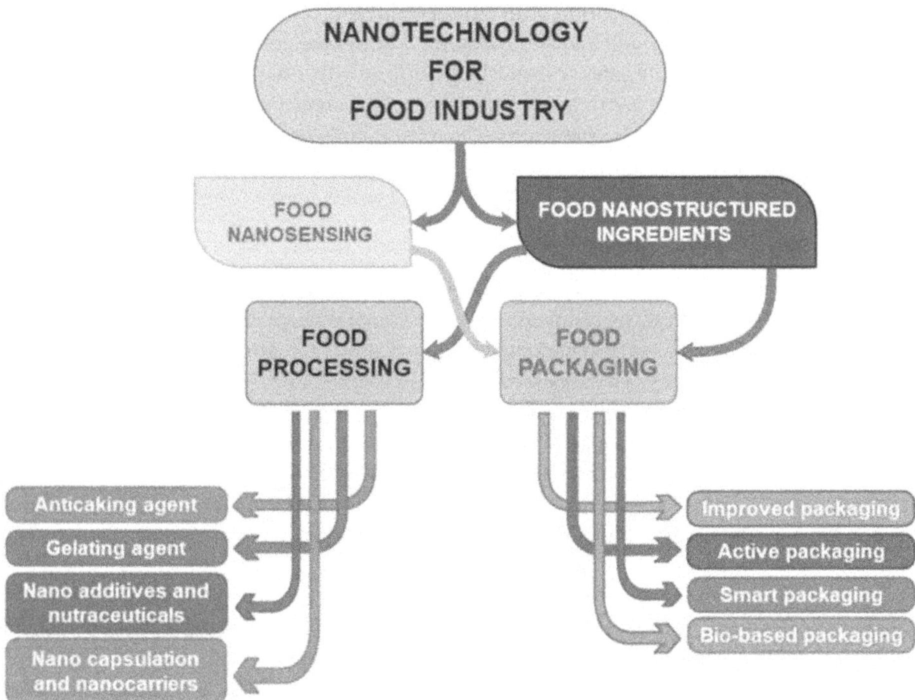

FIGURE 29.1   Application of nanotechnology in different fields of the food industry.

SOURCE: [2]

29.2.1 Metal and Metal Oxide Nanoparticles

29.2.1.1 Silver Nanoparticles

Researches on silver nanoparticles have gone in-depth and are frequently used particularly in the improvement of the performance that can be seen in coatings or films utilised when packaging food. Due to their incredibly small size, these particles, which range from 1 to 100 nanometres in size, have higher catalytic reactivity and antibacterial efficacy compared to bulk silver that is conventional. Many researchers have conducted numerous studies attesting to the fact that by adding silver nanoparticles to polymer packaging materials, a substantial decrease in the risk of food spoilage and contamination is observed. Numerous microorganisms that are associated with food deterioration are found to be greatly affected by these small nanoparticles due to their strong antibacterial activity, including major foodborne pathogens such as *Salmonella*, *E. coli*, Listeria, and *S. aureus*. It is believed that the antibacterial activity of nano-silver on microorganisms is caused by contact with their cell walls, the disruption of molecular transport within the cell, and the release of silver ions that can harm DNA and enzymes.

An effective method to impart antimicrobial properties to package surfaces is by incorporating a thin layer of silver nanoparticles onto plastic films. Polymer nanocomposites with dispersed silver nanoparticles throughout the bulk matrix offer longer-lasting and more uniform antibacterial film performance. Plastic-based nanocomposites, such as low-density polyethylene (LDPE), polypropylene, and polystyrene, containing silver nanoparticles, have demonstrated the ability to inhibit the growth of food pathogens in laboratory testing and food models.

Antimicrobial packaging with nano-silver integration shows promise in enhancing the shelf life and safety of high-risk foods, such as raw and processed meats, ready-to-eat foods, fresh produce, bakery, and dairy items. These applications offer added protection against recontamination in cases where packaging damage or improper handling may allow bacteria penetration. Moreover, the use of nano-silver in packaging contributes to sustainability by reducing product losses, waste generation, and energy consumption in cold supply chains.

While some plastic food packaging films with ultra-thin silver layers or coatings have successfully provided antimicrobial functionality and barrier properties, widespread commercialisation faces regulatory concerns and uncertainty regarding the potential migration of silver nanoparticles into food. Questions persist about the long-term effects of chronic low-dose silver ingestion on consumer health. Ongoing research priorities aim to definitively characterise the release kinetics of nano-silver particles into foods, establishing appropriate safety limits and conducting risk assessments [5, 6].

29.2.1.2 Titanium Dioxide Nanoparticles

Titanium dioxide nanoparticles (TiO_2 NPs) are emerging as versatile additives for food packaging films, offering improvements in optical, mechanical, and antimicrobial properties. When pigment grade TiO_2 particles are reduced to sizes below 100 nanometres, they exhibit high ultraviolet light absorption capacity and visible light transparency. This has spurred research into using TiO_2 NPs to enhance polymeric packaging materials, aiming to improve UV blocking, film opacity, and resistance to aging.

While bulk titanium dioxide has a robust safety record as a common food contact substance, concerns arise when nano-sized particles are considered due to their potential toxicity. Hence, understanding the migratory behaviour of nano-TiO_2 from packaging films is crucial for evaluating consumer risk. Research on migration kinetics and bioavailability of TiO_2 NPs following packaging release scenarios is currently limited.

Most development efforts have concentrated on creating LDPE nanocomposite films with embedded TiO_2 nanoparticles, typically at concentrations up to 10% by weight. The incorporation of TiO_2 NPs has demonstrated benefits, including improved film tensile and puncture strength, increased opacity, water contact angles, and UV shielding. These properties can prevent flavour loss and vitamin degradation in packaged contents. Additionally, antimicrobial effects suggest that TiO_2 NPs can help limit bacterial growth on package surfaces and nearby surroundings through the generation of photocatalytic reactive oxygen species upon light activation.

Recent studies have explored nano-TiO_2 migration from various polymer food packaging structures, such as PE and PET, into food stimulants or actual food samples. However, research quantifying true particle transfer levels for comprehensive risk analysis remains limited. Some studies, particularly in chicken meat, have reported the migration of only ionic titanium from films, rather than intact TiO_2 nanoparticles. Evidence also suggests that particle migration into non-fatty and acidic foods is minimal when TiO_2 NP loading in PE is kept at 10% or lower. Additionally, limited titanium absorption from packaging into blood/tissues in rodent oral toxicity studies provides further reassurance.

Titanium dioxide nanoparticles hold technical promise for enhancing the functionality of food packaging films. Nevertheless, uncertainties surrounding the safety, migration, and toxicology of nano-TiO_2 have hindered regulatory approval for many active packaging applications. Addressing these concerns can guide appropriate usage levels to balance desirable benefits against potential exposure risks. Ongoing research also indicates that layered films with nanoparticles trapped between outer polymer layers may offer a reliable construct to prevent direct food contact [7, 8].

29.2.1.3 Zinc Oxide Nanoparticles

Similar to titanium dioxide, zinc oxide is a versatile nanoparticulate additive for food packaging materials that has antibacterial and UV-blocking properties. Zinc oxide nanoparticles, or ZnO NPs, are generally regarded by the FDA as safe. Compared to TiO_2, they have the advantage of not having photocatalytic activity, which can hasten the degradation of polymeric packaging.

ZnO nanoparticles have been shown in studies to have a substantial UV absorption capacity and to preserve or increase optical clarity when incorporated into food-contact polymers like low-density polyethylene. This capability rises in direct proportion to ZnO concentrations, up to a maximum of 10% w/w. Improved UV blocking has demonstrated further advantages in preventing food photooxidation, which helps to preserve freshness.

ZnO nanoparticles coated in thin layers on polyethylene and polypropylene films have antibacterial action against major foodborne pathogens such as Salmonella, Listeria, and *E. coli* in addition to improving UV barrier qualities. The suggested antimicrobial methods combine the toxicity of reactive oxygen species produced following UV activation with direct contact to ZnONPs, which damages intra-cellular proteins and cell membranes.

ZnONPs have effects at far lower doses than bulk ZnO, despite the fact that bulk ZnO is typically thought to be benign. This suggests that there may be nano-specific toxicity. Exposure issues are brought up by this, particularly with regard to ingesting. Therefore, in order to determine safety limits, concerns pertaining to ZnO NPs' stability and bioavailability must be characterised by studies of simulated intestinal destiny. Initial findings in artificial stomach fluid point to quick ionisation, which releases zinc ions as the main migratory species. To examine ZnO NP dissolution over a range of digestion model parameters, including transit time, pH, temperature, enzymatic breakdown, food matrices, and more, much research is still required. Evaluations of bioaccumulation are also essential for identifying actual tissue uptake levels and possible toxicological effects.

As a material that has been used in food contact, ZnONPs provide a promising mix of low cost, scalable enhancement for food packaging films, strong and long-lasting antibacterial and UV-blocking characteristics, and general safety assurance. However, before ZnO nano-enabled active packaging may advance towards commercial and regulatory adoption, further research is required due to uncertainty regarding exposure hazards and destiny [9, 10].

29.2.1.4 Iron Oxide Nanoparticles

Iron oxide nanoparticles exhibit advantageous light-blocking capabilities when incorporated into polymer matrices such as LDPE for food packaging purposes. These nanoparticles, specifically magnetite and maghemite particles with diameters less than 50 nm, are cost-effective and generally considered non-toxic at typical exposure levels. Their ability to strongly absorb visible light while maintaining high transparency in the visible spectrum makes them appealing as film enhancers for improved UV resistance, oxidative stability, and pigmentation capacity in food packaging.

While there is limited research on the impact of composite films with nano-iron oxides on food quality and safety attributes, early studies involving LDPE films incorporating low concentrations of iron oxide nanoparticles (around 1%) indicate preserved optical clarity, enhanced UV opacity, and improved tensile strength. Moreover, these studies showed no significant effects on overall migration levels into food simulants compared to pure LDPE films.

The anticipated low migration of iron oxide nanoparticles from well-distributed polymer composites, due to strong interactions and a large surface area contact with the host matrix, may be influenced by acids and liquid lipids, potentially increasing release potential. Specific studies on migration across various food stimulants are needed. Assessing the bioavailability of iron after any migration is crucial, as free iron can contribute to oxidative damage.

In addition to functional enhancements, iron oxide nanoparticles present promising applications for intelligent and active food packaging roles. The radio frequency heating of iron oxide nanocomposite films offers a potential external trigger mechanism for controlled temperature adjustments, and the integration of iron oxide may enable various sensor platforms based on changes in polymer optical, mechanical, or magnetic properties.

Iron oxide nanoparticles hold notable potential for the development of multifunctional food packaging. However, comprehensive research confirming the safety and migration aspects of this novel application area is limited. Key priorities for facilitating commercial adoption include conducting evaluations of iron bioavailability and toxicity following nanoparticle introduction into different food systems. Migration testing is also crucial across a range of food oil, alcohol, and acid simulants. Addressing these considerations can better position iron oxide nanocomposites as approved next-generation food packaging film platforms [11, 12].

29.2.1.5 Magnesium Oxide Nanoparticles

Nanoscale magnesium oxide (MgO) particles offer advantageous characteristics such as additives for food packaging, serving as antimicrobial agents, and enhancing barriers. Similar to other metal oxide nanoparticles, nano-MgO demonstrates increased reactivity compared to traditional bulk particles. It also possesses strong microwave absorption potential, making it suitable for active heating applications. The adhesion of nanoscale MgO onto polymer films, such as LDPE, may provide reversible antioxidant, moisture scavenging, and gas barrier properties responsive to environmental triggers.

Early studies on LDPE nanocomposites incorporating up to 5% loading of MgO nanoparticles indicate maintained optical transparency and flexibility. Additionally, these studies show improvements in thermal stability, tensile strength, and oxygen barrier properties with the incorporation of nano-MgO platelets [13, 14].

29.2.2 Carbon-Based Nanomaterials

29.2.2.1 Graphene

Graphene is an emerging advanced nanomaterial with tremendous potential to enhance various properties of food packaging structures, including mechanical strength, barrier resistance, and durability. Composed of a single-layer lattice of carbon atoms, graphene's unique two-dimensional configuration provides exceptional strength, gas impermeability, and lightweight characteristics, making it highly advantageous for packaging applications.

Even minor percentages of graphene inclusion as a filler reinforcement in common food packaging polymers, such as polyethylene and PET, have demonstrated the ability to significantly improve material strength; enhance barrier resistance to gases, liquids, and vapours; and increase temperature resilience. Graphene nanocomposites also exhibit excellent optical transparency and electrical conductivity, making them suitable for intelligent packaging applications [15].

Ultrathin graphene oxide films showcase high natural antibacterial activity, inhibiting pathogen growth on package surfaces and extending food shelf life. These multifunctional enhancements, achieved with minimal graphene loading, make graphene an attractive nanomaterial filler for advanced packaging performance while minimising safety or leaching concerns.

However, challenges in consistently producing high-quality graphene at an affordable, mass production scale have hindered extensive commercialisation in food packaging. Difficulties in homogeneously dispersing hydrophobic graphene sheets into polymer matrices have also limited the development of composite materials. While most published studies focus on laboratory-prepared graphene nanopackaging constructs, active efforts are underway to address these manufacturing and integration challenges [16].

Promising techniques for scalable graphene production from waste sources or lower-cost graphitic feedstocks are emerging. New compatibilization chemistries that alter surface properties aim to overcome dispersion difficulties, achieving stable composite uniformity. Major packaging producers are initiating large-scale research and development programs dedicated to graphene-based materials.

In the near term, the most significant promise lies in ultrabarrier packaging applications that utilise standalone graphene or hybrid films to prevent gases, moisture, aromas, and oils from permeating container walls. Graphene oxide coating solutions could serve as thin, flexible, and durable liners for existing packaging structures, enhancing performance as a value-add. Integration of intelligent sensor technologies with such constructs could further enhance graphene-enabled active and smart packaging.

Graphene has the potential to revolutionise food packaging performance while reducing environmental impacts. Ongoing efforts to refine manufacturing and tailor graphene composite integration will facilitate industry utilisation. Realistic commercialisation timeframes suggest widespread adoption of graphene packaging within 5–10 years, contingent upon regulatory clearances supported by benchmark safety studies [17].

29.2.2.2 Carbon Nanotubes

Carbon nanotubes (CNTs) present another promising carbon-based nanostructure that can enhance the mechanical strength, thermal conductivity, gas barrier, and other properties of food packaging polymers when used as filler materials. CNTs are composed of rolled graphene sheets, forming tiny cylindrical tubes with diameters less than 100 nm and lengths that can extend up to centimetres, resulting in remarkable material tensile strength and resilience.

Early composites incorporating small percentages of CNTs into packaging films, such as polyethylene, have shown significant increases in strength and modulus, surpassing unmodified films by over 50%, potentially allowing for downgauging. These composites also exhibit improved resistance against gas diffusion through films, suggesting the potential for extending the shelf life of food products. Inclusion of CNTs in composites offers additional benefits such as enhanced thermal and electrical conductivity, making them useful for active and intelligent packaging applications. Some studies also indicate antioxidative properties [18].

Moreover, CNTs exhibit broad antimicrobial activity attributed to direct contact with nanoparticles and disruption of cell walls. These multifunctional enhancements from minor CNT filler fractions underscore their potential as specialty additives for packaging. However, challenges in achieving uniform dispersion and stable intermolecular bonding of CNTs within target polymer matrices have hindered the consistent and high-performance composite constructs needed for widespread commercial adoption [19].

Recent research progress has improved understanding of surface treatment methodologies, using surfactants or chemical functionalisation, to balance adequate polymer–CNT adhesion against the risk of additive dispersion aggregation. The development of masterbatches, consisting of premixed CNT–polymer formulations, has further facilitated progress in product manufacturing. With these technical refinements, CNT nanocomposites are transitioning from academic curiosity to commercially investigational platforms in major food packaging companies globally.

Early applications focus on thin CNT-based films or coatings to enhance the barrier properties, temperature resilience, and shelf life of perishable products like meat, seafood, produce, and prepared meals. Flexible nano-enabled structures also show promise for increasing the impact resistance and robustness of food shipping crates and outer protective packaging.

Simultaneously, toxicological assessments analysing CNT leaching potential and bioaccumulation behaviours need expansion to confirm safety specifications for intended packaging applications. However, initial cytotoxicity and animal testing studies continue to provide reasonable safety assurances as long as CNT nanotubes remain fully encapsulated within stable polymer matrices.

CNTs are overcoming technical development hurdles and are now becoming commercially investable solutions for enhancing barrier, mechanical, and active functionalities in food packaging. Achieving consistent, scalable manufacturing of pure, high-quality CNTs tailored for food contact is crucial. Advancements in composite interface engineering and migration testing will strengthen regulatory confidence, paving the way for widespread commercial adoption opportunities over the next decade [20].

29.2.2.3 Fullerenes

Fullerenes represent another class of carbon allotropes with the potential to enhance the properties of food packaging at the nano-scale. These carbon molecules, which can be spherical, ellipsoidal, or cylindrical, have exceptionally high surface area to volume ratios, making them effective in improving diffusion pathway barriers when incorporated into packaging polymers like LDPE [21].

Among fullerenes, C60 buckminsterfullerene is the most extensively studied variant for materials applications. Early research on LDPE nanocomposites incorporating C60 fullerenes has shown improved thermal and oxidation resistance of films. The inclusion of low percentages of C60 has also resulted in enhancements in mechanical strength, glass transition temperatures, and reduced oxygen permeability [22].

Fullerenes with antimicrobial properties could offer another promising application in packaging films or coatings. Chemical derivatisation of fullerenes, such as appending antimicrobial enzymes or metals, might enable responsive activity against food surface pathogens. Metallofullerene derivatives may also provide controlled release of nutraceutical supplements in active packaging constructs [22, 23].

Despite these potentials, high production costs and purification inconsistencies have hindered extensive research and commercialisation of fullerene-enabled food packaging. Technical challenges in dispersing strongly hydrophobic fullerenes into polymer matrices further impede composite development. Overcoming these challenges, such as demonstrating cost-effective, scalable fullerene derivation and tailoring chemistry suitable for food contact applications, remains a hurdle [24].

Efforts to address manufacturing and compositing challenges are advancing fullerene applications for specialty packaging. Techniques utilising nano-scale fullerene dispersion and extrusion in enhanced film production show potential for achieving consistent, high-performance composite properties needed for commercial investment. Hybrid chemical treatment approaches may also enhance fullerene interfaces for improved miscibility with polar polymer matrices.

With progress in these areas, the unique size, surface chemistry, and barrier augmentation properties of fullerenes could soon lead to new possibilities for performance enhancement and active packaging. Particularly, antimicrobial fullerene nanocomposite constructs offer promising solutions for extending the shelf life of perishable foods. While current regulatory frameworks do not specifically address fullerene food contact usages, existing evidence of low oral toxicity and lack of dermal penetration supports reasonable safety expectations.

Fullerenes are yet to realise their commercial promise as multifunctional packaging reinforcement additives. However, early-stage principles established in lab settings continue to showcase the vast potential of tuned fullerene properties for advanced food packaging designs. Overcoming remaining manufacturing hurdles can position fullerenes as competitive next-generation materials for extending food quality, safety, and supply chain sustainability [25].

29.2.3 Biopolymer Nanoparticles

29.2.3.1 Protein Nanoparticles

The nano-processing of proteins such as zein, whey, soy, or albumen into spherical particles or encapsulating nano-shells offers improved delivery functionalities for antioxidants, antimicrobials, nutraceuticals, and other bioactive supplements relevant to food packaging systems. Reducing the dimensions of protein carriers to below 1 micron enhances dispersion stability, improves mobility, and facilitates targeted interactions, thereby increasing the efficacy of active ingredients [26].

Zein protein is particularly suitable for nano-encapsulation and controlled release applications in food packaging due to its excellent film-forming capacity, hydrophobicity for protecting sensitive compounds, and generally recognised safe status. The preparation of zein nanoparticles with omega fatty acids, minerals, vitamins, and other nutraceuticals has demonstrated 2–3 times enhancement in supplement stability and retention over 15+ days of storage in lab models. Nano-encapsulation also reinforces antioxidants like curcumin to better maintain activity after exposure to light, oxygen, and temperature fluctuations [27].

Incorporating these protein-based nano-carriers into active food packaging film layers enables sustained and targeted release of bioactives into foods over time. This nano-enabled dosing approach helps achieve nutritional augmentation goals while minimising sensory impacts and risks of ingredient oxidation compared to direct food supplementation [27, 28].

Additionally, ultrafine protein nano-platelets can serve as a renewable reinforcement to strengthen and enhance the barrier properties of biomaterial films when combined with other biopolymers such as starch or alginate. Soy protein nano-platelets, with diameters around 150 nm, have demonstrated reinforced thermomechanical film performance at high humidity when incorporated into various starch composites [29].

In conclusion, protein-based nano-architectures offer versatile performance enhancement and controlled delivery functionalities that are beneficial for active and bio-based food packaging designs. The refinement of nano-protein fabrication protocols for consistent size and purity is crucial for scaling up production. Emerging techniques, such as streamlined nano-processing using microfluidics and membrane technologies, indicate promising pathways for commercialisation.

29.2.3.2 Polysaccharide Nanoparticles

Polysaccharides such as chitosan, starch, and alginate can undergo self-assembly or chemical cross-linking to form nanoscale hydrogel particles or crystalline reinforcements [30]. When applied to food packaging structures, these materials offer beneficial modulation of stability, release kinetics, and mechanical and barrier properties:

Chitosan nano-gels exhibit selective permeability, allowing for controlled diffusion kinetics influenced by crosslink density. This property enables applications like nano-reinforced antimicrobial films, contributing to shelf-life extension. The safety of oral administration supports its usage in food packaging [31].

Ultrasmall starch nanocrystals, derived from corn, potato, and rice starches, significantly enhance the strength and barrier properties of traditional starch biopolymer matrices with minimal additions (3–5%). Additionally, these nanocrystals contribute to increased biodegradability [32].

Alginate-based nanofibers, produced through gel spinning techniques, provide rheological benefits when blended into conventional alginate food packaging films. This inclusion enhances rigidity and water resistance [33].

Cellulose nanofibers/nanocrystals are valuable for reinforcing the mechanical performance of recycled cellulose packaging composites, such as moulded egg carton foams/trays [34].

In summary, the nano-engineering of polysaccharides improves the mechanical and release properties of bio-based food packaging, contributing to increased biodegradation. Advancements in nano-processing methods towards commercially scalable platforms will drive further adoption of these innovative materials.

29.2.3.3 Solid Lipid Nanoparticles

Solid lipid nanoparticles (SLNs) with sizes smaller than 100 nm offer controlled and tuneable degradation and release kinetics, enhancing the delivery efficacy, bioavailability, and stability of sensitive ingredients like essential oils, antimicrobial peptides, and lipophilic nutraceuticals. The protective capacity of SLNs increases as their nano-scale dimensions decrease.

Compositions made from food-grade lipids, such as monoglycerides, waxes, or emulsifiers, allow precise customisation of SLN dispersibility, melting points, crystallinity, and hydrophobic thickness parameters. These factors influence both controlled release capacities and long-term stability profiles during storage. The resulting enhancements support antimicrobial, antioxidant, and other active functionalities, making them valuable for bioactive food packaging applications.

Although challenges remain in cost-effectively producing SLNs at commercial scales, advancements in microfluidic manufacturing methods show promise for scalable nano-emulsion templating techniques using food-grade lipids. Demonstrating the efficacy improvements compared to traditional encapsulation forms will further incentivise the adoption of SLNs as innovative, bio-based delivery platforms [35].

In summary, the nano-sizing of lipid carriers simplifies diffusion barriers, enhancing stability and improving timed release properties, which are beneficial for next-generation food packaging. Progress in fabrication methods and component tailoring will eventually enable commercially viable antimicrobial, antioxidant, and nutraceutical delivery systems.

29.2.4 Nanoclays

Nanoclays are naturally occurring hydrated aluminium silicates like bentonite and hectorite, featuring a fine layered crystal lattice structure. They can be easily dispersed into polymer matrices at low filler loadings, enhancing the tortuosity of gas diffusion pathways and significantly improving barrier properties for moisture, oils, and gases such as oxygen and carbon dioxide. Nanoclays find common applications in food packaging, particularly in enhancing the oxygen barrier properties of PET bottles and trays. Additionally, nanoclays offer attributes like heat resistance, flame retardancy, and mechanical improvements. Various types of nanoclays have received FDA approval for use in food packaging, contributing to their increasing global adoption [36].

29.2.5 Liposomes

Liposomes, which are spherical lipid vesicles with single or multiple bilayer membranes, enable the encapsulation and protection of active ingredients such as antimicrobial enzymes and nutraceuticals. These ingredients can be selectively released based on triggers like enzymes, pH, temperature, or other stimuli in foods and packaging headspace. Liposomes also demonstrate oxygen scavenging capabilities through the incorporation of antioxidants. The ability to precisely control size, composition, and release kinetics enhances the utility of liposomes for nano-enabled smart packaging [37].

29.2.6 Dendrimers

Dendrimers, synthetically engineered macromolecules with a branching tree-like architecture, offer precise tailoring of shape, size, and surface reactivity. Dendrimer films chemically grafted onto packaging exhibit antimicrobial and oxygen-scavenging functionalities suitable for food packaging. However, potential toxicity and migration concerns need investigation before widespread commercialisation [38].

29.2.7 Electro Spun

Electro-spun nanofibers, produced by applying electrostatic fields to spin nano-scale fibres from biopolymer and polymer solutions, create nonwoven mats and membranes with high surface area and tuneable porosity. These nanofibers, made from polysaccharides, proteins, and composites, show excellent oxygen barrier properties and transparency. They can also provide antioxidant and antimicrobial functionalities. Despite technical challenges in consistent production and mechanical integrity, research and development in this field are progressing rapidly [39].

29.3 Key Functionalities Enabled by Nanomaterials

Engineered nanomaterials offer unique properties that enable various functionalities in food packaging.

29.3.1 Barrier Enhancement

Nanoparticles, nanoclays, nanofibers, and nanosheets create highly tortuous pathways with multiple interfaces, slowing down the transmission rates of gases, moisture, oils, volatiles, and other compounds across packaging materials [40]. This results in:

- Preservation of food quality and freshness by restricting oxygen ingress.
- Reduction of moisture loss to maintain texture.
- Hindrance of odours and aromas uptake.
- Slowing down oxidation and rancidity of fats/lipids.
- Improvement of shelf life and overall quality.

29.3.2 Mechanical Property Augmentation

Nanoscale fillers and structures like graphene, CNTs, and cellulose nanocrystals bring exceptional strength capabilities to reinforce packaging polymer matrices, composite liners, and biopolymer films. Even with low filler additions (<5%), significant improvements in tensile strength, flexibility, impact resistance, and puncture resistance are achieved. This enables the potential for thinner and lighter packaging. The reinforcing effects arise from effective stress transfer between the crystalline nanomaterial structure and the surrounding matrix [41].

29.3.3 Antimicrobial Effect:

Nanoparticles of silver, copper, ZnO, MgO, and titanium dioxide exhibit toxicity against various food spoilage microbes and pathogens, including bacteria like Listeria, *Escherichia coli*, and Salmonella, as well as moulds. They achieve this by disrupting membrane permeability, hindering enzyme function, and damaging cellular proteins and DNA. The high surface area to volume ratio enhances their reactivity and potency. These nanomaterials can also suppress undesirable yeasts, fungi, and biofilm formation. Encapsulated natural extracts such as essential oils, bacteriocins, enzymes, and metabolites enable controlled release of additional antimicrobials with nano-enabled regulators and triggers responding to time, moisture, temperature, or microbiological stimuli [42].

29.3.4 Antioxidant Effects

Antioxidants like natural polyphenols can be embedded and released in a controlled and targeted manner using nanocarriers to restrict oxidation reactions in foods. TiO_2 nanoparticles also provide intrinsic antioxidant capacity through photocatalytic superoxide and hydroxyl radical scavenging.

Essentially, nanomaterials play a crucial role in elevating the performance of food packaging by enhancing barrier properties, reinforcing mechanical strength, and providing antimicrobial/antioxidant benefits, thereby contributing to prolonged shelf life and the preservation of product quality [43].

29.4 Conclusion

The integration of nanotechnology offers a highly promising opportunity to address crucial challenges related to food quality, safety, and supply chain efficiency through the development of next-generation "smart" packaging designs. As highlighted, the use of nano-engineered materials such as antimicrobial silver, barrier-enhancing graphene, and controlled-release protein capsules demonstrates remarkable capabilities in extending the shelf life of food, improving quality, enabling intelligent sensing, enhancing mechanical strength, and more. Although there are lingering questions about the scalable manufacturing and lifecycle impacts of certain novel nanomaterials, the increasing momentum in research and the early adoption in commercial settings suggest that significant breakthroughs in nano-enabled packaging are expected to emerge within the next decade. Ongoing technical efforts to refine fabrication processes, maximising efficacy while minimising material usage, will further enhance sustainability benefits and instil confidence in safety. Simultaneously, advancing risk characterisation and toxicology research becomes crucial for establishing appropriate regulatory policies in this rapidly evolving packaging landscape. In summary, the purposeful and science-based application of nanotechnology presents a highly promising set of customisable solutions for addressing interconnected goals such as reducing food waste, minimising packaging impacts, and delivering fresh and safe foods to consumers worldwide.

REFERENCES

[1] V. Katiyar, T. Ghosh, T. Ghosh, V. Katiyar, Edible food packaging: An introduction, Nanotechnology in edible food packaging. *Food preservation practices for a sustainable future*, (2021) 1–23.

[2] M. Primožič, Ž. Knez, M. Leitgeb, (Bio) Nanotechnology in food science—food packaging, *Nanomaterials, 11* (2021) 292.

[3] K. Cooley, V. Lynn, *Nanotechnology: Principles and Practices*, Scientific e-Resources, 2018.

[4] N. Baig, I. Kammakakam, W. Falath, Nanomaterials: A review of synthesis methods, properties, recent progress, and challenges, *Materials Advances, 2* (2021) 1821–1871.

[5] B.S. Sekhon, Food nanotechnology–an overview, *Nanotechnology, science and applications*, (2010) 1–15.

[6] Y. Echegoyen, C. Nerín, Nanoparticle release from nano-silver antimicrobial food containers, *Food and chemical toxicology, 62* (2013) 16–22.

[7] Y. Xing, X. Li, Q. Xu, X. Bi, X. Liu, *TiO2 Nanoparticles and Composite Materials: Antimicrobial Activity, Antimicrobial Mechanism and Applications, Interfaces Between Nanomaterials and Microbes*, CRC Press, 2021, pp. 114–136.

[8] Z.P. Tang, C.W. Chen, J. Xie, Development of antimicrobial active films based on poly (vinyl alcohol) containing nano-TiO2 and its application in macrobrachium rosenbergii packaging, *Journal of food processing and preservation, 42* (2018) e13702.

[9] A. Sirelkhatim, S. Mahmud, A. Seeni, N.H.M. Kaus, L.C. Ann, S.K.M. Bakhori, H. Hasan, D. Mohamad, Review on zinc oxide nanoparticles: antibacterial activity and toxicity mechanism, *Nano-micro letters, 7* (2015) 219–242.

[10] J. Jiang, J. Pi, J. Cai, The advancing of zinc oxide nanoparticles for biomedical applications, *Bioinorganic chemistry and applications, 2018* (2018).

[11] A. Ali, H. Zafar, M. Zia, I. ul Haq, A.R. Phull, J.S. Ali, A. Hussain, Synthesis, characterization, applications, and challenges of iron oxide nanoparticles, *Nanotechnology, science and applications*, (2016) 49–67.

[12] S. F. Hasany, N. H. Abdurahman, A. R. Sunarti, R. Jose, Magnetic iron oxide nanoparticles: chemical synthesis and applications review, *Current nanoscience*, 9 (2013) 561–575.

[13] J. Hornak, P. Trnka, P. Kadlec, O. Michal, V. Mentlík, P. Šutta, G.M. Csányi, Z.Á. Tamus, Magnesium oxide nanoparticles: dielectric properties, surface functionalization and improvement of epoxy-based composites insulating properties, *Nanomaterials*, 8 (2018) 381.

[14] J. Hornak, Synthesis, properties, and selected technical applications of magnesium oxide nanoparticles: a review, *International Journal of Molecular Sciences*, 22 (2021) 12752.

[15] B.M. Yoo, H.J. Shin, H.W. Yoon, H.B. Park, Graphene and graphene oxide and their uses in barrier polymers, *Journal of Applied Polymer Science*, 131 (2014).

[16] H. Chen, T. Filleter, Effect of structure on the tribology of ultrathin graphene and graphene oxide films, *Nanotechnology*, 26 (2015) 135702.

[17] B.S. Sekhon, Nanotechnology in agri-food production: an overview, *Nanotechnology, science and applications*, (2014) 31–53.

[18] A.H. Dar, G.A. Nayik, *Nanotechnology Interventions in Food Packaging and Shelf Life*, (2022).

[19] I. Armentano, M. Gigli, F. Morena, C. Argentati, L. Torre, S. Martino, Recent advances in nanocomposites based on aliphatic polyesters: Design, synthesis, and applications in regenerative medicine, *Applied Sciences*, 8 (2018) 1452.

[20] M.P. Motloung, T.G. Mofokeng, V. Ojijo, S.S. Ray, A review on the processing–morphology–property relationship in biodegradable polymer composites containing carbon nanotubes and nanofibers, *Polymer Engineering & Science*, 61 (2021) 2719–2756.

[21] N. Aich, *Environmental implications of higher order fullerenes and conjugated nanostructures*, 2015.

[22] X. Fan, N. Soin, H. Li, H. Li, X. Xia, J. Geng, Fullerene (C60) nanowires: The preparation, characterization, and potential applications, *Energy & Environmental Materials*, 3 (2020) 469–491.

[23] M. Azizi-Lalabadi, H. Hashemi, J. Feng, S.M. Jafari, Carbon nanomaterials against pathogens; the antimicrobial activity of carbon nanotubes, graphene/graphene oxide, fullerenes, and their nanocomposites, *Advances in Colloid and Interface Science*, 284 (2020) 102250.

[24] M. Naffakh, A.M. Díez-Pascual, C. Marco, G.J. Ellis, M.A. Gomez-Fatou, Opportunities and challenges in the use of inorganic fullerene-like nanoparticles to produce advanced polymer nanocomposites, *Progress in Polymer Science*, 38 (2013) 1163–1231.

[25] M.A. Tofighy, T. Mohammadi, *Barrier, diffusion, and transport properties of rubber nanocomposites containing carbon nanofillers, Carbon-based nanofillers and their rubber nanocomposites*, Elsevier, 2019, pp. 253–285.

[26] M. Alizadeh Sani, A. Khezerlou, M. Tavassoli, K. Mohammadi, S. Hassani, A. Ehsani, D.J. McClements, Bionanocomposite active packaging material based on soy protein Isolate/Persian Gum/Silver nanoparticles; fabrication and characteristics, *Colloids and Interfaces*, 6 (2022) 57.

[27] S.R. Ahmad, P. Ghosh, Application of nanoprotein in food industry and its potential toxicity related health issues, *Indian Journal of Pure & Applied Biosciences*, 8 (2020) 678–689.

[28] E. Assadpour, S.M. Jafari, An overview of biopolymer nanostructures for encapsulation of food ingredients, *Biopolymer nanostructures for food encapsulation purposes*, (2019) 1–35.

[29] H. Tian, G. Guo, X. Luo, Advances in Soy Protein-Based Nanocomposites, Soy Protein-Based Blends, Composites and Nanocomposites, (2017) 39–66.

[30] N. Lin, J. Huang, A. Dufresne, Preparation, properties and applications of polysaccharide nanocrystals in advanced functional nanomaterials: a review, *Nanoscale*, 4 (2012) 3274–3294.

[31] P.I. Siafaka, E. Özcan Bülbül, M.E. Okur, I.D. Karantas, N. Üstündağ Okur, The application of nanogels as efficient drug delivery platforms for dermal/transdermal delivery, *Gels*, 9 (2023) 753.

[32] N.A. Hassan, O.M. Darwesh, S.S. Smuda, A.B. Altemimi, A. Hu, F. Cacciola, I. Haoujar, T.G. Abedelmaksoud, Recent trends in the preparation of nano-starch particles, *Molecules*, 27 (2022) 5497.

[33] F. Xie, C. Gao, L. Avérous, Alginate-based materials: Enhancing properties through multiphase formulation design and processing innovation, *Materials Science and Engineering: R: Reports*, 159 (2024) 100799.

[34] K. Neenu, C.M. Dominic, P.S. Begum, J. Parameswaranpillai, B.P. Kanoth, D.A. David, S.M. Sajadi, P. Dhanyasree, T. Ajithkumar, M. Badawi, Effect of oxalic acid and sulphuric acid hydrolysis on

the preparation and properties of pineapple pomace derived cellulose nanofibers and nanopapers, *International Journal of Biological Macromolecules, 209* (2022) 1745–1759.

[35] C. Viegas, A.B. Patrício, J.M. Prata, A. Nadhman, P.K. Chintamaneni, P. Fonte, Solid lipid nanoparticles vs. nanostructured lipid carriers: a comparative review, *Pharmaceutics, 15* (2023) 1593.

[36] K.Y. Perera, M. Hopkins, A.K. Jaiswal, S. Jaiswal, Nanoclays-containing bio-based packaging materials: Properties, applications, safety, and regulatory issues, *Journal of Nanostructure in Chemistry, 14* (2024) 71–93.

[37] C. Muñoz-Shugulí, C.P. Vidal, P. Cantero-López, J. Lopez-Polo, Encapsulation of plant extract compounds using cyclodextrin inclusion complexes, liposomes, electrospinning and their combinations for food purposes, *Trends in Food Science & Technology, 108* (2021) 177–186.

[38] J. Tang, W. Chen, W. Su, W. Li, J. Deng, Dendrimer-encapsulated silver nanoparticles and antibacterial activity on cotton fabric, *Journal of Nanoscience and Nanotechnology, 13* (2013) 2128–2135.

[39] L. Zhao, G. Duan, G. Zhang, H. Yang, S. He, S. Jiang, Electrospun functional materials toward food packaging applications: A review, *Nanomaterials, 10* (2020) 150.

[40] T.V. Duncan, Applications of nanotechnology in food packaging and food safety: barrier materials, antimicrobials and sensors, *Journal of colloid and interface science, 363* (2011) 1–24.

[41] S. Jafarzadeh, S.M. Jafari, Impact of metal nanoparticles on the mechanical, barrier, optical and thermal properties of biodegradable food packaging materials, *Critical reviews in food science and nutrition, 61* (2021) 2640–2658.

[42] A.A. Anvar, H. Ahari, M. Ataee, Antimicrobial properties of food nanopackaging: A new focus on food-borne pathogens, *Frontiers in microbiology, 12* (2021) 690706.

[43] C. Vasile, M. Baican, Progresses in food packaging, food quality, and safety—controlled-release antioxidant and/or antimicrobial packaging, *Molecules, 26* (2021) 1263.

30

Development of Smart Textiles Using Functional Nanofibres for Wearable Technology

R. M. Devarajaiah
Acharya Institute of Technology, Bengaluru, India

M. K. Chaanthini
E.G.S. Pillay Engineering College, Nagapattinam, India

D. R. Srinivasan
JNTUA College of Engineering, Anantapur, India

Nidamanuri Sreenivasa Babu
University of Technology and Applied Sciences-Shinas, College of Engineering and Technology, Engineering Department, Sultanate of Oman

Amos Gamaleal David
Panimalar Engineering College, Chennai, India

30.1 Nanofibre Synthesis Techniques for Smart Textiles

30.1.1 Nanofibre Synthesis Techniques

Nanofibre synthesis techniques are crucial for the development of smart textiles, which are gaining prominence in various fields such as healthcare, sports, and fashion. These techniques enable the fabrication of nanoscale fibres with unique properties and functionalities suitable for integration into textile materials. In this chapter, we will explore three primary nanofibre synthesis techniques: Electrospinning, solution blow spinning, and Forcespinning™ technology. Each technique offers distinct advantages and challenges, making them suitable for different applications in smart textiles [1].

30.1.2 Electrospinning: A Versatile Method for Nanofibre Production

Electrospinning is a widely used and versatile technique for producing nanofibres. It involves the application of a high-voltage electric field to a polymer solution or melt, causing the polymer to stretch and form ultrafine fibres that are collected on a grounded substrate [2]. Electrospinning offers precise control over fibre morphology, diameter, and alignment, making it suitable for a wide range of applications in smart textiles. However, electrospinning also has limitations such as low production rates and difficulties in scaling up for industrial applications.

30.1.3 Solution Blow Spinning: Fabrication of Functional Nanofibres for Wearable Technology

Solution blow spinning is an emerging technique for the fabrication of functional nanofibres for wearable technology applications. It involves the extrusion of a polymer solution through a nozzle at high velocity, followed by rapid evaporation of the solvent, resulting in the formation of nanofibres. Solution blow

DOI: 10.1201/9781003495437-30

spinning offers advantages such as high throughput, scalability, and the ability to incorporate functional additives or nanoparticles into the nanofibres during the spinning process. This technique shows promise for the development of smart textiles with enhanced functionalities such as sensing, actuation, and energy harvesting.

30.1.4 Forcespinning™ Technology: Innovations in Nanofibre Synthesis for Smart Textiles

Forcespinning™ technology is a novel approach to nanofibre synthesis that offers several advantages over traditional spinning methods. Unlike electrospinning and solution blow spinning, which rely on electric or pneumatic forces, Forcespinning™ technology utilises mechanical forces generated by a rotating spinneret to stretch and draw out nanofibres from a polymer solution. This technique offers high production rates, uniform fibre morphology, and the ability to process a wide range of polymers and functional materials. Forcespinning™ technology holds great promise for the scalable and cost-effective production of nanofibres with precisely controlled properties for smart textile applications [3].

30.2 Functionalisation of Nanofibres for Enhanced Properties

30.2.1 Nanofibres in Smart Textiles

Nanofibres play a pivotal role in the development of smart textiles, offering unique properties such as high surface area-to-volume ratio, flexibility, and tuneable porosity. However, to fully realise their potential in smart textile applications, it is essential to enhance their properties through functionalisation. In this chapter, we will explore various techniques for functionalising nanofibres to tailor their properties for specific applications in smart textiles [4]. We will discuss surface modification techniques, incorporation of nanoparticles, and chemical and biological functionalisation methods, highlighting their significance in enhancing the functionality and performance of nanofibres for wearable technology.

30.2.2 Surface Modification Techniques for Tailoring Nanofibre Properties

Surface modification techniques are widely used to tailor the properties of nanofibres for specific applications in smart textiles. These techniques involve modifying the surface chemistry or morphology of nanofibres to impart desired functionalities such as hydrophobicity, antimicrobial properties, or biocompatibility. Common surface modification methods include plasma treatment, chemical grafting, and physical vapour deposition. Plasma treatment involves exposing the nanofibre surface to a plasma discharge to introduce functional groups or alter surface morphology. Chemical grafting involves covalently attaching functional molecules or polymers to the nanofibre surface, while physical vapour deposition involves depositing thin films of functional materials onto the nanofibre surface [5]. These surface modification techniques offer precise control over the surface properties of nanofibres, enabling the design of smart textiles with tailored functionalities for various applications.

30.2.3 Incorporation of Nanoparticles for Added Functionality in Smart Textiles

Incorporating nanoparticles into nanofibres is another effective strategy for enhancing their functionality in smart textiles. Nanoparticles offer unique properties such as high surface area, quantum size effects, and enhanced mechanical, electrical, or optical properties, making them ideal candidates for imparting additional functionalities to nanofibres. Common nanoparticles used in smart textiles include metal nanoparticles (e.g., silver and gold), metal oxide nanoparticles (e.g., titanium dioxide and zinc oxide), carbon-based nanoparticles (e.g., carbon nanotubes and graphene), and quantum dots. These nanoparticles can be incorporated into nanofibres through various methods such as electrospinning, solution blending, or in situ synthesis [6]. The incorporation of nanoparticles enables the development of smart textiles with advanced functionalities such as antimicrobial activity, UV protection, conductive properties, or optical sensing capabilities, expanding their potential applications in healthcare, sports, and fashion.

30.2.4 Chemical and Biological Functionalisation of Nanofibres for Wearable Sensors

Chemical and biological functionalisation of nanofibres is a promising approach for developing wearable sensors with enhanced sensing capabilities. These functionalisation methods involve immobilising sensing molecules or biomolecules onto the nanofibre surface to enable selective detection of target analytes such as biomarkers, gases, or environmental pollutants. Chemical functionalisation techniques include covalent immobilisation, physical adsorption, and molecular imprinting, while biological functionalisation techniques involve immobilising enzymes, antibodies, or DNA aptamers onto the nanofibre surface. These functionalised nanofibres can be incorporated into wearable sensor devices for applications such as health monitoring, environmental sensing, and food safety. By leveraging the unique properties of nanofibres and the selectivity of functionalised sensing elements, wearable sensors based on functionalised nanofibres offer high sensitivity, specificity, and reliability, paving the way for personalised and real-time monitoring of physiological and environmental parameters.

30.3 Integration of Nanofibres into Smart Textiles

30.3.1 Nanofibre Coating and Lamination Techniques for Textile Integration

Nanofibres offer unique properties such as high surface area, flexibility, and tuneable porosity, making them ideal candidates for integration into smart textiles. One of the key challenges in incorporating nanofibres into textiles is ensuring seamless integration while maintaining the desired properties of both materials. Nanofibre coating and lamination techniques provide effective solutions for achieving this integration.

Coating involves depositing a layer of nanofibres onto the surface of textile substrates using techniques such as dip coating, spray coating, or electrostatic deposition. This process allows for precise control over the thickness and distribution of nanofibres on the textile surface, enabling the enhancement of specific functionalities such as water repellency, breathability, or antibacterial properties. Lamination, on the other hand, involves bonding nanofibre membranes onto textile substrates using techniques such as heat pressing, ultrasonic welding, or adhesive bonding. This approach provides a more robust and durable integration of nanofibres into textiles, making it suitable for applications requiring higher mechanical strength and stability.

30.3.2 3D Printing of Smart Textiles Using Nanofibres

3D printing technologies offer a versatile and customisable approach for integrating nanofibres into smart textiles. Additive manufacturing techniques such as fused deposition modelling (FDM), selective laser sintering (SLS), and inkjet printing allow for the precise deposition of nanofibre-based inks onto textile substrates, enabling the creation of complex 3D structures with tailored functionalities [7].

In FDM, nanofibre-based filaments are extruded through a heated nozzle and deposited layer by layer onto a textile substrate, allowing for the creation of intricate patterns and designs. SLS involves using a laser to sinter nanofibre powders onto the textile surface, providing high resolution and accuracy in the deposition process. Inkjet printing utilises inkjet printheads to deposit nanofibre-based inks onto textile substrates in a controlled manner, offering flexibility in design and material selection.

3D printing of smart textiles using nanofibres enables the integration of functional elements such as sensors, actuators, and energy harvesting devices directly into the textile structure, providing enhanced functionalities such as sensing, actuation, and energy storage [8]. This approach allows for the development of smart textiles with customisable properties tailored to specific applications in healthcare, sports, fashion, and beyond.

30.3.3 Seamless Integration of Nanofibres for Enhanced Comfort and Performance

Seamless integration of nanofibres into smart textiles is essential for ensuring comfort, flexibility, and performance in wearable applications. Traditional textile manufacturing processes such as weaving, knitting, and braiding often result in seams or joints that can compromise the integrity and functionality of nanofibre-based materials.

To address this challenge, advanced manufacturing techniques such as seamless knitting, seamless weaving, and 3D knitting have been developed to seamlessly integrate nanofibres into textile structures without the need for additional stitching or joining processes. Seamless knitting involves using computer-controlled knitting machines to produce seamless garments with integrated nanofibre components, offering superior comfort and flexibility compared to traditional knitting methods. Seamless weaving utilises specialised looms to create seamless fabrics with embedded nanofibres, providing enhanced strength and durability [9]. 3D knitting combines knitting and additive manufacturing techniques to create complex 3D structures with integrated nanofibre elements, allowing for the production of customised and functional textiles with seamless integration of nanofibres.

The seamless integration of nanofibres into smart textiles enhances comfort, flexibility, and performance, making them suitable for a wide range of applications in healthcare, sports, fashion, and beyond. By leveraging advanced manufacturing techniques, researchers and engineers can develop smart textiles with superior properties and functionalities, paving the way for the next generation of wearable technology.

30.4 Applications of Smart Textiles in Wearable Technology

In recent years, smart textiles have emerged as a promising platform for integrating advanced functionalities into wearable devices, revolutionising various fields such as healthcare, fitness, and lifestyle. This chapter explores the diverse applications of smart textiles in wearable technology, focusing on wearable sensors, energy harvesting and storage technologies, and textile electronics for health monitoring and wellness.

30.4.1 Wearable Sensors and Biosensors Based on Functional Nanofibres

Wearable sensors play a crucial role in monitoring various physiological parameters and environmental conditions in realtime. Functional nanofibres offer unique advantages for fabricating wearable sensors due to their high surface area, flexibility, and biocompatibility. These nanofibres can be functionalised with sensing elements such as conductive polymers, metal nanoparticles, and biomolecules to detect a wide range of analytes, including temperature, humidity, pressure, and biochemical markers.

One of the key applications of wearable sensors is in healthcare, where they are used for continuous monitoring of vital signs such as heart rate, respiration rate, and blood glucose levels. Nanofibre-based biosensors enable non-invasive and real-time monitoring of biomarkers in sweat, saliva, and tears, providing valuable insights into the wearer's health status. Additionally, wearable sensors integrated into clothing and accessories offer seamless and unobtrusive monitoring, enhancing user comfort and compliance [10].

30.4.2 Energy Harvesting and Storage Technologies in Smart Textiles

Energy harvesting and storage technologies play a vital role in powering wearable devices and extending their operating time. Smart textiles offer a versatile platform for integrating energy harvesting devices such as thermoelectric generators, piezoelectric nanogenerators, and solar cells into clothing and accessories, enabling the conversion of ambient energy sources into electrical power.

Nanofibre-based materials are well-suited for energy harvesting applications due to their lightweight, flexible, and scalable nature. Nanofibre-based piezoelectric materials can generate electrical energy from mechanical vibrations and movements, while nanofibre-based thermoelectric materials can convert temperature gradients into electrical power. Furthermore, nanofibre-based supercapacitors and batteries provide efficient energy storage solutions for storing harvested energy and powering wearable devices.

30.4.3 Nanofibre-Based Textile Electronics for Health Monitoring and Wellness

Textile electronics, or e-textiles, refer to fabrics and garments integrated with electronic components such as sensors, actuators, and communication modules. Nanofibres play a crucial role in fabricating textile-based electronic devices due to their lightweight, flexible, and breathable properties. Nanofibre-based conductive materials, such as conductive polymers, carbon nanotubes, and metal nanowires, enable the fabrication of flexible and stretchable electronic circuits and electrodes for health monitoring and wellness applications. One of the key applications of nanofibre-based textile electronics is in monitoring physiological parameters such as heart rate variability, skin temperature, and muscle activity. Wearable garments embedded with nanofibre-based sensors and electrodes can provide continuous and non-invasive monitoring of the wearer's health status, facilitating early detection of medical conditions and personalised healthcare management [11]. Additionally, nanofibre-based textile electronics can be integrated into smart clothing and accessories for tracking physical activity, promoting fitness, and enhancing overall wellness.

In summary, smart textiles offer a versatile platform for integrating advanced functionalities into wearable devices, enabling applications such as wearable sensors for health monitoring, energy harvesting and storage technologies for powering wearable devices, and textile electronics for enhancing user comfort and performance. Nanofibres play a crucial role in enabling these applications by providing lightweight, flexible, and functional materials for fabricating wearable technology solutions.

30.5 Exploring Novel Functionalities Enabled by Nanofibre Integration

Nanofibres, with their high surface area-to-volume ratio and unique mechanical properties, offer a versatile platform for integrating advanced functionalities into wearable devices. One of the key areas of exploration is the development of smart textiles capable of sensing and responding to external stimuli. Nanofibre-based sensors embedded within textiles can detect various parameters such as temperature, humidity, and pressure, enabling real-time monitoring of environmental conditions and user health metrics. Additionally, nanofibre membranes with selective permeability properties can be incorporated into wearable devices for applications such as moisture management, odour control, and protection against harmful UV radiation [1].

Furthermore, nanofibre-based composites are being investigated for their potential in energy harvesting and storage applications. By incorporating piezoelectric nanofibres or nanogenerators into wearable textiles, it becomes possible to convert mechanical energy from body movements into electrical energy, thus powering the devices or storing it for later use. This opens up new possibilities for self-powered wearable electronics that eliminate the need for external batteries or charging ports.

30.5.1 Enhancing Wearable Technology with Nanofibre-Based Sensors and Actuators

Nanofibre-based sensors play a pivotal role in enhancing the functionality and usability of wearable technology. These sensors can be designed to detect a wide range of physiological parameters, including heart rate, blood pressure, muscle activity, and even biochemical markers present in sweat. By integrating these sensors seamlessly into garments or accessories, such as shirts, socks, or wristbands, it becomes possible to monitor an individual's health status in realtime, providing valuable insights for healthcare professionals and users alike.

Moreover, nanofibre-based actuators offer new possibilities for creating interactive and adaptive wearable devices shown in Figure 30.1. By incorporating shape memory polymers or electroactive nanofibres into textiles, it becomes possible to engineer fabrics that can change their shape, stiffness, or porosity in response to external stimuli such as temperature, electrical voltage, or mechanical stress. This opens up avenues for developing self-adjusting garments that can adapt to the wearer's needs or environmental conditions, providing enhanced comfort and performance in various situations.

30.5.2 Future Trends in Nanofibre-Enabled Wearable Devices

Looking ahead, the field of nanofibre-based wearable technology is poised for significant advancements and innovations. One of the key trends is the continued miniaturisation and integration of functional components into textiles, leading to the development of lightweight, flexible, and unobtrusive wearable devices that seamlessly blend into everyday clothing. This trend is driven by advancements in nanofibre

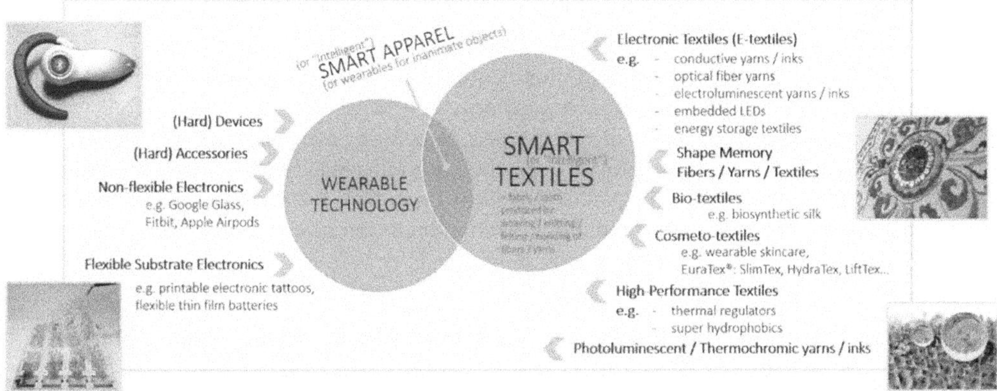

FIGURE 30.1 Enhancing wearable technology with nanofibre-based sensors and actuators. (https://www.intechopen.com/chapters/68403)

fabrication techniques, such as electrospinning and solution blowing, which enable precise control over fibre properties and morphology [12].

Additionally, there is growing interest in the development of smart textiles with autonomous capabilities, such as self-healing materials that can repair damage or degradation caused by wear and tear. By incorporating self-healing polymers or nanocomposites into textiles, it becomes possible to extend the lifespan of wearable devices, reducing the need for frequent replacements and contributing to sustainability.

Furthermore, the integration of artificial intelligence (AI) and machine learning algorithms into nanofibre-enabled wearable devices is expected to unlock new functionalities and applications. AI-powered wearable systems can analyse data collected from sensors in realtime, providing personalised insights and recommendations to users based on their unique physiology and activity levels. This personalised approach to healthcare and wellness monitoring has the potential to revolutionise preventive medicine and lifestyle management, empowering individuals to take control of their health and well-being.

30.6 Biomedical Applications of Nanofibre-Based Smart Textiles

Nanofibre-based smart textiles have garnered significant interest in biomedical applications due to their unique properties, including high surface area, porosity, and tuneable mechanical characteristics. This chapter explores the various biomedical applications of nanofibre-based smart textiles, focusing on biosensors for healthcare monitoring, drug delivery systems, and advancements in medical textiles for therapeutic applications.

30.6.1 Nanofibre Biosensors for Healthcare Monitoring and Diagnostics

Nanofibre-based biosensors have emerged as powerful tools for real-time healthcare monitoring and diagnostics. These biosensors leverage the high surface area and excellent biocompatibility of nanofibres to detect specific biomarkers with high sensitivity and selectivity. By functionalising nanofibres with biomolecules or nanoparticles, researchers can create biosensors capable of detecting a wide range of analytes, including glucose, proteins, and nucleic acids, in bodily fluids such as blood, sweat, and saliva. These biosensors can be integrated into wearable textiles, providing continuous monitoring of physiological parameters and early detection of health conditions such as diabetes, cardiovascular diseases, and infectious diseases [13].

30.6.2 Drug Delivery Systems Integrated into Nanofibre-Based Textiles

Nanofibre-based smart textiles offer a promising platform for controlled drug delivery systems with applications in wound healing, transdermal drug delivery, and localised therapy. By loading drugs or therapeutic agents into nanofibre matrices, researchers can achieve sustained release profiles, targeted delivery, and enhanced bioavailability of therapeutics. Nanofibre-based drug delivery systems can be incorporated into wound dressings, bandages, or patches, providing localised treatment while minimising systemic side effects. Furthermore, the porous structure of nanofibre scaffolds allows for the encapsulation of both hydrophobic and hydrophilic drugs, making them versatile platforms for a wide range of therapeutic applications.

30.6.3 Advancements in Nanofibre-Based Medical Textiles for Therapeutic Applications

Recent advancements in nanofibre-based medical textiles have led to the development of innovative solutions for various therapeutic applications, including tissue engineering, regenerative medicine, and implantable medical devices. Nanofibre scaffolds mimicking the extracellular matrix of tissues provide an ideal environment for cell growth, proliferation, and differentiation, making them valuable tools for tissue regeneration and repair. These scaffolds can be engineered to exhibit specific mechanical properties, surface topographies, and bioactive functionalities to promote tissue regeneration in various anatomical locations, including bone, cartilage, skin, and vascular tissues.

Moreover, nanofibre-based medical textiles offer unique opportunities for the development of implantable medical devices with enhanced biocompatibility and functionality. By incorporating nanofibres into implants such as vascular grafts, stents, and orthopaedic implants, researchers can improve the integration of implants with host tissues, reduce the risk of infection, and enhance the overall performance and longevity of the devices. Furthermore, nanofibre coatings can be applied to medical devices to provide drug-eluting or antimicrobial properties, further enhancing their therapeutic efficacy and safety.

30.7 Challenges and Future Directions in Smart Textiles Development

Smart textiles hold immense promise for revolutionising various industries by integrating advanced functionalities into everyday fabrics and garments. However, the development and widespread adoption of smart textiles face several challenges, ranging from durability and washability issues to scalability and commercialisation hurdles. This chapter examines these challenges in detail and explores future trends and innovations in the field of smart textiles using functional nanofibres.

30.7.1 Durability and Washability Issues in Nanofibre-Based Smart Textiles

One of the primary challenges in the development of smart textiles is ensuring their durability and washability, especially when incorporating nanofibres into fabric structures. Nanofibres, while offering unique properties such as flexibility and breathability, can be fragile and prone to damage during washing and regular wear. Moreover, the integration of electronic components, such as sensors and conductive materials, into textile substrates further complicates the durability aspect.

To address these challenges, researchers are exploring various strategies, including the development of protective coatings for nanofibres, reinforcement techniques to enhance mechanical strength, and novel fabrication methods to improve washability. Additionally, advances in nanomaterial synthesis and textile manufacturing processes are facilitating the production of durable and washable smart textiles that can withstand repeated use and laundering without compromising functionality.

30.7.2 Scalability and Commercialisation Challenges of Nanofibre Production

Another significant challenge in the development of smart textiles is the scalability and commercialisation of nanofibre production processes. While electrospinning, solution blow spinning, and other techniques offer precise control over nanofibre morphology and properties, they often suffer from limitations in terms of production scale and throughput. Scaling up these processes to meet industrial demand while maintaining cost-effectiveness remains a significant challenge.

Furthermore, the commercialisation of nanofibre-based smart textiles requires addressing regulatory requirements, ensuring product safety and compliance with industry standards, and establishing reliable supply chains for raw materials and manufacturing equipment. Collaborations between academic researchers, industry partners, and government agencies are essential for overcoming these challenges and accelerating the adoption of smart textiles in various market sectors.

30.7.3 Future Trends and Innovations in the Field of Smart Textiles Using Functional Nanofibres

Despite the challenges, the field of smart textiles using functional nanofibres is poised for significant growth and innovation in the coming years. Several trends and advancements are shaping the future of smart textiles, including the following:

1. Advanced Functionalities: Future smart textiles are expected to offer a wide range of advanced functionalities beyond sensing and actuation, including self-healing, shape memory, and adaptive thermal regulation. Nanofibres will play a crucial role in enabling these functionalities by serving as versatile building blocks for fabricating multifunctional textile structures.

2. Sustainable Materials: With growing concerns about environmental sustainability, there is a rising demand for smart textiles made from eco-friendly and biodegradable materials. Nanofibres derived from renewable sources, such as cellulose and chitosan, are gaining attention as sustainable alternatives to traditional synthetic polymers.

3. Wearable Healthcare Technologies: The integration of smart textiles with healthcare monitoring and diagnostic systems holds promise for revolutionising personalised healthcare delivery. Future smart textiles will incorporate advanced biosensors, drug delivery systems, and physiological monitoring devices for continuous health monitoring and disease management.

4. Customisation and Personalisation: Advances in digital fabrication technologies, such as 3D printing and additive manufacturing, are enabling the customisation and personalisation of smart textiles to meet individual user preferences and requirements. Nanofibre-based smart textiles will offer tailored solutions for diverse applications, ranging from sports performance optimisation to medical rehabilitation.

5. Internet-of-Things (IoT) Integration: Smart textiles will increasingly be integrated into the IoT ecosystem, enabling seamless connectivity and data exchange with other smart devices and systems. Nanofibre-based sensors and actuators will play a crucial role in enabling real-time communication and interaction between wearable textiles and IoT platforms.

30.8 Conclusion

In conclusion, the development of smart textiles using functional nanofibres has ushered in a new era of innovation in wearable technology. Throughout this study, we have explored the various facets of this emerging field, from the synthesis and functionalisation of nanofibres to their integration into wearable devices for diverse applications. The integration of nanofibres has enabled the creation of wearable sensors, drug delivery systems, and energy harvesting devices with enhanced functionality and performance. Despite the progress made, challenges such as durability, scalability, and environmental sustainability remain areas of concern that require further attention. However, the future looks promising, with ongoing research and development efforts focusing on addressing these challenges and unlocking new possibilities for nanofibre-based smart textiles. As we continue to push the boundaries of innovation, the convergence of nanotechnology and wearable technology holds immense potential to revolutionise healthcare, environmental monitoring, and consumer electronics, ultimately improving lives and shaping the future of wearable technology.

REFERENCES

[1] Syduzzaman, M., Hassan, A., Anik, H.R., Akter, M. and Islam, M.R., 2023. Nanotechnology for high-performance textiles: A promising frontier for innovation. *ChemNanoMat*, 9(9), p. e202300205.

[2] Bhattarai, P., Thapa, K.B., Basnet, R.B. and Sharma, S., 2014. Electrospinning: how to produce nanofibers using most inexpensive technique? An insight into the real challenges of electrospinning such nanofibers and its application areas. *International Journal of Biomedical and Advance Research*, 5(9), pp. 401–405.

[3] Deravi, L.F., Sinatra, N.R., Chantre, C.O., Nesmith, A.P., Yuan, H., Deravi, S.K., Goss, J.A., MacQueen, L.A., Badrossamy, M.R., Gonzalez, G.M. and Phillips, M.D., 2017. Design and fabrication of fibrous nanomaterials using pull spinning. *Macromolecular Materials and Engineering*, 302(3), p. 1600404.

[4] Chamanehpour, E., Thouti, S., Rubahn, H.G., Dolatshahi-Pirouz, A. and Mishra, Y.K., 2024. Smart nanofibers: Synthesis, properties, and scopes in future advanced technologies. *Advanced Materials Technologies*, 9(3), p. 2301392.

[5] Gopal, R., Zuwei, M., Kaur, S. and Ramakrishna, S., 2007. Surface modification and application of functionalized polymer nanofibers. In G. Ali Mansoori, Thomas F. George, Lahsen Assoufid, Guoping Zhang (eds.) *Molecular Building Blocks for Nanotechnology: From Diamondoids to Nanoscale Materials and Applications* (pp. 72–91). New York, NY: Springer New York.

[6] Dolez, P.I., 2015. Nanomaterials definitions, classifications, and applications. In Patricia I. Dolez (ed.) *Nanoengineering* (pp. 3–40). Elsevier: QC, Canada.

[7] Arockiam, A.J., Subramanian, K., Padmanabhan, R.G., Selvaraj, R., Bagal, D.K. and Rajesh, S., 2022. A review on PLA with different fillers used as a filament in 3D printing. *Materials Today: Proceedings*, 50, pp. 2057–2064.

[8] Joseph Arockiam, A., Rajesh, S. and Karthikeyan, S., 2023. Development of fish scale particle reinforced PLA filaments for 3D printing applications. *Journal of Applied Polymer Science*, 141, p. e55132.

[9] Arockiam, A.J., Rajesh, S., Karthikeyan, S., Thiagamani, S.M.K., Padmanabhan, R.G., Hashem, M., Fouad, H. and Ansari, A., 2023. Mechanical and thermal characterization of additive manufactured fish scale powder reinforced PLA biocomposites. *Materials Research Express*, 10(7), p. 075504.

[10] Jacobsen, M., Dembek, T.A., Kobbe, G., Gaidzik, P.W. and Heinemann, L., 2021. Noninvasive continuous monitoring of vital signs with wearables: fit for medical use?*Journal of Diabetes Science and Technology*, 15(1), pp. 34–43.

[11] Yang, G., Pang, G., Pang, Z., Gu, Y., Mäntysalo, M. and Yang, H., 2018. Non-invasive flexible and stretchable wearable sensors with nano-based enhancement for chronic disease care. *IEEE Reviews in Biomedical Engineering*, 12, pp. 34–71.

[12] Barhoum, A., Pal, K., Rahier, H., Uludag, H., Kim, I.S. and Bechelany, M., 2019. Nanofibers as new-generation materials: From spinning and nano-spinning fabrication techniques to emerging applications. *Applied Materials Today*, 17, pp. 1–35.

[13] Guk, K., Han, G., Lim, J., Jeong, K., Kang, T., Lim, E.K. and Jung, J., 2019. Evolution of wearable devices with real-time disease monitoring for personalized healthcare. *Nanomaterials*, 9(6), p. 813.

31

Novel Energy Harvesting Systems for Nanorobots

D. Elayaraja and M. Anusuya
Indra Ganesan College of Engineering, Trichy, Tamil Nadu, India

R. Thenmozhi
Indra Ganesan College of Arts and Science, Trichy, India

V. Saravanan
Indra Ganesan College of Engineering, Trichy, India

R. Venkatesh
PSNA College of Engineering and Technology, Dindugal, India

31.1 Introduction

Research on green energy, sustainable energy, and renewable energy is the greatest challenge to human civilisation in sustainable development. The energy sources fuelling the world in large scale are nuclear energy, petroleum, natural gas, coal, and hydraulic energy. Research and development activities are also carried in non-conventional energy sources such as solar wind, biomass, geothermal, wind, and hydrogen [1].

Researches are also carried in the much smaller operation of micro electrochemical systems, ultrasonic chemical and biomolecular sensors, personal electronics and mobile environment sensors, and nanorobots. A nanorobot is a smart machine that performs complex function, manipulates objects, is able to feel, and cooperateswith the environment. But the tough task is to get a source of power for the robot with lesser weight [2].

Conventional energy harvesting systems are mechanical, optical, and vibration. The novel energy harvesting techniques are nanowires from piezoelectric materials, microbial fuel cells, and adenosine triphosphate (ATP) energy converter.

31.2 Conventional Energy Harvesting Systems

31.2.1 Mechanical Energy Harvesting

Electric power could be produced by combining mechanical energy with several types of energy transducers. Piezoelectric materials are one possibility; they display a charge separation under mechanical stress resulting from motion or periodic vibration. They can be installed directly on a surface that exhibits mechanical strain or constructed as a resonating cantilever structure. Because of this, this idea is challenging for applications whose vibration frequencies are irregular [4].

Although they are far more complicated, tuneable vibration harvesters can be adjusted to the stimulating frequencies. Piezoelectric materials include PVDF (polyvinylidene fluoride) and PZT (lead–zirconate–titanate). PZT generates extremely high voltages, yet is extremely fragile. Although PVDF has a lower voltage output, it is quite versatile. Another option is electro-dynamic or inductive generators, which use a coil and a magnet as the primary components to generate electricity. One component moves in relation to the other and causes an electrical current to flow through the coil if the entire combination is subjected to vibration [5].

DOI: 10.1201/9781003495437-31

31.2.2 Optical Energy Harvesting

An energy harvesting system opts best for energy harvesting in micro and nanoscale is light. Optical energy harvesting can be done using photovoltaic cells designed for indoor and outdoor conditions.

Optical power generation is a power generation method that gathers light energy from lighting sources such as solar light, incandescent light, fluorescent light, and LED to generate power. A solar cell can be made of many different materials, including organic thin films, dye-sensitised silicon, silicon amorphous, and silicon crystal type [6].

31.2.3 Vibration Energy Generation

Three different types of electromechanical sensors are used, namely piezoelectric, electrostatic, and electromagnetic vibration-based generators.

The electromagnetic microgenerator creates an alternating electric current in a closed circuit by means of a moving magnet or coil. Nevertheless, a few microgenerators the size of micro electromechanical systems have been produced. This method is suitable for large structure having oscillations. The oscillation ranges from 0.5 mm to 1 mm and the range of power output is 10 μW to 1 kW. Vibration power generation is the process of producing electricity by harvesting vibration energy from many sources, including motors, engines, and other machinery, as well as from the construction of bridges and roads. While some generate power by electromagnetic or electrostatic induction, others use piezoelectric elements [7].

31.2.4 Thermal Energy Harvesting

Thermoelectric power generation is a method of producing electricity by using thermal energy as a source of energy. It generates power by collecting thermal energy from various sources, such as engines, motors, and machinery as well as from the pipes in buildings and factories. The primary one uses a Peltier element to generate power via temperature differential [8].

31.2.5 Radio Waves Energy Harvesting

The process of creating electricity by collecting radio wave energy from many sources, including wireless local area networks (LANs), televisions, radios, cell phones, and more, is known as radio waves power production. To convert radio wave energy into direct current, a device called rectenna is used [9].

31.3 Novel Energy Harvesting System

There is a lot of promise in using nanoscale mechanical energy harvesting to power small circuits and produce self-sufficient electronic devices. Big battery and energy storage components can be left out because electronic device packages can function without an external power source. After the invention of nanogenerators, it became a good power source for nanogenerators. Piezoelectric nanogenerators occupied an important role in nanogenerators.

Power scavenging can power small wireless computer devices due to very large-scale industry (VLSI) design and wireless sensor nodes [10]. First, a detailed examination of potential power scavenging techniques as well as conventional energy sources is given. Extensive study is conducted on low-frequency vibrations that may be a potential source of electricity and that can happen in ordinary home and working environments. Rather than arguing that vibration conversion is the most efficient or versatile technique to scavenge ambient electricity [11].

The flow field drives the waterwheel which is the special design of vibration energy harvesting which is of fluid solid energy harvesting nature. By changing the magnetic field, this waterwheel generates power by turning a wheel that is equipped with magnets. The flow field is also responsible for driving connecting rods and gears. A piezoelectric patch (PZT) is introduced and magnetised upstream of the pipeline [12].

Daochun et al. [13] examined vibration energy-based energy harvesting techniques which cover creation, experiment, and application. The initial stage is to offer numerous energy harvester designs designed to exploit a wide class of flow-induced vibrations. Special attention was given to study the operational characteristics of unmanned aerial vehicle. We investigate several phenomena exhibiting flow-induced vibrations, including air turbulence and gusts, limit cycle oscillations of plates and wing sections, vortex-induced vibrations of downstream structures, and vortex-induced and galloping oscillations of bluff materials [14].

Christian Falconi et al. [3] provided a report on high-efficiency nanogenerators and developed high-efficiency piezoelectric generator,whichwas compared with type II contact piezoelectric nanogenerators because its size, location, and direction of the input force can all be very important.

Photovoltaic (PV) technology and the type of illumination source determine the average voltage that PV cells can produce, which is 0.5 V.PV cells linked in series cell design can increase capacity. Optical antennas are another option for PV energy collection. The optical antenna's size is on the order of nanoscale. It has theoretical efficiency above 60%. With plasmon, a mean of optical coupling close to sub-wavelength dimensions, optical antennas are coupled with PV technology.

31.3.1 Piezoelectric Nanowires for Energy Harvesting Applications

Wang et al. [2] started their research on piezoelectric energy harvesting system in the year 2006. They studied design, modelling, operation mechanism, and performance optimisation. Various flexible nanogenerators were developed to extract vibration energy from human body or environment. The generators will run LED and LCD.

31.3.1.1 Vertical-Aligned Nanowire Array Energy Harvesting

Vertical compression and lateral bending are two different models of vertical aligned nanowire.

Because of different mechanical and electrical configuration, the working mechanism is also different in different configurations to certain degrees, but with one consistent basis: The coupling of the semiconductor behaviour and the piezoelectric property of the piezoelectric nanowire. A Schottky barrier was built up between the nanowire and the AFM tip due to the difference in working function and electron affinity.

31.3.1.2 Lateral-Aligned Nanowire Networks

The nanowire bends laterally because the pressure acts on the radial direction and uniform bending takes place in it. It is considered as lateral stretching due to ultra high aspect ratio.

Falconi et al. [3] studied the energy conversion performance of vertically compared and laterally stretched on the wires and it has low power out. The wires are packed using soft polymers. Package devices receive the outside load and not on the tube. Hence the design of the nanotube is very important.

The variant which is vertically compared and had silicone packing stood in vertical posture. In the axial direction of the rod, outside force is applied on the rod. The rod underwent compression and produced electric potential. The rod is placed between 2 electrodes made up of Au. The rod was developed on the rod's upper surface in radial direction.

The nanorod in the vertically compressed variant was packaged by silicone, fixed at the end, and stood vertically. On the silicone's upper surface, an external force was applied along the nanorod's axial direction. The nanorod underwent vertical compression under the compressing force, resulting in the generation of a piezoelectric potential of approximately 350 mV between its top and bottom surfaces. The nanorod with the same piezoelectric characteristics was packaged by silicone and laterally positioned between two Au electrodes in the lateral stretching model. The pressure was delivered to the nanorod's upper surface in a radial manner.

31.3.2 Microbial Fuel Cells

The capacity of microorganisms to catalyse the conversion of various complex substrates and carbohydrates into electricity is helpful. The direct glucose fuel cell power device like pacemaker. Direct glucose fuel cells have demonstrated a 150-day lifespan in animal trials. All eukaryotic species, including humans, have cells and tissues that contain both oxygen and glucose. One can utilise the body's inherent resources.

31.3.3 Adenosine Triphosphate (ATP) Energy Converter

An alternative for powering next nanomechanical devices is enzymes. The utilisation of RNA, polymerase, myosin, adenosine, ribose, kinesin, and three covalently bound phosphate groups are examples of nanoscale biological motor enzymes. Adenosine diphosphate is formed after the first phosphate group is eliminated.

31.4 Conclusion

In this chapter, conventional energy harvesting systems like thermal energy harvesting, mechanical energy harvesting, optical radio energy, and vibration energy harvesting systems were discussed. Novel energy harvesting systems such as nanowires from piezoelectric structure, microbial fuel cells, and ATP energy converter were discussed. The review will be useful in the field of nanorobotics.

REFERENCES

[1] Basma El Zein, Self-sufficient energy harvesting in robots using nanotechnology, *Advanced Robotics and Automation 2*, 2013, 3.

[2] Zhao Wang, Xumin Pan, Yahua He, Yongming Hu, Haoshuang Gu, and Yu Wang, Piezoelectric nanowires in energy harvesting applications, *Advances in Materials Science and Engineering 2015*(1), 2023, 165631.

[3] Christian Falconi, Giulia Mantini, Arnaldo D'Amicoa, and Zhong Lin Wang, Studying piezoelectric nanowires and nanowalls for energy harvesting, *Sensors and Actuators B 139*, 2009, 511–519.

[4] Yaokun Pang, Yunteng Cao, Masoud Derakhshani, Yuhui Fang, Zhong Lin Wang, and Changyong Cao, Hybrid energy-harvesting systems based on triboelectric nanogenerators, *Matter 4*, January 6, 2021, 116–143.

[5] E. S. Nour, Azam Khan, Omer Nur, and A. Magnus Willander, A flexible sandwich nanogenerator for harvesting piezoelectric potential from single crystalline zinc oxide nanowires, *Nanomaterials and Nanotechnology. 4*, 2014, 24

[6] P. Gambier, S. R. Anton, N. Kong, A. Erturk, and D. J. Inman, Piezoelectric, solar and thermal energy harvesting for hybrid low-power generator systems with thin-film batteries, *Measurement Science and Technology 23*, 2012, 015101.

[7] Z. L. Wang, Energy harvesting for self-powered nanosystems, *Nano Research 1*, 2008, 1–8.

[8] Xudong Wang, Piezoelectric nanogenerators—Harvesting ambient mechanical energy at the nanometer scale, *Nano Energy, 1*(1), January 2012, 13–24.

[9] Shad Roundy, Paul K. Wright, and Jan Rabaey, A study of low level vibrations as a power source for wireless sensor nodes, *Computer Communications 26*(11), July 1, 2003, 1131–1144.

[10] Yi-Ren Wang, and Pin-Tung Chen, Energy harvesting analysis of the magneto-electric and fluid-structure interaction parametric excited system, *Journal of Sound and Vibration 569*(20), January 2024, 118087.

[11] Daochun Li, Yining Wu, Andrea D. Ronch, and Jinwu Xiang, Energy harvesting by means of flow-induced vibrations on aerospace vehicles, *Progress in Aerospace Sciences 86*, October 2016, 28–62.

[12] James P. Thomas, Muhammad A. Qidwai, and James C. Kellogg, Energy scavenging for small-scale unmanned systems. *Journal of Power Sources 159*(2), September 22 2006, 1494–1509.

[13] R. Dauksevicius, G. Kulvietis, V. Ostasevicius, and I. Milasauskaite, Finite element analysis of piezoelectric microgenerator—Towards optimal configuration, *Procedia Engineering 5*, 2010, 1312–1315.

[14] Shlomi Dolev, Sergey Frenkel, Michel Rodrnblit, Ram Prasath Narayanan, and K. Muni Venkateswaralu, In-vivo energy harvesting nano robots. In 2016 *IEEE International Conference on the Science of Electrical Engineering, ICSEE 2016* Article 7806107 (2016 IEEE International Conference on the Science of Electrical Engineering, ICSEE 2016). Institute of Electrical and Electronics Engineers, 2016.

32

Aerodynamic Shape Memory Alloy-Based Actuators for Robotics Applications

A. Anu Kuttan
Noorul Islam Centre for Higher Education, Kanyakumari, India

Barla Madhavi
IFHE, Faculty of Science and Technology, Hyderabad, India

Balkeshwar Singh
Adama Science and Technology University, Adama, Ethiopia

D. R. Srinivasan
JNTUA college of Engineering, Anantapur, India

Amos Gamaleal David
Panimalar Engineering College, Chennai, India

32.1 Introduction

Shape memory alloys (SMAs) represent a class of innovative materials that have garnered significant attention in various engineering fields, particularly in the realm of robotics. This introduction aims to provide a comprehensive overview of SMAs, encompassing their definition, historical background, development, and properties that render them highly suitable for actuation in robotics.

32.1.1 Definition and Characteristics of SMAs

SMAs are a unique category of metallic materials that possess the remarkable ability to recover their original shape after undergoing substantial deformation when subjected to specific external stimuli, most commonly changes in temperature or mechanical stress [1]. This phenomenon, known as the shape memory effect (SME), distinguishes SMAs from conventional materials and facilitates their use in diverse applications, including robotics.

In addition to the SME, SMAs exhibit several other distinctive characteristics that contribute to their utility in robotics. These include high specific strength and stiffness, excellent fatigue resistance, and the ability to generate large strains during actuation [2]. These attributes make SMAs highly desirable for applications requiring precise and efficient motion control, such as robotic manipulators and prosthetic devices.

32.1.2 Historical Background and Development

The concept of shape memory was first observed in the 1930s by Swedish metallurgist Arne Ölander, who discovered the unique properties of an alloy composed of nickel and titanium (NiTi) [3]. However, it was not until the 1960s that researchers began to systematically investigate and exploit the potential of SMAs for practical applications.

Throughout the latter half of the 20th century, significant advancements were made in the understanding and development of SMAs, driven by research efforts in materials science, metallurgy, and engineering [4]. The discovery of new SMA compositions and the refinement of manufacturing processes have contributed to expanding the range of applications for these materials, with robotics emerging as a prominent domain benefiting from SMA technology.

32.1.3 Properties That Make SMAs Suitable for Actuation in Robotics

Several key properties inherent to SMAs make them particularly well-suited for actuation in robotics. These include their reversible phase transformation behaviour, which enables precise control of motion and facilitates complex shape changes in robotic components [5]. Furthermore, SMAs offer high energy density, rapid response times, and silent operation, making them ideal candidates for applications requiring efficient and quiet actuation, such as robotic prosthetics and exoskeletons.

In summary, SMAs represent a compelling class of materials with unique properties that make them highly suitable for actuation in robotics. Through a combination of their SME, historical development, and inherent properties, SMAs have emerged as indispensable tools for advancing the capabilities of robotic systems, paving the way for innovative solutions in fields ranging from industrial automation to medical robotics.

32.2 Fundamentals of Shape Memory Effect (SME)

The SME is a unique property exhibited by SMAs, defining their remarkable ability to recover their original shape after undergoing deformation upon exposure to specific external stimuli. This effect has garnered significant interest in various engineering applications, particularly in robotics, due to its potential for enabling precise and adaptive motion control.

32.2.1 The Shape Memory Effect

At the heart of the SME lies the reversible phase transformation between two distinct crystal structures within SMAs: The martensitic phase and the austenitic phase [1]. When SMAs are in their martensitic phase, they exhibit a lower symmetry crystal structure characterised by a distorted lattice. In contrast, the austenitic phase corresponds to a higher symmetry crystal structure with a more regular lattice arrangement.

Upon deformation in the martensitic phase, such as bending or stretching, the SMA retains this altered shape even after the applied stress is removed. However, when subjected to an appropriate stimulus, typically a change in temperature or mechanical stress, the SMA can undergo a reversible phase transition back to its original austenitic phase. This transition triggers the recovery of the SMA's initial shape, effectively exhibiting the SME.

32.2.2 Martensitic and Austenitic Phases

The martensitic and austenitic phases play a crucial role in determining the behaviour of SMAs and their response to external stimuli. The martensitic phase is stable at lower temperatures, where the SMA material exhibits high ductility and can be easily deformed. In contrast, the austenitic phase becomes stable at higher temperatures, leading to the recovery of the SMA's original shape.

The transition between these phases is governed by temperature and stress conditions, with specific temperature ranges known as transformation temperatures marking the boundaries between martensitic and austenitic behaviours. Understanding and controlling these phase transformations are essential for harnessing the SME in practical applications.

32.2.3 Factors Affecting SME in SMAs

Several factors influence the efficacy and reliability of the SME in SMAs, including composition, processing techniques, and external conditions such as temperature and stress. The composition of SMAs, particularly the ratio of alloying elements such as nickel and titanium in NiTi alloys, significantly impacts their transformation temperatures and mechanical properties [2]. Additionally, the processing methods used to fabricate SMAs, such as heat treatment and alloying, play a crucial role in tailoring their microstructure and properties.

External conditions, such as temperature and stress levels, also affect the phase transformation behaviour of SMAs. Changes in temperature can induce phase transitions between martensitic and austenitic phases, while mechanical stress can influence the onset and completion of these transformations. Optimising these factors is essential for maximising the performance and reliability of SMAs in practical applications.

In summary, the SME represents a fundamental phenomenon in SMAs, enabling them to exhibit unique and desirable properties for applications in robotics and beyond. Understanding the underlying principles of the SME, including phase transformations and influencing factors, is essential for harnessing the full potential of SMAs in engineering applications.

32.3 Actuation Mechanisms in Robotics

Actuation mechanisms play a pivotal role in enabling motion and manipulation in robotic systems. This section provides an overview of actuation methods commonly employed in robotics, compares traditional actuators with SMA-based actuators, and discusses the advantages and limitations of SMAs in this context.

32.3.1 Overview of Actuation Methods in Robotics

Robotic systems utilise various actuation methods to achieve motion, including electrical, hydraulic, pneumatic, and mechanical actuators. Electrical actuators, such as DC motors and stepper motors, are widely used for their precise control and ease of integration with electronic systems. Hydraulic actuators offer high power density and are suitable for applications requiring significant force output, while pneumatic actuators provide lightweight and rapid motion capabilities. Mechanical actuators, such as gears and linkages, transmit motion through mechanical components, offering simplicity and reliability in certain applications.

32.3.2 Comparison between Traditional Actuators and SMAs

Traditional actuators and SMAs differ significantly in their operating principles, performance characteristics, and suitability for robotic applications. Traditional actuators, while well established and widely used, often exhibit limitations such as high-power consumption, limited stroke length, and complex control requirements. In contrast, SMAs offer unique advantages that make them increasingly attractive for actuation in robotics. SMAs possess the ability to generate large deformations with low energy input, enabling compact and lightweight actuation systems. Additionally, SMAs exhibit silent operation, rapid response times, and the capability to withstand harsh environmental conditions, making them suitable for various robotic applications where traditional actuators may fall short.

32.3.3 Advantages and Limitations of SMA-Based Actuators

SMA-based actuators offer several advantages over traditional actuators in robotics. One significant advantage is their intrinsic compliance, which allows for adaptive and flexible motion control, particularly in applications requiring interaction with uncertain environments or delicate objects [4]. Furthermore, SMAs enable precise control of motion through the exploitation of the SME, offering potential advancements in robotics applications such as soft robotics, minimally invasive surgery, and human–robot interaction.

However, SMA-based actuators also present certain limitations that must be considered. These include limited force output compared to traditional actuators, slower actuation speeds, and the need for precise temperature control to induce phase transformations effectively [3]. Additionally, SMAs may exhibit hysteresis and fatigue behaviour over extended use, necessitating careful design considerations to mitigate these effects in practical applications.

32.4 Design Considerations for SMA-Based Actuators

The successful implementation of SMA actuators in robotics necessitates careful consideration of various design aspects to ensure optimal performance, reliability, and efficiency. This section explores key design considerations, including material selection criteria, mechanical design considerations, and thermal management strategies, essential for the effective integration of SMA actuators into robotic systems.

32.4.1 Material Selection Criteria

Selecting the appropriate SMA material is crucial for the performance and functionality of SMA actuators. Nickel–Titanium (NiTi) alloys are the most commonly used SMAs due to their excellent shape memory and super elastic properties, as well as their biocompatibility [6]. When choosing an SMA material, factors such as transformation temperatures, mechanical properties (e.g., stiffness, strength, and ductility), fatigue resistance, and corrosion resistance must be carefully evaluated to ensure compatibility with the intended application requirements.

32.4.2 Mechanical Design Considerations

The mechanical design of SMA actuators plays a significant role in determining their functionality, durability, and efficiency. Design considerations include the selection of appropriate actuator geometries, such as wire, coil, or sheet configurations, to achieve desired motion profiles and force outputs [7]. Additionally, structural considerations, such as ensuring sufficient rigidity and stiffness while minimising weight and volume, are essential for maintaining performance and minimising energy consumption. Furthermore, the integration of SMA actuators with other mechanical components, such as linkages, gears, and bearings, requires careful attention to ensure compatibility and optimise system performance.

32.4.3 Thermal Management Strategies

Effective thermal management is critical for maximising the performance and longevity of SMA actuators. Thermal cycling between martensitic and austenitic phases can induce fatigue and degradation in SMA materials over time [8]. Therefore, implementing thermal insulation to minimise heat loss and maintaining precise temperature control within the desired operating range are essential. Additionally, thermal cycling can be minimised by utilising efficient heat transfer mechanisms, such as embedded heat sinks or active cooling systems, to dissipate excess heat generated during actuation [9]. Furthermore, integrating temperature sensors and feedback control mechanisms enables real-time monitoring and adjustment of operating temperatures, ensuring consistent and reliable actuator performance.

In summary, the design considerations of SMA actuators encompass a multidisciplinary approach, integrating material science, mechanical engineering, and thermal management principles. By carefully selecting SMA materials, optimising mechanical designs, and implementing effective thermal management strategies, engineers can realise the full potential of SMA actuators in robotics applications, achieving enhanced performance, reliability, and functionality.

32.5 Applications of SMA-Based Actuators in Robotics

SMA-based actuators have garnered significant interest in robotics due to their unique properties and capabilities. This section explores the diverse range of applications of SMA-based actuators, providing an overview of current applications, showcasing case studies and examples, and discussing emerging trends shaping the future of SMA technology in robotics.

32.5.1 Overview of Current Applications

SMA-based actuators find application across various domains within robotics, ranging from industrial automation to biomedical devices. In industrial robotics, SMA actuators are utilised in precision positioning systems, grippers, and end-effectors due to their high force density, compact size, and silent operation. In the field of medical robotics, SMA-based actuators are employed in minimally invasive surgical instruments, prosthetic devices, and rehabilitation exoskeletons, enabling precise and adaptive motion control in confined spaces [10]. Furthermore, SMA actuators are increasingly being integrated into soft robotics platforms, offering compliant and biomimetic motion capabilities for applications such as human–robot interaction and assistive devices.

32.5.2 Case Studies and Examples

Several case studies and examples highlight the versatility and effectiveness of SMA-based actuators in practical robotic applications. For instance, in industrial robotics, SMA actuators are employed in gripper systems for handling fragile objects with delicate manipulation requirements. By leveraging the SME, SMA grippers can adapt their grip force and contact pressure dynamically, enabling gentle handling of objects with varying shapes and sizes. In medical robotics, SMA actuators are utilised in robotic surgical instruments for minimally invasive procedures, providing surgeons with precise control and dexterity in confined surgical environments. The flexibility and adaptability of SMA actuators allow for complex motion trajectories and force modulation, enhancing surgical outcomes and patient safety [11].

32.5.3 Emerging Trends

Emerging trends in SMA-based actuators are shaping the future of robotics by enabling new capabilities and applications. One emerging trend is the integration of SMA actuators with advanced sensing and control technologies, enabling closed-loop feedback control and autonomous operation in robotic systems [12]. By incorporating sensors to monitor actuator performance and environmental conditions, SMA-based robots can adapt their behaviour in realtime to changing operating conditions, improving reliability and robustness. Another emerging trend is the development of multifunctional SMA materials capable of sensing, actuation, and energy harvesting, opening up new possibilities for self-powered and autonomous robotic systems [13]. Additionally, advancements in manufacturing techniques, such as additive manufacturing and microfabrication, are facilitating the design and fabrication of complex SMA-based actuators with tailored performance characteristics for specific applications.

In summary, SMA-based actuators are finding widespread application in robotics, offering unique capabilities and enabling innovative solutions across various domains. Through case studies, examples, and emerging trends, it is evident that SMA technology holds great promise for advancing the capabilities of robotic systems, driving new developments and applications in the field.

32.6 Challenges and Future Directions

Despite the promising applications and capabilities of SMA-based actuators in robotics, several challenges and opportunities for future development exist. This section discusses the key challenges facing the widespread adoption of SMA actuators in robotics and outlines potential future directions for research and innovation in this field.

32.6.1 Challenges

Performance Optimisation: One of the primary challenges in SMA-based actuation is optimising performance parameters such as actuation speed, force output, and energy efficiency. Achieving fast and precise motion control while minimising energy consumption remains a significant challenge due to the inherent material properties and operating principles of SMAs [3, 14].

Reliability and Durability: SMA actuators are susceptible to fatigue and degradation over time, particularly during cyclic loading and thermal cycling. Ensuring long-term reliability and durability under real-world operating conditions is crucial for the practical deployment of SMA-based robotic systems [4].

Temperature Management: SMA actuators rely on precise temperature control to induce phase transformations and achieve SMEs. Managing temperature variations and thermal cycling within the desired operating range poses challenges, especially in dynamic and unpredictable environments.

Integration Complexity: Integrating SMA actuators into robotic systems requires careful consideration of mechanical, electrical, and control aspects. Addressing the complexity of system integration, including interfacing with sensors, actuators, and control electronics, is essential for achieving seamless operation and compatibility with existing robotic platforms.

32.6.2 Future Directions

Advanced Material Development: Continued research into advanced SMA materials, such as multifunctional alloys and composites, holds promise for enhancing performance and functionality. Developing new SMA compositions with tailored properties, such as enhanced fatigue resistance and improved actuation characteristics, can unlock new possibilities for robotics applications.

Advanced Manufacturing Techniques: Advancements in manufacturing technologies, such as additive manufacturing and microfabrication, offer opportunities for fabricating complex SMA-based actuators with precise geometries and customisable features. Leveraging advanced manufacturing techniques can streamline production processes and enable the rapid prototyping of novel SMA-based robotic systems.

Smart Sensing and Control: Integrating smart sensing and control algorithms into SMA-based robotic systems can enhance autonomy, adaptability, and robustness. Developing advanced feedback control strategies and sensor fusion techniques enables real-time monitoring of actuator performance and environmental conditions, improving system reliability and responsiveness.

Interdisciplinary Collaboration: Encouraging interdisciplinary collaboration between materials scientists, mechanical engineers, control theorists, and roboticists is essential for addressing the multifaceted challenges of SMA-based actuation. Collaborative research efforts can facilitate knowledge exchange, accelerate innovation, and drive progress towards practical solutions for real-world robotics applications.

In conclusion, addressing the challenges and pursuing future directions outlined in this section are crucial for advancing the field of SMA-based actuation in robotics. By overcoming technical barriers, fostering innovation, and promoting interdisciplinary collaboration, researchers can unlock the full potential of SMA technology and realise transformative advancements in robotic systems.

REFERENCES

[1] Otsuka, K., & Wayman, C. M. (1998). *Shape Memory Materials*. Cambridge University Press.

[2] Lagoudas, D. C. (2008). *Shape Memory Alloys: Modeling and Engineering Applications*. Springer Science & Business Media.

[3] Boyd, J. G. (2017). *Shape Memory Alloys for Aerospace Applications*. Woodhead Publishing.

[4] Bhattacharyya, D., & Majumdar, B. (2015). *Shape Memory Alloys: Processing, Characterisation and Applications*. Woodhead Publishing.

[5] Tiwari, R. K. (Ed.). (2020). *Shape Memory Alloys: Fundamentals, Design, and Applications*. CRC Press.

[6] Zhang, X., Zhang, X., Wang, Z., & Zheng, Y. (2020). Shape Memory Alloys and Their Applications in Actuators. *Actuators*, *9*(4), 87.

[7] Chen, X., He, Y., Zhao, Y., & Zhao, H. (2021). Shape Memory Alloy Actuators: A Review of Materials, Technologies, and Applications. *Materials*, *14*(7), 1834.

[8] Wang, H., Huang, W., & Huang, Z. (2018). Recent Advances in Shape Memory Alloy Actuators: Materials, Mechanisms, and Applications. *Micromachines*, *9*(1), 34.

[9] Lai, K. W. C., & Li, W. J. (2020). Shape Memory Alloy Actuators: A Comprehensive Review. *Smart Materials and Structures*, *29*(7), 073001.

[10] Kim, H., et al. (2019). Design and Performance of Shape Memory Alloy Actuators: A Review of Recent Advances. *Journal of Intelligent Material Systems and Structures, 30*(20), 3030–3054.

[11] Yuan, W., Li, J., He, Y., & Ma, J. (2022). Challenges and Opportunities of Shape Memory Alloy Actuators in Robotics: A Review. *IEEE/ASME Transactions on Mechatronics, 27*(1), 123–136.

[12] Zhao, Y., Chen, X., He, Y., & Wang, Y. (2023). Advanced Manufacturing Techniques for Shape Memory Alloy Actuators: Current Status and Future Perspectives. *Journal of Manufacturing Processes, 59*, 180–195.

[13] Li, X., Li, X., Wang, Y., & Cheng, Y. (2022). Advances in Shape Memory Alloy Actuator Design and Integration: A Review. *Robotics, 11*(2), 66.

[14] Gao, S., Li, X., & Zhang, X. (2024). Smart Sensing and Control Strategies for Shape Memory Alloy Actuators in Robotics: A Comprehensive Review. *Robotics and Autonomous Systems, 148*, 102–118.

33

Functional Materials for Energy Conversion and Storage

Gitanjali Jothiprakash
Agricultural Engineering College and Research Institute, Tamil Nadu Agricultural University, Coimbatore, India

Deepa Jaganathan
Hindustan College of Engineering and Technology, Coimbatore, India

Sriramajayam Srinivasan
Agricultural College and Research Institute, Tamil Nadu Agricultural University, Killikulam, Vallanad, India

Ramesh Desikan and Karthikeyan Subburamu
Agricultural Engineering College and Research Institute, Tamil Nadu Agricultural University, Coimbatore, India

33.1 Introduction

Recent researchers are reinvigorated to develop maintenance-free, high-performance energy conversion and energy storage systems, as a result of the hasty commercialisation of a range of electronic devices and artificial intelligence systems. It can be challenging to reduce energy demand by developing an efficient method for energy harvesting and then storing it [1]. The storage and conversion of energy with high efficiency and purity have a strong influence on the long-term growth of the economy and the ecology of the industry. Secondary/rechargeable batteries and fuel cells are promising alternatives to replace conventional fossil fuels.

33.2 Energy Conversion Systems

Energy conversion systems are the collection of techniques used to extract useful energy from resources and transform it into a usable form (Figure 33.1). The most significant topic of concern in the current situation is harvesting energy in any useful form. To successfully capture the energy, the introduction of an efficient energy conversion system has become crucial [2]. Energy conversion technology will stay vital in the future, as evidenced using the growing challenge of energy shortage in current duration. Functional materials are utilised in thermal energy and photovoltaic cells to harness waste heat, magnetic energy, and mechanical power. Functional materials with a purpose hold promise for thermoelectric devices. Materials used with purpose can improve the efficiency of converting heat energy into electrical energy [3].

DOI: 10.1201/9781003495437-33

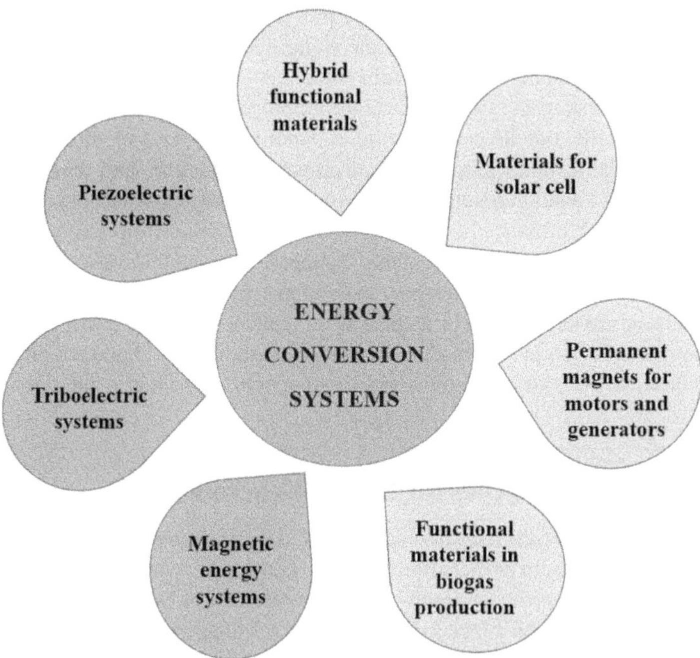

FIGURE 33.1   Energy conversion systems.

33.3 Piezoelectric Systems

A double-clamped micro-electromechanical system resonator with nonlinear stiffness and a very wide bandwidth is used for piezoelectric energy harvesting. This structure was found to increase power density and represents a promising direction for future research. Researchers reported 0.8 V open circuit voltage at 1.3 kHz using an ideal forward resistive load of 22 W. This corresponds to the estimated power that can be extracted.

Lead zirconate titanate nanofibres can be produced by electrospinning. These nanofibres are deposited on prefabricated interconnected platinum wire array electrodes (50 μm diameter) mounted on a silicon substrate [4].

33.4 Triboelectric Systems

To capture energy from environmental mechanical motions including sound, water waves, and mechanical vibrations, a technology known as a triboelectric nano-generator or Wang generator has been developed. Triboelectric nano-generators can be used for several different measuring sensors and energy harvesters. Applications of these systems have distinct requirements for production processes and materials. For instance, flexible materials are needed for portable triboelectric nano-generators. Nanoscale materials are unable to fulfil many of the requirements, but macro-scale materials can. They feature several distinctive characteristics, including a straightforward structure, a wide range of material choices, and a high conversion efficiency. They also show promising uses in huge blue energy, self-sufficient sensors, micro/nano-power sources, and direct power sources with high voltage [5].

As a result, each example had a conversion efficiency of >50% and an area power of 500 Wm^{-2}. Triboelectric nano-generators can be a promising alternative energy source for power generation when combined with electrochemistry for electrochemical manipulation considering these benefits. The selection of electrodes is crucial because triboelectric materials have different physical and chemical properties.

Gold or copper nanofilms that have been thermally formed on the reverse of triboelectric materials are the most frequently employed nanomaterials as electrode materials. This nanofilm deposition method increases the electrostatic induction contact surface [6].

When extremely rough materials, like textiles, are employed as triboelectric materials, these nanofilms are crucial. Additionally, the direct deposition of nanofilms does away with the requirement for an adhesive layer, shortens the distance required for electrostatic induction, and thus improves the current performance of triboelectric nano-generators. The application of inorganic nanomaterials to enhance triboelectric nano-generator performance has garnered the most attention from scientists.

On flat surfaces, surfaces containing inorganic nanoparticles exhibit various charge distributions. Inorganic nanoparticles' high surface energies may improve charge transport between triboelectric surfaces. Inorganic nanoparticles' dielectric properties enhance charge transport between surfaces. Additionally, nanomaterials can give triboelectric nano-generators new characteristics that will make it easier for them to be used. Of course, nanomaterials also benefit triboelectric nano-generators.

33.5 Hybrid Functional Materials for Energy Generation

The hybrid functional nanomaterials have a wide range of uses, including enhanced catalysis and energy conversion/storage. The production of multifunctional nanostructures with much better optical, electrical, mechanical, and magnetic properties is possible via nanofabrication, which includes the self-assembly of mineral nanomaterials based on block copolymers. Block copolymers of metal nanoparticles like gold, palladium, and platinum can be used to self-assemble nanostructures into macroscopically structured domains, which have potential applications. The extremely high surface-to-volume ratio of these nanoparticles considerably improves surface properties. To detect alterations in the particle's environment, such as the presence of additional compounds or ligands, optical sensors can therefore be built [7].

Inorganic nanoparticles and polystyrene-*b*-poly(4-vinylpyridine) are combined to create a hybrid material that allows for the selective incorporation of functional components like metal and semiconductor nanoparticles. It contains a copolymer. Block copolymer self-assembly can therefore be improved. A potential method for creating well-defined hybrid nanostructured material that can be used in a variety of techniques is by altering its nature [8].

33.6 Magnetic Energy Systems

A combination of magnetoelectric and piezoelectric effects is frequently used in multi-modal systems for the simultaneous generation of power from magnetic and mechanical stray energies. These systems typically consist of a magnetoelectric beam attached to the centre of a cantilever beam with maximal mass. A gadget with a huge magnetostrictive material is connected to a copper substrate-based sensing coil and an energy-harvesting circuit. This system utilises the magnetostriction effect. Vibrational stress brought on by bending results in a change in magnetisation. A rotating energy harvester makes use of a cantilever beam and laminated magnetostrictive/piezoelectric transducers [9].

33.7 Materials for Solar Cell Applications

Semiconductor materials are used in photovoltaics to transform solar energy into electrical energy. The process of producing energy often starts with complicated molecules absorbing light. After that, the electron is energised and moved to the acceptor. Numerous nanomaterials are utilised to make solar cells, which produce electricity. These include nanocrystals, quantum dots, and nanotubes, among others. Innovative solar energy conversion designs may result from improvements in the production of nanoparticles with clearly specified geometries. Thermoelectric solar power generation is quickly catching up to utility power in price. Porous titanium dioxide microspheres' improved ability to scatter light is responsible for this rise. The hydrolysis and condensation reaction that results from the conversion of titanium glycolate microspheres to titanium dioxide produces mesopores inside the spheres, but the particles preserve their form. After a hydrothermal process, titanium dioxide microspheres were produced. Titanium glycolate spheres have a smooth surface and a diameter of 400–600 nm. The hydrothermal treatment caused the spheres' diameter to significantly drop to 300–500 nm. Compared to their bulk gold counterparts, they efficiently trap light and prevent unwanted energy transfer on the gold surface.

33.8 Permanent Magnets for Motors and Generators

Magnetic nanoparticles with diameters less than approximately 15 nm and high magneto crystallinity can be employed as building blocks of new generation permanent magnets. A series of processes, including self-assembly, straightforward axis alignment, and appropriate nanostructuring, are outlined which allow for the creation of densely packed nanoparticle arrays with better permanent magnetic characteristics. To boost the energy density of permanent magnets based on hard magnetic nanoparticles, these particles can be mixed with the appropriate soft phases to create exchange-bonded nanocomposites. Super-paramagnetism and surface modification by oxidation or surfactant molecules have an effect on nanoparticles due to its size.

New materials that can be utilised to produce lightweight, high-performance electric motors and replace copper and iron in conventional motors include carbon nanotubes and superparamagnetic nanoparticles with a magnetic core, also known as carbon nanomaterials. A cutting-edge example of a future energy conversion technology could be a motor. It has been demonstrated that the conductivity of the best carbon nanotubes is far higher than that of the best metals. Therefore, compared to contemporary copper windings, windings composed of such materials may have twice as much conductivity (100 μS/m). As an alternative, hard magnetic nanoparticles are non-critical, accessible, stable, and made of a 3D metal that is simple to recycle.

New materials that can replace copper, iron, and rare earth permanent magnets in traditional motors include carbon nanotube wires made of carbon nanotubes, magnetic cores made of superparamagnetic nanoparticles, and permanent magnets built of hard magnetic nanoparticles. This will make it possible to create new types of energy converters, such as robust, lightweight electric motors. Unquestionably, the quick acceleration, great torque, and capacity to operate at high temperatures, and high speeds are other benefits of carbon engines. Modern engines must be created by integrating various elements, pushing the limits of already existing materials, and utilising the capabilities of new materials.

33.9 Functional Materials in Biogas Production

The effect of titanium dioxide functional materials (1120 mg/L, 7.5 nm) on the anaerobic digestion of sewage sludge over 50 days in mesophilic and thermophilic environments is a 10% increase in biogas generation. Fe_3O_4 functional materials (7 nm) at a concentration of 100 ppm boosted biogas output by 180% and methane production by 234% in a waste anaerobic digestion tank at mesogenic temperature (37°C) for 60 days [10].

33.10 Energy Storage Systems

Due to the accelerating economic growth and growing need for energy-based products, there is an alarming and dramatic rise in energy demand. Environmental research concerns and social difficulties related to the use of conventional fossil fuels are growing in importance because of rising energy consumption. As a cutting-edge strategy, nanotechnology creates new materials for efficient energy storage and conversion, especially at micro/nanometre scale. Recently, micro/nanomaterials have been widely investigated for use in energy storage systems (Figure 33.2) and conversion processes [11]. Cost and competence issues are the main reasons for the actual use of micro/nanomaterials. Therefore, the development of low-cost, highly effective micro/nanomaterials is still crucial for energy storage systems.

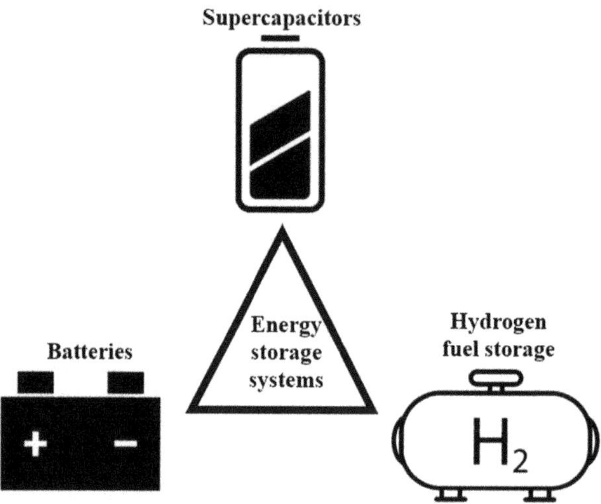

FIGURE 33.2   Energy storage systems.

33.11 Batteries

Advanced nanotechnological materials having internal or external dimensions in the nanometre range are known as nanostructured functional materials. These materials are special and have potential for lithium batteries because of their incredibly small dimensions. The electrochemical performance of suggested alternative electrode materials can be significantly enhanced by nanostructured materials. More focus has been placed on composites made of graphene and active anode materials. The active material's volume expansion during the lithiation process is also considered [12].

One-dimensional tubular forms composed of graphitic carbon called 'graphene coils' are unique and show promise for use in electrochemical energy storage. With a lower discharge potential of 370 mV than Li/Li, silicon is regarded as a viable anode material. This has a ten times better strength than regular graphite and is cycling-stress resistant. The synthesis of nested silicon nanospheres, as well as their excellent high-speed capabilities and extremely reversible lithium deposition, is possible. The solvothermal method was used to create nested silicon nanospheres [13].

Silicon nanospheres have a specific capacity of 3052 mAh/g initially and after 48 cycles. The silicon nanoparticle-based composite anodes are rapidly expanding. These can lessen considerable Si expansion while preserving the electrode's structural integrity and boosting stability by lowering electrochemical sintering and silicon agglomeration. Composite electrodes made of silicon are combined with conductive fillers such as carbon, graphite, carbon nanotubes, biographene, as well as a binder. A hydrothermal technique is used to create cobalt oxide/nickel oxide core nanofunctional materials before covering them with nickel oxide via a chemical bath deposition. After 6000 cycles, the materials and nanostructural characteristics generate a high specific capacity. By hydrolysing aqueous nickel chloride solution in hexangular zinc oxide nanorod patterns and heating it to change nickel hydroxide to nickel oxide, self-assembled porous nickel oxide-coated zinc oxide nanorod electrodes were created on stainless steel substrates.

Due to the layered structure of the crystal, layered transition metal di-chalcogenides (2D) have been shown to offer good lithium-ion storage capacity. Metal sulphides can be employed as cathode materials, depending on the potential value. Metals are used as anode in lithium batteries. Carbon functional materials containing molybdenum disulphide nanocomposite-based functional materials are regarded as potential battery materials since they have good cycling stability and reversible capacity values of 800–1000 mAh/g.

This is because the electrode surface has a large surface area, a short route length, and high degree of freedom for volume change, all of which may result in dropped overpotentials and advantageous reaction kinetics. It was revealed to use triethanolamine to fabricate hollow microspheres doped with zinc germanite. One 1D single-crystal nanorod with dimensions of around 2 mm in length and 100 nm in diameter was used to self-assemble the hollow microsphere shell.

A high capacity of 882 mAh/g at 75% at 1.0 A/g is maintained even after 100 discharge cycles, and a capacity of 464 mAh/g can be maintained even at a high current density of 5.0 A/g. Three-layer cathodes are an additional possible cathode material. Among the many cathodes under investigation, nanocomposites combine the advantages of lithium-based materials. Nanocomposites have high specific capacitance, high operating voltage, cycle stability, and structural stability.

33.12 Supercapacitors

Supercapacitors still fall short of real-world commercial applications in flexible transparent electronics, despite numerous significant advancements. Consideration should be given to growth electrode materials with excellent electronic conductivity. This is so that it does not require additional current collectors [14]. Another topic worth looking into is the rational design of electrode materials, which can significantly enhance mechanical flexibility, electrochemical characteristics, and optoelectronic capabilities.

The long-range supercapacitors' triaxial structure, porous structure, hierarchical structure, core–shell coaxial storage, and cycling stability allow for fast dynamics and stable energy.

Device stability depends on the electrode materials being thermodynamically stable. The supercapacitors shelf life can be extended by stored under ambient circumstances. As flexible, transparent, conducting substrates for interconnection with non-conducting active materials, ultrathin metal networks offer a representative sample. Supercapacitors perform more electrochemically when the active material's large specific surface area expands the active sites and reduces the distance that electrons and ions must travel. However, for the mentioned electrode materials, this is still insufficient.

To improve the specific surface part of the active material in supercapacitors, particular procedures should be investigated. This entails creating wrinkly nanosheet building blocks and porous 3D structures. The fundamental mechanism was quickly determined. The link between the structural and electrochemical properties of electrode materials makes theoretical models and computations essential. The experimental results can also be examined in further depth. Supercapacitors have grown in surface area, per theoretical models. Supercapacitors expand a device's ability to store energy.

Because the operating voltage of portable electronic devices is frequently above 1.8 V, the creation of asymmetric supercapacitors with a broad voltage window also minimises the number of supercapacitors that must be linked in series to obtain high energy and power densities. A more precise method of evaluating the mechanical stability of supercapacitors is to compare the capacitance values at different radii of curvature during continuous bending or stretching processes. Flexibility is a key indicator of mechanical stability, inconsistent shapes (e.g., sheet supercapacitors and micro-supercapacitors), and dimensions (height, breadth, etc.). Even with the same bending angle and tensile stress, it is challenging to compare the elasticity of different parts in supercapacitors. A precise quantitative comparison of flexibility is also still lacking, as is a clear quantitative definition of flexibility. Future applications requiring commercial use will urgently require a consistent quantitative flexibility analysis. The focus of next-generation supercapacitor breakthrough is on a variety of multifunctional devices, including solar hybrid supercapacitors, self-healing supercapacitors, self-charging photonic supercapacitors, and electrochromic supercapacitors.

For integrated, self-healing, flexible, and transparent portable electronics applications, curable supercapacitors are preferred. Most of today's self-healing supercapacitors are opaque and transparent and possess self-healing qualities. Each component's transparency and capacity for self-healing must be assessed. Additionally, much work is required to overcome the mutual exclusion of high performance, transparency, and self-healing in supercapacitors.

Universal criteria and technical indicators must be standardised for commercialisation even in the early stages of laboratory research. Supercapacitors require the establishment of numerous technical standards, such as optoelectronic activity test procedures, standards for assessing mechanical flexibility, and indices for assessing electrochemical performance. Applications today strive to dispose of recycling in an eco-friendly manner with minimal consumption.

33.13 Hydrogen Fuel Storage

Research on hydrogen energy has a high priority for developing strategies for safe and inexpensive storage of H_2. For instance, it takes energy to compress hydrogen to 700 bar, and additional improvements in volume and weight can only be made by employing higher pressures. Due to the high pressure needed, the tank geometry is restricted to cumbersome, pricey carbon fibre-reinforced composite cylinders. To prevent overheating, hydrogen must be precooled to 233 K due to the isenthalpic expansion of hydrogen during charging, which raises the temperature.

Although it must be held at an extremely low temperature below 30 K, hydrogen can also be kept as a liquid. This scale of cooling requires highly insulated tanks for long-term storage, which is very expensive and consumes more energy than compressing hydrogen to 700 bar. Cryogenic compression, a third choice, combines cooling and compression to produce high bulk densities that resemble liquid hydrogen in its gaseous state. To give large gravimetric and volumetric capacities at more realistic pressures and temperatures, materials that absorb or adsorb hydrogen are frequently used as an alternative to storing hydrogen in compressed, liquid, or cryogenically compressed form. This special edition discusses a variety of possibilities, each having pros and cons. Metal and complex hydrides fall within this category [15].

High gravimetric and volumetric capacities are possible with complex hydrides; however, they are not always reversible, frequently operating at high temperatures, and when they are, they typically exhibit delayed hydrogen absorption and desorption. However, interstitial metal hydrides typically have poor gravimetric capacities of less than 2% in weight, can function at temperatures close to ambient, and have better bulk capacities than liquid hydrogen. Furthermore, the sorption and desorption enthalpies involved are high in each scenario because hydrogen is chemically linked through ionic, metallic, or covalent interactions, making temperature management essential for successful functioning as a component. The issue of storing hydrogen can be solved physically, as opposed to chemically, by adsorbing hydrogen on nanoporous materials.

Molecular hydrogen is either physically adsorbed or trapped in the pores of materials having huge internal surfaces and consequently extensive gas–solid interfaces, such as zeolites, activated carbons, and metal–organic frameworks to be adsorbed [16]. This method allows for the storage of hydrogen at densities greater than compressed gas, at pressures lower than 100 bar, and temperatures greater than liquid storage. The practical issue is that, to obtain high capacity, low temperatures around 77 K and pressures as high as 100 bar are still required. In contrast to chemical storage on hydrides, it also has benefits like quick adsorption kinetics and low adsorption heat.

Another crucial benefit of storing hydrogen in nanoporous materials is that it provides a temporary solution that can provide excellent gravimetric and bulk density storage at pressures lower than 100 bar, minimising compression or liquefaction losses. It indicates that it is possible to avoid the losses and issues in handling and distributing hydrogen under high-pressure gas or cryogenic liquid at 700 bar.

33.14 Conclusion

Economic progress and national development are directly related to improvements in energy storage and conversion technologies and sustainable energy research. Fossil fuels are a major component of the world's energy consumption, resulting in severe pollution and climate change. The growth of green fuel energy generation alternatives depends on the creation of inexpensive materials for renewable energy conversion and storage systems.

REFERENCES

[1] Kumar, D. Praveen, S. Pugalendhi, P. Subramanian, and S. Karthikeyan. 2018. "Interactive Effect of Plasma Power and Equivalence Ratio upon Plasma Gasification of Coconut Shell.", *Madras Agricultural Journal 105* (March): 109–112.

[2] Jothiprakash, Gitanjali, and Venkatachalam Palaniappan. 2014. "Development and Optimization of Pyrolysis Unit for Producing Charcoal." *International Journal of Agriculture, Environment and Biotechnology 7* (4): 863. https://doi.org/10.5958/2230-732x.2014.01397.7

[3] Liu, Jun, Guozhong Cao, Zhenguo Yang, Donghai Wang, Dan Dubois, Xiaodong Zhou, Gordon L. Graff, Larry R. Pederson, and Ji Guang Zhang. 2008. "Oriented Nanostructures for Energy Conversion and Storage." *ChemSusChem 1* (8–9): 676–697. https://doi.org/10.1002/cssc.200800087

[4] Hajati, Arman, and Sang Gook Kim. 2011. "Ultra-Wide Bandwidth Piezoelectric Energy Harvesting." *Applied Physics Letters 99* (8): 1–4. https://doi.org/10.1063/1.3629551

[5] Aazem, Irthasa, Dhanu Treasa Mathew, Sithara Radhakrishnan, K. V. Vijoy, Honey John, Daniel M. Mulvihill, and Suresh C. Pillai. 2022. "Electrode Materials for Stretchable Triboelectric Nanogenerator in Wearable Electronics." *RSC Advances 12* (17): 10545–10572. https://doi.org/10.1039/d2ra01088g

[6] Zhang, Renyun, and Håkan Olin. 2022. "Advances in Inorganic Nanomaterials for Triboelectric Nanogenerators." *ACS Nanoscience Au 2* (1): 12–31. https://doi.org/10.1021/acsnanoscienceau.1c00026

[7] Radousky, H. B., and H. Liang. 2012. "Energy Harvesting: An Integrated View of Materials, Devices and Applications." *Nanotechnology 23* (50). https://doi.org/10.1088/0957-4484/23/50/502001

[8] Fahmi, Amir, Torsten Pietsch, Cesar Mendoza, and Nicolas Cheval. 2009. "Functional Hybrid Materials." *Materials Today 12* (5): 44–50. https://doi.org/10.1016/S1369-7021(09)70159-2

[9] Balamurugan, B., D. J. Sellmyer, G. C. Hadjipanayis, and R. Skomski. 2012. "Prospects for Nanoparticle-Based Permanent Magnets." *Scripta Materialia 67* (6): 542–547. https://doi.org/10.1016/j.scriptamat.2012.03.034

[10] Manidurai, Paulraj, and Ramkumar Sekar. 2017. "Nanomaterials in Energy Generation." *Handbook of Composites from Renewable Materials 1–8* (July): 207–228. https://doi.org/10.1002/9781119441632.ch155

[11] Zhou, Linglin, Di Liu, Li Liu, Lixia He, Xia Cao, Jie Wang, and Zhong Lin Wang. 2021. "Recent Advances in Self-Powered Electrochemical Systems." *Research 2021*. https://doi.org/10.34133/2021/4673028

[12] Ramar, Alagar, and Fu Ming Wang. 2020. "Emerging Anode and Cathode Functional Materials for Lithium-Ion Batteries." *Nanostructured, Functional, and Flexible Materials for Energy Conversion and Storage Systems*. ElsevierInc.https://doi.org/10.1016/B978-0-12-819552-9.00015-4

[13] Sahito, Iftikhar Ali, Chengyu Hong, and Sung Hoon Jeong. 2018. "Advanced Hybrid Functional Materials for Energy Applications." *Journal of Nanomaterials 2018*: 1–2. https://doi.org/10.1155/2018/7814784

[14] Zhao, Weiwei, Mengyue Jiang, Weikang Wang, Shujuan Liu, Wei Huang, and Qiang Zhao. 2021. "Flexible Transparent Supercapacitors: Materials and Devices." *Advanced Functional Materials 31* (11): 1–30. https://doi.org/10.1002/adfm.202009136

[15] Zhang, Linda, Mark D. Allendorf, Rafael Balderas-Xicohténcatl, Darren P. Broom, George S. Fanourgakis, George E. Froudakis, Thomas Gennett, et al. 2022. "Fundamentals of Hydrogen Storage in Nanoporous Materials." *Progress in Energy 4* (4): 12–28. https://doi.org/10.1088/2516-1083/ac8d44

[16] Gitanjali, J., P. Venkatachalam, and P. Subramanian. 2014. "Development of high efficient combustion system for turmeric boiling." *Journal of Environmental Research And Development 9* (01): 67–74.

34

Exploring Perovskite Solar Cells for Enhanced Photovoltaic Performance

Sarvani Jowhar Khanam
University of Hyderabad, Hyderabad, India

34.1 Introduction

The inexhaustible reserve of solar energy presents a beacon of hope in our quest for sustainable energy solutions amidst the dual challenges of climate change and finite fossil fuel resources. As nations strive to transition towards cleaner energy sources, the imperative for advanced photovoltaic technologies capable of efficiently harnessing solar power becomes increasingly evident.

The trajectory of global energy consumption underscores the pivotal role of solar photovoltaics (PV) in meeting burgeoning electricity demand. Projections from the International Energy Agency (IEA) forecast that solar PV installations will dominate new electricity capacity additions in the years ahead, surpassing conventional energy sources [1]. This surge in solar deployment underscores the urgency for enhancing the efficiency and affordability of PV technologies.

Traditional silicon-based solar cells, while effective, are constrained by resource-intensive manufacturing processes and associated costs, inhibiting their widespread adoption. In this context, PSCs have emerged as a compelling alternative, captivating the attention of researchers and industry stakeholders alike.

Perovskite materials exhibit remarkable optoelectronic properties, including high light absorption coefficients, long charge carrier diffusion lengths, and tuneable bandgaps, rendering them highly promising for PV applications. Moreover, their compatibility with solution-processing techniques offers the prospect of low-cost and scalable fabrication, addressing key barriers to solar technology deployment.

This chapter embarks on a comprehensive exploration of PSCs as a disruptive force in advancing PV technology. By scrutinising their intricate structural design, diverse fabrication methodologies, and performance metrics, we seek to unravel the underlying mechanisms driving their extraordinary potential.

Through a critical appraisal of the current state of PSC research, alongside recent breakthroughs and persistent challenges, we aim to provide nuanced insights into their transformative capacity. Furthermore, by examining their role in shaping a sustainable energy future, we endeavour to elucidate pathways for accelerating their integration into mainstream energy systems.

34.2 Fundamentals of Perovskite Solar Cells

PSCs have emerged as a frontrunner in the realm of PV technology owing to their exceptional optoelectronic properties and the promise of cost-effective manufacturing processes. To understand the workings of PSCs, it is crucial to delve into their fundamental aspects, encompassing their structure, working principles, and unique advantages for PV applications.

34.2.1 Structure and Composition of Perovskite Materials

Perovskite materials, named after the mineral perovskite (calcium titanate), adopt a distinctive crystal structure characterised by the general formula ABX_3, where A and B represent cations, and X denotes

DOI: 10.1201/9781003495437-34

anions. In the context of PSCs, the most widely studied perovskite composition is represented as $APbX_3$, where A commonly refers to an organic or inorganic cation, such as methylammonium (MA) or formamidinium (FA), while X typically represents halide ions like iodide (I), bromide (Br), or chloride (Cl) [2].

These perovskite materials can be synthesised using various methods, including solution-processing techniques like spin-coating, doctor-blading, or vapour deposition methods such as thermal evaporation or sputtering. The flexibility in material composition and fabrication methods allows for tailoring the properties of perovskite films to optimise device performance [3, 4].

34.2.2 Working Principles of Perovskite Solar Cells

PSCs operate based on the principles of the PV effect; wherein incident photons generate electron–hole pairs (excitons) within the perovskite absorber layer. The excitons subsequently undergo charge separation at the perovskite/electron transport layer (ETL) interface, with electrons moving towards the ETL and holes migrating towards the hole transport layer (HTL).

The generated charge carriers are then collected at the respective electrodes, creating an electric current that can be utilised as electrical energy. Notably, the high carrier mobility and long carrier diffusion lengths in perovskite materials facilitate efficient charge transport and collection, contributing to the high PV performance of PSCs [5].

34.2.3 Key Characteristics and Advantages of Perovskite Materials for Photovoltaic Applications

Perovskite materials exhibit several advantageous properties that make them highly attractive for PV applications. These include high absorption coefficients across a broad spectral range, enabling efficient light harvesting even with thin absorber layers. Additionally, perovskite films can be fabricated using low-cost solution-processing techniques, offering the potential for scalable and economically viable production of solar cells.

Furthermore, the tuneable bandgap of perovskite materials allows for bandgap engineering to match the solar spectrum, maximising energy conversion efficiency. Their defect tolerance and facile defect passivation strategies contribute to the high open-circuit voltage and reduced non-radiative recombination losses in PSCs, further enhancing their performance.

34.3 Fabrication Methods of Perovskite Solar Cells

The fabrication of PSCs encompasses a variety of techniques tailored to depositing high-quality perovskite films with controlled morphology and composition. These methods can be broadly categorised into solution-processing techniques and vapour deposition methods, each offering distinct advantages and challenges.

34.3.1 Solution-Processing Techniques

Solution-processing techniques are widely employed for their simplicity, scalability, and compatibility with large-scale manufacturing processes. Among the most common solution-based methods are spin-coating, doctor-blading, and spray-coating.

Spin-coating: In spin-coating, a perovskite precursor solution is dispensed onto a substrate and spun at high speeds to spread the solution uniformly, resulting in the formation of a thin film. This method allows for precise control over film thickness and morphology, making it a preferred choice for research and development purposes [6].

Doctor-blading: Doctor-blading involves spreading a viscous perovskite precursor solution onto a substrate using a blade or a doctor blade. This method is advantageous for its potential to scale up production and its compatibility with roll-to-roll processing, making it suitable for large-area module fabrication [7].

Spray-coating: Spray-coating entails atomising the perovskite precursor solution into fine droplets, which are then deposited onto a substrate to form a uniform film. This technique offers the advantage of high throughput and can be easily adapted for continuous processing, making it suitable for industrial-scale production [8].

34.3.2 Vapour Deposition Methods

Vapour deposition methods involve the deposition of perovskite films by vaporising precursor materials and condensing them onto a substrate. Common vapour deposition techniques include thermal evaporation, sputtering, and atomic layer deposition (ALD).

Thermal evaporation: In thermal evaporation, perovskite precursor materials are heated to high temperatures in a vacuum chamber, causing them to vaporise and deposit onto a substrate, where they subsequently crystallise into a perovskite film. This method offers excellent control over film thickness and composition but is typically limited to small-area deposition due to its batch-processing nature [9].

34.3.3 Sputtering

Sputtering involves bombarding a target material with energetic ions in a vacuum chamber, causing atoms to be ejected and deposited onto a substrate to form a thin film. While less commonly used for perovskite deposition, sputtering offers advantages in terms of film uniformity, composition control, and compatibility with various substrates [10].

34.3.4 Atomic Layer Deposition (ALD)

ALD is a precise and controlled thin-film deposition technique that relies on sequential, self-limiting surface reactions to deposit atomic layers of material onto a substrate. While less frequently employed for PSC fabrication, ALD offers advantages in terms of thickness control, uniformity, and conformal coating, making it suitable for certain applications requiring precise film properties [11].

34.3.5 Role of Fabrication Parameters on Device Performance and Reproducibility

The choice of fabrication method and parameters significantly impacts the performance and reproducibility of PSCs. Parameters such as precursor concentration, solvent type, annealing temperature, and processing atmosphere influence the crystallinity, morphology, and defect density of perovskite films, thereby affecting device efficiency, stability, and hysteresis behaviour [12].

Optimisation of fabrication parameters is essential to achieving high-performance PSCs with reproducible characteristics. Furthermore, process scalability, cost-effectiveness, and environmental considerations are important factors to consider in the development of commercial-scale manufacturing processes for PSCs.

34.4 Factors Influencing Photovoltaic Performance

Several key factors play pivotal roles in determining the PV performance of PSCs, encompassing material composition, device architecture, and processing conditions. Understanding and optimising these factors are essential for achieving high-efficiency and stable PSCs.

34.4.1 Material Composition

The choice of halide ions, organic cations, and additives profoundly influences the optoelectronic properties and stability of perovskite materials. Variation in halide composition (e.g., iodide, bromide, or chloride) enables bandgap engineering, facilitating the absorption of a broader spectrum of sunlight and enhancing device efficiency. Moreover, the selection of organic cations (e.g., MA and FA) influences the perovskite crystal structure, affecting charge carrier dynamics and device performance. Additionally, the incorporation of additives, such as passivation agents or dopants, can mitigate defect densities, suppress non-radiative recombination, and enhance device stability [13].

34.4.2 Device Architecture

The architectural design of PSCs, including planar and mesoporous structures, significantly impacts charge transport, light absorption, and interfacial charge extraction. Planar heterojunction architectures offer advantages in terms of simplified fabrication processes and reduced hysteresis effects but may suffer from limited charge extraction efficiency and light absorption compared to mesoporous counterparts [14]. Mesoporous architectures, on the other hand, provide enhanced surface area for perovskite deposition, facilitating efficient charge transport and collection. Furthermore, interfacial engineering, such as modifying the interfaces between the perovskite layer and charge transport layers, is critical for minimising energy barriers, reducing charge recombination, and improving overall device performance.

34.4.3 Impact of Processing Conditions, Film Morphology, and Crystallinity

Processing conditions during perovskite film deposition, including precursor concentration, solvent type, temperature, and annealing conditions, profoundly influence film morphology, crystallinity, and defect density. Optimisation of these parameters is essential to achieve uniform, pinhole-free films with optimal crystalline quality, thereby enhancing charge carrier mobility and reducing recombination losses [4, 13]. Moreover, controlling the morphology and crystallinity of the perovskite film is crucial for achieving efficient charge transport and collection, as well as long-term device stability under operating conditions.

In summary, the PV performance of PSCs is intricately governed by the interplay of material composition, device architecture, and processing conditions. By carefully tuning these factors and optimising device fabrication processes, researchers can advance the development of high-efficiency and stable PSCs for practical applications.

34.5 Recent Advantages and Challenges

In recent years, the field of PSC research has witnessed significant advancements alongside persistent challenges, shaping the trajectory of this promising PV technology. This section provides an overview of recent breakthroughs, highlights existing challenges, and discusses strategies for overcoming these obstacles to accelerate the development and commercialisation of PSCs.

34.5.1 Review of Recent Breakthroughs in Perovskite Solar Cell Research

The past few years have seen remarkable progress in enhancing the efficiency, stability, and scalability of PSCs. Breakthroughs include the achievement of power conversion efficiencies (PCEs) exceeding 25%, facilitated by innovative materials engineering, device architectures, and fabrication techniques. Moreover, advancements in tandem and perovskite–silicon hybrid solar cells have demonstrated the potential for achieving even higher efficiencies by leveraging complementary absorption spectra and optimised charge transport pathways [15].

Furthermore, research efforts have focused on improving the long-term stability of PSCs against environmental factors such as moisture, heat, and light exposure. Strategies such as encapsulation, interface engineering, and the development of novel materials with enhanced stability have shown promising results, with PSCs demonstrating extended operational lifetimes under accelerated aging conditions [16].

34.5.2 Challenges Related to Toxicity, Scalability, and Commercialisation

Despite the significant progress, several challenges hinder the widespread adoption of PSCs in commercial applications. One such challenge is the toxicity of lead-based perovskite materials, raising concerns regarding environmental and health hazards during manufacturing, disposal, and end-of-life recycling processes [4]. Efforts are underway to develop lead-free alternatives and mitigate the environmental impact of PSCs through recycling and sustainable manufacturing practices.

Scalability remains another key challenge, as the transition from laboratory-scale prototypes to large-area module production requires addressing issues related to reproducibility, uniformity, and cost-effectiveness. Scalable deposition techniques, such as roll-to-roll processing and printing technologies, are being explored to enable mass production of PSCs at competitive costs.

Additionally, the commercialisation of PSCs faces regulatory hurdles, market acceptance challenges, and competition from established solar technologies. Addressing these challenges requires concerted efforts from academia, industry, and policymakers to establish standards, ensure product safety, and foster public trust in the reliability and sustainability of PSCs.

34.5.3 Strategies for Addressing Current Limitations and Accelerating Technology Development

To overcome existing limitations and expedite the commercialisation of PSCs, interdisciplinary research collaborations and strategic investments are essential. Research initiatives aimed at developing lead-free perovskite materials, improving manufacturing processes, and advancing device stability are critical for realising the full potential of PSCs as a clean and sustainable energy solution.

Furthermore, partnerships between academia, industry, and government entities can facilitate technology transfer, scaleup production, and drive innovation in PSCs. Public–private collaborations, funding initiatives, and supportive policies are instrumental in creating an enabling environment for the growth of the perovskite solar industry, driving down costs, and accelerating market penetration.

In conclusion, while recent advances in PSC research have been promising, significant challenges remain on the path to commercialisation. By addressing issues related to toxicity, scalability, and market acceptance through collaborative efforts and targeted investments, PSCs hold the potential to revolutionise the solar energy landscape and contribute to a sustainable future.

34.6 Conclusion

In summary, this chapter has provided a comprehensive exploration of perovskite solar cells (PSCs), delving into their fundamental aspects, recent advancements, challenges, and future prospects. Through an examination of material composition, fabrication methods, factors influencing photovoltaic performance, recent breakthroughs, and challenges, valuable insights have been gleaned regarding the potential of PSCs as a transformative technology in the renewable energy landscape.

Key findings from this chapter underscore the remarkable optoelectronic properties of perovskite materials, their compatibility with solution-processing techniques, and the pivotal role of material composition and device architecture in determining photovoltaic performance. Recent breakthroughs in PSC research, including record-breaking efficiencies, enhanced stability, and scalability, highlight the immense progress made towards realising the full potential of this technology.

However, challenges such as toxicity, scalability, and commercialisation hurdles persist, posing barriers to the widespread adoption of PSCs. Addressing these challenges requires collaborative efforts from academia, industry, and policymakers to develop sustainable, safe, and cost-effective solutions.

Despite these challenges, the significance and potential impact of perovskite solar cells in the field of renewable energy cannot be overstated. With their high efficiency, low-cost fabrication processes, and compatibility with flexible and tandem architectures, PSCs hold promise as a viable alternative to conventional silicon-based solar cells. Moreover, their potential applications in building-integrated photovoltaics, wearable devices, and beyond underscore their versatility and versatility in meeting diverse energy needs.

In conclusion, perovskite solar cells represent a disruptive force in the quest for sustainable energy solutions. Through continued research, innovation, and collaboration, PSCs have the potential to revolutionise the solar energy landscape, paving the way towards a cleaner, greener, and more sustainable future.

REFERENCES

[1] International Energy Agency (IEA). (2021). *Renewables 2021: Analysis and Forecast to 2026*, IEA, Paris.

[2] Kojima, A., et al. (2009). Organometal Halide Perovskites as Visible-Light Sensitizers for Photovoltaic Cells. *Journal of the American Chemical Society*, *131*(17), 6050–6051.

[3] Yang, W. S., et al. (2012). Iodide Management in Formamidinium-Lead-Halide–Based Perovskite Layers for Efficient Solar Cells. *Science*, *356*(6345), 1376–1379.

[4] Stranks, S. D., & Snaith, H. J. (2015). Metal-Halide Perovskites for Photovoltaic and Light-Emitting Devices. *Nature Nanotechnology*, *10*(5), 391–402.

[5] Green, M. A., et al. (2014). Solar Cell Efficiency Tables (Version 42). *Progress in Photovoltaics: Research and Applications*, *22*(7), 701–710.

[6] Kim, H. S., et al. (2016). Lead Iodide Perovskite Sensitized All-Solid-State Submicron Thin Film Mesoscopic Solar Cell with Efficiency Exceeding 9%. *Scientific Reports*, *2*(3), 1–7.

[7] Chen, Q., et al. (2015). Planar Heterojunction Perovskite Solar Cells via Vapor-Assisted Solution Process. *Journal of the American Chemical Society*, *137*(1), 40–43.

[8] Hu, X., et al. (2017). Scalable Fabrication of Efficient Perovskite Solar Cells via Doctor-Blade Coating. *Advanced Materials*, *29*(24), 1–7.

[9] Jeong, S., & Seo, J. (2020). Spray-Coating Techniques for Organic and Perovskite Solar Cells. *Advanced Materials Interfaces*, *7*(9), 1–18.

[10] Jeon, N. J., et al. (2014). Solvent engineering for high-performance inorganic–organic hybrid perovskite solar cells. *Nature Materials*, *13*(9), 897–903.

[11] Docampo, P., et al. (2014). Sputtered Thin Films of Tin(II) Sulfide and Their Application in Thin-Film Transistors. *Chemistry of Materials*, *26*(6), 2063–2068.

[12] Lee, K. T., et al. (2019). Atomic Layer Deposition of Lead Halide Perovskite Materials: A Room-Temperature, Solution-Free Synthesis Method for Efficient Solar Cells. *Nano Letters*, *19*(3), 2210–2217.

[13] Momblona, C., & Snaith, H. J. (2016). Perovskite Photovoltaics: Slowly but Surely. *ACS Energy Letters*, *1*(4), 949–955.

[14] Abdi-Jalebi, M., et al. (2018). Impact of Stoichiometry on the Stability of Perovskite Solar Cells. *Nature Communications*, *9*(1), 2133.

[15] Bush, K. A., et al. (2020). 23.6%-Efficient Monolithic Perovskite/Silicon Tandem Solar Cells with Improved Stability. *Nature Energy*, *5*(7), 1–9.

[16] Li, N., et al. (2021). Towards Stable and Scalable Perovskite Solar Cells: Strategies for Interface Engineering and Encapsulation. *Energy & Environmental Science*, *14*(3), 1330–1365.

35

Nanofunctional Materials in Food Processing

Hariharan Thangavel
Tamil Nadu Agricultural University, Madurai, India

M. PrasannaBlessy
Dr. Mahalingam College of Engineering & Tech., Pollachi, Tamil Nadu, India

Rohit Kumar
Coer University, Roorkee, India

R. Halima
Sir M Visvesvaraya Institute of Technology, Bangalore, India

M. Anusuya
Indra Ganesan College of Engineering, Trichy, Tamil Nadu, India

35.1 Introduction

Nanotechnology is a groundbreaking and encouraging technology that has been combined into an assortment of areas, including prescription, horticulture, and food fabricating. This section gives extra bits of knowledge about the expected transformative impacts of nanofunctional materials on the standard food industry. The content features significant advantages that can be accomplished by utilising nanomaterials in food lattices on account of their extraordinary physicochemical properties including tiny size, enormous surface region, and cooperation capacities at the subatomic level. Key uses featured incorporate improved supplement flexibly and bioavailability, keen packaging with nanosensors to guarantee food well-being, expanded timeframe of realistic usability, improved quality, new handling choices, and supportability. Notwithstanding, concerns, for example, possible poisonousness, absence of administrative normalisation, and buyer incredulity, are appropriately recognised. To guarantee the moral and responsible mix of nanotechnology into the food industry, it is critical to set up comprehensive well-being examinations and strong worldwide administrative structures.

Nanofunctional materials have made a disruptive wave in the food area, significantly changing a few components, like horticultural and domesticated animals creation, flavour and nourishment improvement, and developments in food packaging [1]. Their tiny nanoscale size gives them one of the kind properties that makes them significant for some cooking applications. This has started incredible interest in the logical local area as these materials can possibly fundamentally work on the nature of food, well-being, and timeframe of realistic usability. The goal of the presentation is to give a far-reaching outline of nanofunctional materials with regard to food handling. Nanomaterials like nanodispersions and nanocapsules are described by giving functional segments through encapsulation, focused on delivery, controlled delivery, and improved steadiness. Also, nanotechnology assumes a vital part in separating supplements from crude materials, supplanting customary strategies, and working on food creation, handling, protection, and circulation. This comprehensive methodology guarantees better food quality, less waste, and, more noteworthy, food security.

In spite of the expected advantages, it is vital to perceive the likely risks of nanotechnology for human and creature well-being [2]. The ingestion, circulation, digestion, and evacuation of nanoparticles after oral utilisation raise concerns about negative effects. Standardised tests are viewed as essential to assess

DOI: 10.1201/9781003495437-35

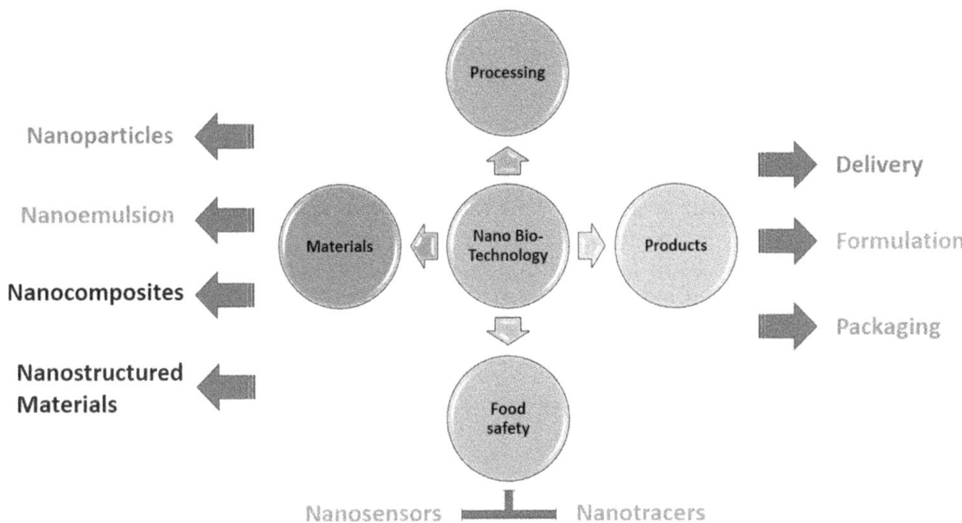

FIGURE 35.1 Several sectors of the agricultural industry that are dependent on nanotechnology.

hazardous impacts on the climate and human well-being. Precise and far-reaching poisonous substance appraisals require tough administrative guidelines worldwide, just as the utilisation of instrumentation and informatics to guarantee protected joining of nanomaterials into food items [3]. The story features the developing interest for the utilisation of nanofunctional materials in food handling and their significance in tackling modern difficulties. Specifically, its capacity to expand the bioavailability of nutraceuticals and bioactive substances expands the advantages for the shopper. The interesting properties of nanoparticles empower an ideal model shift in business food measures, prompting improved quality, decreased waste, and expanded well-being.

Given these empowering possibilities, there is a need to painstakingly analyse the potential risks related with nanotechnology in foods. Further examinations on ingestion, circulation, digestion, and end are required, zeroing in on conceivable unfriendly impacts on human well-being. Standardised testing, worldwide administrative models, and current logical methods are suggested to guarantee protected joining of nanomaterials into food handling. The fragment explicitly includes nanofunctional materials and features their basic part in tackling different difficulties in the developing climate of the horticultural area. Its rise is portrayed as an upheaval power that drives development and opens the way to new potential outcomes. Exploiting their extraordinary properties, nanofunctional materials are portrayed as basic for fostering cooking applications and advancing modern turn of events.

Nanofunctional materials have set up themselves as drivers of progress in food handling. Their remarkable qualities and steady responsibility to taking care of complex issues put them at the bleeding edge of logical revelation. As their importance keeps on captivating the logical local area, they end up being key drivers for a future described by better food quality, improved well-being measures, and maintainable practices (Figure 35.1).

35.1.1 Applications of Nanofunctional Materials in Food Processing

Nutrient enhancers: Nanofunctional materials can be utilised to work on the wholesome substance of cooking items. For instance, nanoencapsulation can expand the bioavailability of nutrients and minerals [4].

Food packaging materials: Enhanced barrier properties in food packaging composed of nanofunctional materials could possibly broaden the freshness of wares [4].

Bottom-up processing technology: In bottom-up processing technology, nanostructures are made from singular atoms or particles. This methodology can be utilised to produce innovative food parts with higher usefulness [4].

Top-down processing technology: Top-down processing technology includes the decrease of mass materials to the nanoscale. This strategy can be utilised to make nanoemulsions equipped for expanding the steadiness and bioavailability of lipophilic substances [4].

35.1.1.1 Advantages and Disadvantages

The utilisation of nanofunctional materials in food handling offers various advantages, including higher wholesome substance, more extended timeframe of realistic usability, and improved usefulness. In any case, there are additional concerns in regard to the well-being of these mixtures, as their impacts on human well-being are not yet completely perceived [5].

35.2 Nanomaterial Production Process

As displayed in Figure 35.2, currently there are two essential techniques for obtaining nanomaterials: A top-down methodology and a base-up procedure. The "top-down" strategy includes actually cutting nanoscale materials using lithography, carving, granulating, and processing. On the other hand, bottom-up nanotechnology in the food industry has been influenced by biologically inspired concepts like self-assembly and self-assembly. Building larger structures molecule by molecule or atom by atom is one example of a bottom-up strategy. These processes involve self-assembly, chemical synthesis, and situational organization [6].

35.2.1 Production of Nanomaterials for Food and Bioactive Substances

When making nanoparticles, an assortment of conveyance frameworks is utilised, like restricting colloids, biopolymer nanoparticles, nanoemulsions, nanofibres, nanocapsules, and so forth. These frameworks fill in as methods for shipping functional components, ensuring the functional component from degradation, and controlling the dispersal of dynamic substances [5].

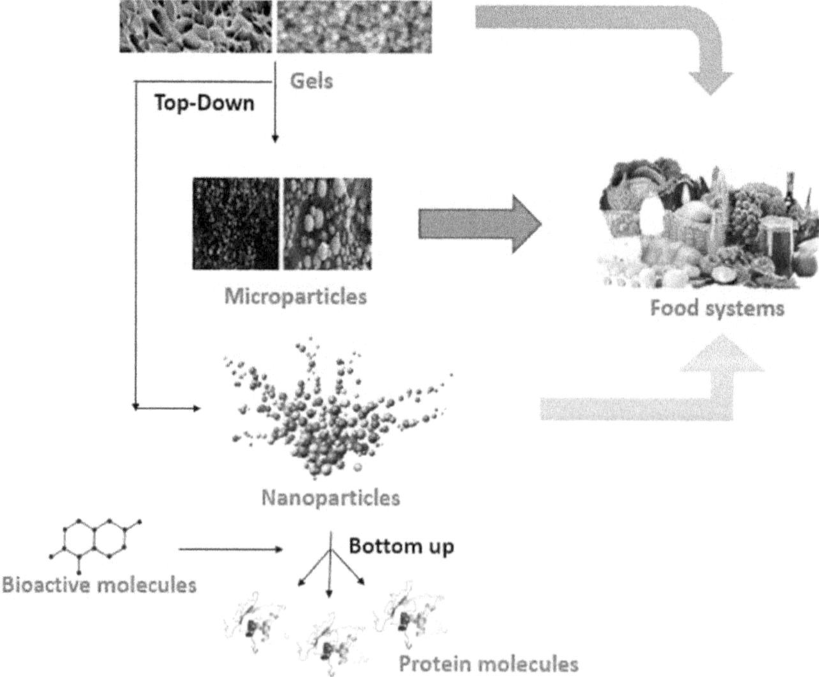

FIGURE 35.2 Nanomaterial production process.

35.2.2 Associative Colloids

For many years, restricting colloids, such as vesicles, bilayers, switch micelles, surface actuated micelles, and fluid gems, have been utilised to encapsulate and ship polar, non-polar, and/or amphibious functional parts. A non-polar functional component, for instance, might be appropriated inside the micellar film structure or the hydrophobic centre of a surfactant micelle; subsequently, it very well may be controlled in a water-based arrangement dependent on the particular application necessities. Relationship colloids are frameworks that are thermodynamically useful and are regularly determined to foster because of the hydrophobic impact, which is the decrease of the contact region between the non-polar gatherings of the surfactant and water in the colloidal relationship. The sorts of following designs and the kinds of restricting colloids that are created are controlled by the subatomic qualities and fixations of the surfactant and cosurfactant that are applied, notwithstanding the environmental conditions (like pH, particle strength, and temperature). Importantly, the location of the functional component encapsulated in the binding colloid (e.g., in the hydrophobic core or as part of the binding colloid membrane) is of fundamental consequence for the functioning of the self-assembled system. The diameters of many binding colloids are between 5 and 100 nm, consequently these structures are dubbed nanoparticles.

Colloidal restricting frameworks have three primary benefits: They are thermodynamically helpful, spontaneously foster, and frequently give novel arrangements. In any case, the primary impediment is that creation requires a lot of surfactant (and now and again cosurfactant), which may cause issues with taste, cost, or lawfulness. Also, weakening of liquids containing colloids might trigger their unconstrained dissociation since the arrangement of associative colloids is focus subordinate. Accordingly, keeping up with the viability of colloids under an assortment of ecological conditions relies on the choice of surfactants and cosurfactants.

35.2.3 Nanoemulsions

Nanoemulsions are thermodynamically unaffected emulsions compared to conventional emulsions in a spectrum of various conditions. This is owing to their modest size (usually 50 to 500 nm vs 1200 nm) and their monodispersity. They may be diluted with water without influencing the particulate size distribution. The strength of the last emulsion relies on the sort of surfactant utilised to make the nanoemulsion. The nanoemulsion arrangements might be utilised to encapsulate functional food parts at oil/water interfaces or in the persistent stage of the interaction (Weiss et al. [5]).

35.2.4 Some Uses for Nanoemulsions

- Delivery of active ingredients to the body.
- Stabilisation of biologically active ingredients.
- Longer life due to better stability and reduced concentrations of the oil phase.

Nanotechnology has a wide scope of expected uses in the cooking area. In any case, a large number of these thoughts could be challenging to execute monetarily since they are either exorbitant or impractical to utilise on a huge scale. The accompanying areas centre around a restricted number of nanotechnology applications that may, before long, show financial worth. As nanomanufacturing innovations become more reasonable, it is very expected that the restricted utilisation of nanotechnology in the food industry will change.

35.2.5 Nanoencapsulation

The cycle of fusing substances into minuscule vesicles or shut materials with nanometric (or submicron) measurements is known as nanoencapsulation. These nanomaterials have various benefits, like acting as a vehicle for the delivery of fat-dissolvable fixings, protecting them from contamination in the processing framework or during handling, permitting controlled delivery at a particular area, being viable with other food fixings, having a more drawn out home time, and working on retention [7]. This strategy makes it simpler to safeguard bioactive fixings including proteins, lipids, nutrients, cancer prevention agents, and carbs, which prompt more steady and functional dinners. Since nanoencapsulation lessens the measure of dynamic fixings needed, it might assist formulators with saving cash.

35.2.6 Nanofibres

The numerous capacities of nanofibers show their unique highlights. The capacity to make filaments with an estimated distance of under 100 nm gratitude to the electrospinning process has prompted huge headways in assembling innovation. This process delivers solid, slender polymer strands from an answer by applying a high electric field to a spinneret. Different electrospun polymer strands with distances across going from 10 to 1000 nm serve various purposes as far as their warm, electrical, and mechanical properties. These filaments are utilised in numerous modern areas for the creation of materials, electrical gear, defensive hardware, and prosthetics.

The possibility of utilising electrostatic powers for fibre union returns to patent applications recorded somewhere between 1934 and 1944. In this technique, a solution between terminals that are inversely charged structures polymer strands. This arrangement makes a fluid charge on the fibre's surface, which prompts the arrangement of a Taylor cone. With the assistance of this cone, an accused fluid polymer stream might be delivered, which, when the solvent dissipates, shapes strong strands. The mechanical and useful properties of these strands are impacted by the fly's direction and speed, which might bring about twisting precariousness that improves the mechanical properties.

Current advancements like microchips utilise composite nanofibers, particularly those strengthened with carbon nanotubes. However, most patents and research focus on life sciences and show applications such as biocompatible nanofibers for wound healing, regeneration of blood vessels and nerves, and matrices for drug delivery. Electrospun strands have a higher surface-to-volume proportion, which works on their functional properties.

Other than its utilisation in science, electrospinning finds use in the filtration of gases and fluids, military safeguard frameworks, and the improvement of exceptionally touchy sensors. While electrospun strands find numerous new uses, their relevance in the food and farming businesses is yet generally restricted. This is clarified by the way that engineered polymers are frequently utilised in electrospinning instead of biopolymers got from food and agribusiness. The food industry is relied upon to utilise biopolymer nanofibers as advances in the creation of nanofibers from food-grade biopolymers are made.

35.2.7 Biopolymer Nanoparticles

Food-grade biopolymers, including proteins or polysaccharides, might be utilised to make nanoscale particles by self-affiliation, total, or stage partition in blended biopolymer frameworks [8]. At first, egg whites and non-biodegradable engineered polymers like methyl polyacrylate and polyacrylamide were utilised to make biopolymer nanoparticles. One famous biodegradable nanoparticle that is broadly used to encapsulate and transport drugs and micronutrients including iron, nutrients, proteins, and so forth is polylactic acid (PLA). It has been demonstrated that PLA requires an associative synthetic, like polyethylene glycol, to give great outcomes. Also, functional parts might be encapsulated in nanoparticles and delivered in the light of specific natural boosts [9].

35.2.8 Nanotubes

In nanotechnology, carbon nanotubes are regularly utilised for uses other than in food industry. Among different uses, these designs have been utilised as impetus measure holders or as low-obstruction conductors. Under certain natural conditions, hydrolysed α-lactalbumin and other globular milk proteins have been displayed to self-organize into structurally comparable nanotubes. This method has been considered to deliver imitates of muscle fibre design or to help in the immobilisation of compounds. It is additionally relevant to different proteins.

35.2.9 Nanoliposomes

Watery stage and other particles coupled to lipid or phospholipid bilayers set up the establishment of liposomes, which are shut, persistent phospholipid esters. Similar to liposomes in terms of chemistry, physics, and thermodynamics, nanoliposomes may survive aggregation and hold onto their nanoscale dimension while being stored. The bifunctional nanoliposome was built to ship cancer prevention agents with contrasting solubilities all the while inside a solitary lipid vesicle, like water-dissolvable glutathione and lipid-dissolvable α-tocopherol. The essential benefit of nanoliposomes is in their capacity to catch, transmit, and release water-solvent lipids and bioactive mixtures at the fitting moment and spot because they are amphiphilic. Nanoliposomes might be used to change unpredictable, receptive, or delicate particles (like nutrients, compounds, cancer prevention agents, and thinning specialists) into steady ones (Figure 35.3 and Table 35.1).

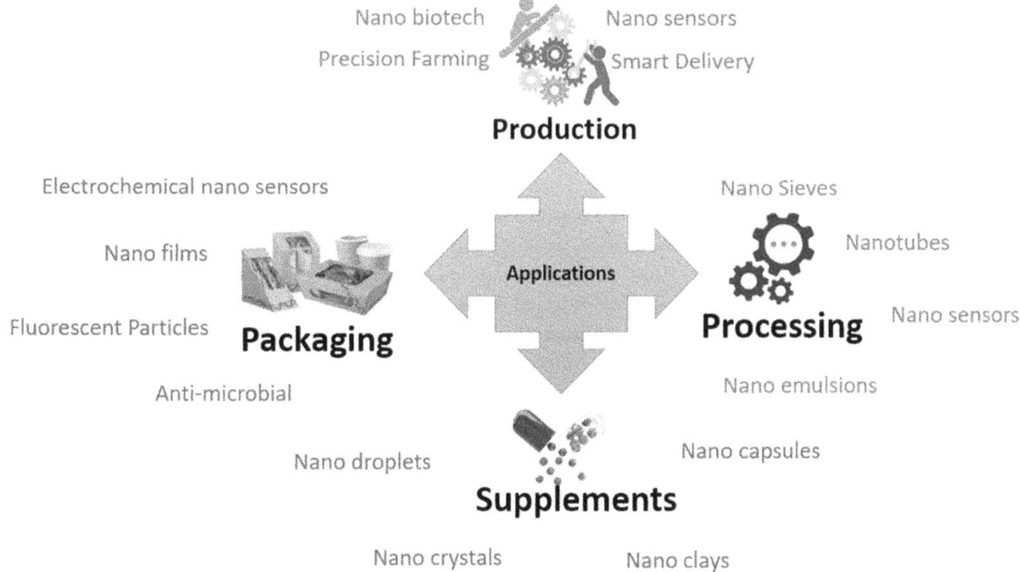

FIGURE 35.3 Summary of various uses of nanomaterials in the food industry.

TABLE 35.1

A Summary of the Many Applications of Nanoparticles in the Food Industry

Nanomaterials	Types	Uses in the Food Industry
Nanoparticles	Ag, ZnO, Mg, and SiO_2	Food packaging, oxidation of contaminants, and antibacterials
Nanosieve	Specific nanoparticles	Elimination of infections or pollutants
Nanocapsules	Bioactive compounds	Increased potency, solubility in water, and controlled and localised release
Nanoemulsions	Extensions or interpolations; gum Arabic or altered caseinate, soy, or starch	Food processing, food encapsulation, stability, colouring, and antimicrobials and preservation
Nanospheres	Starch nanosphere	Food encapsulation and synthetic adhesives
Nanosensors	Aptasensors	Detection of microorganisms and control of food spoilage
Nanomicelles	Acquanova and Novasol	Liquid vehicle and improved solubility
Nanocomposite	$Fe–Cr/Al_2O_3$ and Ni/Al_2O_3	Improve shelf life and protection and packaging of food

35.3 Food Packaging

The cutting-edge food industry relies intensely on bundling. While it is fundamental for food conservation, the primary motivation behind packaging is to work on the nature of food all through its whole life pattern, from assembling to utilisation. As Figure 35.4 shows, the utilisation of nanotechnology to food packaging offers huge open doors to work on food packaging adequacy. By exploiting their ability to encapsulate dynamic substances and help usefulness, steadiness, and bioavailability, an assortment of nanoparticles is created and utilised in items. Also, food packaging exploits nanofillers, which fill explicit needs and help the development of this area. As Figure 35.4 shows, nanofillers in the field of biosensors not just ensure food from ecological impacts but additionally give novel attributes to the packaging materials, making a plenty of chances for the food packaging area [10].

Carbon dioxide radiators and oxygen scroungers are extra significantly added substances used in nano-sized packaging materials and utilised to manage the microbiological development and air oxidation. These forms of active packaging materials are implemented use packing of livestock, fish, milk and dairy commodities. The deterioration of food items causes change in pH and temperature of containers over time. Thus, the integration pH indicator in packaging film provides particular features to functions as time and temperature indicators to monitors distinct characteristics of packaged times.

Zinc oxide is an essential cancer prevention agent. For the food industry, it is a nanocomposite material utilised in the active packaging of food parts. Oxidation is the principle debasement process that occurs in consumables and abbreviates their time span of usability. Cancer prevention agents are incorporated into the packaging material as a feature of active packaging technique, which works on food conservation [11]. Food packaging regularly utilises nanocomposites related to nanolaminates notwithstanding nanoparticles for antimicrobial insurance. Scientists broaden food's timeframe of realistic usability by ensuring it from outrageous heat and mechanical pressure. More extended time span of usability and worked on food quality are accomplished by adding nanoparticles to food fixing packaging [12].

Shrewd and proactive packaging is known as smart food packaging. By and large, active packaging broadens time span of usability and works on packaged products. By providing packaged goods with products critical medical advantages, bioactive packaging decidedly affects purchaser well-being.

Food packaging is one of the most essential business uses of nanotechnology in the cooking business. The main usage of nanotechnology in the food industry is the utilisation of nanoparticles to work on the adaptability and gas boundary properties of food packaging. The active packaging incorporates

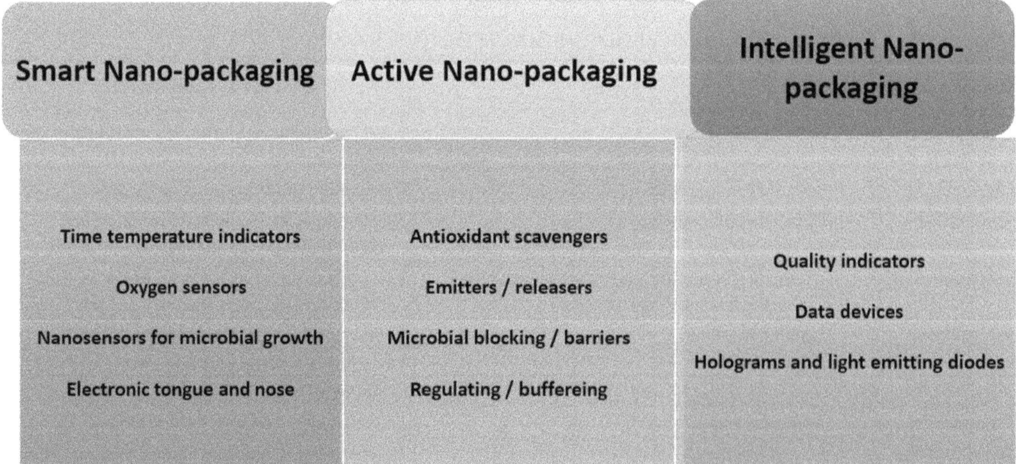

FIGURE 35.4 Classification of food nanopackaging.

nanoparticles that have antibacterial and oxygen-scrounging properties just as brilliant packaging and brilliant food packaging that incorporate nanosensors intended to recognise and send signals on microbial and biochemical changes [13].

Nanotechnology is remembered for the packaging of numerous cooking products. One of the special characteristics of the new Living Nano Packs is their capacity to annihilate infections present in food. When utilised for food packaging, nanomaterials bring out the flavours of food without changing the food in an unfavourable manner. Food components may also be nanopackaged using nanoparticles. The use of nanotechnology in food packaging offers technological answers by altering the film's penetration activities and improving its mechanical, chemical, and microbiological barriers, as well as its heat resistance.

The utilisation of nanotechnology to make biodegradable packaging diminishes ecological contamination. In the culinary industry, a silicate layer is a kind of nanostructure used in food packaging. Advancements in sensor technology used in food items' smart packaging provide data on material half-lives, safety, and quality. Nutritious products are packaged using nanocomposites as well. The remarkable opposition to warm pressure in food planning, transportation, and capacity. At this point, nanocomposites are utilised in capacity compartments, broadening their time span of usability by up to a half year.

Due to their heat resistance, carbon-based graphene nanoplatelets may find use in the food industry for food packaging. The method the analysts are utilising to make covering materials utilising ultrasound is called sonochemical covering, and it offers a dexterous and straightforward approach to apply this innovation.

It has been shown that the covering attempts to counteract an assortment of bacterial species, including Gram-positive *Escherichia coli* microbes and Gram-negative *Staphylococcus aureus*. It has been checked that this cycle works on the creation of parts that can endure long. Silver nanoparticles have been utilised for quite a while and are believed to be vital particles for material encapsulation. For instance, objects covered with these particles are shielded from tainting. Although just a couple of uses of silver nanoparticles have been endorsed by the European Food Safety Authority (EFSA) for reuse in food conservation and bundling, various analysts have expressed that silver nanoparticles are huge in the food packaging explore and food conservation industry. Zinc oxide is a food added substance that has been endorsed by the FDA and is by and large thought about protected. Different antibacterial mixtures with particular properties are made conceivable by nanotechnology, including silver nanoparticles, zinc oxide, magnesium oxide, and nickel oxide. At the nanoscale, these nanoparticles have shown intriguing uses and antibacterial properties. These nanoparticles are added to polymeric lattices to furnish them with benefits including worked on packaging properties and antibacterial action.

35.3.1 Nanosensor

Identifying ecological changes is a significant capacity of nanosensors. varieties in temperature, mugginess, and gas creation; additionally, metabolites from microbial development and results of food contamination might be distinguished. For these purposes, an assortment of nanosensors is utilised, like exhibit biosensors, sensors in light of carbon nanotubes, electronic tongue or nose, microfluidic gadgets, and nanoelectromechanical systems (NEMS). Chemical and counteracting agent recognising biosensors are vital for following the nature of food well-being. To identify *Staphylococcus enterotoxin* B in dairy items, for instance, a microfluidic immunodetection framework utilising biotin-functionalised antibodies was created. Application on fluid films showed that the location limit of the method is 0.5 ng/mL. Streptavidin-reinforced double-layer wafers covered in polyoxidized material (dimethylsiloxane).

35.4 Advances in Food Technology: Nanofunctional Materials

Nanotechnologies, the control of matter at the subatomic and nuclear level, have brought about tremendous changes in different fields, whose upsetting impact additionally reaches out to the arrangement of food. This developing discipline has fostered an assortment of nanofunctional materials that offer the potential to change food handling, conservation, and utilisation [1]. These materials, working at the nanoscale, offer various advantages in the food industry, going from more extended time span of usability, better taste and surface, higher wholesome worth, and diminished gamble of contamination. Also, they show the capability of progressive packaging arrangements that add to the quality of food and reduction of waste.

In the field of food handling, the presentation of nanofunctional materials helps work on the general nature of food sources and supportability. By consolidating nanoparticles, analysts can foster dry spell and irritant safe yields, expanding yields and horticultural creation, particularly in poor nations. Meanwhile, the utilisation of these assets in developed nations empowers the creation of fresher and more advantageous food sources according to purchaser requests. The utilisation of nanoparticles in food science is developing as an upsetting power empowering developed and developing nations to effectively address horticultural difficulties.

In spite of the potential advantages, public assessment stays divided on the utilisation of nanotechnology in the food area. The requirement for future studies to research the expected unfriendly impacts of nanoparticles on human well-being and the climate is perceived [14]. The principle objective is to track down a trade-off between misusing the upsetting expected of nanofunctional materials and tending to expected risks and moral results.

The mix of nanofunctional materials into food handling offers transformative potential to work on food well-being and quality by lessening pollutants, expanding time span of usability, and working on in general creation effectiveness and manageability. With a worldwide populace of more than 6 billion individuals and quick advancement representing difficulties, the utilisation of nanofunctional materials turns out to be basic to address food deficiencies and guarantee security in developing and developed nations. In any case, given the double idea of the expected risks and advantages, there is a need to direct further exploration and set up far-reaching administrative systems to direct the protected and liable utilisation of functional nanomaterials in the food industry.

Incorporation of nanomaterials into food production and packaging not just offers modern change potential but additionally offers an economical arrangement by reducing waste, broadening time span of usability, and working on item well-being and food quality. With a developing worldwide populace, the key utilisation of nanofunctional materials shows up to be a vital factor in tackling issues related to food deficiencies and guaranteeing food security for shoppers on an overall scale.

35.5 The Role of Nanomaterials in Modern Food Processing Technology

Nanofunctional materials play a crucial role in establishing modern food processing methods and act as excellent carriers of functional ingredients due to their optimal performance in encapsulation, emulsification, and dispersion activities [15]. Food scientists may produce functional meals with extra health benefits, including better digestion and immune support, by using nanoparticles to increase the bioavailability and controlled release of nutraceuticals and bioactive substances. Furthermore, the use of nanofunctional materials improves food preservation by slowing down the development of microorganisms responsible for spoilage and extending the useful life of products. These materials also help improve the texture and sensory qualities of foods, thus, improving consumers' overall dining experience. In particular, its implementation contributes to improving the efficiency and sustainability of food processing, which translates into a reduction in energy consumption and waste production. In summary, the use of nanofunctional materials in food processing shows enormous potential to improve food quality, safety, and sustainability [16].

In the field of nanotechnology in food processing, the expected benefits are better distribution of functional ingredients, better food preservation, and greater efficiency and sustainability. However, it is important to recognise that despite various advantages, the use of nanofunctional materials presents potential dangers and problems [14]. These issues range from the likely toxicity of nanoparticles to human health to the need for comprehensive regulation to adequately control associated risks [3]. Therefore, to ensure the safe and responsible integration of nanotechnology into the food processing sector, more cooperation between industry players and regulatory organisations is needed [14].

Appropriate safety laws and procedures in the field of nanotechnology are essential to reduce potential dangers and protect consumer safety. Rigorous criteria and controls are essential to regulate the use of nanomaterials in food production. This includes rigorous testing and evaluation of the safety and potential health effects of nanoparticles used in foods. Furthermore, simple labelling is essential as it provides customers with information on the incorporation of nanotechnology into food production and allows them to make informed decisions.

Compliance with safety measures and standards not only protects the health of consumers but also preserves the reputation of food manufacturers, thus increasing customer trust. It is essential that food industry professionals stay abreast of the latest regulatory developments and follow best practices for the safe use of nanotechnology. By focusing on safety and regulatory compliance, the food industry can leverage the benefits of nanofunctional materials while ensuring customer well-being and long-term business survival.

35.6 New Technologies: Nanofunctional Materials in the Food Industry

The food sector is undergoing a metamorphosis with the incorporation of edge technologies, and, among these, the incorporation of nanofunctional materials stands out as a revolutionary force in food processing. These advanced materials, including nanodispersions and nanocapsules, serve as optimal vehicles for the delivery of functional compounds to food products, as Nile et al. [2]. Nanotechnology in food processing allows companies to leverage their capabilities to efficiently extract nutrients from raw materials, increase production rates, and address challenges related to food loss due to microbial contamination and spoilage.

The use of nanotechnology in food processing goes beyond efficiency benefits. It plays a crucial role in improving food quality, maintaining safety, and extending the shelf life of products. This, in turn, contributes to global food security and solves multiple problems caused by population growth and environmental concerns. The relevance of nanofunctional materials not only is limited to solving challenges at the macro level but also has a profound impact on food properties at the micro level [14]. Its smaller size and larger surface area allow for greater dispersion and diffusion within food matrices, resulting in better sensory qualities and greater nutritional value [15].

With the integration of nanotechnology, the food sector is well positioned to meet consumer expectations in developed countries for fresher and healthier food alternatives. At the same time, this enables the development of sustainable and efficient food production technologies suited to the urgent problem of food shortages in developing countries. The current global population of more than 6 billion people highlights the need for more efficient, safe, and sustainable food production systems that are protected from the harmful effects of pathogenic organisms. The application of nanofunctional materials in food processing is proving to be a multidimensional solution that addresses the challenges of nutrient extraction, shelf-life extension, sensory enhancement, and maintaining food safety in various global contexts.

In summary, the use of nanofunctional materials in food processing not only represents a solution to critical challenges in the food sector but also paves the way for sustainable and efficient food production systems. This innovative technology responds to the need to produce food not only more productively but also in a safer and more environmentally sustainable way.

The benefits of using nanofunctional materials in food processing are numerous and varied. The main advantages include the following:

Improved nutritional value: Foods may have their nutritional content increased by using nanofunctional materials to increase the bioavailability of vitamins, minerals, and other nutrients [17].
Extended shelf life: Food packaging may use nanomaterials to improve barrier properties, which may help extend product expiry dates [17].
Improved food quality: Nanostructured additives can improve the taste, texture, and consistency of foods, leading to an overall increase in food quality [18].
Food safety: Nanotechnology can be used to detect diseases and pathogens in foods, helping improve food safety [17].
Innovative food processing: Nanofunctional materials enable the creation of innovative food components and processing methods, resulting in improved taste, colour, and nutritional quality of foods [19].

Potential risks associated with the use of nanofunctional materials in food processing include the following:

Toxicity: Nanoparticles may pose a potential toxicity risk to human health because they have different properties than their bulk components, such as a massive surface area [17].

Penetration of cells and membranes: Because nanoparticles may be able to cross membrane and cell barriers, the use of nanomaterials in food packaging may be concerning [17].

Conversion into hazardous forms: Nanoparticles can be converted into hazardous forms or vice versa when present in foods, which could lead to significant health effects [20].

Environmental risks: The use of nanotechnology in the food sector may raise potential environmental concerns, as the release of nanoparticles into the environment could have unknown effects [21].

Regulatory challenges: Since further study is needed to examine the potentially harmful effects of nanoparticles on human health and the environment, it is imperative to establish an appropriate regulatory framework to control the possible risks and toxicity of nanoparticles in food [17].

Consumer awareness: Consumers are increasingly concerned about the safety of nanotechnology in foods, which could lead to a decline in consumer acceptance and trust in these products [20].

While nanofunctional materials offer many benefits in food processing, it is important to consider the potential risks associated with their use, including toxicity, cell and membrane penetration, conversion to harmful forms, environmental risks, regulatory challenges, and consumer awareness.

35.7 Innovations in the Field of Food Safety: The Application of Nanofunctional Materials

Ensuring food safety remains a major challenge and stimulates the development of new methods, such as the incorporation of nanofunctional materials into food processing [15]. These materials have significant potential to improve food safety by minimising contamination issues, extending shelf life, and overall improving food quality [14].

Nanofunctional materials, characterised by their unique properties, appear to be particularly suitable for use in culinary preparation. Its small size allows penetration and interaction with food matrices at the molecular level, resulting in greater functionality. For example, nanoparticles can act as effective antibacterial agents and limit the growth of harmful pathogens in foods. Furthermore, they can act as antioxidants, limiting oxidative damage and preserving the freshness of foods. Integrating nanofunctional elements into food packaging appears to be a way to significantly extend the shelf life of products. Furthermore, these materials help create environmentally friendly packaging solutions, reducing waste, and limiting environmental impacts associated with food processing methods [3].

Despite these potential benefits, public opinion remains divided on the use of nanotechnology in the food sector. The need for future studies to investigate the potential adverse effects of nanoparticles on human health and the environment is recognised [14]. The main goal is to find a compromise between exploiting the revolutionary potential of nanofunctional materials and addressing potential dangers and ethical issues (Primožič et al., 2021).

The integration of nanofunctional materials into food processing offers transformative potential to improve food safety and quality by reducing contaminants, increasing shelf life, and improving overall production efficiency and sustainability. With a global population of over 6 billion people and rapid development posing challenges, the use of nanofunctional materials becomes critical to address food shortages and ensure security in emerging and developed countries. However, given the dual nature of the potential risks and benefits, there is a need to conduct further research and establish comprehensive regulatory frameworks to guide the safe and responsible use of functional nanomaterials in the food industry.

The introduction of nanomaterials into food production and packaging offers not only industrial transformation potential but also a sustainable solution by reducing waste, extending shelf life and improving product safety and food quality. With a growing global population, the strategic use of nanofunctional materials appears to be a crucial factor in solving problems associated with food shortages and ensuring food security for consumers on a global scale. The underlying theme illustrates that while nanotechnologies in food processing offer opportunities for innovation and development, their implementation requires a comprehensive and careful approach, taking into account potential dangers and ethical consequences.

35.8 Safety Issues

The extensive application of nanotechnology in the food industry, ranging from testing to packaging and ingredient selection, has raised serious questions regarding potential impacts on animal and human health. While the potential uses of nanotechnology seem endless, the interaction of products containing nanoparticles with the food chain poses a danger of poisoning to both plants and animals. Notably, there is no formal legislation controlling its use in the farming industry. The absence of regulatory oversight highlights the critical need for strict guidelines and regulations to guarantee the secure use of nanoparticles in the food industry.

Concerns over the safety of nanotechnologies and their potential impact on human health have been raised by the growing use of these technologies in the agriculture sector. Concerns about possible risks to human health are raised by the special qualities of nanomaterials, such as their enormous surface area. This worry also extends to the use of nanotechnology in food packaging, with a focus on the ability of nanoparticles to breach membrane and cellular barriers because of their minuscule size. Notwithstanding these obstacles, applications of nanotechnology are already being carried out, as seen by the use of nanoparticles in food packaging. Although these nanoparticles do not directly harm people, if they find their way into food, that may be a problem [22].

Some nations do not have total dietary limitations because people are not aware of the toxicity, availability, and exposure to specific foods. Some nations have called for the creation of a regulatory framework to address the risks associated with nanofoods due to the growing regulatory challenges. The proper use of nanotechnology in the food industry necessitates the development of stringent legal frameworks, regulations, and thorough toxicity testing procedures. An globally acknowledged regulatory framework is very necessary to control the use of nanoparticles in food production.

The use of nanoprocessed food and human products highlights how important it is to adhere to risk assessment procedures when processing food. There are still barriers to a sustainable and nutritious food system, even with the advancements made by nanotechnology in the food industry. Public education on the potential harm that nanoparticles may pose to human health and the environment is becoming more and more important as nanotechnology makes its way into the food industry. Regulations have been established in a number of EU and non-EU governments to ensure the safety of nanomaterials used in feed, agriculture, and food. The criteria for marketable food items have been changed by regulatory bodies like the Food and Drug Administration (FDA) and the Environmental Protection Agency (EPA), with a focus on quality, safety, and health.

35.9 Conclusion

In summary, nanofunctional materials have great potential to bring about a paradigm shift in the food industry by improving quality, safety, sustainability, and processing capabilities. However, a balanced approach is needed, balancing the benefits with potential concerns for human health and the environment. Safety considerations related to the different properties of nanoparticles require further toxicological studies and the development of detection technologies. Additionally, it is critical to establish stringent laws and labelling requirements through a collaborative effort between regulators and industry stakeholders. This holistic strategy, focused on safety, transparency, and responsible oversight, will enable the food industry to leverage the important benefits of nanofunctional materials to improve food technology, prioritising ethics and human well-being.

REFERENCES

[1] Prasad, V., Felix, S., Srikanta, S., Biswas, P. P., & Bose, S. (2018). Implication of nanoscience in the food processing and agricultural industries. In *Impact of Nanoscience in the Food Industry* (pp. 57–85). Academic Press.

[2] Nile, S. H., Baskar, V., Selvaraj, D., Nile, A., Xiao, J., & Kai, G. (2020). Nanotechnologies in food science: applications, recent trends, and future perspectives. *Nano-Micro Letters*, *12*, 1–34.

[3] Kumar, P., Mahajan, P., Kaur, R., & Gautam, S. (2020). Nanotechnology and its challenges in the food sector: A review. *Materials Today Chemistry*, *17*, 100332.

[4] Su, Q., Zhao, X., Zhang, X., Wang, Y., Zeng, Z., Cui, H., & Wang, C. (2022). Nano functional food: Opportunities, development, and future perspectives. *International Journal of Molecular Sciences*, *24*(1), 234.

[5] Weiss, J., Takehito, P., & McClements, D. J. (2006). Functional materials in food nanotechnology. *Journal of Food Science*, *71*(9), R107–R116.

[6] Sozer, N., & Kokini, J. L. (2008). Nanotechnology and its applications in the food sector. *Trends in Biotechnology*, *27*, 82–88.

[7] Chen, L. (2006). Food protein based materials as nutraceuticals delivery systems. *Trends in Food Science and Technology*, *17*, 272–283.

[8] Gupta, A. K., & Gupta, M. (2005). Synthesis and surface engineering of iron oxide nanoparticles for biomedical applications. *Biomaterials*, *26*(18), 3995–4021.

[9] Riley, T., Govender, T., Stolnik, S., Xiong, C. D., Garnett, M. C. L., & IllumandDavis, S. S. (1999). Colloidal stability and drug incorporation aspects of micellar like PLA-PEG nanoparticles. *Colloids and Surfaces B*, *16*, 47–159.

[10] Arroyo, B. J., Santos, A. P., de Melo, E. D. A., Campos, A., Lins, L., & Boyano-Orozco, L.C. (2019). Bioactive compounds and their potential use as ingredients for food and its application in food packaging. In *Bioactive Compounds* (pp. 143–156). Woodhead Publishing.

[11] Clemente, J. C., Pehrsson, E. C., Blaser, M. J., Sandhu, K., Gao, Z., Wang, B., Magris, M., Hidalgo, G., Contreras, M., & Noya-Alarcón, Ó. (2015). The microbiome of uncontacted Amerindians. *Science Advances*, *1*, e1500183.

[12] Couch, L. M., Wien, M., Brown, J. L., & Davidson, P. (2016). Food nanotechnology: Proposed uses, safety concerns and regulations. *Agro Food Industry Hi Tech*, *27*, 36–39.

[13] Jafarizadeh-Malmiri, H., Sayyar, Z., Anarjan, N., & Berenjian, A. (2019). In *Nanobiotechnology in Food: Concepts, Applications and Perspectives*, (pp. 81–94). Springer: Berlin, Germany.

[14] Shafiq, M., Anjum, S., Hano, C., Anjum, I., & Abbasi, B. H. (2020). An overview of the applications of nanomaterials and nanodevices in the food industry. *Foods*, *9*(2), 148.

[15] Ameta, S. K., Rai, A. K., Hiran, D., Ameta, R., & Ameta, S. C. (2020). Use of nanomaterials in food science. *Biogenic Nano-Particles and Their Use in Agro-Ecosystems*, Springer; Singapore (pp. 457–488).

[16] Ponce, A. G., Ayala-Zavala, J. F., Marcovich, N. E., Vázquez, F. J., & Ansorena, M. R. (2018). Nanotechnology trends in the food industry: Recent developments, risks, and regulation. *Impact of Nanoscience in the Food Industry*, 113–141.

[17] Mohammad, Z. H., Ahmad, F., Ibrahim, S. A., & Zaidi, S. (2022). Application of nanotechnology in different aspects of the food industry. *Discover Food*, *2*(1), 12.

[18] Singh, T., Shukla, S., Kumar, P., Wahla, V., Bajpai, V. K., & Rather, I. A. (2017). Application of nanotechnology in food science: Perception and overview. *Frontiers in Microbiology*, *8*, 1501.

[19] Biswas, R., Alam, M., Sarkar, A., Haque, M. I., Hasan, M. M., & Hoque, M. (2022). Application of nanotechnology in food: Processing, preservation, packaging and safety assessment. *Heliyon.*, *8*(11).

[20] Naseer, B., Srivastava, G., Qadri, O., Faridi, S., Islam, R. & Younis, K. (2018). Importance and health hazards of nanoparticles used in the food industry. *Nanotechnology Reviews*, *7*(6), 623–641. https://doi.org/10.1515/ntrev-2018-0076

[21] Cushen, M., Kerry, J., Morris, M., Cruz-Romero, M., & Cummins, E. (2012). Nanotechnologies in the food industry–Recent developments, risks and regulation. *Trends in Food Science & Technology*, *24*(1), 30–46.

[22] Huang, Y., Chen, S., Bing, X., Gao, C., Wang, T., & Yuan, B. (2011). Nanosilver migrated into food-simulating solutions from commercially available food fresh containers. *Packaging Technology and Science*, *24*(5), 291–297.

36

Design and Optimisation of Flexible Electronics Using Organic Semiconductors

M. Sharanya
Malla Reddy College of Engineering and Technology, Secunderabad, Telangana, India

R. Premalatha
S.A. Engineering College, Poonamallee, India

Srinivasa Acharya
Aditya Institute of Technology and Management, Tekkali, Andhra Pradesh, India

M. Monisha
Vels Institute of Science Technology and Advanced Studies VISTAS, Chennai, India

36.1 Synthesis and Characterisation of Organic Semiconductors

Organic semiconductors have garnered significant attention in the field of flexible electronics due to their unique properties, such as mechanical flexibility, low-cost fabrication, and tuneable optoelectronic properties. This chapter explores the various techniques employed for synthesising organic semiconductor materials and the characterisation methods used to assess their properties [1].

36.1.1 Techniques for Synthesising Organic Semiconductor Materials

Several synthesis techniques are utilised to fabricate organic semiconductor materials with desired properties tailored for specific applications in flexible electronics. One common method is chemical synthesis, which involves the reaction of organic precursors to form conjugated polymers or small molecules. This approach offers precise control over molecular structure and allows for the incorporation of functional groups to modulate electronic properties [2]. Additionally, physical vapour deposition techniques such as vacuum evaporation and molecular beam epitaxy are employed to deposit thin films of organic semiconductors with high purity and uniformity. Solution-based methods like spin coating, inkjet printing, and doctor-blading offer scalability and versatility in depositing organic semiconductor layers over large areas.

36.1.2 Characterisation Methods for Assessing Semiconductor Properties

Characterisation plays a crucial role in evaluating the performance and properties of organic semiconductors for flexible electronics applications as shown in Figure 36.1. Various analytical techniques are employed to assess parameters such as molecular structure, charge transport properties, and morphological characteristics. Spectroscopic methods such as UV–vis absorption spectroscopy, fluorescence spectroscopy, and infrared spectroscopy provide insights into the electronic structure and optical properties of organic semiconductors. X-ray diffraction and electron microscopy techniques offer information on crystal structure, grain morphology, and film morphology, which influence charge transport behaviour. Electrical measurements, including field-effect transistor (FET) measurements and impedance spectroscopy, are utilised to evaluate charge carrier mobility, conductivity, and device performance [3].

DOI: 10.1201/9781003495437-36

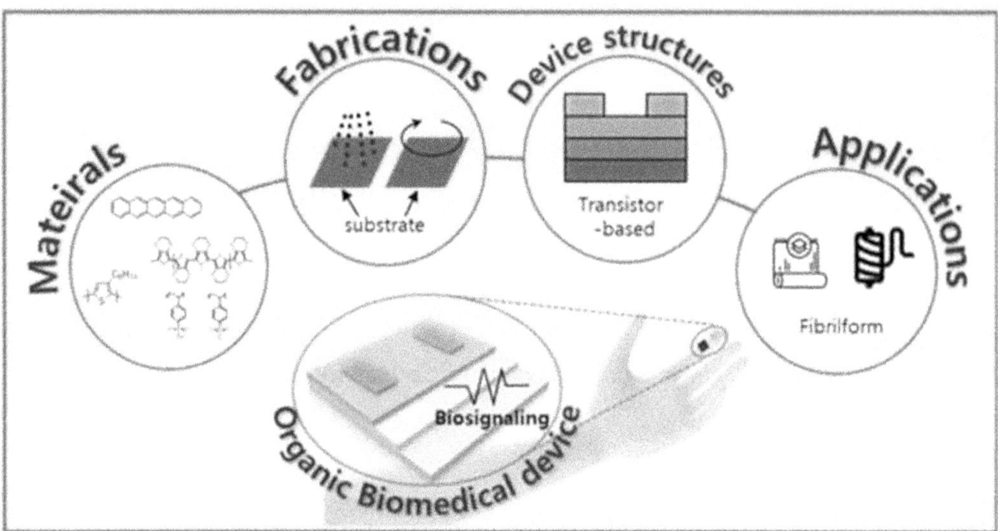

FIGURE 36.1 Graphical method of producing and application of organic semiconductors. (https://pub.mdpi-res.com/polymers/polymers-14-02960/article_deploy/html/images/polymers-14-02960-ag-550.jpg?1658744728).

Additionally, techniques like atomic force microscopy (AFM) and scanning tunnelling microscopy (STM) offer nanoscale insights into surface morphology and molecular packing arrangement, aiding in understanding structure–property relationships.

36.2 Device Fabrication Techniques for Flexible Electronics

Flexible electronics represent a paradigm shift in device design and fabrication, enabling the development of conformable, lightweight, and bendable electronic systems for various applications. This chapter explores state-of-the-art device fabrication techniques tailored for organic semiconductors, focusing on printing technologies and roll-to-roll (R2R) manufacturing processes.

36.2.1 Printing Technologies for Deposition of Organic Semiconductors

Printing technologies offer cost-effective and scalable methods for depositing organic semiconductors onto flexible substrates, enabling the fabrication of large-area electronic devices with high throughput and precision. Inkjet printing, one of the most versatile printing techniques, employs droplet ejection mechanisms to deposit semiconductor inks onto substrates, allowing for the precise patterning of organic semiconductor layers. Screen printing, another widely used technique, utilises a stencil-based approach to deposit thick-film semiconductor layers, making it suitable for applications requiring high throughput and low-cost production. Additionally, gravure printing and flexographic printing offer continuous and high-speed deposition of organic semiconductor inks, making them suitable for R2R processing [4].

36.2.2 Roll-to-Roll Manufacturing Processes for Scalable Device Production

R2R manufacturing processes enable the continuous fabrication of flexible electronic devices on flexible substrates, offering scalability and cost-effectiveness for large-scale production. R2R processing involves feeding a flexible substrate through a series of processing stations, where various deposition and patterning steps are performed sequentially. Slot-die coating is a common R2R technique for depositing uniform layers of organic semiconductor materials onto flexible substrates, offering high material utilisation and control over film thickness [5]. Additionally, rotary screen printing and flexographic printing are employed in R2R processes for large-area patterning of organic semiconductor layers, allowing for the fabrication of flexible electronic devices with intricate designs.

36.3 Optimisation Strategies for Organic Semiconductor Devices

Organic semiconductor devices hold immense promise for flexible electronics applications due to their unique properties such as mechanical flexibility, low-cost fabrication, and compatibility with large-area manufacturing processes. However, to realise their full potential, optimisation strategies are essential to enhance charge transport, stability, and overall device performance. This chapter explores various techniques and approaches for optimising organic semiconductor devices, focusing on design optimisation for improved charge transport and strategies to enhance device stability and performance.

36.3.1 Design Optimisation for Enhanced Charge Transport in Flexible Electronics

Design optimisation plays a critical role in improving charge transport properties in organic semiconductor devices, ultimately leading to enhanced device performance. Molecular engineering techniques enable the precise tuning of organic semiconductor materials' electronic structure, energy levels, and intermolecular interactions to facilitate efficient charge carrier transport. By designing conjugated polymer backbones with specific molecular architectures and side-chain modifications, researchers can tailor materials with optimised charge mobility, enabling faster charge transport and higher device efficiency [6]. Moreover, optimising the morphology of organic semiconductor thin films through techniques such as solvent engineering, annealing, and surface modification facilitates the formation of ordered molecular packing structures, minimising charge trapping and improving charge carrier mobility.

a. Electrical characterisation: Ohm's law, $V=IR$, where V is the voltage, I is the current, and R is the resistance. Ohm's law describes the relationship between voltage, current, and resistance in a material.

b. Optimisation algorithms: Gradient descent update rule, $\theta t+1 = \theta t - \alpha \nabla J(\theta t)$, where θt is the parameter vector at iteration t, α is the learning rate, and $\nabla J(\theta t)$ is the gradient of the cost function J with respect to θt. The gradient descent algorithm is an iterative optimisation technique used to minimise a cost function by updating the parameters in the direction of the steepest descent of the cost function.

c. Material design: Density functional theory (DFT) energy functional, $E[n]=Ts[n]+Eee[n]+Exc[n]+Eext[n]$, where $E[n]$ is the total energy, $Ts[n]$ is the kinetic energy of non-interacting electrons, $Eee[n]$ is the electron–electron interaction energy, $Exc[n]$ is the exchange-correlation energy, and $Eext[n]$ is the external potential energy. DFT is a quantum mechanical modelling method used to calculate the electronic structure and properties of materials. The total energy functional is derived from the Hohenberg–Kohn theorems and the Kohn–Sham equations.

d. Statistical analysis: Regression equation, $y = mx+c$, where y is the dependent variable, x is the independent variable, m is the slope, and c is the y-intercept. The regression equation describes the linear relationship between two variables, with the slope m representing the rate of change of the dependent variable with respect to the independent variable.

These mathematical concepts and derivations play a crucial role in understanding and optimising the design of flexible electronics using organic semiconductors.

36.3.2 Strategies for Improving Device Stability and Performance

Enhancing the stability and performance of organic semiconductor devices is crucial for their practical applications in flexible electronics. Several strategies have been developed to address challenges such as material degradation, environmental sensitivity, and device reliability. Encapsulation techniques, such as thin-film barriers and conformal coatings, protect organic semiconductor devices from moisture, oxygen, and other environmental factors, prolonging their operational lifetime and stability. Additionally, interface engineering approaches involving the incorporation of interlayers and interface modifiers help optimise charge injection, extraction, and transport at organic semiconductor/electrode interfaces, thereby improving device efficiency and stability. Furthermore, the development of novel device architectures, such as inverted and tandem structures, enables improved charge balance, reduced recombination losses, and enhanced device performance under various operating conditions.

36.4 Interface Engineering and Simulation Modelling in Organic Semiconductor Devices

Interface engineering and simulation modelling play crucial roles in optimising the performance of organic semiconductor devices. This chapter delves into the methodologies and techniques employed in interface engineering to enhance charge injection and transport, as well as the computational approaches used for simulating and modelling organic semiconductor devices [7].

36.4.1 Interface Engineering for Improved Charge Injection and Transport

Interface engineering focuses on manipulating the properties of interfaces between organic semiconductors and electrodes to facilitate efficient charge injection and transport. Various techniques, such as surface treatments, interfacial layers, and molecular doping, are utilised to tailor the energy level alignment and charge carrier dynamics at the interface [8]. By optimising the interface morphology and energy level alignment, interface engineering minimises charge carrier barriers and traps, leading to improved charge injection and transport efficiency in organic semiconductor devices.

36.4.2 Surface Modification Techniques for Enhancing Device Interfaces

Surface modification techniques involve chemical and physical treatments applied to electrode surfaces or organic semiconductor layers to modify their properties and improve device interfaces. Surface functionalisation with self-assembled monolayers (SAMs), polymer brushes, or interfacial layers enhances surface wettability, adhesion, and charge carrier injection efficiency. Additionally, surface passivation and interface modification with dielectric layers or interfacial dopants mitigate charge trapping and interface recombination, contributing to enhanced device performance and stability [9].

36.4.3 Computational Approaches for Predicting Device Performance

Computational approaches offer valuable insights into the performance of organic semiconductor devices by simulating their electronic and optoelectronic properties. DFT and molecular dynamics (MD) simulations are employed to predict the structural, electronic, and vibrational properties of organic semiconductor materials at the atomic level [10]. These simulations provide information on energy levels, band structures, and charge carrier mobility, aiding in material selection and design optimisation for improved device performance.

36.4.4 Multiscale Modelling Techniques for Understanding Charge Transport

Multiscale modelling techniques bridge the gap between atomistic and macroscopic scales, enabling a comprehensive understanding of charge transport mechanisms in organic semiconductor devices. Kinetic Monte Carlo (KMC) simulations and coarse-grained models capture the stochastic nature of charge carrier motion and hopping dynamics within organic semiconductor layers. Continuum models, such as drift-diffusion and Poisson equations, describe charge transport phenomena at larger length and time scales, facilitating device-level simulations and performance predictions.

36.5 Applications and Flexible Sensor Technologies of Organic Semiconductor-Based Electronics

Organic semiconductor-based flexible electronics have garnered significant attention due to their potential applications in various fields. This chapter explores the diverse applications of organic semiconductor-based flexible electronics, including flexible displays and lighting devices, organic photovoltaics (OPVs), energy harvesting devices, and flexible sensor technologies [11].

36.5.1 Flexible Displays and Lighting Devices Using Organic Semiconductors

Flexible displays and lighting devices represent one of the most promising applications of organic semiconductor-based flexible electronics. Organic light-emitting diodes (OLEDs) are widely used in flexible displays due to their lightweight, thin form factor, and vibrant colours. These OLED displays can be bent, rolled, or even folded, enabling the development of innovative display concepts such as rollable and foldable screens. Furthermore, organic light-emitting transistors (OLETs) offer additional functionalities by combining the switching capabilities of transistors with the light-emitting properties of OLEDs, paving the way for flexible lighting panels and signage applications.

36.5.2 Organic Photovoltaics and Energy-Harvesting Devices

OPVs have emerged as promising alternatives to traditional silicon-based solar cells for flexible energy harvesting applications. OPVs utilise organic semiconductor materials to convert sunlight into electricity, offering advantages such as lightweight, flexibility, and low-cost fabrication. These flexible solar cells can be integrated into various surfaces, including clothing, windows, and building facades, enabling the generation of renewable energy in diverse environments. Additionally, organic thermoelectric generators (OTEGs) harness waste heat from electronic devices or industrial processes to generate electricity, providing a sustainable solution for energy harvesting in IoT devices, wearable electronics, and remote sensing systems [12].

36.5.3 Flexible Sensor Technologies Using Organic Semiconductors

Flexible sensor technologies represent another exciting application area for organic semiconductor-based electronics, offering opportunities for real-time monitoring and diagnostics in various fields. The development of flexible sensor platforms involves the integration of organic semiconductors with flexible substrates and sensor components to create lightweight, conformable, and stretchable sensor devices. These sensors can be employed in a wide range of applications, including healthcare monitoring, environmental sensing, human–machine interfaces, and structural health monitoring. Integration of organic semiconductors in wearable and healthcare sensors enables the development of comfortable, non-invasive sensing solutions for continuous health monitoring, disease detection, and wellness tracking [13]. Additionally, flexible sensors based on organic materials exhibit excellent mechanical properties, allowing them to withstand bending, stretching, and deformation without compromising their sensing performance.

36.6 Advancements in Organic Semiconductor Device Integration

In recent years, significant progress has been made in the integration of organic semiconductors with other materials to enhance device performance and functionality. This chapter explores novel approaches and hybrid organic–inorganic device architectures that leverage the unique properties of organic semiconductors for various applications [14].

36.6.1 Novel Approaches for Integrating Organic Semiconductors with Other Materials

One of the key advancements in organic semiconductor device integration is the development of novel approaches that enable seamless integration with other materials while preserving the unique properties of organic semiconductors. One such approach is the use of interfacial engineering techniques to modify the surface properties of organic semiconductor films, allowing for improved charge transport and interface compatibility with other functional layers. For example, SAMs can be employed to modify the surface energy and chemical properties of organic semiconductor films, leading to enhanced device performance and stability [15].

Another promising approach is the development of hybrid organic–inorganic composites, where organic semiconductors are combined with inorganic materials to create multifunctional device architectures. By carefully selecting and optimising the composition of the hybrid materials, researchers can tailor the electronic and optical properties of the devices to meet specific application requirements. For instance, hybrid organic–inorganic perovskite materials have shown great promise for use in photovoltaic devices, where they exhibit high carrier mobility and absorption coefficients, leading to improved power conversion efficiencies.

36.6.2 Hybrid Organic–Inorganic Device Architectures for Enhanced Functionality

Hybrid organic–inorganic device architectures offer a versatile platform for integrating organic semiconductors with other materials to achieve enhanced functionality. One example is the integration of organic semiconductors with metallic nanoparticles or quantum dots to create plasmonic or quantum-enhanced devices. These hybrid architectures leverage the unique optical and electronic properties of metallic nanoparticles or quantum dots to enhance light absorption, charge generation, and transport in organic semiconductor devices.

Another approach is the incorporation of organic semiconductors into flexible or stretchable substrates to create bendable, conformable, and wearable electronic devices. By integrating organic semiconductors with flexible substrates such as polymers or elastomers, researchers can develop devices that can withstand mechanical deformations without compromising their electrical performance. These flexible hybrid devices are ideal for applications in wearable electronics, medical devices, and electronic skins.

Furthermore, hybrid organic–inorganic device architectures can be tailored for specific applications by controlling the morphology, composition, and interface properties of the hybrid materials. For example, by tuning the nanoscale morphology of hybrid perovskite materials, researchers can optimise the light absorption and charge transport properties of photovoltaic devices, leading to improved power conversion efficiencies. Similarly, by engineering the interface between organic semiconductors and inorganic charge transport layers, researchers can enhance the charge injection and extraction properties of OLEDs, leading to improved device performance and stability.

36.7 Challenges, Opportunities, and Future Directions in Organic Semiconductor-Based Electronics

Organic semiconductor-based electronics hold great promise for a wide range of applications, but they also face significant challenges. This chapter delves into the current research landscape, addressing challenges, exploring emerging opportunities, and discussing future directions in the field.

36.7.1 Addressing Challenges in Device Stability and Reliability

One of the primary challenges in organic semiconductor-based electronics is achieving device stability and reliability. Organic materials are inherently susceptible to degradation when exposed to environmental factors such as moisture, oxygen, and light. This can lead to device performance degradation and limited operational lifetimes. Researchers are actively working to address these challenges through various approaches, including the development of encapsulation techniques to protect organic devices from environmental factors. Additionally, the design of stable organic semiconductor materials with improved chemical and thermal stability is a key area of research. By enhancing the stability and reliability of organic devices, researchers aim to unlock their full potential for practical applications.

36.7.2 Exploring Emerging Opportunities for Organic Semiconductor Applications

Despite the challenges, organic semiconductor-based electronics offer numerous opportunities for innovative applications. One emerging opportunity lies in the development of flexible and stretchable electronics. Organic semiconductors can be deposited onto flexible substrates, allowing for the creation of bendable and conformable electronic devices. These flexible electronics have applications in wearable technology, healthcare monitoring, and flexible displays. Another opportunity lies in the field of OPVs, where organic semiconductors are used to convert sunlight into electricity. OPVs offer advantages such as lightweight and low-cost production, making them attractive for renewable energy applications. Furthermore, organic semiconductor-based electronics have the potential to revolutionise the field of Internet of Things (IoT). Organic sensors and actuators can be integrated into everyday objects, enabling them to sense and respond to their environment. For example, organic sensors can be used in smart homes to monitor air quality, temperature, and humidity, while organic actuators can control lighting, heating, and ventilation systems. These IoT applications have the potential to enhance convenience, efficiency, and sustainability in various domains.

36.7.3 Future Directions in Flexible Electronics Using Organic Semiconductors

Looking ahead, several future directions hold promise for the advancement of flexible electronics using organic semiconductors. One emerging trend is the development of multifunctional flexible devices that integrate multiple functionalities into a single platform. For example, researchers are exploring the integration of energy harvesting, sensing, and communication capabilities into flexible electronic devices. This convergence of functionalities enables the creation of smart, self-powered systems with applications in smart cities, healthcare, and environmental monitoring.

Another future direction is the exploration of novel materials and device designs to enhance the performance and functionality of organic semiconductor-based electronics. Researchers are investigating new organic semiconductor materials with improved charge transport properties, stability, and process ability. Additionally, the development of novel device architectures, such as organic thin-film transistors (OTFTs) and OLETs, opens up new possibilities for flexible electronics. By pushing the boundaries of materials and device design, researchers aim to overcome existing limitations and unlock new capabilities for organic semiconductor-based electronics.

Finally, opportunities for commercialisation and market adoption are essential for the widespread adoption of flexible electronics using organic semiconductors. As research continues to advance, it is crucial to bridge the gap between academia and industry to facilitate technology transfer and commercialisation. Collaborative efforts between researchers, industry partners, and policymakers are needed to address manufacturing challenges, scale up production, and bring innovative products to market. By fostering collaboration and innovation, the field of flexible electronics using organic semiconductors can realise its full potential and revolutionise the electronics industry.

36.8 Conclusion

The development and optimisation of flexible electronics using organic semiconductors represent a significant advancement in the field of electronic devices. Throughout this exploration, we have delved into various aspects of designing and optimising these flexible electronics, considering the synthesis and characterisation of organic semiconductors, device fabrication techniques, optimisation strategies, challenges, opportunities, and future directions. Organic semiconductor materials offer unique advantages, including flexibility, lightweight, and low-cost production, making them well-suited for flexible electronic applications. Techniques for synthesising organic semiconductor materials have been developed, allowing for the precise control of their properties such as molecular structure, energy levels, and charge transport characteristics. Characterisation methods enable researchers to assess the performance of these materials accurately and optimise their properties for specific applications. In device fabrication, printing technologies and roll-to-roll manufacturing processes have emerged as key techniques for depositing organic semiconductors onto flexible substrates. These techniques enable scalable and cost-effective production of flexible electronic devices, paving the way for large-scale commercialisation. Additionally, optimisation strategies focusing on design optimisation, stability improvement, and performance enhancement have been implemented to address challenges associated with organic semiconductor-based devices. Despite significant progress, challenges remain in achieving the desired stability, reliability, and performance of organic semiconductor-based flexible electronics. Issues such as device degradation due to environmental factors, limited charge transport properties, and scalability concerns pose obstacles to widespread adoption. However, ongoing research efforts aim to overcome these challenges through the development of encapsulation techniques, stable organic semiconductor materials, and innovative device architectures. Looking ahead, the future of flexible electronics using organic semiconductors is promising, with emerging opportunities and future directions driving innovation in the field. Multifunctional flexible devices that integrate energy harvesting, sensing, and communication capabilities are envisioned, enabling the creation of smart, self-powered systems for various applications. Novel materials and device designs are being explored to enhance device performance and functionality, while opportunities for commercialisation and market adoption are being pursued to bridge the gap between research and industry. In conclusion, the design and optimisation of flexible electronics using organic semiconductors hold immense potential for revolutionising electronic devices. By leveraging the unique properties of organic semiconductor materials and employing advanced fabrication techniques and optimisation strategies, researchers aim to overcome existing challenges and unlock new capabilities for flexible electronics. Collaboration between academia, industry, and policymakers will be crucial in realising the full potential of organic semiconductor-based flexible electronics and driving innovation in the field.

REFERENCES

[1] Liu, K., Ouyang, B., Guo, X., Guo, Y. and Liu, Y., 2022. Advances in flexible organic field-effect transistors and their applications for flexible electronics. *npj Flexible Electronics*, 6(1), p. 1.

[2] Liao, C., Zhang, M., Yao, M.Y., Hua, T., Li, L. and Yan, F., 2015. Flexible organic electronics in biology: Materials and devices. *Advanced Materials*, 27(46), pp. 7493–7527.

[3] Mitta, S.B., Choi, M.S., Nipane, A., Ali, F., Kim, C., Teherani, J.T., Hone, J. and Yoo, W.J., 2020. Electrical characterization of 2D materials-based field-effect transistors. *2D Materials*, 8(1), p. 012002.

[4] Cruz, S.M.F., Rocha, L.A. and Viana, J.C., 2018. Printing technologies on flexible substrates for printed electronics. In Simas Rackauskas (ed.) *Flexible Electronics*. IntechOpen.

[5] Amruth, C., Pahlevani, M. and Welch, G.C., 2021. Organic light emitting diodes (OLEDs) with slot-die coated functional layers. *Materials Advances*, 2(2), pp. 628–645.

[6] Kukhta, N.A., Marks, A. and Luscombe, C.K., 2021. Molecular design strategies toward improvement of charge injection and ionic conduction in organic mixed ionic–electronic conductors for organic electrochemical transistors. *Chemical Reviews*, 122(4), pp. 4325–4355.

[7] Zampetti, A., Minotto, A. and Cacialli, F., 2019. Near-infrared (NIR) organic light-emitting diodes (OLEDs): Challenges and opportunities. *Advanced Functional Materials*, 29(21), p. 1807623.

[8] Li, P. and Lu, Z.H., 2021. Interface engineering in organic electronics: Energy-level alignment and charge transport. *Small Science*, 1(1), p. 2000015.

[9] Vilan, A. and Cahen, D., 2017. Chemical modification of semiconductor surfaces for molecular electronics. *Chemical Reviews*, 117(5), pp. 4624–4666.

[10] Beljonne, D., Cornil, J., Muccioli, L., Zannoni, C., Brédas, J.L. and Castet, F., 2011. Electronic processes at organic–organic interfaces: Insight from modeling and implications for opto-electronic devices. *Chemistry of Materials*, 23(3), pp. 591–609.

[11] Yuvaraja, S., Nawaz, A., Liu, Q., Dubal, D., Surya, S.G., Salama, K.N. and Sonar, P., 2020. Organic field-effect transistor-based flexible sensors. *Chemical Society Reviews*, 49(11), pp. 3423–3460.

[12] Kazem, H.A., Al-Waeli, A.H., Chaichan, M.T., Sopian, K., Al Busaidi, A.S. and Gholami, A., 2023. Photovoltaic-thermal systems applications as dryer for agriculture sector: A review. *Case Studies in Thermal Engineering*, 47, p. 103047.

[13] Kazanskiy, N.L., Khonina, S.N. and Butt, M.A., 2024. A review on flexible wearables-Recent developments in non-invasive continuous health monitoring. *Sensors and Actuators A: Physical*, 366, p. 114993.

[14] Wang, H. and Yu, C., 2019. Organic thermoelectrics: Materials preparation, performance optimization, and device integration. *Joule*, *3*(1), pp. 53–80.

[15] Miozzo, L., Yassar, A. and Horowitz, G., 2010. Surface engineering for high performance organic electronic devices: The chemical approach. *Journal of Materials Chemistry*, *20*(13), pp. 2513–2538.

37

Nanofunctional Material in Food Analytical Sensors

Venkatasami Murugesan
Agricultural Engineering College and Research Institute, Tamil Nadu Agricultural University, Coimbatore, India

Arulmari Ramaraj
Agricultural Engineering College and Research Institute, Tamil Nadu Agricultural University, Tiruchirappalli, India

Gitanjali Jothiprakash
Agricultural Engineering College and Research Institute, Tamil Nadu Agricultural University, Coimbatore, India

Deepa Jaganathan
Hindustan College of Engineering and Technology, Coimbatore, India

Amuthaselvi Gopal
Agricultural Engineering College and Research Institute, Tamil Nadu Agricultural University, Coimbatore, India

37.1 Introduction

Food sensors detect and transmit information about food quality, safety, origin, or degree of contamination [1]. They can help reduce food waste and improve food management. Common types of food sensors include time–temperature indicators, humidity sensors, pH indicators, gas sensors, pesticide detectors, and pathogen detectors [2]. They often use optical or electrical transducers. Recent advances have focused on incorporating sensors into materials and devices for food packaging and processing to help translate into commercial products. However, few food sensor technologies have achieved widespread practical use so far. Challenges for food sensor development include complexity and natural variation in food products, different causes of food spoilage, difficulties with the food–sensor interface and signal processing, and issues related to legislation and public perception [3]. Opportunities for improvement include better accounting for food variability during sensor design, developing universal systems or simple binary sensors, advances in incorporating sensors into packaging, leveraging signal processing tech like smartphones, expanding food standards and regulations, and addressing safety concerns [4].

37.2 Nanofunctional Materials in Food Analytical Sensors

Food sensors typically consist of four components – the target analyte, recognition element, signal transducer, and processor [5]. The recognition element binds to and detects the analyte, the transducer converts this into a signal, and the processor interprets the signal. Common analytes targeted by food sensors include temperature, humidity, pH, gases, pesticides, and pathogens [6]. Strategies for detection include monitoring storage conditions like temperature and humidity which indirectly indicate food spoilage, detecting microbial metabolites like lactic acid that generally indicate spoilage, or detecting specific

DOI: 10.1201/9781003495437-37

pathogens and contaminants. Many sensors incorporate naturally occurring compounds like anthocyanins or synthetic polymers like polydiacetylenes as recognition elements [7]. These change colour or electrical properties when reacting with target analytes. Transduction methods include optical/colorimetric, electronic, and acoustic [8]. Optical methods like colour changes are convenient for consumer applications while electronic methods integrate better with automated food processing systems. Signal processing ranges from simple (e.g., visible colour change) to complex (e.g., machine learning interpretation of electronic nose datasets) [9]. Smartphones and computer vision are enabling more complex optical signal analysis.

Nanomaterials like metal nanoparticles, carbon nanotubes (CNTs), graphene, and quantum dots have unique optical, electrical, and chemical properties that make them useful for sensing applications [10]. Their high surface area-to-volume ratio improves sensitivity. Common nanomaterials used as recognition elements in food sensors include gold nanoparticles, silicon nanowires, titanium dioxide nanoparticles, and nanocarbons [11]. These can be functionalised to bind to specific food analytes. Nanoparticle-based sensors using techniques like surface-enhanced Raman spectroscopy (SERS) offer high sensitivity for detecting food contaminants like pesticides, toxins, and pathogens down to very low concentrations [12]. Nanomaterial-based electronic noses and tongue sensors with cross-reactive sensor arrays allow the classification of food samples based on overall volatile compound profiles. Despite their promise, safety concerns around nanomaterials regarding toxicity, accumulation in the body, and environmental impact are barriers to commercial use in food. Strict regulations are in place for approval of nanotechnology uses in food products [13]. Recent advances focus on using food-safe materials like proteins and cellulose to make nanosensors, developing immobilisation and encapsulation strategies to prevent nanoparticles from contacting food, and integrating nanosensors with microfluidics for on-site food analysis.

Along with nanomaterials, nanocomposites play an important role in the development of sensors. Three main classes of hybrid nanocomposites are presented – binary, ternary, and multiple. Binary nanocomposites like metal oxide nanoparticles combined with ionic liquids can lower over-potential and increase sensitivity for compounds. Ternary nanocomposites like CNTs with imprinted polymers and polythionine enhance electrocatalytic performance [14]. Multiple nanocomposites with three or more components provide further versatility. Several types of prime for biointegration are noted. Binary composites of metal nanoparticles like gold or iron oxide with antibodies, aptamers, or enzymes help capture targets and amplify signals [15]. Ternary composites of nanotubes, conducting polymers, and metal nanoparticles bind elements like glucose oxidase effectively [16]. Multiple composites with dendrimers, graphene, and nanoparticles enable covalent enzyme immobilisation [17]. The nanocomposites have advantages like catalysing redox reactions, high strength, stability, surface area, and electronic transfer ability. Their biocompatibility maintains biological activity. Versatile multi-component composites avoid the limitations of single materials. Some composites have dual uses for ensuring food functionality (nutrients) and safety (contaminants). Metal/metal oxide composites are applied for antioxidants and preservatives. Graphene composites sense amino acids and pesticides. Metal/CNT composites determine vitamins and sweeteners. Metal/graphene types analyse pigments and pathogens [14].

37.3 Carbon Dots

Carbon dots (CDs) are a novel class of fluorescent carbon nanomaterials with sizes below 10 nm. They have unique optical and physicochemical properties and great potential for sensing applications [18]. CDs can be synthesised via top-down methods like laser ablation, arc discharge, and electrochemical approaches or bottom-up methods using hydrothermal/solvothermal treatment of organic precursors. Surface passivation and heteroatom doping can enhance the properties of CDs [19]. CDs exhibit tuneable photoluminescence (PL), high photostability, pH-dependent fluorescence, and low cytotoxicity. Their fluorescence can be quenched or recovered for sensing applications. CD sensors have been developed based on fluorescence quenching or recovery mechanisms using interactions with metal ions, organic compounds, enzymes, etc. [20]. Techniques like molecular imprinting, aptamers, and Fluorescence Resonance Energy Transfer (FRET) with nanoparticles further improve selectivity. CDs show promise for sensing applications in food analysis to detect nutrients, pesticides, additives, pathogens, and toxins, with detection limits down to ppb/ppm levels. Challenges remain in terms of selectivity in complex food matrixes [21, 22].

CDs can act as fluorescent labels or probes in sensors based on fluorescence quenching or recovery mechanisms. Their fluorescence is quenched in the presence of certain substances like metal ions, and nanoparticles, or recovers in the presence of other target analytes [23]. CDs have unique optical absorption and PL properties which can be tuned via doping or surface modifications. This allows their use in optical sensors for things like pH and temperature [24]. CDs conduct electricity and can participate in electron transfer reactions. This allows their use in electrochemical sensors and biosensors. The low toxicity and good biocompatibility of CDs allow their use in biosensors. They can be integrated with biomolecules like antibodies, aptamers, and DNA to enable selective binding to target analytes [25]. Due to their small size, CDs can be combined with structures like quantum dots, gold nanoparticles, and graphene to create nanosensors with unique properties. As highlighted, CDs have promising applications in fluorescent sensors to detect a range of targets relevant to food quality and safety [25].

CDs are used to detect and quantify ascorbic acid, tannic acid, etc., in fruits, vegetables, juices, and wine; used to detect pesticide residues like methyl parathion in agricultural produce [26]; used to detect banned food colorants like tartrazine in foods like buns, honey, and candy and bacterial toxins like aflatoxin B1 in corn, peanuts, and other susceptible commodities [27]. Additionally, they are used to detect foodborne pathogens like *Salmonella typhimurium* in eggs and water. One application is detecting heavy metal contaminants like lead, mercury, and chromium in water [28] (Figure 37.1).

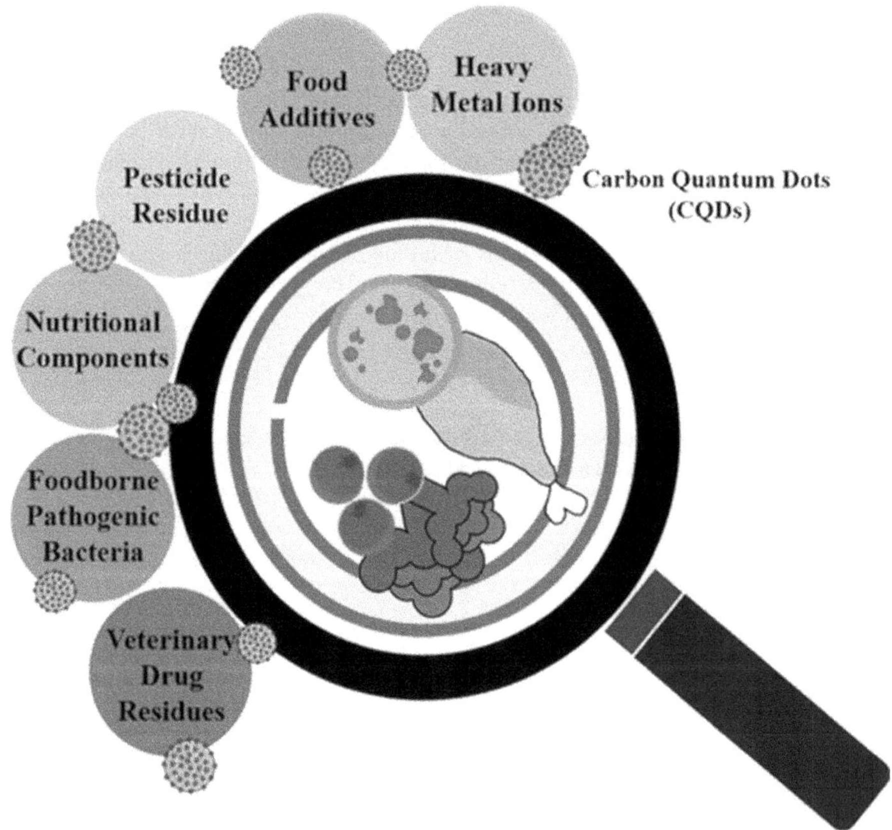

FIGURE 37.1   Carbon dot nanomaterial-based biosensors in food analysis [22].

37.4 Silicon Nanowire

Silicon nanowires leverage silicon's semiconducting properties at the nanoscale, where unique quantum effects emerge. Their small size (below 100 nm in diameter) means they have a very large surface area to volume ratio [29]. This allows excellent sensitivity when they are used as chemical or biological sensors. They can exhibit changes in electrical conductance based on surface effects and charge accumulation/depletion. This enables silicon nanowire devices to directly transduce binding events into measurable electrical signals. Their semiconducting nature combined with nanoscale effects allows control of conductive properties through doping, surface charges, or gate voltage [30]. This tunability expands their potential applications.

Nanowires grow from silicon seed particles using vapour–liquid–solid techniques. This involves a metal catalyst and exposure to silicon precursor gases at high heat [31]. Electron beam or UV lithography defines nanowire patterns on a silicon wafer surface. Anisotropic etching then removes surrounding material, leaving nanowires of the desired width and placement behind [32]. Nanowires made using either bottom-up or top-down methods can be harvested/released and deposited between conductive electrodes to complete nanowire devices. Alignment and placement can use electric fields [33].

Silicon nanowires were fabricated using a top-down approach of electron beam lithography and plasma etching to create highly uniform and reproducible structures [34]. The nanowires were 60–100 nm in width and integrated with electrodes using photolithography to create electrical devices [35]. The silicon nanowire devices were functionalised with aminopropyltriethoxysilane (APTES) and glutaraldehyde and showed promising results as highly sensitive pH sensors. The 60 nm width nanowire device exhibited the best pH response. The optimised 60 nm width silicon nanowire device was further modified to create a biosensor for detecting DNA hybridisation [36]. An immobilised DNA probe could selectively bind to complementary target DNA strands, causing measurable resistance changes. The silicon nanowire DNA sensor demonstrated a detection limit down to the pM level, on par with or better than other electrical and electrochemical biosensors. It also showed good specificity in discriminating mismatched DNA [37]. Key advantages of the silicon nanowire platform include ease of fabrication, direct electrical readout without labels, rapid response, and the ability to regenerate the sensor surface [38]. The results indicate promise for diagnostic applications (Figure 37.2).

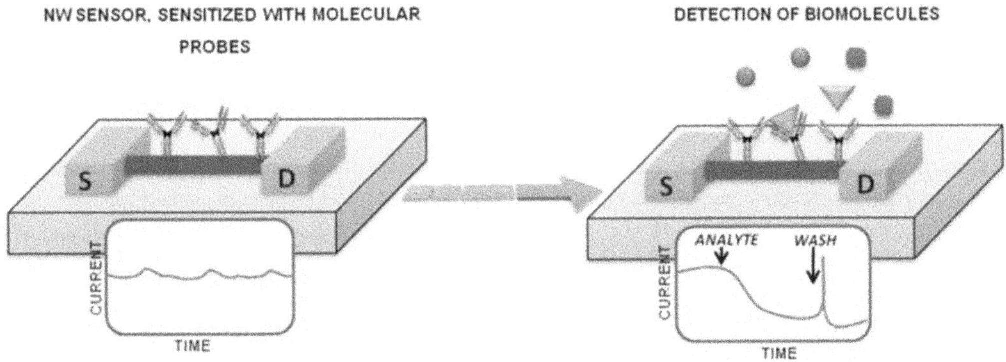

FIGURE 37.2   The Si-NW-sensors arrangement and depiction of detecting biomolecules using nanowires. S –source and D – drain [36].

37.5 Silicon Quantum Dots (Si QDs)

Si QDs are nanostructures of silicon with dimensions in the quantum confinement regime (typically below 10 nm). At these tiny sizes, the electronic and optical properties of silicon become size dependent as the laws of quantum mechanics start to dominate [39]. Some key properties and interests in Si QDs include size-tuneable PL across the visible and near-infrared spectra. This makes Si QDs promising materials for optoelectronic applications (Wang, Li, *et al.*, 2015). Long carrier lifetimes and high stability are compared to other quantum dot systems. This gives potential for applications in electronics such as memory devices. Silicon-based device technologies are eco-friendly, abundant, and compatible with nano applications. This gives Si QDs an advantage for scalable nano- and opto-electronic applications. The quantum confinement effects open up new possibilities to precisely engineer the electronic band structure and properties.

A wide variety of techniques have been investigated for preparing Si QDs, which can broadly be divided into physical, chemical, and biological approaches [40]. Key preparation methods include ion implantation –high-energy Si+ ions are implanted into SiO_2 and then annealed, allowing precipitation of Si QDs. Plasma synthesis – SiH4 plasma produces SiHx species that can deposit as Si QDs [41]. Laser ablation –pulsed laser ablation of silicon sources creates a Si vapour that condenses into QDs [42]. Electrochemical etching –anodic etching of silicon wafers under illumination can produce surface Si QDs [43]. Sol–gel methods –processing silicon alkoxide precursors via supercritical fluids or reverse micelles [44]. Reduction of silicon halides – $SiCl_4$ is reduced by reducing agents like sodium or magnesium [45]. Solution phase reduction –reduction of Si salts by metal particles or heat produces colloidal Si QDs [46]. Use of bacteria, fungi, or plant extracts to reduce Si salts or oxides to nanostructured Si. Environmentally friendly approach [47]. Key differences between these preparation routes include production rate, control over size distributions, and surface chemistry. Combined approaches are also adopted, for example, using electrochemical etching or plasma methods to liberate Si QDs which are then size selected or surface modified via solution chemistry [48].

The fundamental origin of the optoelectronic properties of silicon quantum arises from quantum confinement effects that emerge at the nanoscale when silicon structures have dimensions smaller than the Bohr radius of silicon (around 5 nm). As the size of the Si nanostructure becomes smaller than about 5 nm, the electronic band structure changes from continuous parabolic bands to discrete, atomic-like electronic levels [49]. This reduces the density of state function, impacting optical and electronic properties. Quantum confinement causes the conduction band edge to shift up in energy while the valence band edge shifts down. This leads to an effective expansion of the fundamental bandgap – allowing the tuning of light emission across visible wavelengths. The precise dependence of confinement energies depends on the exact size, shape, and surface properties of the Si QDs in a quantum mechanical way [50]. This enables the engineering of electronic and optical properties by precise control over Si QD properties. The discrepancy between slow phonon relaxation times and fast radiative recombination means excited carriers remain "trapped" in the dot and enhance light emission efficiency. This phonon bottleneck effect improves optical properties [51]. Given the intricate link between Si QD properties and their nanoscale dimensions, morphology, and chemistry, there are several important characterisation techniques: Transmission electron microscopy (TEM) directly images Si QD dimensions and morphology [52]. X-ray diffraction (XRD) can determine the crystalline nature and presence of different Si phases. Raman spectroscopy detects confinement-modified phonon modes. PL spectroscopy reveals light emission and electronic structure. Absorption spectroscopy confirms transitions related to bandgap expansion. X-ray photoelectron spectroscopy (XPS) provides information on chemical composition and surface bonding [53].

The ongoing improvements in controlled preparation and fundamental understanding of Si QD properties point towards a range of potential applications across optoelectronics, photovoltaics, electronics, sensing, and imaging: emitting layers, optical gain media, and luminescent solar concentrators based on engineered Si QD PL [54]. Biological imaging and sensing exploit the intrinsic non-toxicity and bright stability of Si QD fluorescence. Tandem photovoltaic devices with Si QDs as luminescent top layers enable solar energy conversion via multiple stages [55]. Continued fundamental studies open up new frontiers linking nanoscale manipulation of matter to quantum mechanical effects (Figure 37.3).

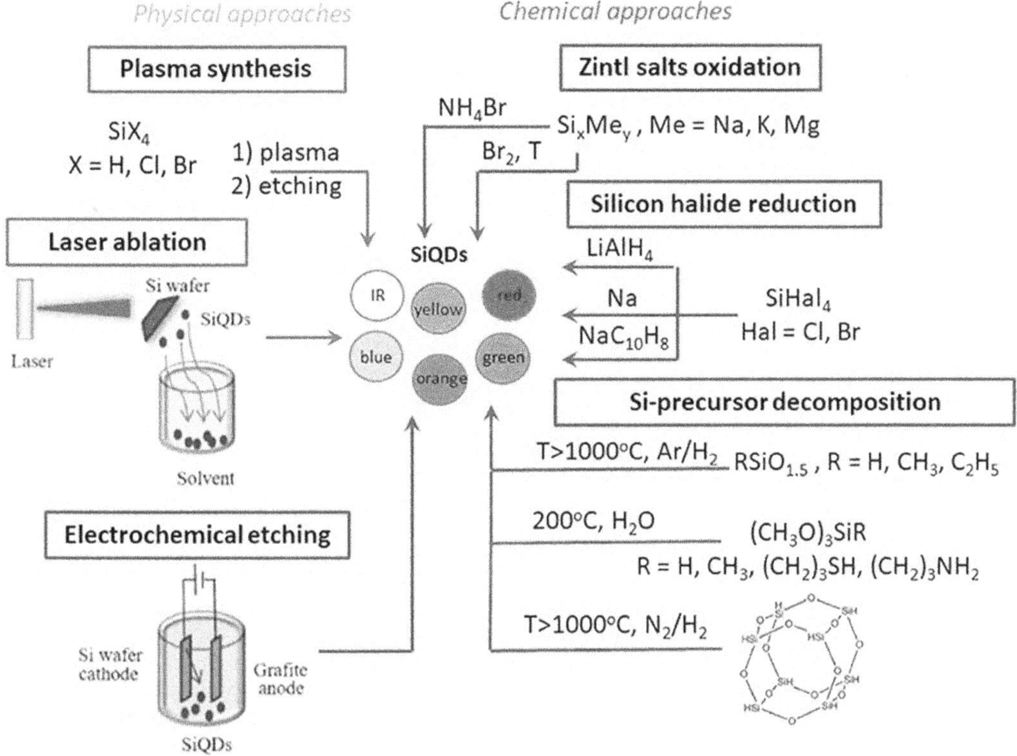

FIGURE 37.3 Various methods for synthesising SiQDs [40].

37.6 Carbon Nanotubes

CNTs are tiny cylindrical structures made up of rolled graphene sheets [56]. They can be either single-walled CNTs (SWCNTs) composed of a single cylinder of graphene or multi-walled CNTs (MWCNTs) consisting of multiple concentric cylinders [56]. CNTs possess remarkable electrical, mechanical, and thermal properties due to the sp^2-hybridised carbon atom configuration, making them valuable for sensors and other applications. CNTs change electrical resistance when exposed to certain molecules, allowing them to detect gases like NH$_3$ and NO$_2$ down to parts per billion sensitivity. Their small size also enables detection of glucose, DNA nucleotides, proteins, and other biomolecules. CNT films, transistors, and hybrid devices with polymers or nanoparticles provide inexpensive, portable chemical and biological sensors [57].

CNT growth involves transition metal nanoparticle catalysts like iron, cobalt, or nickel to catalyse the cracking of carbon feedstocks into nanotube structures [58]. Common techniques include chemical vapor deposition, utilising hydrocarbons like methane at 700–1200°C, and arc discharge, applying high voltage across carbon electrodes in inert gas environments [59]. CNT chirality, diameter, defect density, and other properties depend strongly on synthesis parameters like temperature, catalyst, and feedstock gases (Wang, Yuan, *et al.*, 2015). The excellent electrical conductivity and stability of CNTs allow integration into devices like conductive films, transistors, sensors, and probes. Adding CNTs at low-weight fractions can enhance the mechanical properties of polymer composites several-fold, enabling lightweight, durable materials. High thermal conductivity suits CNTs to remove heat in electronics. Large surface area combined with biocompatibility fosters applications like drug delivery and implant coatings [60]. However, higher CNT loading often impairs composite properties due to poor dispersion and interfacial issues with the matrix. Surface functionalisation can improve dispersion and compatibility. CNTs possess exceptional properties for sensors, conductive composites, electronics, and other uses [61]. Controlled growth conditions and surface modifications allow tailoring CNT characteristics for different applications. While

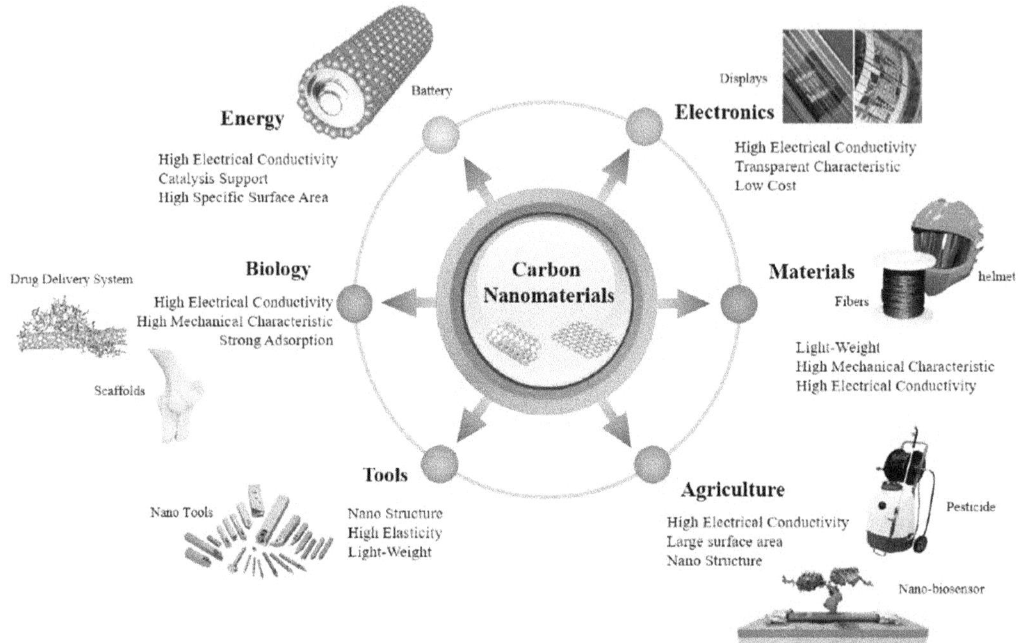

FIGURE 37.4 The characteristics of carbon nanotubes (CNTs) and their applications in both synthetic and transdermal fields [63].

high loadings degrade polymer properties, small CNT additions enable substantial improvements in mechanical, electrical, and thermal performance [62]. Continued research aims to leverage their combination of nanoscale dimensions and excellent properties for next-generation devices and multifunctional materials [63] (Figure 37.4).

37.7 Nano Inorganic Structures Material

Nano inorganic hybrid structures are combinations of carbon nanomaterials like CNTs or graphene with inorganic nanoparticles like gold, platinum, or quantum dots [64]. These hybrid structures display enhanced electrical, thermal, chemical, and mechanical properties compared to their components. They can be produced using various ex situ and in situ techniques. The properties of nanoinorganic hybrids depend on the purity of CNTs, the modification of their surface chemistry, and the type of inorganic nanoparticles used. Typically, the inorganic nanoparticles are deposited on the CNT surface to form the hybrid structure. The synthesis methods can be categorised as ex situ, where nanoparticles are first synthesised separately, or in situ, where nanoparticles are directly grown on the CNTs [65]. Ionic liquids or chitosan can aid solubilisation and manipulation of CNTs and subsequent functionalisation [64].

Hybrid nanostructures of CNTs and inorganic nanoparticles have shown useful applications in biosensing and bioanalysis. Examples include highly sensitive immunoassays for cancer biomarkers, enhanced amperometric enzyme electrodes for monitoring glucose levels, and detection of DNA or proteins [66]. Key factors for enhanced performance include the use of advanced nanolabels that amplify signals: gold-wrapped graphene nanolabels lowered the detection limit of tumour markers like carcinoembryonic antigen (CEA) to 100 pg/mL [67]. Gold nanoparticle-coated CNTs provided signal enhancement via redox cycling down to 0.8 fg/mL for anti-fingerprint (AFP). Modification of electrode transducer + nanolabelling: gold nanoparticles modified carbon electrode transducer for immobilising antibodies + CNTs as nanocarriers for enzyme tags provided dual amplification for detecting cancer biomarkers [68]. Carbon-nanostructured transducer + inorganic nanolabels: gold nanoflowers/graphene-modified gold electrode transducer + gold nanoflowers as labels for signal amplification detected 10 pg/mL CEA. Hybrid electrode transducer: gold nanoparticle/graphene electrode transducer surface offers efficient immobilisation of capture antibodies. The combination of enhanced electrical properties of graphene and the catalytic ability of gold particles synergistically enhances the performance of the resulting immunosensor [69]. The rapid development of nanoinorganic hybrids has led to major advancements in electroanalysis techniques, especially diverse biosensing applications relevant to clinical diagnosis and healthcare monitoring. Future efforts aim to tailor the hybrid synthesis to meet application-specific electrochemical biosensing needs.

37.8 Chemical Nanosensors

Nanosensors work at the nanoscale, around 10^{-9} metres, and are capable of carrying information about particles at that scale up to the macroscopic level. Due to their tiny size, nanosensors offer enormous improvements in selectivity, speed, and sensitivity over traditional chemical and biological methods of detection [70]. Some methods used for detection in chemical sensors are fluorescence, colorimetric, and electrochemical. Fluorescence offers very high sensitivity and is widely applicable [71]. Colorimetric is similar but less sensitive, while electrochemistry requires simpler equipment but works for only certain target species.

An anionic sensor using nano-silica particles is functionalised with organic molecules to detect fluoride ions through fluorescence change. The fluorescence shifts only in the presence of fluoride, not other anions like bromide or sulphate. A sensor detects phosphate anions through the release of fluorescein derivatives loaded inside mesoporous nano-silica [72]. The fluorescein release causes a colour change only when phosphate is present. A sensor functionalising silica nanotube with organic molecules detects copper cations by a colorimetric change. The colour change occurs specifically in response to copper [73]. Nanosensors constructed from magnetic nanoparticles functionalised with organic molecules can selectively identify and isolate target molecules from complex biological and environmental samples. For example, they have been applied to separate lead from blood samples or detect trace levels of uranium in water through fluorescence or colour signals.

Key applications of nanosensors include detecting microbes, contaminants, heavy metals, pesticides, and other toxins in water systems [74]. They can also monitor parameters like pH and dissolved oxygen. Nanomaterials like silicon nanotubes, gold nanoparticles, graphene, and quantum dots are often used to build these sensors [75]. The tiny size of nanosensors allows detection of species even at cellular levels inside living organisms. Their ease of use means sampling, dilution, and other processing may not be required. Fields utilising nanosensors include medical diagnostics, environmental monitoring, food safety, and industrial process control. With recent advances in precise organic synthesis, sensors can now be designed even for analytes without natural receptors. When combined with developments in electronics and optics, nanosensors promise continued improvements in rapid, accurate detection across many critical fields. Their expanding use will enable catching contaminants before they can pose significant threats (Figure 37.5).

37.9 Nano Energy Storage

The nanostructuring of materials enables faster charging due to shorter diffusion distances. It also accommodates the large volume changes during cycling that lead to failure and capacity loss in microparticle materials. This allows the use of high-capacity materials like silicon anodes in batteries. Nanomaterials can enable batteries using multivalent ions like Mg^{2+} and Al^{3+}, which otherwise diffuse too slowly in bulk materials to be practical. Understanding nanoscale processes at electrochemical interfaces is essential to uncover the fundamental mechanisms of charge storage. This knowledge can lead to new concepts and revolutionary advances in battery and capacitor technology.

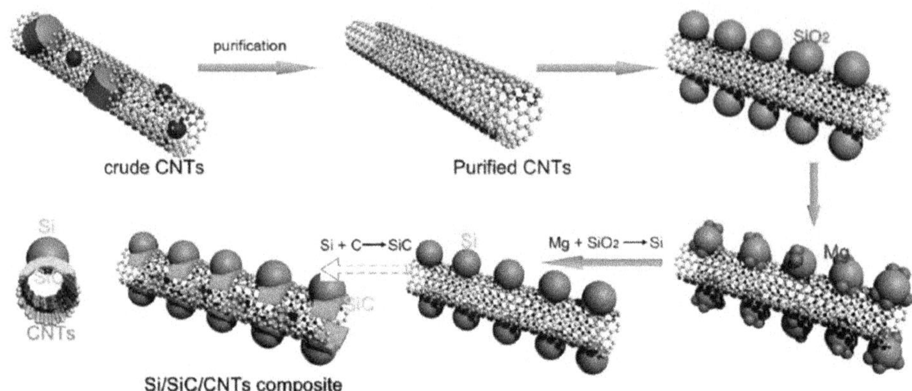

FIGURE 37.5 A diagram representing the process of preparing a nanocomposite involving silicon (Si), silicon carbide (SiC), and carbon nanotubes (CNTs) [65].

Nanomaterials and nanostructured architectures offer routes to achieve high energy density along with power density and cycle life approaching that of supercapacitors. Examples are tuning pore size in hierarchical carbon materials and conformally coating graphene or polymers onto porous scaffolds. In situ studies of energy storage processes at the nanoscale using advanced microscopy and spectroscopy can provide critical insight. New techniques for such characterisation are highly valuable. Modelling at the quantum and molecular levels facilitates the evaluation and even design of new nanoscale materials assemblies for electrical energy storage applications [76].

The researchers developed a triboelectric nanogenerator-based wireless gas sensor system (TWGSS) that integrates a triboelectric nanogenerator (TENG) and gas sensor to realise selective, sensitive, and real-time wireless assessment of food quality in a cold supply chain [77, 78]. The TENG uses porous, conductive wood and a fluorinated ethylene propylene (FEP) film as the triboelectric materials. The wood is chemically treated to make it conductive by incorporating CNTs into its porous 3D structure, which provides more surface area and sites for gas absorption while maintaining wood's natural flexibility [79]. Nanoscale design, engineering, production, and characterisation enable tailoring of structure, morphology, and interfaces in electrodes to optimise performance and provide key knowledge of charge storage mechanisms to enable transformative innovation in the field.

Some potential connections between microfluidics and nanotechnology for energy storage could be: using microfluidic platforms for synthesis and assembly of nanostructured battery electrode materials; employing microfluidics for precision deposition/patterning of nanomaterials onto substrates; integrating nanoscale battery materials and devices with microfluidic systems for characterisation or testing; leveraging advantages of both microfluidics (fluid control) and nanomaterials (high surface area) for improved electrochemical energy storage systems (Figure 37.6).

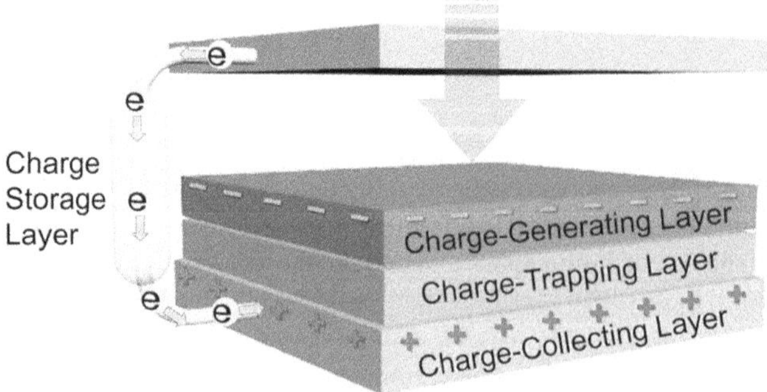

FIGURE 37.6 Key elements of a triboelectric nanogenerator (TENG) influencing the generation of triboelectric power [78].

37.10 Microfluidics

Microfluidics refers to the technology of manipulating and controlling fluids in channels with dimensions of tens to hundreds of micrometres. Microfluidics has promising applications in the food industry, including for food safety, processing, and production [80]. For food safety, microfluidics enables rapid, sensitive, and high-throughput detection of pathogens, toxins, and contaminants in food samples [81]. Examples include microfluidic devices to capture and detect *Escherichia coli* or Salmonella on gold electrodes using magnetic nanoparticles and a microfluidic sensor to detect botulinum toxin using enzymatic cleavage of fluorescent substrates and beads in a microchannel for glutamate detection [82]. Key advantages are portability, automation, and minimal reagent and sample requirements. Limitations currently include a lack of integration of different functionality and difficulty bridging microfluidic systems with readout electronics.

For food processing, microfluidics allows precise control over the formation of emulsions, bubbles, and other structures. This includes devices for the formation of monodisperse double emulsions, capsules, and lipid microparticles, which may allow the production of novel food textures and encapsulation systems. Scaled-up microchannel emulsification systems can generate higher productivity of uniform droplets [83]. In addition, mixers and reactors are being developed for applications like nanoparticle production and biodiesel synthesis. Limitations are meeting production-scale throughput and effective infusion/mixing control. Microfluidics also supports areas like animal health monitoring with portable analysis of biomarkers in saliva, purification of enzymes for biofuels using partitioned microfluidic extraction, and plant metabolomics through integrating concentration gradients and cell culture [84].

Microfluidics is an emerging enabling technology for food analysis, production, and process engineering. Key advantages are exquisite fluid control, automation, and portability. While specialised expertise is still needed to fabricate and operate devices, self-contained commercial systems are becoming available. With progress in overcoming throughput limitations and developing user-friendly interfaces, microfluidics promises a paradigm shift in food manufacturing through rapid, tailored production of specialised structures and formulations. This can lead to advances in food quality, safety, functionality, and sustainability (Figure 37.7).

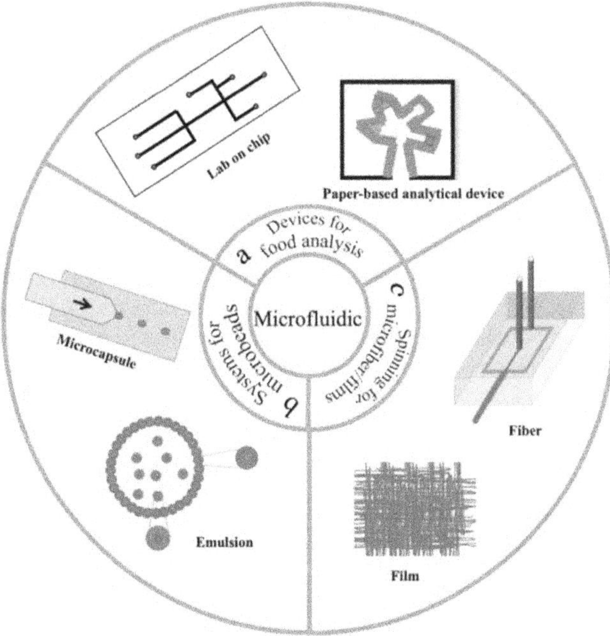

FIGURE 37.7 Possible uses of microfluidics in the field of food science and technology [80].

37.11 Conclusion

Nanotechnology offers tremendous advances in food analytical sensors by leveraging unique properties at the nanoscale. Fluorescence, colorimetric, and electrochemical transduction methods prevail, with extensive efforts to impart selectivity using nanoscale interactions, biomolecule recognition, and chemical functionalisation [85]. Integration with electronics and wireless communication will enable networked systems to track parameters like time, temperature, humidity, and gas profiles to continually assess food quality during storage and transport [86]. The field aims to balance improved performance with safety assurances before translation into commercial monitoring systems and packaging. Precision in designing nanomaterials and tailoring them to specific detection needs will allow these technologies to eventually realise routine, widespread application in the food industry.

REFERENCES

[1] Weston, M, S Geng, and R Chandrawati. 2021. "Food sensors: Challenges and opportunities." *Advanced Materials Technologies 6* (5):2001242.

[2] Antonacci, A, F Arduini, D Moscone, G Palleschi, and V Scognamiglio. 2016. "Commercially available (bio) sensors in the agri food sector." In Biosensors for Sustainable Food – New Opportunities and Technical Challenges *Comprehensive Analytical Chemistry*, 315–340. Elsevier.

[3] Sridhar, A, M Ponnuchamy, PS Kumar, A Kapoor, D-VN Vo, and G Rangasamy. 2023. "Digitalization of the agro-food sector for achieving sustainable development goals: Areview." *Sustainable Food Technology.* (*1*):783–802.

[4] Lee, MH, and HR Nicholls. 1999. "Review article tactile sensing for mechatronics—A state of the art survey." *Mechatronics 9* (1):1–31.

[5] Terry, LA, SF White, and LJ Tigwell. 2005. "The application of biosensors to fresh produce and the wider food industry." *J Agric Food Chem 53* (5):1309–1316.

[6] Adley, CC. 2014. "Past, present and future of sensors in food production." *Foods 3* (3):491–510.

[7] Weston, M. 2020. *Colourimetric Sensors for Rapid Detection of Contaminants and Food Quality.* UNSW Sydney.

[8 El Kazzy, M, JS Weerakkody, C Hurot, R Mathey, A Buhot, N Scaramozzino, and Y Hou. 2021. "An overview of artificial olfaction systems with a focus on surface plasmon resonance for the analysis of volatile organic compounds." *Biosensors 11* (8):244.

[9 Ghasemi-Varnamkhasti, M, SS Mohtasebi, M Siadat, H Ahmadi, and SH Razavi. 2015. "From simple classification methods to machine learning for the binary discrimination of beers using electronic nose data." *Engineering in Agriculture, Environment and Food 8* (1):44–51.

[10] Mansuriya, BD, and Z Altintas. 2020. "Applications of graphene quantum dots in biomedical sensors." *Sensors 20* (4):1072.

[11] Khan, Y, H Sadia, SZ Ali Shah, MN Khan, AA Shah, N Ullah, MF Ullah, H Bibi, OT Bafakeeh, and NB Khedher. 2022. "Classification, synthetic, and characterization approaches to nanoparticles, and their applications in various fields of nanotechnology: A review." *Catalysts 12* (11):1386.

[12] Guo, Y, M Girmatsion, H-W Li, Y Xie, W Yao, H Qian, B Abraha, and A Mahmud. 2021. "Rapid and ultrasensitive detection of food contaminants using surface-enhanced Raman spectroscopy-based methods." *Critical Reviews in Food Science and Nutrition 61* (21):3555–3568.

[13] Chaudhry, Q, M Scotter, J Blackburn, B Ross, A Boxall, L Castle, R Aitken, and R Watkins. 2008. "Applications and implications of nanotechnologies for the food sector." *Food Additives and Contaminants 25* (3):241–258.

[14] Lu, L, Z Zhu, and X Hu. 2019. "Hybrid nanocomposites modified on sensors and biosensors for the analysis of food functionality and safety." *Trends in Food Science & Technology 90*:100–110.

[15] Neupane, D, and KJ Stine. 2021. "Electrochemical sandwich assays for biomarkers incorporating aptamers, antibodies and nanomaterials for detection of specific protein biomarkers." *Applied Sciences 11* (15):7087.

[16] Luong, JH, T Narayan, S Solanki, and BD Malhotra. 2020. "Recent advances of conducting polymers and their composites for electrochemical biosensing applications." *Journal of Functional Biomaterials 11* (4):71.

[17] Das, R, A Dwevedi, and AM Kayastha. 2021. "Current and future trends on polymer-based enzyme immobilization." *Polymeric Supports for Enzyme Immobilization: Opportunities and Applications 1*: 150–180.

[18] Jorns, M, and D Pappas. 2021. "A review of fluorescent carbon dots, their synthesis, physical and chemical characteristics, and applications." *Nanomaterials 11* (6):1448.

[19] Modi, PD, VN Mehta, VS Prajapati, S Patel, and JV Rohit. 2023. "Bottom-up approaches for the preparation of carbon dots." In Suresh Kumar Kailasa and Chaudhery Mustansar Hussain *Carbon Dots in Analytical Chemistry*, 15–29. Elsevier.

[20] Shamsipur, M, A Barati, and Z Nematifar. 2019. "Fluorescent pH nanosensors: Design strategies and applications." *Journal of Photochemistry and Photobiology C: Photochemistry Reviews 39*:76–141.

[21] Luo, X, Y Han, X Chen, W Tang, T Yue, and Z Li. 2020. "Carbon dots derived fluorescent nanosensors as versatile tools for food quality and safety assessment: A review." *Trends in Food Science & Technology 95*:149–161.

[22] Tian, J, M An, X Zhao, Y Wang, and M Hasan. 2023. "Advances in fluorescent sensing carbon dots: An account of food analysis." *ACSOmega 8* (10):9031–9039.

[23] Sun, X, and Y Lei. 2017. "Fluorescent carbon dots and their sensing applications." *TrAC Trends in Analytical Chemistry 89*:163–180.

[24] Dhenadhayalan, N, KC Lin, and TA Saleh. 2020. "Recent advances in functionalized carbon dots toward the design of efficient materials for sensing and catalysis applications." *Small 16* (1):1905767.

[25] Pirsaheb, M, S Mohammadi, and A Salimi. 2019. "Current advances of carbon dots based biosensors for tumor marker detection, cancer cells analysis and bioimaging." *TrAC Trends in Analytical Chemistry 115*:83–99.

[26] Huang, L, DW Sun, H Pu, and Q Wei. 2019. "Development of nanozymes for food quality and safety detection: Principles and recent applications." *Comprehensive Reviews in Food Science and Food Safety 18* (5):1496–1513.

[27] Barciela, P, A Perez-Vazquez, and M Prieto. 2023. "Azo dyes in the food industry: Features, classification, toxicity, alternatives, and regulation." *Food and Chemical Toxicology 178*:113935.

[28] Landa, SDT, NKR Bogireddy, I Kaur, V Batra, and V Agarwal. 2022. "Heavy metal ion detection using green precursor derived carbon dots." *Iscience 25* (2).

[29] Doerk, GS. 2010. *Synthesis, Characterization, and Integration of Silicon Nanowires for Nanosystems Technology*. University of California.

[30] Riederer, F, T Grap, S Fischer, MR Mueller, D Yamaoka, B Sun, C Gupta, KT Kallis, and J Knoch. 2018. "Alternatives for doping in nanoscale field-effect transistors." *Physica Status Solidi (a) 215* (7):1700969.

[31] Schmidt, V, J Wittemann, and U Gosele. 2010. "Growth, thermodynamics, and electrical properties of silicon nanowires." *Chemical Reviews 110* (1):361–388.

[32] Yin, BA. 2015. "Fabrication of silicon nanowires with controlled nano-scale shapes using wet anisotropic etching." *UT Electronic Theses and Dissertations*. 2152:46753

[33] Hobbs, RG, N Petkov, and JD Holmes. 2012. "Semiconductor nanowire fabrication by bottom-up and top-down paradigms." *Chemistry of Materials 24* (11):1975–1991.

[34] Kim, K, JK Lee, SJ Han, and S Lee. 2020a. "A novel top-down fabrication process for vertically-stacked silicon-nanowire array." *Applied Sciences 10* (3):1146.

[35] Ko, H-S, Y Lee, S-Y Min, S-J Kwon, and T-W Lee. 2017. "Large-scale metal nanoelectrode arrays based on printed nanowire lithography for nanowire complementary inverters." *Nanoscale 9* (41):15766–15772.

[36] Ivanov, YD, TS Romanova, KA Malsagova, TO Pleshakova, and AI Archakov. 2021. "Use of silicon nanowire sensors for early cancer diagnosis." *Molecules 26* (12):3734.

[37] Nuzaihan, M, U Hashim, MM Arshad, S Kasjoo, S Rahman, A Ruslinda, M Fathil, R Adzhri, and M Shahimin. 2016. "Electrical detection of dengue virus (DENV) DNA oligomer using silicon nanowire biosensor with novel molecular gate control." *Biosensors and Bioelectronics 83*:106–114.

[38] Luan, E, H Shoman, DM Ratner, KC Cheung, and L Chrostowski. 2018. "Silicon photonic biosensors using label-free detection." *Sensors 18* (10):3519.

[39] Pucker, G, E Serra, and Y Jestin. 2012. "Silicon quantum dots for photovoltaics: Areview." *Quantum Dots—A Variety of New Applications*, 59–92. Intech Open Publishers.

[40] Morozova, S, M Alikina, A Vinogradov, and M Pagliaro. 2020. "Silicon quantum dots: Synthesis, encapsulation, and application in light-emitting diodes." *Frontiers in Chemistry 8*:191.

[41] Shimizu-Iwayama, T, DE Hole, and IW Boyd. 1999. "Mechanism of photoluminescence of Si nanocrystals in SiO$_2$ fabricated by ion implantation: The role of interactions of nanocrystals and oxygen." *Journal of Physics: Condensed matter 11* (34):6595.

[42] Kabashin, AV, A Singh, MT Swihart, IN Zavestovskaya, and PN Prasad. 2019. "Laser-processed nanosilicon: A multifunctional nanomaterial for energy and healthcare." *ACS Nano 13* (9):9841–9867.

[43] Rashid, M, NM Ahmed, NAM Noor, and MZ Pakhuruddin. 2020. "Silicon quantum dot/black silicon hybrid nanostructure for broadband reflection reduction." *Materials Science in Semiconductor Processing 115*:105113.

[44] O'Neil, A, and JJ Watkins. 2005. "Fabrication of device nanostructures using supercritical fluids." *MRSBulletin 30* (12):967–975.

[45] Yasuda, K, and TH Okabe. 2010. "Solar-grade silicon production by metallothermic reduction." *Jom 62*:94–101.

[46] Mulvaney, P, L Liz-Marzan, M Giersig, and T Ung. 2000. "Silica encapsulation of quantum dots and metal clusters." *Journal of Materials Chemistry 10* (6):1259–1270.

[47] Agarwal, H, SV Kumar, and S Rajeshkumar. 2017. "A review on green synthesis of zinc oxide nanoparticles–An eco-friendly approach." *Resource-Efficient Technologies 3* (4):406–413.

[48] Chinnathambi, S, S Chen, S Ganesan, and N Hanagata. 2014. "Silicon quantum dots for biological applications." *Advanced Healthcare Materials 3* (1):10–29.

[49] Saar, A. 2009. "Photoluminescence from silicon nanostructures: The mutual role of quantum confinement and surface chemistry." *Journal of Nanophotonics 3* (1):032501.

[50] Barbagiovanni, EG, DJ Lockwood, PJ Simpson, and LV Goncharova. 2014. "Quantum confinement in Si and Ge nanostructures: Theory and experiment." *Applied Physics Reviews 1* (1).

[51] Zhang, Y, G Conibeer, S Liu, J Zhang, and JF Guillemoles. 2022. "Review of the mechanisms for the phonon bottleneck effect in III–V semiconductors and their application for efficient hot carrier solar cells." *Progress in Photovoltaics: Research and Applications 30* (6):581–596.

[52] Mahfoud, Z, AT Dijksman, C Javaux, P Bassoul, A-L Baudrion, Jrm Plain, B Dubertret, and M Kociak. 2013. "Cathodoluminescence in a scanning transmission electron microscope: A nanometer-scale counterpart of photoluminescence for the study of ii–vi quantum dots." *The Journal of Physical Chemistry Letters 4* (23):4090–4094.

[53] Mustafeez, W. 2012. *Silicon Nanocrystals, Dust to Gold: Progress in Material, Devices and Synthesis.* Stanford University.

[54] Jariwala, D, VK Sangwan, LJ Lauhon, TJ Marks, and MC Hersam. 2013. "Carbon nanomaterials for electronics, optoelectronics, photovoltaics, and sensing." *Chemical Society Reviews 42* (7):2824–2860.

[55] Ramachandran, AM, AS Thampi, M Singh, and A Asok. 2022. "A comprehensive review on optics and optical materials for planar waveguide-based compact concentrated solar photovoltaics." *Results in Engineering 16*:100665.

[56] Malik, S, M Maqbool, S Hussain, and H Irfan. 2008. "Carbon nanotubes: Description, properties and applications." *Journal of the Pakistan Materials Society 2* (1):21–26.

[57] Murugasenapathi, NK, R Ghosh, S Ramanathan, S Ghosh, A Chinnappan, SAJ Mohamed, KA Esther Jebakumari, SC Gopinath, S Ramakrishna, and T Palanisamy. 2023. "Transistor-based biomolecule sensors: Recent technological advancements and future prospects." *Critical Reviews in Analytical Chemistry 53* (5):1044–1065.

[58] Jourdain, V, and C Bichara. 2013. "Current understanding of the growth of carbon nanotubes in catalytic chemical vapour deposition." *Carbon 58*:2–39.

[59] Shahzad, H. 2014. "Carbon nanotubes deposited by hot wire plasma CVD and water assisted CVD for energetic and environmental applications." Thesis, Universitat de Barcelona. Departament de Física Aplicada i Òptica.

[60] Cardenas, JA, JB Andrews, SG Noyce, and AD Franklin. 2020. "Carbon nanotube electronics for IoT sensors." *Nano Futures 4* (1):012001.

[61] Ibrahim, KS. 2013. "Carbon nanotubes? properties and applications: A review." *Carbon Letters 14* (3):131–144.

[62] Spitalsky, Z, D Tasis, K Papagelis, and C Galiotis. 2010. "Carbon nanotube–polymer composites: Chemistry, processing, mechanical and electrical properties." *Progress in Polymer Science 35* (3):357–401.

[63] Patel, DK, H-B Kim, SD Dutta, K Ganguly, and K-T Lim. 2020. "Carbon nanotubes-based nanomaterials and their agricultural and biotechnological applications." *Materials 13* (7):1679.

[64] Singh, VV, and AK Malik. 2023. "Approaches to produce hybrid nanomaterials for engineering applications." In Janardhan Reddy Koduru, Rama Rao Karri, Nabisab Mujawar Mubarak *Hybrid Nanomaterials for Sustainable Applications* 55–69. Elsevier.

[65] Zhang, Y, K Hu, Y Zhou, Y Xia, N Yu, G Wu, Y Zhu, Y Wu, and H Huang. 2019. "A facile, one-step synthesis of silicon/silicon carbide/carbon nanotube nanocomposite as a cycling-stable anode for lithium ion batteries." *Nanomaterials 9* (11):1624.

[66] Li, J, S Li, and CF Yang. 2012. "Electrochemical biosensors for cancer biomarker detection." *Electroanalysis 24* (12):2213–2229.

[67] Laraib, U, S Sargazi, A Rahdar, M Khatami, and S Pandey. 2022. "Nanotechnology-based approaches for effective detection of tumor markers: A comprehensive state-of-the-art review." *International Journal of Biological Macromolecules 195*:356–383.

[68] Yáñez-Sedeño, P, S Campuzano, and JM Pingarrón. 2017. "Carbon nanostructures for tagging in electrochemical biosensing: Areview." *Journal of Carbon research 3* (1):3.

[69] Khaniani, Y, Y Ma, M Ghadiri, J Zeng, D Wishart, S Babiuk, C Charlton, JN Kanji, and J Chen. 2022. "A gold nanoparticle-protein G electrochemical affinity biosensor for the detection of SARS-CoV-2 antibodies: A surface modification approach." *Scientific Reports 12* (1):12850.

[70] Mahfuz, MU, and K Ahmed. 2005. "A review of micro-nano-scale wireless sensor networks for environmental protection: Prospects and challenges." *Science and Technology of Advanced Materials 6* (3–4):302–306.

[71] Ajay Piriya V S, P Joseph, SCG Kiruba Daniel, S Lakshmanan, T Kinoshita, and S Muthusamy. 2017. "Colorimetric sensors for rapid detection of various analytes." *Materials Science and Engineering: C 78*:1231–1245.

[72] Candel Busquets, I. 2014. *"Functional Silica Materials for Controlled Release, Sensing and Elimination of Target Molecules."* Universitat Politècnica de València.

[73] Deuermeyer, H. 2020. *"Design and Synthesis of Fluorescent Silica Nanoparticle Conjugates for Metal Ion Sensing Applications."* Master's Theses. 1324. https://repository.usfca.edu/thes/1324

[74] Salouti, M, and F Khadivi Derakhshan. 2020. "Biosensors and nanobiosensors in environmental applications." *Biogenic Nano-Particles and Their Use in Agro-Ecosystems*, 515–591. Springer Singapore

[75] Anas, NAA, YW Fen, NAS Omar, WMEMM Daniyal, NSM Ramdzan, and S Saleviter. 2019. "Development of graphene quantum dots-based optical sensor for toxic metal ion detection." *Sensors 19* (18):3850.

[76] Gogotsi, Y. 2014. *What Nano Can Do for Energy Storage.* ACS Publications.

[77] Cai, C, J Mo, Y Lu, N Zhang, Z Wu, S Wang, and S Nie. 2021. "Integration of a porous wood-based triboelectric nanogenerator and gas sensor for real-time wireless food-quality assessment." *Nano Energy 83*:105833.

[78] Kim, DW, JH Lee, JK Kim, and U Jeong. 2020b. "Material aspects of triboelectric energy generation and sensors." *NPG Asia Materials 12* (1):6.

[79] Farid, T, MI Rafiq, A Ali, and W Tang. 2022. "Transforming wood as next-generation structural and functional materials for a sustainable future." *Eco Mat 4* (1):e12154.

[80] Mu, R, N Bu, J Pang, L Wang, and Y Zhang. 2022. "Recent trends of microfluidics in food science and technology: Fabrications and applications." *Foods 11* (22):3727.

[81] Nilghaz, A, SM Mousavi, M Li, J Tian, R Cao, and X Wang. 2021. "based microfluidics for food safety and quality analysis." *Trends in Food Science & Technology 118*:273–284.

[82] Vidic, J, P Vizzini, M Manzano, D Kavanaugh, N Ramarao, M Zivkovic, V Radonic, N Knezevic, I Giouroudi, and I Gadjanski. 2019. "Point-of-need DNA testing for detection of foodborne pathogenic bacteria." *Sensors 19* (5):1100.

[83] Kobayashi, I, Y Wada, K Uemura, and M Nakajima. 2010. "Microchannel emulsification for mass production of uniform fine droplets: Integration of microchannel arrays on a chip." *Microfluidics and Nanofluidics 8*:255–262.

[84] Arya, SS, SB Dias, HF Jelinek, LJ Hadjileontiadis, and A-M Pappa. 2023. "The convergence of traditional and digital biomarkers through AI-assisted biosensing: A new era in translational diagnostics?" *Biosensors and Bioelectronics* (235):115387.

[85] Wang, H, Y Yuan, L Wei, K Goh, D Yu, and Y Chen. 2015a. "Catalysts for chirality selective synthesis of single-walled carbon nanotubes." *Carbon 81*:1–19.

[86] Wang, L, Q Li, H-Y Wang, J-C Huang, R Zhang, Q-D Chen, H-L Xu, W Han, Z-Z Shao, and H-B Sun. 2015b. "Ultrafast optical spectroscopy of surface-modified silicon quantum dots: Unraveling the underlying mechanism of the ultrabright and color-tunablephotoluminescence." *Light: Science & Applications 4* (1):e245–e245.

38

Application Studies: Synthesis, Characterisation, Antimicrobial Activity of LSNN Crystal

M. Renugadevi
St. Joseph's College (Autonomous), Affiliated to Bharathidasan University, Tiruchirappalli, Tamil Nadu, India

P. Lalitha
Indra Ganesan College of Engineering, Affiliated to Anna University, Tiruchirappalli, Tamil Nadu, India

A. Sinthiya
St. Joseph's College (Autonomous), Affiliated to Bharathidasan University, Tiruchirappalli, Tamil Nadu, India

38.1 Introduction

Amino acids are chemical compounds made up of the carboxylic acid (COOH) and the amino acid group (NH_2), which are both attached to the same single carbon atom, known as the α-carbon. There are more than 700 different kinds of amino acids in nature. They are almost all α-amino acids. They are discovered in antimicrobial, fungal, algal, and distinct plant species [1, 2]. Three categories of amino acids exist: basic, neutral, and acidic. Certain amino acids have to be consumed through diet because the body cannot produce them on its own. There are many classes of fundamental amino acids [3]. The body can produce a number of amino acids, but only in sufficient amount. These amino acids are referred to as semi-fundamental amino acids. A few amino acids are needed for the synthesis of enzymes. L-Serine belongs to the class of chemical compounds that make up amino acids. It belongs to proteinogenic amino acids that are found naturally. L-Serine is a zwitterionic molecule that can mix with cationic, anionic, and neutral substances in general. Research has shown that transition metal mixed with ligand complexes of different amino acids have cytotoxic, antifungal, and antibacterial properties [4]. According to studies, there has been a substantial advancement in the use of the transition metal materials' complexes as pharmaceuticals to indulge a number of human illnesses, including diabetes, cancer, and neurological problems. [5]. They have also been used as anti-inflammatory and infection control medicines. Amino acid complexes with mixed binders are beneficial instances of enzyme suppressors [6, 7]. Understanding certain physiological functions requires an understanding of the shape, quantity, atoms from donors, and type of connecting these ligands [8]. In comparison to the parent complexes, several complexes of amino acids with mixed ligands have demonstrated increased antibacterial activity [9, 10]. Copper(II) compounds of the form $[M(Q)(L)] \cdot 2H_2O$ have been prepared using an 8-hydroxylquinoline (HQ) as the major ligand and the N-and O-donor amino acids (HL) such as L-thronine, L-hydroxyproline, L-proline, L-serine, and L-isoleucine as secondary ligands [11].

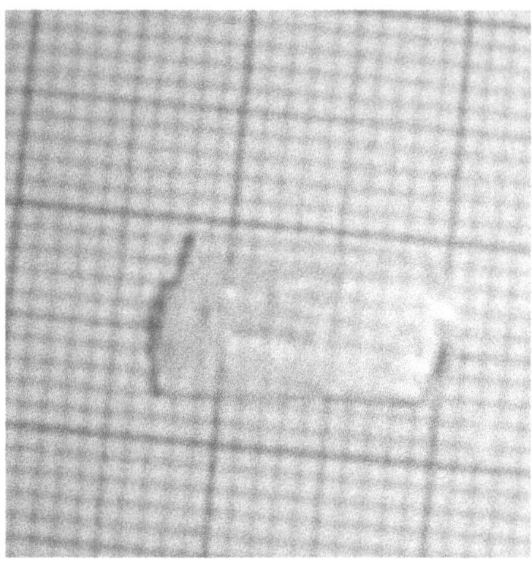

FIGURE 38.1   Crystal photograph of LSNN single crystal.

38.2 Experimental Process

38.2.1 Materials and Methods

Using water as the solvent, the solution growth evaporation method was utilised to successfully develop the crystals of L-serine-doped nickel nitrate (LSNN) at atmospheric temperature. The L-serine and nickel nitrate (1:1) solution was made by dissolving the mixture into the water solvent and swirling constantly at room temperature (35°C) using a magnetic stirrer. Gravimetric analysis was used to determine the solute's equilibrium concentration once saturation was reached. The solution-filled beaker was properly closed to allow for evaporation. After 21 days, transparent single crystals measuring 1 mm were produced. In Figure 38.1, the growing crystal is seen.

38.3 Results and Discussion

38.3.1 Analysis of Powder X-Ray Diffraction

An X-ray diffraction (XRD) experiment employing powder was conducted on LSNN crystals. The experiment was conducted at atmospheric temperature using CuKα radiation ($\lambda = 0.15418$ nm) and required a current supply of 100–240V/15A. The step size was set to 0.01°, and the angle range was 2θ–80°θ. One gram of finely ground sample powder was stored. The resolution of 2D HPAD (High Pixel Array Detector) detector Hypix is 3000. Figure 38.2 displays the powder XRD pattern for the formed crystal. The presence of prominent, pointed peaks indicates that the grown sample has good crystallinity.

L-serine peaks are represented by ◆ peaks, and nickel nitrate peaks by ♠ peaks. The peak positions obtained are in match with the information found in [JCPDS card number. 27-1989] and [JCPDS card no. 78-0643]. The XRD analysis of the grown crystal verifies that it is the grown LSNN crystal in its crystalline form.

38.3.2 Spectroscopy Analysis of FT-IR

The crystalline powder obtained from the functional groups confirmation study was examined. Pellets based on potassium bromide (KBr) were created by applying compression for around 30 seconds at a rate of 10 kg/cm². The FTIR spectrum is collected at room temperature using a Perkin Elmer FTIR spectrophotometer, covering the range of 500 cm⁻¹–4000 cm⁻¹. The resulting FTIR spectrum is shown

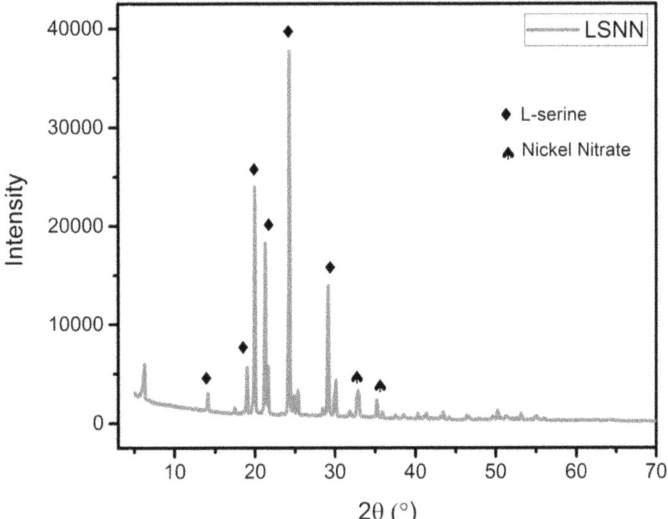

FIGURE 38.2 Powder X-ray diffraction pattern of LSNN.

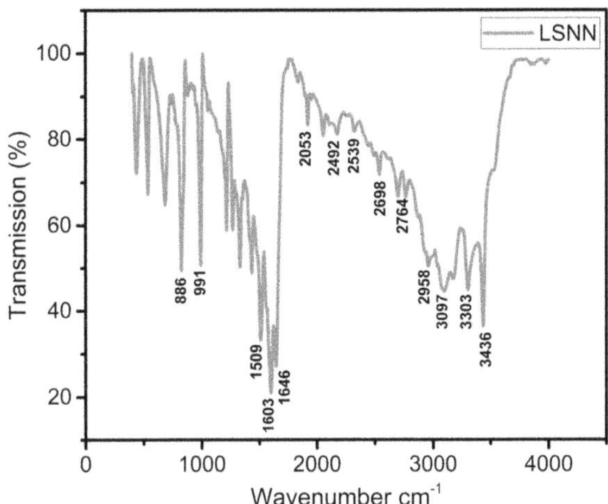

FIGURE 38.3 FT-IR spectrum of LSNN.

in Figure 38.3. In this spectrum, the broadband at 3436 cm^{-1}, 3303 cm^{-1}, 2958 cm^{-1} is due to the presence of OH. Frequencies observed at 3097 cm^{-1} and 2764 cm^{-1} are attributed to NH^{3+} stretching and CH stretching, respectively [12]. COO-antisymmetric stretching vibration is observed at 1509 cm^{-1}. The CH$_2$ bending of vibrations is at 2764 and 2698 cm^{-1} [13]. The bending and rocking vibrations formed due to the CH group have been observed at 2539 and 2492 cm^{-1}, respectively [14]. The CN-stretching vibrations present the peaks around 991 cm^{-1}. The C–N groups are revealed by the peaks observed at1603 cm^{-1} and 1646 cm^{-1}. The peaks observed at 886 cm^{-1} indicate the presence of the CH$_2$ group [15].

38.3.3 UV-Vis Spectroscopy

UV-V is spectroscopy, a wavelength peaks in the range of (200–1400) nm, a medium scan speed, a sampling interval of 2.0, the ability to disable auto sampling intervals, a single scanning mode, measuring mode of absorbance, a slit width of 2.0 nm, a light source change wavelength of 340.8 nm, normal S/R exchange, and software names UV probe, digital electric balance, and micropipette. There is a cut-off

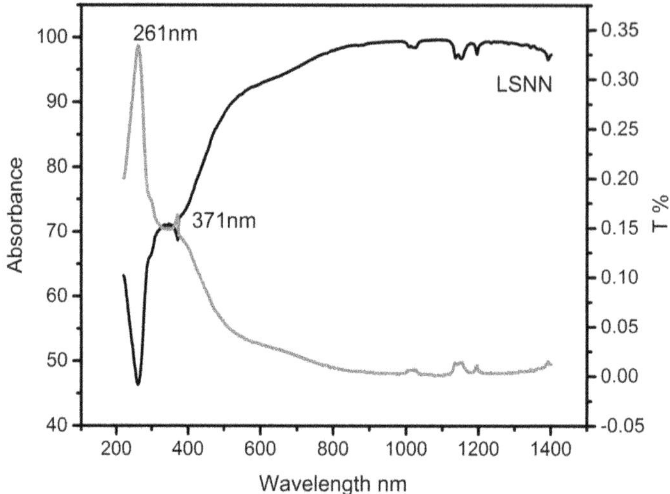

FIGURE 38.4   UV-visible spectrum of LSNN.

wavelength of approximately 261 nm and 371 nm as shown in Figure 38.4. Over the whole visible spectrum, the crystal's transmittance is determined to be 98%.

38.3.4 Fluorescence Spectroscopy

Using Perkin Elmer LS 45 fluorescence spectrometer finds a wide range of applications in the fields of biochemical, medical, and chemical research fields for analysing organic-based compounds. The emission ranges of spectra for LSNN were recorded using a spectrofluorometer at 200–900 nm. The sample was emitted at 482 nm as shown in Figure 38.5. In the emission spectrum, a peak at about 539 nm also was observed. The results are indicating that LSNN crystals have a green fluorescence emission (16).

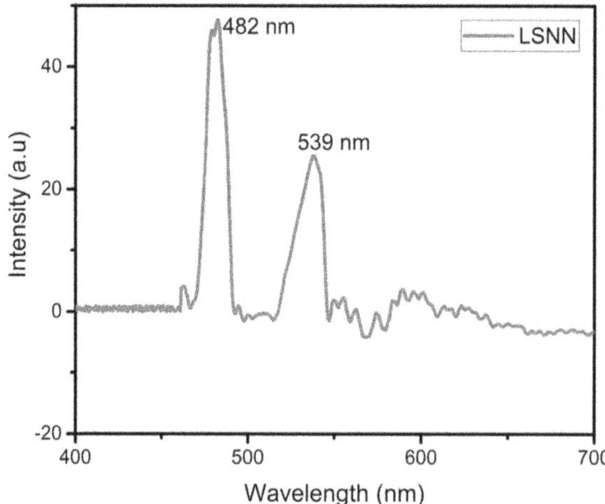

FIGURE 38.5   Fluorescence spectrum of LSNN.

38.3.5 Anti-microbial Activity

38.3.5.1 Determination of Anti-microbial Activity

The disc diffusion method, which was recommended by NCCLS [17] and Awoyinka et al. [16], was used to carry out the antimicrobial activity.

38.3.5.1.1 Setting up the Media

NA-Himedia Nutrient Agar Sample Media for Bacteria: Media Composition

Tissue from the animal: 5.00 g
Chloride sodium: 5.00 g
Extract from beef: 1.50 g
Extract of yeast: 1.50 g
Agar weight: 15.0 g

38.3.5.1.2 Process of the Media

Composition of the Media

Infusion of potatoes: 200.00 g
Weight of dextrose: 20.00 g
Weight of agar: 15.00 g

38.3.5.1.3 Setting up the medium

Agar (28.0 grams) rich in nutrients should be suspended in 1000 millilitres of purified water. Simmer until totally dissolved in the liquid. Sterilise via autoclaving for 15 minutes at 15 Ibs of pressure (121°C). Blend and then transfer to sterilised Petri dishes. Potato dextrose agar often abbreviated as PDA. Most mycologists widely use it as a general-purpose medium for yeasts and molds that can be identified to some extent based on their morphological features and their pigmentation in culture often being important for identification of cultures.

38.3.5.1.4 Microorganisms

In the biological experiments, *Aspergillus flavus* (MTCC 1783) and *Aspergillus niger* (MTCC 1783) were the fungal strains used, and *E. coli* (MTCC 732) and *Staphylococcus aureus* (MTCC 3160) were the strains of bacteria derived from the Institute of Microbial Technology (IMTECH), located in Chandigarh, India, through the Microbial Type Culture Collection (MTCC).

38.3.5.1.5 Cultivating a 24-hour pure culture

Each bacterium was suspended in a loop that held approximately 10 millilitres of physiological salinity within a Roux bottle. All of them were sprayed onto the appropriate culture slopes and then maintained at 37 degrees for 24 hours, with the exception of the fungal, which had been kept at 25 degrees over 48 hours. Once the incubation period was over and development was evident, the tubes were stored between 2 and 8°C until needed. Plant extract and nanoparticles (NPs) solutions were made ready for the experiment.

The standard solution contains the following: 10 mg of the samples was weighed and dissolved in 10 millilitres of distilled water, 25 mg of chloramphenicol for bacteria, and 25 mg of fluconazole (30 µl) for fungus. Unless they were needed for the experiment, they were stored in a refrigerator.

38.3.5.1.6 Setting up the discs of filter paper

Using Whatman filter paper (No. 1), four discs with a diameter of about 6 mm were created and then sterilised in hot air. Following sterilisation, discs were filled with 50 µl, 100 µl, and 150 µl of sample. A standard solution containing 30 µl of fluconazole and chloramphenicol, as well as a control solution of 30 µl, was utilised for contrasting the test result. Unless they were needed for the experiment, they were stored in a refrigerator.

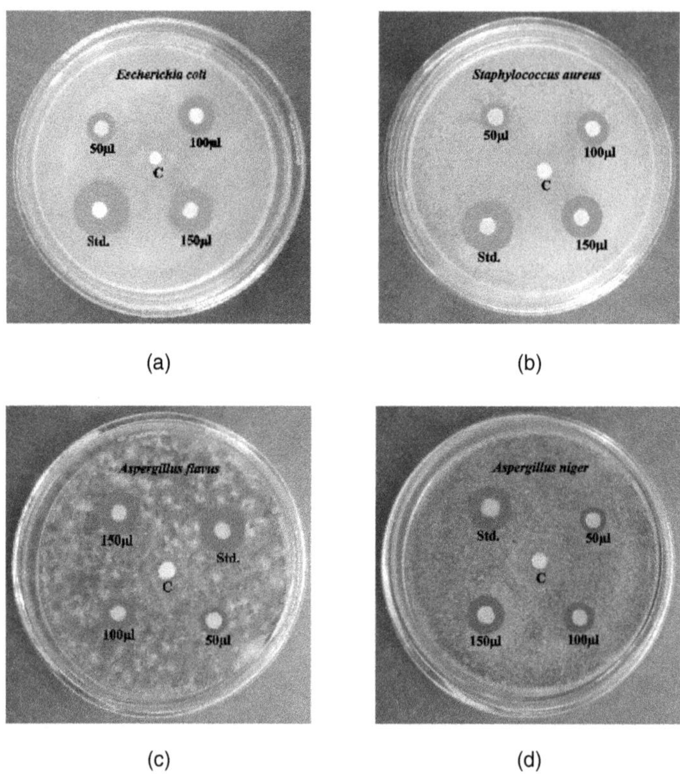

FIGURE 38.6   Plates: anti-microbial activity of LSNN. (a) Escherichia coli; (b) Staphylococcus aureus; (c) Aspergillusflavus; (d) Aspergillusniger.

38.3.5.1.7 Antimicrobial assay

The antibiogram was conducted using samples and the disc diffusion method [16, 17]. The Narcotics Anonymous/*Potato dextrose agar* (NA/PDA) medium was poured into 30 millilitres of petri dishes. In this, the organism was spread out on an agar plate that had been toughened and allowed to dry for 10 minutes using a micropipette. From a fluid culture, microorganisms were injected into the surfaces of the medium. Using a sterile cotton swab wiped in a standardised microbiological test suspension, the whole the PDA/nutrient agar plates' surface is equally inoculated. To put it briefly, microbiological strains were added to PDA/nutrient agar plates.

38.3.5.1.8 Scaling of zone of suppression

The average width of the suppression area around the disc, measured in inches, was used to assess the antibacterial properties of the test compounds. The microbes analysed by the samples had graded zones of suppression (Figure 38.6 and Table 38.1).

38.4 Conclusion

Clear and well developed by using the slow evaporation approach, LSNN single crystals were produced. PXRD analysis validated the crystalline nature. By using FTIR analysis, the functional groups were found. By utilising UV-VIS-NIR analysis to examine the optical behaviour, it was discovered that there is no absorption between 400 nm and that the crystal's transmittance is present. The LSNN's green fluorescence emission was revealed via the fluorescence spectrum. *Aspergillus flavus, Aspergillus niger, Staphylococcus aureus*, and *E. coli* were among the bacterial strains against which LSNN showed good antibacterial efficacy.

TABLE 38.1

Anti-microbial activity of LSNN

Microbial Strains	Control	50 µl	100 µl	150 µl	Std. (30 µl)
Bacterial strains					
Escherichia coli (mm)	Nil	2.00 ± 0.14	4.60 ± 0.32	7.00 ± 0.49	8.50 ± 0.59
Staphylococcus aureus (mm)	Nil	1.70 ± 0.11	4.00 ± 0.28	6.70 ± 0.46	8.20 ± 0.57
Fungal strains					
Aspergillus flavus (mm)	Nil	1.00 ± 0.07	3.10 ± 0.21	6.30 ± 0.44	7.10 ± 0.49
Aspergillus niger (mm)	Nil	0.80±0.05	2.40±0.16	5.80±0.40	7.00 ± 0.49

Notes: For triplicate values, they are expressed as mean ± SD.
Bacterial standard: chloramphenicol.
Fungal standard: fluconazole.
Control: methanol.

REFERENCES

[1] M. Spiliopoulou, A. Valmas, D-P. Triandafillidis, Christos Kosinas, Applications of X-ray Powder Diffraction in Protein Crystallography and Drug Screening (2020). *Crystals 2020*, 10, 54.

[2] R. Vinayagamoorthy, A. Albert Irudayaraj, A. Dhayal Raj, S. Karthick, G. Jayakumar. Comparative Study of Properties of L-Histidine and L-Histidine Nickel Nitrate Hexahydrate (2017). *Mechanics, Materials Science & Engineering Journal*, 9. https://doi.org/10.2412/mmse.83.74.689ff.ffhal-015036

[3] R. Vinayagamoorthy, A. Albert Irudayaraj, A. Dhayal Raj, S. Karthick, G. Jayakumar, Comparative Study of Properties of L-Histidine and L-Histidine Nickel Nitrate Hexahydrate Crystals Grown by Slow Evaporation, *Mechanics, Materials Science & Engineering Journal*, (2017) (9), 92–97. ISSN 2412-5954.

[4] Ahmed A. Soliman. Spectral and Thermal Study of the Ternary Complexes of Nickel with Sulfasalazine and Some Amino Acids (2006). *Spectrochimica Acta Part A: Molecular and Biomolecular Spectroscopy* 65, 5 1180–1185.

[5] E.J. Waheed. Synthesis and characterization of some metals complexes of {N-[(BenzoylAmino)-ThioxoMethyl] Proline} (2012). *Al-Nahrain Journal of Science*, 15, 4, 1–10.

[6] E.J. Waheed. Synthesis and characterization of complexes with some amino acid derivatives of the glycine with some metal salts. Msc. thesis. University of Baghdad, Collage of Education for Pure Sciences Ibn Al Haitham, (2008).

[7] T. Schradera, G. Bitanb, F.G. Klarnera. Molecular Tweezers for Lysine and Arginine – Powerful inhibitors of pathologic protein aggregation (2016). *Chemical Communication*, 52, 76:11318–11334.

[8] S.P.J. Melomedov, A. Wunsche, V. Leupoldt, K. Heinze. Gold(III) Tetra Aryl Porphyry in Amino Acid Derivatives: Ligand or Metal Centered Red Ox Chemistry (2016). *Chemical Science*, 7, 596–610.

[9] O. Sovagok, K. Varnagy. Copper(II) Complexes of Amino Acids and Peptides Containing Chelating bis (imidazolyl) Residues (2003). *Bioinorganic Chemical Applications*, 12, 123–139.

[10] N.S. Buttrus. Synthesis and Characterization of Some Cr^{+3}, Fe^{+3}, Co^{+2}, Ni^{+2}, Cu^{+2} and Zn^{+2} Complexes with N-Phthalyl Amino Acid Ligands (2014). *Research Journal of Chemical Sciences*, 4, 5, 41–47.

[11] Sunil S. Patil, Ganesh A. Thakur and Manzoor M. Shaikh Synthesis, Characterization and Antibacterial Studies of Mixed Ligand Dioxouranium Complexes with 8-Hydroxy Quinoline and Some Ano Acids (2011). *ISRN Pharmaceutics*, 10.

[12] M. Nageshwari, C. Rathikathaya Kumari, G. Vinitha, S. Muthu, M. Lydia Caroline. Growth and Characterization of L-Serine: A Promising a Centric Organic Crystal. (2018). *Physica B: Condensed Matter*, 541, 32–42.

[13] O.E. Olasomi, O.F. Akinyele, E.O. Akinkunmi, D.A. Isabirye, Synthesis Characterization and Antimicrobial Studies of Coordination Compounds of L-Serine and Their Mixed Ligand Complexes with Aspartic Acid (2017). *Asian Journal of Chemistry*, 29, 371–374.

[14] B. Lal, K.K. Bamzai, P.N. Kotru, Mechanical Characteristics of Melt Grown Doped $KMgF_3$ Crystals (2003). *Material Chemistry and Physics*, 78, 202–207.

[15] H.H. Willard, L.L. Merritt Jr., J.A. Dean, F.A. Settle Jr., *Instrumental Methods of Analysis*, Sixth ed., Wads worth Publishing Company, USA, (1986), 609.

[16] O.A. Awoyinka, I.O. Balogun, A.A. Ogunnow. Phytochemical Screening and *in vitro* Bioactivity of *Cnidoscolusaconitifolius* (Euphorbiaceae) (2007). *Journal of Medicinal Plant Research, 1*, 63–95.

[17] NCCLS. *National Committee for Clinical Laboratory Standards. Performance Standards for Antimicrobial Disc Susceptibility Tests*. NCCLS Publications, PA, (1993), 25.

39

Analysing Wave Propagation Characteristics in Fluid-Conveying Magneto-Thermo-Elastic Zigzag Carbon Nanotubes Subjected to Harmonic Excitation

Mahaveer Sree Jayan Madasamy and Lifeng Wang
Nanjing University of Aeronautics and Astronautics, Nanjing, People's Republic of China

Selvamani Rajendran
Karunya Institute of Technology, Coimbatore, India

Ramya Nandakumar
Saranathan College of Engineering, Trichy, India

39.1 Literature Review

Increased research focus on microscale wave propagation stems from advancements in nano- and micro-electromechanical systems, impacting diverse fields such as medicine, biology, and mechanical engineering. When studying the dynamical responses of nano-sized structures, size-dependent continuum theories are crucial, as classical continuum mechanics, which neglects interatomic forces, may fall short. The mechanical behaviour of carbon nanotubes (CNTs) is a focal point, with theoretical models often substituting for limited and costly nanoscale experimental data. Research interest extends to the structural aspects of fluid-conveying magneto-thermoelastic single-walled CNTs (SWCNTs), engaging communities in exploring the hydroelasticity of nanotubes and nano-electromechanical systems. The coupling of thermomechanical and magnetic fields in thermo-magnetoelastic (TME) nanomaterials has led to extensive studies. The increasing use of TME structures across engineering fields has spurred research into wave propagation in TME nanomaterials subjected to mechanical loads.

Since the introduction of CNTs [1], there has been a surge in interest to explore their mechanical, thermal, chemical, electrical, and electromagnetic properties. Traditional continuum theories struggle to predict these properties at the nanoscale, highlighting the significance of small-scale and nanoscale surface effects in nanotechnology. Unlike local continuum theory, nonlocal theory of elasticity posits that stress at a point depends on strains across the continuum [2, 3]. Numerous studies have explored the static, buckling, and vibration analysis of carbon nanotubes (CNTs) using both local and nonlocal beam theories within the Eringen framework. For instance, frequency equations and mode shapes for various beam configurations have been formulated using a nonlocal Euler-Bernoulli beam (NL-EBB) model. Torsional vibration analysis of double-walled carbon nanotubes (DWCNTs) using NL elasticity was explored by [4]. Investigating buckling under axial compressive load with temperature and surrounding elastic media effects, the authors of [5] employed the NL-EBB model for simply supported SWCNTs. Additionally, the authors of [6] delved into the impact of heat on the ultrasonic wave propagation properties of a nanoplate using NL continuum theory, considering the axial tension induced by the heat effect in their research. This body of work underscores the growing understanding of CNTs behaviour at the nanoscale, emphasising the application of advanced theories for accurate predictions.

The interaction of charged particles in motion with the poles of a magnet produces either an attracting or a repulsive magnetic force, which is a manifestation of the electromagnetic force. The area around a magnet or current-carrying wire is characterised by the magnetic field, a vector quantity expressed in

DOI: 10.1201/9781003495437-39

Tesla. Many aspects of wave propagation in magneto-electro-elastic (MEE) nanotubes have been studied [7], as has the impact of magnetic fields on wave propagation in CNTs in an elastic matrix [8], as well as the dynamic properties of hygro-magneto-thermo-electrical (HMTE) nanobeams with non-ideal boundary conditions [9]. Further studies include vibration responses of DWCNTs under longitudinal magnetic fields using NL [10], bending studies on magneto-electro-piezoelectric nanobeam systems [11], and magnetic field effects on thermally affected acoustical wave propagation in rotary double-nanobeam systems [12]. Both the effect of thermal loading on the free vibration characteristics of CNTs with numerous cracks and nonlinear vibration studies of electro-hygro-thermally actuated embedded nanobeams with different border conditions have been examined [13, 14]. Moreover, for nonlinear vibration analysis of HMTE-piezoelectric nanocomposites resting on an elastic foundation, a non-classical beam model (CBM) based on NL theory has been investigated [15], incorporating Hamilton's principle within the EB model framework with von-Karman-type nonlinearity. With potential implications in a variety of industrial fields, these studies further our knowledge of the complex interplay between magnetic and thermal influences on the mechanical behaviour of nanomaterials.

CNTs, with superior mechanical, chemical, and thermal characteristics and a hollow structure, show promise in advancing nano-biological devices and nanomechanical systems for drug delivery and fluid transport. The impacts of internal moving fluid on the stability and free vibration of CNTs have been studied recently, and the results show that these effects are considerable on resonance frequencies, particularly at larger radii and greater flow velocities. The surrounding elastic medium significantly reduces this impact [16, 17]. Using a CCM, another research examined the impact of temperature on the transverse vibrations of DWCNTs [18]. Examining the thermal–mechanical vibrations and instability of fluid-transporting SWCNTs, a model based on NEL was employed for the investigation [19]. The authors of [20] determined that the vibration frequency and buckling instability of fluid-conveying SWCNTs are notably influenced by temperature, NL value, and elastic medium constant. They assessed the reliability of Euler–Bernoulli and Timoshenko beam models, coupled with stress or strain gradients, for wave propagation analysis in fluid-conveying SWCNTs in a comparative study. They discovered that a more appropriate Timoshenko beam theory for examining the dynamic behaviours of these tubes was the combined strain/inertia gradient theory. The authors of [21] suggested a technique to examine how a longitudinal magnetic field affects the transverse vibration of fluid-conveying, magnetically sensitive SWCNTs. In order to address real-world structural dynamics issues, the authors of [22] used NL beam theories to evaluate tube structures under a moving nanoparticle. The authors of [23–25] used NEL theory to study the vibration analysis of SWCNTs under a moving harmonic load. These investigations provide insightful information on the intricate dynamics of fluid-conveying nanotubes, which directs the development of new applications for nanotechnology.

This work explores the behaviour of NL elastic waves under a moving harmonic load in a fluid-conveying magneto-thermoelastic SWCNT. The study integrates NL components, thermal and Lorenz magnetic forces, and uses Eringen's NLE theory inside the framework of EBM theory to develop governing equations. Dispersion curves, which display the computed dynamic displacement, offer important new perspectives on the intricate behaviour of nanotubes in the presence of fluid, magneto-thermal, and NL factors.

39.2 Modelling of the Problem

39.2.1 Zigzag SWCNT Atomic Structure

The chiral vector $\vec{C_h}$ is shown in Figure 39.1(a). The chiral vector is represented using the basis vectors $\vec{a_1}$ and $\vec{a_2}$:

$$\vec{C_h} = m\vec{a_1} + n\vec{a_2} \tag{39.1}$$

In this scenario, the structure around the perimeter is determined by the translational indices (m, n).

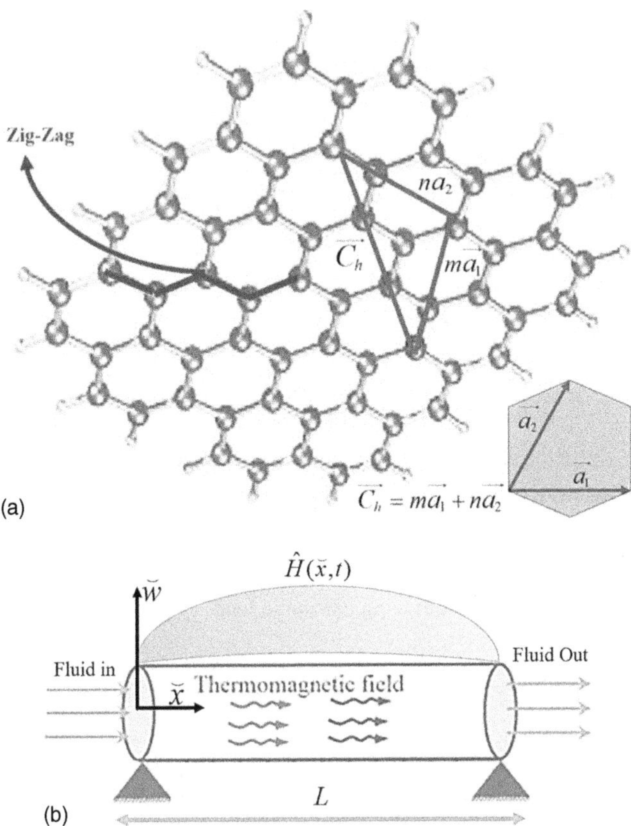

FIGURE 39.1 (a) Graphene sheet hexagonal lattice with base vectors; (b) Geometry of the problem.

The diameter of a SWCNT zigzag configuration where the values of n and m are equal.

$$d = \frac{3ns}{\pi} \tag{39.2}$$

Here s represents the scale of the C–C bands, given as $0.142 \times 10^9 m$. Using a model that equates molecular mechanics to solid mechanics, the authors of [26–28] introduced a method based on energy equivalence to investigate the mechanical characteristics of SWCNTs:

$$P_a = \frac{4\sqrt{3}}{3} \frac{\Re_v \Re_w}{3\Re_w t + 4\Re_v s^2 t \left(\gamma^2_{a1} + \gamma^2_{a2} \right)}, \tag{39.3}$$

$$\text{Radius} = \frac{\left| \vec{C_h} \right|}{2\pi} = b\sqrt{3(m^2 + mn + n^2} / 2\pi \tag{39.4}$$

R_v and R_w denote the force constants and signifies the nanotube thickness, while (γ_{a1}) and (γ_{a2}) are specified parameters:

$$\begin{cases} \gamma_{a1} = \dfrac{4 - \cos^2\left(\pi / 2n\right)}{16 + 2\cos^2\left(\pi / 2n\right)} \\[4mm] \gamma_{a2} = \dfrac{\sqrt{12 - 3\cos^2\left(\pi / 2n\right)}\,\cos\left(\pi / 2n\right)}{32 + 4\cos^2\left(\pi / 2n\right)} \end{cases} \tag{39.5}$$

Letting $n \rightarrow \infty$, Young's modulus (E) expressions for a graphite sheet are given as:

$$E_g = \frac{8\sqrt{3}\mathfrak{R}_v \mathfrak{R}_w}{18\mathfrak{R}_w t + \mathfrak{R}_v s^2 t} \tag{39.6}$$

39.2.1.1 Nonlocal Beam Model

Figure 39.1(b) shows fluid in a SWCNT under magneto-thermo-elastic conditions and harmonic stress. The internal fluid is assumed to be incompressible and stable at zero gravity. The governing equation is a partial differential equation (PDE) for a moving harmonic load.

$$\frac{\partial \Pi}{\partial \breve{x}} + \bar{N}_T \frac{\partial^2 \breve{w}}{\partial \breve{x}^2} + \bar{Q}_{\breve{x}} \frac{\partial^2 \breve{w}}{\partial \breve{x}^2} - m_c \frac{\partial^2 \breve{w}}{\partial \breve{x}^2} + \bar{F}_f = \hat{H}\left(\breve{x},t\right) + \rho A_{\breve{x}\breve{x}} \frac{\partial^2 \breve{w}}{\partial t^2} \tag{39.7}$$

In a beam under a distributed transverse load $\hat{H}\left(\breve{x},t\right)$ along the $\breve{x}$-axis, with ρ being the mass density, $\breve{w}$ representing the transverse bending, and t denoting time, the equilibrium equation defines the shear force Π on the nanotube's cross-section:

$$\Pi = \frac{\partial \bar{M}}{\partial \breve{x}} \tag{39.8}$$

The thermoelastic force can be expressed as

$$\bar{N}_T = -EA_{\breve{x}\breve{x}}\alpha_{\breve{x}}T \tag{39.9}$$

where $\bar{N}_T$ is denoted as thermal term and T is the temperature, while the thermal expansion coefficient in the $\breve{x}$-direction is indicated by $\alpha_{\breve{x}}$.

The fluid force per unit length, while taking into account the plug flow of fluid, is calculated as follows:

$$\bar{F}_f = m_f \left(2v \frac{\partial^2 \breve{w}}{\partial t \partial \breve{x}} + v^2 \frac{\partial^2 \breve{w}}{\partial \breve{x}^2} + \frac{\partial^2 \breve{w}}{\partial t^2} \right) \tag{39.10}$$

Lorentz body force is applied to the CNT:

$$\bar{Q} = \eta A_{\breve{x}\breve{x}} H_{\breve{x}}^2 \frac{\partial^2 \breve{w}}{\partial \breve{x}^2} \tag{39.11}$$

In this case, H_x stands for the magnetic field strength, and η for the magnetic field permeability. Eq. (39.8) gives the following expression for the bending moment $\bar{M}$:

$$\bar{M} = \int_{A_{\breve{x}\breve{x}}} \breve{w}\Theta_{\breve{x}\breve{x}}dA_{\breve{x}\breve{x}}, \tag{39.12}$$

One-dimensional (1D) tube is considered from [2, 29]:

$$\Theta_{\bar{x}\bar{x}} - \left(e_0 a\right)^2 \frac{\partial^2 \Theta_{\bar{x}\bar{x}}}{\partial \bar{x}^2} = E\varepsilon_{\bar{x}\bar{x}} \tag{39.13}$$

The variables involved encompass axial strain $\varepsilon_{\bar{x}\bar{x}}$. In the presence of a temperature environment, the NL relations outlined in Eq. (39.13) can be articulated as follows:

$$\Theta_{\bar{x}\bar{x}} - \left(e_0 a\right)^2 \frac{\partial^2 \Theta_{\bar{x}\bar{x}}}{\partial \bar{x}^2} = E\varepsilon_{\bar{x}\bar{x}} - E\alpha T \tag{39.14}$$

$\varepsilon_{\bar{x}\bar{x}}$ deflection is characterised by the following definition:

$$\varepsilon_{\bar{x}\bar{x}} = -\breve{w}\frac{\partial^2 \breve{w}}{\partial \bar{x}^2} \tag{39.15}$$

Eq. (39.12) may be used to represent the bending moment $\bar{M}$ by combining Eqs. (39.14)– (39.15).

$$\bar{M} - \left(e_0 a\right)^2 \left[\frac{\partial^2 \bar{M}}{\partial \bar{x}^2}\right] = EI\frac{\partial^2 \breve{w}}{\partial \bar{x}^2} \tag{39.16}$$

Expressing the moment of inertia as $I = \int_A \breve{w}^2 dA$, the formulation can be redefined by substituting Eqs. (39.7) through (39.9) into Eq. (39.16). The rewritten expression is as follows:

$$\bar{M} - \left(e_0 a\right)^2 \left[\left(m_c + \rho A_{\bar{x}\bar{x}}\right)\frac{\partial^2 \breve{w}}{\partial t^2} + \bar{F}_f - \bar{Q} + \bar{N}_T + \hat{H}(t)\right] = EI\frac{\partial^2 \breve{w}}{\partial \bar{x}^2} \tag{39.17}$$

$$\Pi - \left(e_0 a\right)^2 \left[\left(m_c + \rho A_{\bar{x}\bar{x}}\right)\frac{\partial^3 \breve{w}}{\partial \bar{x}^2 \partial t^2} + \frac{\partial \bar{F}_f}{\partial \bar{x}} - \frac{\partial^2 \bar{Q}}{\partial \bar{x}^2} + EA_{\bar{x}\bar{x}}\alpha T\right] = EI\frac{\partial^3 \breve{w}}{\partial \bar{x}^3} \tag{39.18}$$

In order to account for dispersed pressure, thermal contact, and harmonic excitation, the equation of motion (Eq. 39.7) for transverse displacement can be written as follows:

$$\hat{H}\left(\bar{x},t\right) = EI\frac{\partial^4 \breve{w}}{\partial \bar{x}^4} + EA_{\bar{x}\bar{x}}\alpha T\frac{\partial^2 \breve{w}}{\partial \bar{x}^2} - \bar{Q}\frac{\partial^2 \breve{w}}{\partial \bar{x}^2} + m_f v^2 \frac{\partial^2 \breve{w}}{\partial \bar{x}^2} + 2m_f v\frac{\partial^2 \breve{w}}{\partial t \partial \bar{x}} +$$
$$\left(\rho A_{\bar{x}\bar{x}} + m_c + m_f\right)\frac{\partial^2 \breve{w}}{\partial t^2} - \left(e_0 a\right)^2 \left(\begin{array}{c} \bar{N}_T \dfrac{\partial^4 \breve{w}}{\partial \bar{x}^4} + m_c \dfrac{\partial^2 \breve{w}}{\partial \bar{x}^2} + m_f v^2 \dfrac{\partial^4 \breve{w}}{\partial \bar{x}^4} \\[2ex] + 2m_f v\dfrac{\partial^4 \breve{w}}{\partial t^2 \partial \bar{x}^2} + m_f \dfrac{\partial^4 \breve{w}}{\partial t^4} - \bar{Q}\dfrac{\partial^4 \breve{w}}{\partial \bar{x}^4} \end{array}\right) \tag{39.19}$$

A moving harmonic load $\hat{h}(t)$ is applied to the SWCNT, translating along the axial direction of the nanotube at a constant velocity v_p:

$$\hat{H}\left(\bar{x},t\right) = \hat{h}(t)\delta\left(\bar{x} - \bar{x}_p\right) \tag{39.20}$$

The expressions $\hat{h}(t) = \hat{h}_o \sin\left(\Omega^h t\right)$ and $\delta\left(\bar{x} - \bar{x}_p\right)$ denote the Dirac delta functions, where $\bar{x}_p$ represents the coordinate of the moving harmonic load. The amplitude of the harmonic load (denoted as $\hat{h}_o$) and the excitation frequency of the moving harmonic load (denoted as Ω) are the key parameters in the system. When Eq. (39.20) is inserted into Eq. (39.19), it results in the particle's differential equation in the NL form, distinguished by constant coefficients:

$$
EI\breve{w}''' - \bar{N}_T\breve{w}'' + m_f v^2 \breve{w}'' + \bar{Q}\breve{w}'' + 2m_f v \frac{\partial^2 \breve{w}}{\partial t \partial \breve{x}}
$$

$$
-\left(e_0 a\right)^2 \left(\begin{array}{l} \left(m_c + m_f + \rho A\right)\dfrac{\partial^4 \breve{w}}{\partial \breve{x}^2 \partial t^2} + \bar{N}_T\breve{w}''' \\[2ex] +\rho A\breve{w}'' + m_f v^2 \breve{w}'''' + 2m_f v\dfrac{\partial^4 \breve{w}}{\partial t^2 \partial \breve{x}^2} \\[2ex] +m_f \dfrac{\partial^4 \breve{w}}{\partial t^4} + \bar{Q}\breve{w}'''' \end{array}\right) = P_0 \sin\Omega t\, \delta\left(\breve{x} - \breve{x}_p\right) - \left(e_0 a\right)\frac{\partial^2 P_0 \sin\Omega t}{\partial \breve{x}^2}
\qquad (39.21)
$$

The subsequent depiction showcases the dynamic displacement of the SWCNT in a direction perpendicular to its axis:

$$
\breve{w}\left(\breve{x},t\right) = \sum_{i=1}^{\infty} \lambda_i^y\left(\breve{x}\right) q_i\left(t\right)
\qquad (39.22)
$$

To determine the unknown time-dependent generalised coordinates (designated as $q_i(t)$) and the Eigenmodes of an undamped, S–S beam (depicted as $\lambda_i^y(\breve{x})$), the selection process can be outlined as follows:

$$
\lambda_i^y\left(\breve{x}\right) = \sin\left(\frac{i\pi\breve{x}}{l}\right), \qquad i = 1, 2, 3, 4 \ldots\ldots\ldots
\qquad (39.23)
$$

Utilising Eq. (39.21) in conjunction with Eqs. (39.22) through (39.23) yields the following results:

$$
EI\left(\sum_{i=1}^{\infty}\sin\left(\frac{i\pi\breve{x}}{l}\right)\left(\frac{\pi}{l}\right)^4 \hat{q}_i(t)\right) - \bar{N}_T\left(\sum_{i=1}^{\infty}\sin\left(\frac{i\pi\breve{x}}{l}\right)\left(\frac{\pi}{l}\right)^2 \hat{q}_i(t)\right) + m_f v^2\left(\sum_{i=1}^{\infty}\sin\left(\frac{i\pi\breve{x}}{l}\right)\left(\frac{\pi}{l}\right)^2 \hat{q}_i(t)\right)
$$

$$
+\bar{Q}\left(\sum_{i=1}^{\infty}\sin\left(\frac{i\pi\breve{x}}{l}\right)\left(\frac{\pi}{l}\right)^2 \hat{q}_i(t)\right) + 2m_f v\left(\sum_{i=1}^{\infty}\sin\left(\frac{i\pi\breve{x}}{l}\right)\left(\frac{\pi}{l}\right)\dot{\hat{q}}_i(t)\right)
$$

$$
-\left(e_0 a\right)^2 \left(\begin{array}{l} \left(m_c + m_f + \rho A\right)\left(\displaystyle\sum_{i=1}^{\infty}\sin\left(\frac{i\pi\breve{x}}{l}\right)\left(\frac{\pi}{l}\right)^2 \ddot{\hat{q}}_i(t)\right) + \\[3ex] \bar{N}_T\left(\displaystyle\sum_{i=1}^{\infty}\sin\left(\frac{i\pi\breve{x}}{l}\right)\left(\frac{\pi}{l}\right)^4 \hat{q}_i(t)\right) + \rho A\left(\displaystyle\sum_{i=1}^{\infty}\sin\left(\frac{i\pi\breve{x}}{l}\right)\left(\frac{\pi}{l}\right)^2 \hat{q}_i(t)\right) + \\[3ex] m_f v^2\left(\displaystyle\sum_{i=1}^{\infty}\sin\left(\frac{i\pi\breve{x}}{l}\right)\left(\frac{\pi}{l}\right)^4 \hat{q}_i(t)\right) + 2m_f v\left(\displaystyle\sum_{i=1}^{\infty}\sin\left(\frac{i\pi\breve{x}}{l}\right)\left(\frac{\pi}{l}\right)^2 \ddot{\hat{q}}_i(t)\right) + \\[3ex] m_f\left(\displaystyle\sum_{i=1}^{\infty}\sin\left(\frac{i\pi\breve{x}}{l}\right)\ddot{\hat{q}}_i(t)\right) - \bar{Q}\left(\displaystyle\sum_{i=1}^{\infty}\sin\left(\frac{i\pi\breve{x}}{l}\right)\left(\frac{\pi}{l}\right)^4 \hat{q}_i(t)\right) \end{array}\right)
$$

$$
= \hat{h}_o \sin\Omega^h t\, \delta(\breve{x} - \breve{x}_p) - (e_o a)\frac{\partial^2 \hat{h}_o \sin\Omega^h t}{\partial \breve{x}^2}
$$

$$
\qquad (39.24)
$$

Indicating the derivative with respect to $\breve{x}$ by a prime, the integration of $\lambda_i^v(\breve{x})$ on both sides allows the above equation to be expressed as

$$\int_0^l \hat{H}(\breve{x},t) - (e_0 a)^2 \frac{\partial^2 \hat{H}(\breve{x},t)}{\partial \breve{x}^2} - Kw\phi_i(\breve{x})d\breve{x} = \sum_{i=1}^\infty \breve{q}_i(t)\int_0^l EI\phi_j(\breve{x})d\breve{x} - \sum_{i=1}^\infty \breve{q}i(t)\int_0^l \bar{N}_T\phi_j(\breve{x})d\breve{x}$$

$$+ \sum_{i=1}^\infty \breve{q}i(t)\int_0^l M_f V^2 \phi_j(\breve{x})d\breve{x} + \sum_{i=1}^\infty qi(t)\int_0^l \bar{Q}\phi_i\phi_j(\breve{x})d\breve{x} + \sum_{i=1}^\infty \breve{q}i(t)\int_0^l 2m_f v\phi_j(\breve{x})d\breve{x}$$

$$- (e_0 a)^2 \begin{bmatrix} \sum_{i=1}^\infty \ddot{\breve{q}}_i(t)\int_0^l (mc+mf+\rho A)\phi_i\phi_j(x)dx + \sum_{i=1}^\infty \breve{q}_i(t)\int_0^l \bar{N}_T\ddot{\phi}_j(\breve{x})d\breve{x} \\[2mm] + \sum_{i=1}^\infty \breve{q}_i(t)\int_0^l \bar{Q}\phi_i\phi_j(\breve{x})d\breve{x} + \sum_{i=1}^\infty \breve{q}_i(t)\int_0^l \rho A\phi_i^*\phi_j(\breve{x})d\breve{x} \\[2mm] + \sum_{i=1}^\infty \breve{q}_i(t)\int_0^l M_f V^2\phi_i\phi_j(\breve{x})d\breve{x} + \sum_{i=1}^\infty \dot{\breve{q}}_i(t)\int_0^l 2m_f v\phi_j(\breve{x})d\breve{x} \\[2mm] + \sum_{i=1}^\infty \dot{q}_i(t)\int_0^l m_f\phi_i\phi_j(\breve{x})d\breve{x} - \sum_{i=1}^\infty \breve{q}_i(t)\int_0^l Kw\phi_i\phi_j(\breve{x})d\breve{x} \end{bmatrix} \tag{39.25}$$

Considering the orthogonal criteria presented in the above equation:

$$\int_0^l \lambda_i^v \lambda_j^v(\breve{x})d\breve{x} = \begin{cases} \dfrac{l}{2} & i=j \\[2mm] 0 & i \neq j \end{cases} \tag{39.26}$$

This is followed by applying the load terms on the right side of the pertinent attribute of the generic Dirac delta function:

$$\int_{x_1}^{x_2} \Gamma(\breve{x})\delta^{(x)}(\breve{x}-\breve{x}_0)d\breve{x} = \begin{cases} (-1)^n g^{(n)}(x_0) & \text{if } \breve{x}_1 < \breve{x}_0 < \breve{x}_2 = j \\[2mm] 0 & \text{otherwise} \end{cases} \tag{39.27}$$

Representing the nth derivative of the Dirac delta function as $\delta^n(\breve{x}-\breve{x}_p)$, the following relations are also taken into consideration:

$$\tilde{k}_j = \int_0^l EI\left(\frac{i\pi}{l}\right)^4 \lambda_i^v \lambda_j^v(\breve{x})d\breve{x} \tag{39.28a}$$

$$\tilde{\alpha}_j = \int_0^l \left(\bar{Q}+mfv^2+\bar{N}_T\right)\left(\frac{i\pi}{l}\right)^2 \lambda_i^v \lambda_j^v(\breve{x})d\breve{x} \tag{39.28b}$$

$$\tilde{M}_j = \int_0^l \rho A_{\breve{x}\breve{x}}\left((e_0 a)^2 - 1 + m_f + m_c\right)\left(\frac{i\pi}{l}\right)^2 \lambda_i^v \lambda_j^v(\breve{x})d\breve{x} \tag{39.28c}$$

$$\tilde{\beta}_j = \int_0^l (e_0 a)^2\left(\bar{N}_T+\bar{Q}+m_f v^2\right)\left(\frac{i\pi}{l}\right)^4 \lambda_i^v \lambda_j^v(\breve{x})d\breve{x} - (e_0 a)\int_0^l (\rho A_{\breve{x}\breve{x}})\left(\frac{i\pi}{l}\right)^2 \lambda_i^v \lambda_j^v(\breve{x})d\breve{x} \tag{39.28d}$$

$$\tilde{D}_j = \left(e_o a\right)^2 \int_0^l \left(\bar{N}_T + 2m_f v + m_f\right)\left(\frac{i\pi}{l}\right)^4 \lambda_i^v \lambda_j^v \left(\check{x}\right) d\check{x} - \left(e_o a\right) \int_0^l \left(\rho A_{\check{x}\check{x}}\right)\left(\frac{i\pi}{l}\right)^2 \lambda_i^v \lambda_j^v \left(\check{x}\right) d\check{x} \quad (39.28\text{e})$$

$$\tilde{f}_i(t) = \hat{H}\left(\check{x},t\right)\int_0^l \left[\delta\left(\check{x}-\check{x}_p\right) - \left(e_o a\right)^2 \frac{\partial^2 \left(\check{x}-\check{x}_p\right)}{\partial \check{x}^2} - Kw\right]\varphi_i d\left(\check{x}\right) \quad (39.28\text{f})$$

The *i*th mode of the generalised deflection is governed by the resulting differential equation:

$$\ddddot{q}_j(t) + \left[\tilde{\omega}_j^2 + \tilde{\zeta}_j^2 + \tilde{\psi}_j^2\right]\ddot{q}_j(t) + \tilde{Z}_j^2 \hat{q}_j(t) = \frac{1}{\tilde{M}_j}\tilde{f}_i(t) \, i, j = 1, 2, 3, 4\ldots\ldots \quad (39.29)$$

in which

$$\tilde{\zeta}_i = \sqrt{\frac{\tilde{D}_i}{\tilde{M}_j}}, \, \tilde{\omega}_j = \sqrt{\frac{\check{k}_i}{\tilde{M}_j}}, \, \tilde{\psi}_j = \sqrt{\frac{\check{\alpha}_i}{\tilde{M}_j}}, \, \tilde{Z}_i = \sqrt{\frac{\tilde{\beta}_i}{\tilde{M}_j}} \text{ and } \tilde{f}_i(t) = 0 \, for \, t > 0, \text{ where } \check{x}_p\left(0 \le \check{x}_P = v_p\right)$$

Consider the direction of the subsequent moving harmonic load. Utilise the following formula for further analysis:

$$\frac{1}{\tilde{M}_j}\tilde{f}_i(t) = \tilde{S}_i(t) \quad (39.30)$$

Additionally, the following equation is present:

$$\tilde{S}_i(t) = \frac{2p(t)}{\rho A_{\check{x}\check{x}}L}\sin\left(\frac{i\pi\check{x}_p t}{L}\right) \quad (39.31)$$

For the homogeneous initial conditions, Eq. (28) may be solved as follows

$$\hat{q}_i(t) = \frac{1}{\tilde{M}_j}\int_0^t \tilde{S}_i(\lambda)\sin\tilde{M}_j\left(t-\lambda\right)d\lambda \quad (39.32)$$

Through the utilisation of integration and the conversion of Eq. (39.32) into Eq. (39.31), the subsequent results are derived:

$$\hat{q}_j(t) = \frac{\left(\left(\rho A_{\check{x}\check{x}} + m_f + m_c\right) - \rho A_{\check{x}\check{x}}\left(e_0 a\right)^2\right)\hat{h}_0}{\rho A_{\check{x}\check{x}}L}\sum_{i=1}^{\infty}\left[\frac{\cos\left(\Omega^h t - \dfrac{i\pi\check{x}_p(t)}{L}\right) - \cos\left(\tilde{M}_j t\right)}{\tilde{M}_j^2 - \left(\Omega^h t - \dfrac{i\pi x_p(t)}{L}\right)^2}\right.$$

$$\left. - \frac{\cos\left(\Omega^h t + \dfrac{i\pi\check{x}_p(t)}{L}\right) - \cos\left(\tilde{M}_j t\right)}{\tilde{M}_j^2 - \left(\Omega^h t + \dfrac{i\pi\check{x}_p(t)}{L}\right)^2}\right] \quad (39.33)$$

By incorporating the expression from Eq. (39.32) into Eq. (39.22), the total dynamic deflection can be computed in the following manner:

$$\breve{w}(\breve{x},t) = \frac{\left(\left(\rho A_{\breve{x}\breve{x}} + m_f + m_c\right) - \rho A_{\breve{x}\breve{x}}\left(e_0 a\right)^2\right)\hat{h}_o}{\rho A_{\breve{x}\breve{x}} L} \sum_{i=1}^{\infty}\left[\breve{\Theta}(\breve{x},t)\right] \times \sin\left(\frac{i\pi\breve{x}}{L}\right) \tag{39.34}$$

here

$$\hat{\Theta}(x,t) = \frac{\cos\left(\Omega^h t - \dfrac{i\pi\breve{x}_p(t)}{L}\right) - \cos\left(\tilde{M}t\right)}{\tilde{M}^2 - \left(\Omega^h t - \dfrac{i\pi\breve{x}_p(t)}{L}\right)^2} - \frac{\cos\left(\Omega^h t + \dfrac{i\pi\breve{x}_p(t)}{L}\right) - \cos\left(\tilde{M}t\right)}{\tilde{M}^2 - \left(\Omega^h t + \dfrac{i\pi x_p(t)}{L}\right)^2}$$

39.2.2 Numerical Results and Discussion

Here, we investigate forced NL wave propagation in the presence of a moving harmonic load in a SWCNT transporting magneto-thermo fluid. Wang and Liew have offered insights and solutions to this problem, addressing structural constraints in estimating effective wall thickness and elastic modulus for the elastic waves of SWCNT [28]. The numerical values of the involved parameters in the problem have been assumed to be $E = 1$ TP_a, $\rho = 2300\ kgm^{-3}$, $t_p = 0.35\ nm$. According to the calculation, the mass of fluid per unit length in the SWCNT is $1.52 \times 10^{-16}\ kgm^{-2}$ and the mass per unit length of the SWCNT is $2.75 \times 10^{-15}\ kgm^{-2}$. The thermal expansion coefficient in room temperature $\alpha^0 = -1.5 \times 10^{-6}C^{-1}$ and $l = 10$ nm, $d = 1\ nm$. The moving harmonic load's non-dimensional velocity parameter is expressed as follows:

$$\xi = \frac{v_p}{v_{cr}}$$

The critical velocity of the SWCNT is defined as:

$$v_{cr} = \frac{\Psi_1}{\pi}$$

In this context, ψ_1 represents the primary resonance frequency of the SWCNT. The effect of the excitation frequency of the mobile harmonic load Ω is delineated by the frequency ratio θ formulated as follows:

$$\theta = \frac{\Omega}{\Psi_1}$$

Denoting dimensionless time as t^*, its equations can be expressed in the following manner:

$$t^* = \frac{\breve{x}_p}{L}$$

Figure 39.2(a) depicts the vibrational displacement variation of the elastic SWCNT with harmonic load velocity, considering $\bar{N}_T = 0.2, 0.5$, as well as $H_{\breve{x}} = 0$, with various NL constants $e_0 a = 0, 0.5, 1.0, 1.5$. These figures reveal that the dynamic deflection of the SWCNT is significantly influenced by the

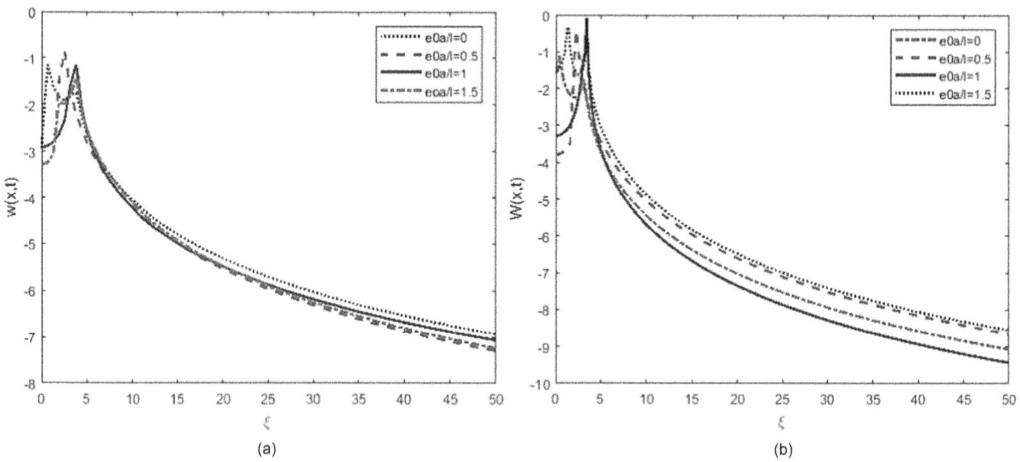

FIGURE 39.2 Graph depicting the relationship between dynamic displacement and harmonic load velocity variation. (a) $\bar{N}_T = 0.2$, $H_{\bar{x}} = 0$. (b) $\bar{N}_T = 0.5$, $H_{\bar{x}} = 0$.

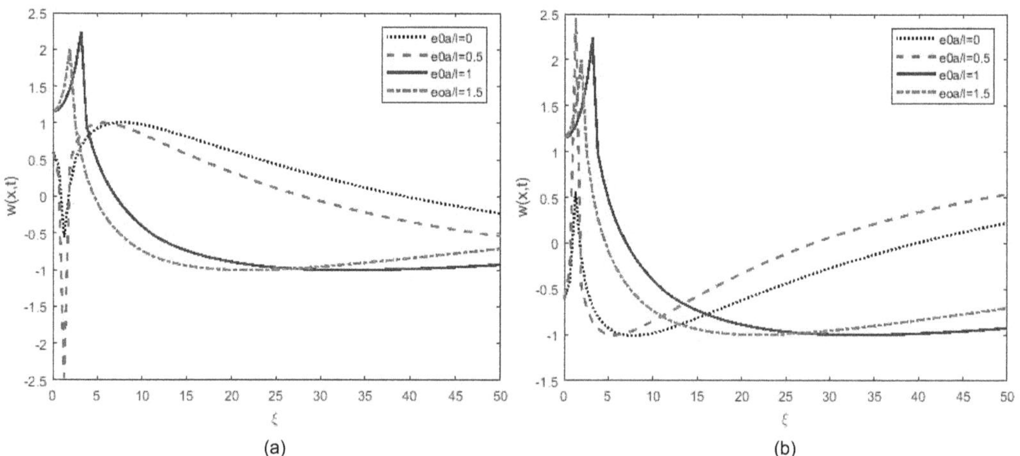

FIGURE 39.3 Comparative analysis of dynamic displacement variations with harmonic load velocities. (a) $\bar{N}_T = 0.2$, $H_{\bar{x}} = 0.5$. (b) $\bar{N}_T = 0.5$, $H_{\bar{x}} = 0.5$.

values of NL, exhibiting higher magnitudes at lower values of the harmonic load velocity. Figure 39.2(b) reveals a dispersion trend in wave propagation attributed to increasing values of thermal parameters. A comparative analysis is presented for the dynamic displacement of the elastic SWCNT under harmonic load velocity, considering $\bar{N}_T = 0.2, 0.5$, and $H_{\bar{x}} = 0.5$, with various NL constants $e_o a = 0$, 0.5, 1.0, 1.5. The results are illustrated in Figure 39.3. Observations from Figures 39.4 and 39.5 highlight that, at the lower range of harmonic velocity, the dynamic displacement of the SWCNT reaches its high range value for both cases of $\bar{N}_T = 0.2$ and $\bar{N}_T = 0.5$, but a deviation in elastic wave behaviour is noticeable when $H_{\bar{x}} = 0.5$ in Figure 39.3. This deviation may be attributed to the influence of the longitudinal magnetic field of the SWCNT.

With respect to the frequency ratio θ, Figure 39.4 investigates the dynamic displacement variation of the elastic SWCNT under various values of $e_0 a = 0.2$, 0.5 and $H_x = 0$, taking into account various values of the temperature parameter $\bar{N}_T$, such as 0.5, 1.0, 1.5 and 2.0. These data demonstrate a significant fluctuation caused by the fluctuating values of the thermal parameters in the SWCNT's dynamic deflection. Furthermore, until θ approaches 35, the dynamic displacement shows a notable increase with the growth

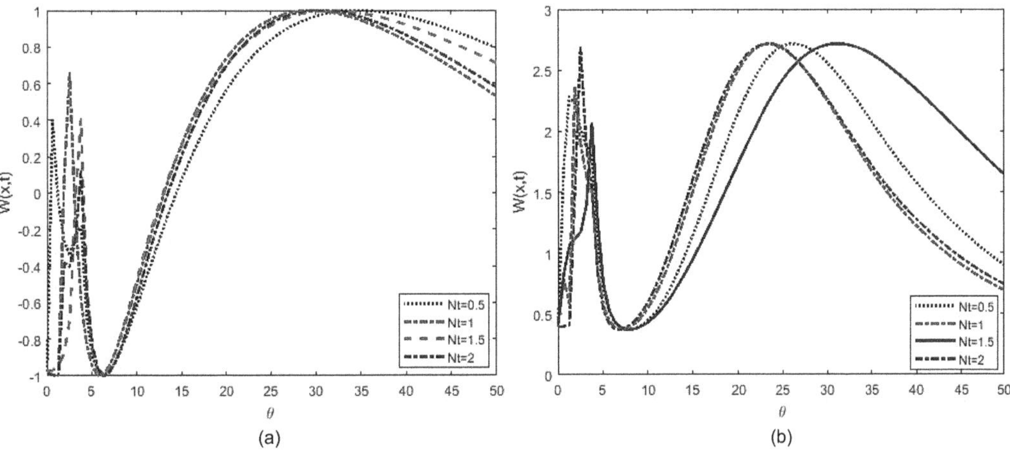

FIGURE 39.4 Variation dynamic displacement versus frequency ratio $H_{\bar{x}} = 0$. (a) $e_0 a = 0.2$. (b) $e_0 a = 0.5$.

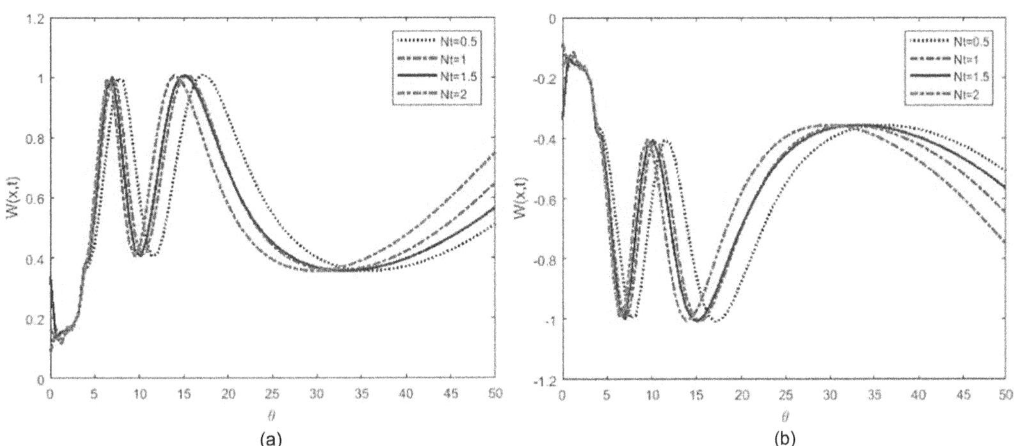

FIGURE 39.5 Relationship between dynamic displacement variation and frequency ratio at $H_{\bar{x}} = 0.5$. (a) $e_0 a = 0.2$. (b) $e_0 a = 0.5$.

in frequency ratio; beyond that, it begins to decline. The pattern of wave propagation in the trend line is illustrated in Figure 39.7, which also highlights the effect of zero magnetic effect and growing nonlocal parameter values.

In Figure 39.5, a comparative analysis of elastic SWCNT deformation with respect to frequency ratio is presented. The chosen parameters are $e_0 a = 0.2$, 0.5, $H_{\bar{x}} = 0.5$, with varying thermal parameter values $\bar{N}_T = 0.5, 1.0, 1.5, 2.0$. As depicted in Figure 39.5(a), the dynamic displacement achieves peak values within the frequency ratio range of $5 \leq \theta \leq 15$. Conversely, in Figure 39.5(b), peak values occur at $\theta = 35$ due to an escalation in the NL effect and the magnetic field value in the context of fluid-conveying SWCNT.

In Figure 39.6, dispersion curves are presented, illustrating the relationship between dynamic displacement and time ratio for a fluid-conveying elastic SWCNT. The specified parameters include $e_0 a = 0.2$, 0.5, $\bar{N}_T = 0$, with different constants $H_{\bar{x}} = 0.5, 1.0, 1.5, 2.0$. The figure reveals that dynamic displacement propagates from negative to maximum positive values as the time ratio increases. Notably, the crossover trend in dispersion curves indicates an exchange of energy among vibrational modes, particularly in scenarios where the thermal parameter value is zero and NL values are on the rise.

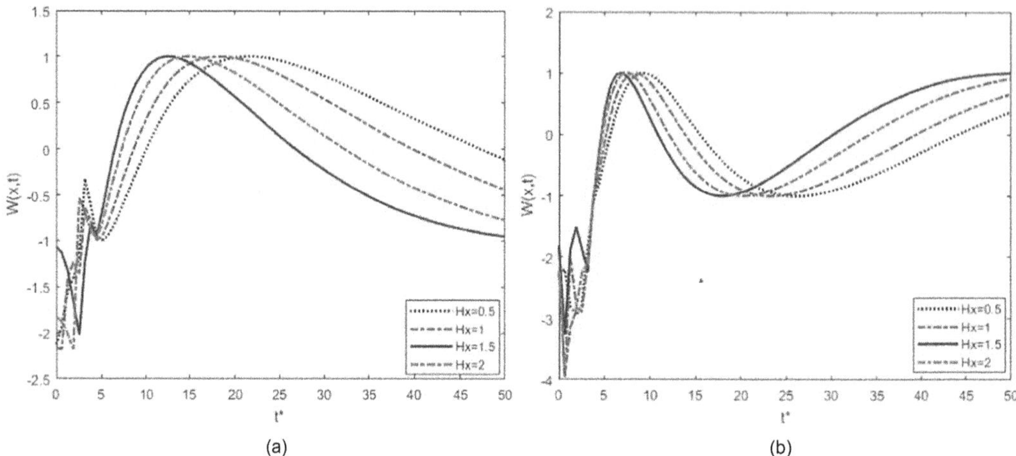

FIGURE 39.6 Relationship between dynamic displacement variation and frequency ratio at $\bar{N}_T = 0$. (a) $e_0 a = 0.2$. (b) $e_0 a = 0.5$.

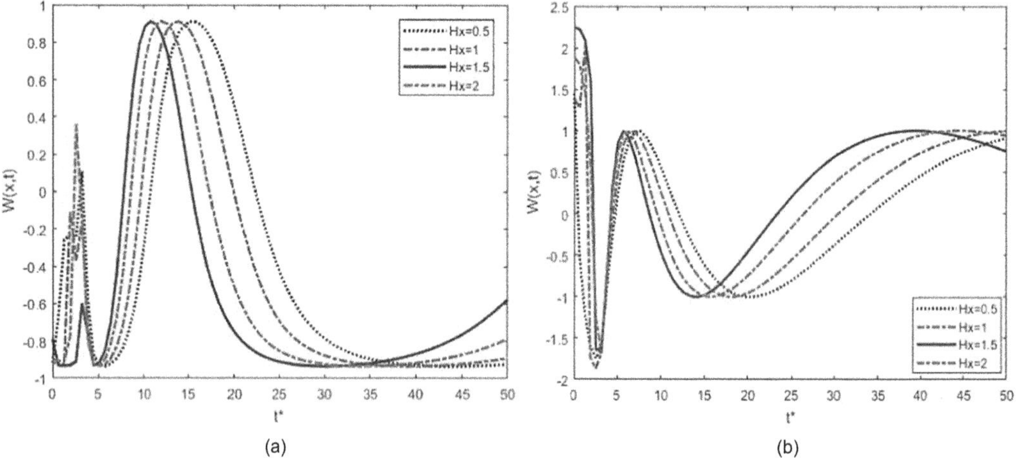

FIGURE 39.7 Relationship between dynamic displacement variation and frequency ratio at (a) $e_0 a = 0.2$, $\bar{N}_T = 0$. (b) $e_0 a = 0.5$, $\bar{N}_T = 0.5$.

Figure 39.8 elucidates the dynamic displacement variation over the time ratio for a fluid-conveying elastic SWCNT with specified parameters: $e_0 a = 0.2$, 0.5, $\bar{N}_t = 0.5$, and magnetic field values $H_{\bar{x}} = 0.5, 1.0, 1.5, 2.0$. The figure suggests that, at lower time ratios, the impact of the magnetic field becomes apparent, particularly with increasing values of the NL parameter and thermal parameter. The observed crossover trend in dispersion curves indicates an energy exchange among vibrational modes, particularly when there is an augmentation in the thermal and NL parameter values.

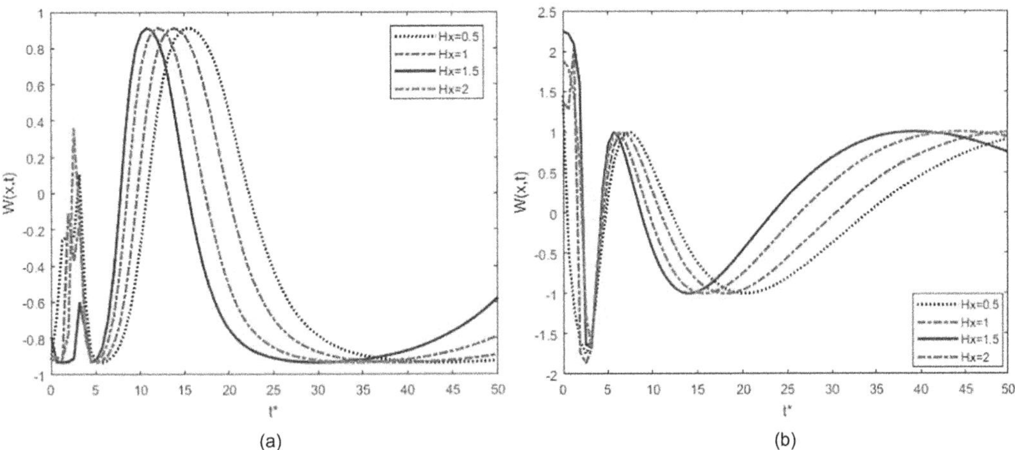

FIGURE 39.8 Deformation variation in relation to frequency ratio with $\bar{N}_T = 0.5$. (a) $e_0 a = 0.2$. (b) $e_0 a = 0.5$.

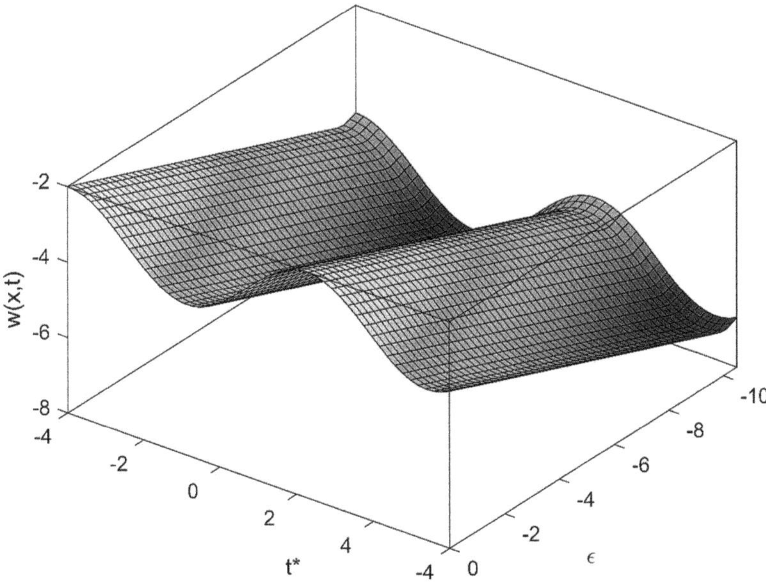

FIGURE 39.9 Spatial distribution of deformation in 3D with t^* and ξ for $e_0 a = 0.5$, $\bar{N}_T = 0$ and $\bar{F}_p = 0$.

Figures 39.9–39.12 showcase 3D curves that illustrate the dynamic displacement's changes over time and velocity ratio. These variations are examined both with and without fluid force, encompassing diverse values for the magnetic parameter, thermal constant, and NL parameter. The curves offer a comprehensive understanding of how dynamic displacement depends on distinct physical parameters $t^*, \xi, \bar{N}_T, H_{\bar{x}}, \bar{F}_p$ – while taking into account different NL parameter values.

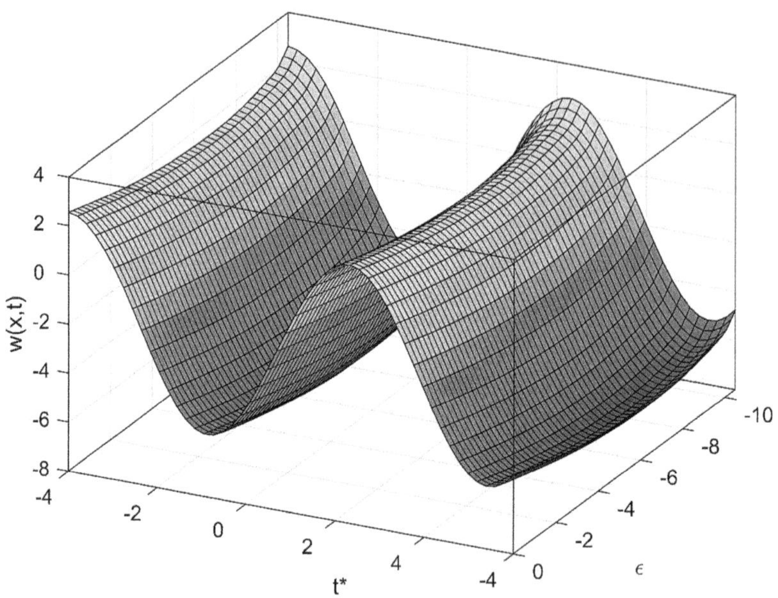

FIGURE 39.10 Spatial distribution of deformation in 3D with t^* and ξ for $e_0a = 0.5$, $\bar{N}_T = 0.5$ and $\bar{F}_p = 0.5$.

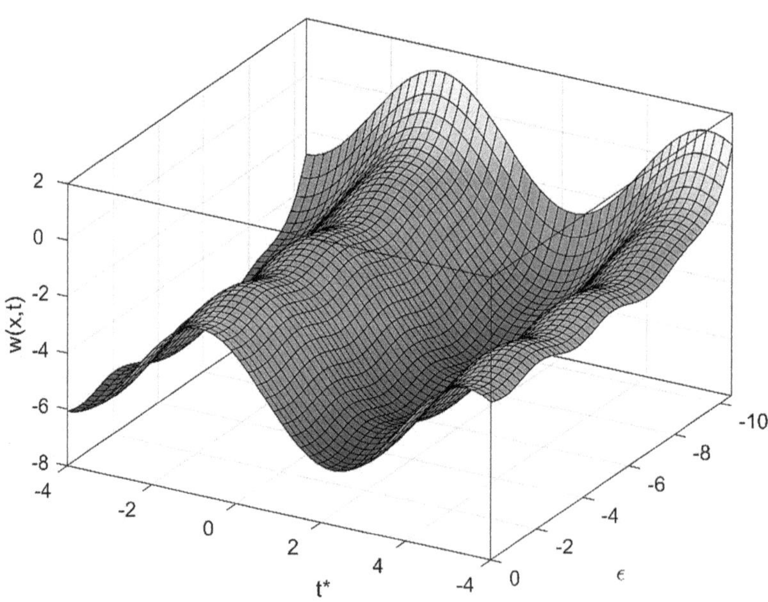

FIGURE 39.11 Spatial distribution of deformation in 3D with t^* and ξ for $e_0a = 0.5$, $H_{\bar{x}} = 0$ and $\bar{F}_p = 0$.

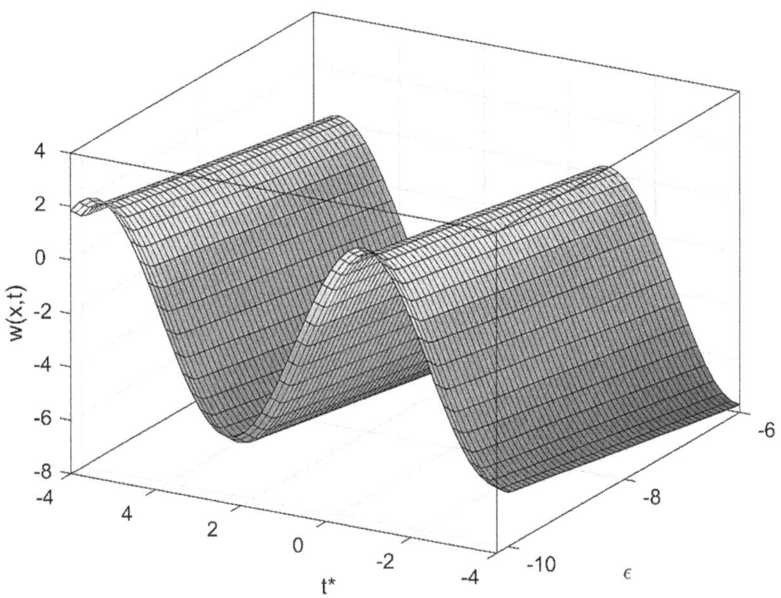

FIGURE 39.12 Spatial distribution of deformation in 3D with t^* and ξ for $e_0 a = 0.5$, $H_{\bar{x}} = 0.5$ and $\bar{F}_p = 0.5$.

39.3 Conclusions

In conclusion, this research has presented a comprehensive analytical model for capturing the two-dimensional dynamics of SWCNTs subjected to harmonic loading. The model, which integrates key parameters including load velocity, excitation frequency, dimensionless time, magnetic field, temperature vector, and nonlocal constants, utilises Eringen's NL theory within the EBB framework. The time-domain responses were calculated employing both modal superposition methodology and Newmark's direct integration method. This study contributes valuable insights into the dynamic behaviour of SWCNTs and provides a foundation for further exploration in nano mechanics and related fields.

The obtained results suggest that:

- In the realm of lower mechanical loads and under the influence of various factors, the specified physical quantities demonstrate a tendency to approach zero, with all functions exhibiting continuity.
- A cyclic pattern is observed in the amplitude of dynamic displacements, exhibiting fluctuations with increased and decreased magnitudes corresponding to higher values of harmonic load velocity and excitation frequency.
- Elevated dynamic displacement values and a discernible trend in wave propagation are observed to correlate with an increase in the non-dimensional time parameter.
- The distribution of dynamic displacement is notably influenced by the magnetic and temperature field vectors, along with the NL scale effect, highlighting their significant impact.

Conflict of Interest

Conflict of interest the authors declare that they have no conflict of interest.

REFERENCES

[1] S. Lijima, "Synthesis of carbon nanotubes," *Nature*, vol. *354*, pp. 56–58, July 1991. https://doi.org/10.5772/intecho.92995

[2] A.C. Eringen, "On differential equations of nonlocal elasticity and solutions of screw dislocation and surface waves," *J. Appl. Phys.*, vol. *54*, pp. 4703–4710, June 1983. https://doi.org/10.1063/1.332803

[3] T. Peddieson, G.R. Buchanan, and R.P. McNitt, "Application of nonlocal continuum model of technology," *Int. J. Eng. Sci.*, vol.*41*, pp. 305–312, July 2003. https://doi.org/10.1016/S0020-7225(02)00210-0

[4] P. Lu, L.P. Lee, C. Lu, and P.Q. Zhang, "Dynamic properties of flexural beams using nonlocal elasticity model," *J. Appl. Phys.*, vol. *99*, pp. 073510–073519, April 2006. https://doi.org/10.1063/1.2189213

[5] T. Murmu, and S.C. Pradhan, "Thermal effect on stability of embedded carbon nanotubes," *Comp. Mater. Sci.*, vol. *47*, pp. 721–726, April 2010. https://doi.org/10.1016/j.commatsci.2009.10.015

[6] S. Narender, and S. Gopalakrishnan, "Temperature effects on wave propagation in nanoplates," *Comp: Part B*, vol. *43*, pp. 1275–1281, April 2012. https://doi.org/10.1016/j.compositesb.2011.11.029

[7] F. Ebrahimi, M. Dehghan, and A. Seyfi, "Eringen nonlocal elasticity theory for wave propagation analysis of magneto-electro-elastic nanotubes", *Adv. Nano. Res.*, vol. *7*, no. 1, pp. 1–11, January 2019. https://doi.org/10.12989/anr.2019.7.1.001

[8] H. Wang, K. Dong, F. Men, Y.J. Yan, and X. Wang, "Influences of longitudinal magnetic field on wave propagation in carbon nanotubes embedded in elastic matrix," *Appl. Math. Model.*, vol. *34*, no. 4, pp. 878–889, April 2014. https://doi.org/10.1016/j.apm.2009.07.005

[9] F. Ebrahimi, M. Kokaba, G. Shaghaghi, and R. Selvamani, "Dynamic characteristics of hygro-magneto-thermo-electrical nanobeam with non-ideal boundary conditions," *Adv. Nano Res.*, vol. *8*, no. 2, pp. 169–182, May 2020. https://doi.org/10.12989/anr.2020.8.2.169

[10] F. Ebrahimi, M. Karimiasl, and R. Selvamani, "Bending analysis of magneto-electro piezoelectric nanobeams system under hygro-thermal loading," *Adv. Nano Res.*, vol. *8*, no. 3, pp. 203–214, April 2020. https://doi.org/10.12989/anr.2020.8.3.203

[11] T. Murmu, and S.C. Pradhan, "Small- scale effect on the vibration of non-uniform nanocantiliever based on nonlocal elasticity theory," *Physica E*, vol. *41*, pp. 1451–1456, August 2009. https://doi.org/10.1016/j.physe.2009.04.015

[12] F. Ebrahimi, and A. Dabbagh, "Magnetic field effects on thermally affected propagation of acoustical waves in rotary double- nanobeam system," *Wave. Random. Complex.*, pp. 1–21, August 2018. https://doi.org/10.1080/17455030.2018.1558308

[13] F. Ebrahimi, and F. Mahmoodi, "Vibration analysis of carbon nanotubes with multiple cracks in thermal environment," *Adv. Nano Res.*, vol. *6*, no. 1, pp. 57–80, March 2018. https://doi.org/10.12989/anr.2018.6.1.057

[14] F. Ebrahimi, and E. Salari, "Thermal buckling and free vibration analysis of size dependent Timoshenko FG Nano beams in thermal environments," *Compos. Struct.*, vol. *128*, pp. 363–380, 2015. https://doi.org/10.1016/j.compstruct.2015.03.023

[15] F. Ebrahimi, M. Boreiry, and G.R. Shaghaghi, "Nonlinear vibration analysis of electro – hygro– thermally actuated embedded nanobeams with various with various boundary conditions," *Micro. Syst. Technol.*, vol. *24*, no. 12, pp. 5037–5054, May 2018. https://doi.org/10.1007/s00542-018-3924-0

[16] J. Yoon, C.Q. Ru, and A. Mioduchowski, "Vibration and instability of carbon nanotubes conveying fluid," *Compos. Sci. Technol.*, vol. *65*, pp. 1326–1336, July 2005. https://doi.org/10.1016/j.compscitech.2004.12.002

[17] J. Yoon, C.Q. Ru, and A. Mioduchowski, "Flow–induced flutter instability of cantilever carbon nanotubes," *Int. J. Solid. Struct.*, vol. *43*, pp. 3337–3349, June 2006. https://doi.org/10.1016/j.ijsolstr.2005.04.039

[18] Y.Q. Zhang, X. Liu, and G.R. Liu, "Thermal effect on transverse vibration of double walled carbon nanotubes," *Nanotechnology*, vol. *18*, pp. 445701(7), October 2007. https://doi.org/10.1088/0957-4484/18/44/445701

[19] T.P. Chang, "Thermal- mechanical vibration and instability of a fluid conveying single walled carbon nanotube embedded in a n elastic medium based on nonlocal elasticity theory," *Appl. Math. Model.*, vol. *36*, pp. 1964–1973, May 2012. https://doi.org/10.1016/j.apm.2011.08.020

[20] L. Wang, "Wave propagation of fluid conveying single walled carbon nanotubes via gradient elasticity theory," *Comp. Mater. Sci.*, vol. *49*, pp. 761–766, October 2010. https://doi.org/10.1016/j.commatsci.2010.06.019

[21] H. Mohammad, and S. Goughari, "Vibration and instability analysis of nanotubes conveying fluid subjected to a longitudinal magnetic field," *Appl. Math. Model.*, vol. *40*, pp. 2560–2576, February 2016. https://doi.org/10.1016/j.apm.2015.09.106

[22] K. Kiani, and B. Mehri, "Assessment of nanotubes structures under a moving nanoparticle using nonlocal beam theories", *J. Sound. Vib.*, vol. *329*, no. 11, pp. 2241–2264, May 2010. https://doi.org/10.1016/j.jsv.2009.12.017

[23] M. Simsek, and T. Kocaturk, "Nonlinear dynamic analysis of an eccentrically prestressed damped beam under a concentrated moving harmonic load," *J. Sound Vib.*, vol. *320*, pp. 235–253, February 2009. https://doi.org/10.1016/j.jsv.2008.07.012

[24] M. Simsek, and T. Kocaturk, "Free and forced vibration of functionally graded beam subjected to concentrated moving harmonic load," *Comp. Struct.*, vol. *90*, pp. 465–473, October 2009. https://doi.org/10.1016/j.compstruct.2009.04.024

[25] M. Simsek, "Vibration of a single walled carbon nanotube under action of a moving harmonic load based on nonlocal elasticity theory", *Physica. E. Low. Dimens. Syst. Nanostruct.*, vol. *43*, no. 1, 182–191, November 2010. https://doi.org/10.1016/j.physe.2010.07.003

[26] Y. Tokio, "Recent development of carbon nanotube", *Synth. Met.*, vol. *70* pp. 1511–8, 1995

[27] D.H. Wu, W.T. Chien, C.S. Chen, and H.H. Chen, "Resonant frequency analysis of fixed-free single-walled carbon nanotube-based mass sensor", *Sens. Actuators A Phys.*, vol. *126*, no. 1, 117–121, January 2006. https://doi.org/10.1016/j.sna.2005.10.005

[28] Q. Wang, and K.M. Liew, "Application of nonlocal continuum mechanics to static analysis of micro and nano structures," *Physics. Lett. A.*, vol. *363*, no. 3, pp. 236–242. March 2006. https://doi.org/10.1016/j.physleta.2006.10.093

[29] M. Aydogdu and M. Arda, "Torsional vibration analysis of double walled carbon nanotubes using nonlocal elasticity", *Int. J. Mech. Mater. Des.*, vol. *12*, no. 1, pp. 71–84, December 2016. https://doi.org/10.1007/s10999-014-9292-8

40

Nano Processing and Preservation in Foods

G. G. Kavitha Shree
Agricultural Engineering College and Research Institute, Tamil Nadu Agricultural University, Coimbatore, India

Arulmari Ramaraj
Agricultural Engineering College and Research Institute, Tamil Nadu Agricultural University, Tiruchirappalli, India

Venkatasami Murugesan
Agricultural Engineering College and Research Institute, Tamil Nadu Agricultural University, Coimbatore, India

Ravi Prakash Balasundaram
Agricultural Engineering College and Research Institute, Tamil Nadu Agricultural University, Tiruchirappalli, India

40.1 Introduction

In the meat industry, nanotechnology applications offer solutions for detecting microbes in packaging, enhancing flavour and colour profiles, and improving barrier properties to enhance safety. However, concerns persist regarding patentable technologies due to potential risks to health and food safety associated with nanoparticle use in cooking. Nanotechnology, the manipulation of nanoparticles for various purposes, is crucial for agriculture and food industries, contributing to improved quality, safety, and human health. The unique physical, chemical, and biological properties of nanoparticles, including their high surface-to-volume ratio and altered solubility and toxicity compared to macroscale counterparts, have sparked interest across sectors like medicine, agriculture, and wastewater treatment. Titanium dioxide (TiO_2), silver (Ag), gold (Au), zinc oxide (ZnO), and carbon nanoparticles find extensive use in air filtration, food packaging, personal care products, paints, and household appliances due to their potential antimicrobial properties. Nanoscale copper oxide is particularly favoured for its strong antimicrobial properties and finds wide application in commercial biocides. Nanomaterials, typically ranging from 1 to 100 nanometres in size, are utilised across various industries, including medicine, food processing, electronics, and beyond, with applications ranging from unknown to biologically determined. In food production, nanotechnologies offer the potential to produce and process healthier, safer, and higher-quality food products. This chapter highlights the significant benefits of employing nanotechnology in food manufacturing, particularly in cooking methods, packaging, quality assurance, and safety measures.

Nanotechnology offers potential benefits such as improved barrier properties, increased mechanical and heat resistance, antimicrobial surfaces, and environmentally friendly practices. By employing mineral delivery systems, nanotechnology enables the creation of nano-formulated agrochemicals, enhanced nutrition, and innovative products via bioactive encapsulation, thus revolutionising the food industry. Additionally, it indirectly contributes to the development of biosensors for detecting pathogens and synthetic pollutants. Concerns about the toxicological effects of nanoparticles in food, particularly bread, highlight the importance of risk assessment and safety considerations. Furthermore, this underscores the need for regulatory frameworks to address the risks associated with nanoparticle use in food science.

DOI: 10.1201/9781003495437-40

40.2 Nano Processing and Preservation in Foods

Nanotechnology involves the manipulation of nanoparticles for diverse applications, playing a crucial role in agriculture and food industries by enhancing agricultural quality, security, and human health through innovative means. Nanoparticles possess distinct physical, chemical, and biological characteristics, including a high surface-to-volume ratio and altered solubility and toxicity compared to larger counterparts, driving interest across fields like medicine, agriculture, and wastewater treatment. Certain nanoparticles, such as TiO_2, Ag, Au, ZnO, and carbon nanoparticles, are utilised in various consumer products like air filters, food packaging, personal care items, and household appliances due to their potential antimicrobial properties. Among these, nanosized copper oxide is widely incorporated into commercial biocides for its potent antimicrobial effects. Nanomaterials, ranging from 1 to 100 nanometres in size, may have unknown or bio-determined characteristics and find application in diverse industries, including agriculture, medicine [9], electronics, and food production. In the food sector, different grades of nanoparticles are employed in nanotechnologies to potentially improve the production and processing of food products, aiming for enhanced healthfulness, reliability, and quality. However, concerns about the safety and regulatory oversight of nanoparticles in food-related applications persist, highlighting the need for comprehensive risk assessment and regulatory frameworks to ensure their responsible use [5]. Despite these challenges, nanotechnology holds significant promise for revolutionising food production and addressing various agricultural and nutritional challenges.

This chapter highlights the significant benefits of integrating nanotechnology into food manufacturing, particularly in cooking, packaging, quality assurance, and safety measures. Nanotechnology offers possibilities such as enhancing the impermeability of packaging materials, strengthening barrier properties, improving mechanical and heat resistance, and creating antimicrobial surfaces. Moreover, it enables the development of environmentally friendly practices. By employing mineral delivery systems, nanotechnology facilitates the production of tailored agrochemicals, enhances nutritional concepts, and introduces innovative products through bioactive encapsulation, potentially revolutionising the snack industry. Nanotechnology is opening up various possibilities for its applications in diverse fields in the food industry (Figure 40.1). Additionally, nanotechnology indirectly contributes to the creation of biosensors for detecting pathogens and synthetic pollutants. However, concerns about the toxicological effects of nanoparticles in bread underscore the importance of rigorous risk assessment and security measures in food science. There is a clear need for a regulatory framework to effectively manage the potential risks associated with nanoparticle use in food production.

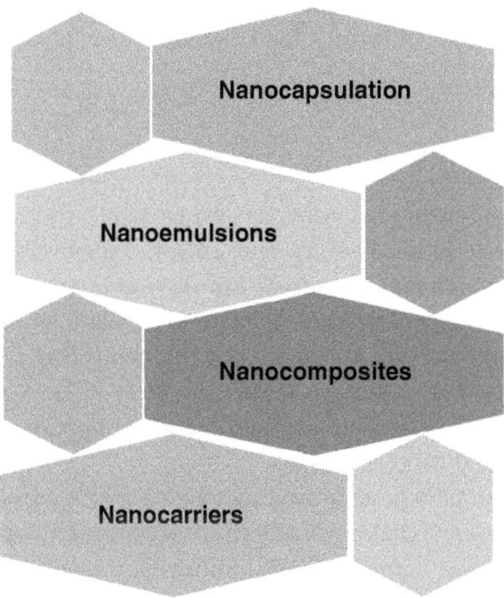

FIGURE 40.1 Application of nanotechnology in food industry.

40.3 Nanotechnology in Food Processing

'Nanofood' refers to food that undergoes processing, preservation, safety enhancement, and packaging using nanotechnology. Nanotechnology shows significant potential in the postharvest food industry, influencing characteristics such as surface charge, particle size, distribution, potential formation of clusters, and texture of food products. Additionally, it enhances bioavailability, taste, composition, and consistency, often by masking undesirable flavours or aromas. The term 'nanofood' specifically denotes food processed with nanotechnology to improve the transformation, outcome, safety, and packaging of bread. Similarly, nanotechnology offers vast opportunities in beverage production post-harvest, where it modifies factors like particle size, categorisation, potential formation of clusters, surface charge, and sensory attributes such as flavour, quality, and viscosity [3, 6]. In both cases, nanotechnology plays a pivotal role in enhancing food products and addressing sensory challenges [4], contributing to improved consumer experiences and product quality.

40.4 Nanocapsulation

Accidental presence of thin edible nano-coatings, approximately 5 nm thick, is observed in various food items like bread, fruits, vegetables, legumes, grains, low-cost quick-to-prepare foods, and bakery products. These coatings serve as moisture and smoke barriers, imparting flavour, colour, antioxidants, enzymes, and anti-browning agents, thereby extending shelf life [7]. Additionally, candy products are coated with appealing antibacterial nano-coatings, enhancing their taste and providing convenience.

Encapsulation of bioactive nutraceutical particles using nano formulations enhances their bioavailability and biodistribution [1]. Nanoencapsulation involves fully enclosing essences within tiny structures, achieved through nano structuration, nano emulsification, or nanocomposites, enabling controlled release of the core ingredients. Various forms of nanoencapsulation, such as micelles, nanoparticles, liposomes, nanospheres, nano cochleates, and nano emulsions, have been developed to cater to specific needs. These encapsulations serve diverse purposes, including acting as digestive supplements, masking unpleasant tastes, improving bioavailability, and efficiently distributing nutrients without requiring emulsifiers or surfactants. For instance, curcumin and quercetin from turmeric extract were encapsulated using polylactic acid (PLA)-based nanoparticles. Stevioside nanoparticles, safe non-caloric sweeteners with antidiabetic properties, have been developed. Additionally, encapsulating polyphenols (catechin and epi-catechin) in bovine serum albumin (BSA) nanoparticles improved their cohesiveness and bioavailability in beverages. Nanoencapsulation presents a promising approach to enhancing the functionality and efficacy of bioactive compounds in various food and beverage applications.

40.5 Nanofillers

Encapsulation of bioactive nutraceutical particles using nano formulations represents a cutting-edge technique to amplify their bioavailability and enhance their distribution throughout the body. Nanoencapsulation involves the complete encapsulation of these bioactive components within minuscule structures, achieved through processes like nano structuration, nano emulsification, or the utilisation of nanocomposites. This encapsulation method facilitates the controlled release of the core ingredients, ensuring their efficient delivery and utilisation within the body [12]. A plethora of nanoencapsulation techniques have been devised to cater to diverse needs, encompassing micelles, nanoparticles, liposomes, nanospheres, nano cochleates, and nano emulsions.

Each of these nanoencapsulation methods serves specific purposes, offering a range of benefits beyond mere delivery. For instance, they can function as digestive supplements, serving to enhance the absorption of essential nutrients and bioactive compounds. Additionally, nanoencapsulation can effectively mask unpleasant tastes or odours associated with certain bioactive ingredients, thereby improving consumer acceptance and compliance with dietary supplements or functional foods. Moreover, this technology aids in optimising the bioavailability of bioactive compounds, ensuring that they are absorbed and utilised more efficiently by the body.

One example of the application of nanoencapsulation is in the encapsulation of curcumin and quercetin, bioactive compounds found in turmeric extract [13]. By encapsulating these compounds using nanoparticles made from PLA, their stability and bioavailability can be significantly enhanced. Similarly, stevioside nanoparticles have been developed as safe and non-caloric sweeteners with additional antidiabetic properties. By encapsulating these sweeteners, their bioavailability and efficacy can be optimised, offering a healthier alternative to traditional sweetening agents.

Another notable application of nanoencapsulation is in the encapsulation of polyphenols, such as catechin and epi-catechin, within BSA nanoparticles. This encapsulation method not only improves the cohesiveness of these polyphenols but also enhances their bioavailability when incorporated into beverages. By encapsulating these bioactive compounds, their functionality and efficacy are maximised, allowing for greater health benefits to be derived from consuming these beverages.

Overall, nanoencapsulation presents a highly promising approach to enhancing the functionality and efficacy of bioactive compounds in various food and beverage applications. By encapsulating these compounds within nanostructures, their stability, bioavailability, and distribution within the body can be significantly improved. This not only enhances the nutritional value of food products but also offers potential health benefits to consumers. Additionally, nanoencapsulation can play a crucial role in the development of novel functional foods and dietary supplements, providing innovative solutions to address various health concerns and dietary needs. As research in this field continues to advance, nanoencapsulation holds immense potential to revolutionise the food and beverage industry, offering new avenues for improving human health and well-being through enhanced nutrition and bioactive compound delivery.

40.6 Nanoemulsions

Nano emulsions are characterised by their colloidal coarse structure, forming lubricant in water emulsions with continuous circles and hazy, lipophilic surfaces, typically with very short bead lengths ranging from 10 to 1000 nm. These unique characteristics make nano emulsions superior carriers for various bioactive substances, offering enhanced material support, improved optical clarity, and increased bioavailability compared to regular emulsions. The small size of nano emulsions is particularly advantageous as it contributes to a large surface area, facilitating the efficient movement of bioactive substances along the gastrointestinal tract.

Compared to conventional emulsions, nano emulsions exhibit a faster rate of digestion due to their increased binding sites for digestive enzymes such as lipase and amylase. Additionally, nano emulsions play a significant role in facilitating the rapid transfer of hydrophobic bioactive substances typically found in functional foods into the lubricant beads. Functional foods, which are designed to provide energy and enhance human strength, include products such as fruit juices fortified with mineral ions like iron and calcium, breads fortified with phytosterols, yogurts containing probiotics, milk products fortified with vitamin D, and cereals enriched with vitamins, minerals, and omega-3 fatty acids.

The concept of functional foods has gained increasing significance in optimising and enhancing the bioavailability of various natural bioactive compounds present in food. Excipient foods, which enhance the bioactivity of co-ingested foods, have become popular choices. The nano emulsion-based technique effectively enhances the bioavailability of biologically active compounds by modifying their structures, arrangements, and features [8]. This technique involves the use of emulsifier-coated lipid beads dispersed in an aqueous medium to create nano emulsion-based formulations.

In summary, nano emulsions exhibit unique characteristics that make them highly effective carriers for bioactive substances in food products. Their small size, colloidal structure, and lipophilic surfaces contribute to enhanced material support, optical clarity, and increased bioavailability compared to regular emulsions. Additionally, nano emulsions facilitate rapid digestion and transfer of hydrophobic bioactive substances, making them valuable tools in the development of functional foods designed to optimise human health and well-being.

40.7 Nanocomposites

Utilising nanomaterials in food packaging offers numerous advantages, including enhanced mechanical barriers, the capacity for detecting microbial contamination, and potentially increased nutrient bioavailability. This application of nanotechnology is widely adopted across the food and food-related industries. Various nanocomposites, comprising polymers integrated with nanoparticles, find extensive use in food packaging and materials that interact with food. Nanoparticles such as ZnO and MgO are commonly employed in food packaging applications. Amorphous silica is also commonly found in both food products and their containers or packaging. Additionally, aerosolised water nanostructures have shown effectiveness in eliminating pathogens like Salmonella, Listeria, and *Escherichia coli* from steel surfaces utilised in food preparation.

40.8 Nanocarrier

The utilisation of nanocapsules aids in enhancing the bioavailability of dietary supplements as they navigate through the digestive system in biological environments. Six methods are employed in the creation of nanocapsules, including layer-by-layer polymer coating, emulsion dispersion, double emulsion, and nanoprecipitation. These nanocapsules find applications in delivering vaccinations, fertilisers, and insecticides to plants, as well as in enhancing the nutritional content of food by delivering lipophilic health supplements like growth hormones, fatty acids, vitamins, and minerals. Encapsulation offers the primary advantage of protecting the encapsulated component and ensuring precise delivery to the target, even under challenging conditions. One type of nanotechnology-based carrier used for nanoencapsulation is the liposome, which facilitates the targeted and controlled distribution of various active components such as vitamins, enzymes, antioxidants, additives, and minerals. A novel encapsulation technique currently under development involves gallic acid-loaded Zein fibre produced via electrospinning, where Zein fibre prevents lipid degradation in the body until the desired dosage is achieved [11]. This innovative approach holds potential for widespread adoption in the food packaging industry due to the selectivity and solubility of components encapsulated in lipid-based systems, making them more effective than other encapsulation methods.

40.9 Anticaking Agent

Handling, processing, and storing food powders often result in clumping, which can diminish product quality and functionality. This caking phenomenon adversely affects the rehydration and dispersibility of powders, leading to a shorter shelf life and compromising the food's sensory attributes. Anticaking agents address this issue by various means, such as coating particles to resist moisture, competing with the powder for moisture, smoothing surfaces to reduce friction between particles, and preventing crystal formation. Silicon dioxide (SiO) is commonly used as an anticaking agent in both food and non-food applications, helping thicken pastes and maintain the flow properties of powdered products like icing sugar, salts, dried milk, spices, and dry mixes.

Synthesised amorphous silica has historically been considered harmless, leading to its widespread use as an anticaking agent without significant concerns. However, numerous studies have highlighted the potential adverse effects of nanomaterials and nanoparticles on bodily molecules. The utilisation of these manufactured nanoparticles has raised concerns regarding consumer product safety and health, underscoring the need for enhanced risk assessment measures. Various chemicals, including sodium bicarbonate, calcium aluminosilicate, sodium silicate, and others, have been incorporated into granular and powdered foods to further prevent caking.

40.10 Gelating Agent

Nanostructured materials serving as gelling agents find application in food processing to enhance food texture. In a study by Amjadi et al., ZnO nanoparticles (ZnONPs), chitosan nanofiber (CHNF), and a gelatine-based nanocomposite were tested for active packaging of chicken fillet and cheese. The gelatine-based nanocomposite, synthesised with ZnONPs and CHNF demonstrated strong antibacterial activity against foodborne pathogens [2]. Through Fourier transform infrared (FT-IR) spectroscopy experiments, Amjadi et al. showcased the interactions and high compatibility between the gelatine matrix, ZnONPs (with a diameter of approximately 30 nm), and CHNF (with a diameter of approximately 28 nm). Furthermore, the nanocomposite exhibited efficient mechanical properties and water barrier capabilities due to its dense structure.

40.11 Antioxidants

Antioxidants are a group of compounds that neutralise free radicals, converting them into harmless substances, which is crucial in reducing oxidative stress and treating related diseases. However, their effectiveness is hindered by challenges such as poor absorption, difficulty crossing cell membranes, and degradation during delivery, leading to limited availability within the body. To address these limitations, antioxidants have been incorporated or covalently linked to various nanoparticle formulations to improve stability, controlled release, biocompatibility, and targeted distribution [10].

40.12 Conclusion

The growing utilisation of nanoscale structures in the food industry has sparked significant interest and efforts in this area of research. Advances in nanobiotechnology have led to the development of more sensitive and compact materials and devices, particularly suitable for applications in packaging and food safety. Additionally, employing nanoparticles for food preservation has shown promising outcomes, as these materials can protect food from lipids, gases, moisture, odours, and flavours, while also serving as effective delivery systems for bioactive compounds. The widespread adoption of nanotechnology is expected to greatly benefit the food industry, impacting various aspects such as bioavailability, shelf life, transportation, processing, packaging, and production. However, concerns regarding human exposure to nanoparticles and their potential health effects have emerged. To ensure consistent manufacturing, product safety, and potential consumer benefits, it is essential to accurately quantify nanomaterials at every stage of the food lifecycle. Establishing a single, international regulatory framework governing nanotechnology in food is crucial for ensuring the acceptance of food and food-related products containing nanomaterials by the public, provided they are deemed safe.

REFERENCES

[1] Ezhilarasi, P. N., Karthik, P., Chhanwal, N., and Anandharamakrishnan, C. (2013). Nanoencapsulation techniques for food bioactive components: A review. *Food Bioprocess Technol. 6*, 628–647. doi: 10.1007/s11947-012-0944-0

[2] Fakhouri, F. M., Casari, A. C. A., Mariano, M., Yamashita, F., Mei, L. I., Soldi, V., et al. (2014). "Effect of a gelatin-based edible coating containing cellulose nanocrystals (CNC) on the quality and nutrient retention of fresh strawberries during storage," in *Proceedings of the IOP Conference Series: Materials Science and Engineering, Conference 1, 2nd International Conference on Structural Nano Composites (NANOSTRUC 2014)*, Vol. *64*, Madrid. doi: 10.1088/1757-899X/64/1/012024

[3] García, M., Aleixandre, M., Gutiérrez, J., and Horrillo, M. C. (2006). Electronic nose for wine discrimination. *Sensors Actuat. B 113*, 911–916. doi: 10.1016/j. snb.2005.03.078

[4] Inbaraj, B. S., and Chen, B. H. (2016). Nanomaterial-based sensors for detection of foodborne bacterial pathogens and toxins as well as pork adulteration in meat products. *J. Food Drug Anal. 24*, 15–28. doi: 10.1016/j.jfda.2015.05.001

[5] Jain, A., Shivendu, R., Nandita, D., and Chidambaram, R. (2018). Nanomaterials in food and agriculture: An overview on their safety concerns and regulatory issues. *Crit. Rev. Food Sci. Nutr.* doi: 10.1080/10408398.2016.1160363

[6] Jianrong, C., Yuqing, M., Nongyue, H., Xiaohua, W., and Sijiao, L. (2004). Nanotechnology and biosensors. *Biotechnol. Adv. 22*, 505–518. doi: 10.1016/j. biotechadv.2004.03.004

[7] Kanazawa, K., and Cho, N. J. (2009). Quartz crystal microbalance as a sensor to characterize macromolecular assembly dynamics. *J. Sens. 6*, 1–17. doi: 10.1155/2009/824947

[8] Kong, M., Chen, X. G., Kweon, D. K., and Park, H. J. (2011). Investigations on skin permeation of hyaluronic acid based nanoemulsion as transdermal carrier. *Carbohydr. Polym. 86*, 837–843. doi: 10.1016/j. carbpol.2011.05.027

[9] Koo, O. M., Rubinstein, I., and Onyuksel, H. (2005). Role of nanotechnology in targeted drug delivery and imaging: A concise review. *Nanomed. Nanotechnol. Biol. Med. 1*, 193–212. doi: 10.1016/j. nano.2005.06.004

[10] Kumar, P., Mahajan, P., Kaur, R, and Gautam, S. (2020). Nanotechnology and its challenges in the food sector: A review. *Mater Today Chem. 17*, 100332. doi: 10.1016/j.mtchem.2020.10033

[11] Lamprecht, A., Saumet, J. L., Roux, J., and Benoit, J. P. (2004). Lipid nanocarriers as drug delivery system for ibuprofen in pain treatment. *Int. J. Pharm. 278*, 407–414. doi: 10.1016/j.ijpharm.2004.03.018

[12] Ravi Kumar, M. N. (2000). Nano and microparticles as controlled drug delivery devices. *Journal of Pharmacy & Pharmaceutical Sciences: A Publication of the Canadian Society for Pharmaceutical Sciences, Societe Canadienne Des Sciences Pharmaceutiques, 3*(2), 234–258.

[13] Ansari, S. H., Islam, F., and Sameem, M. (2012). Influence of nanotechnology on herbal drugs: A review. *J. Adv. Pharm. Techn. Res. 3*, no. 3, 142–146.

Pages in *italics* refer to figures and pages in **bold** refer to tables.

For Product Safety Concerns and Information please contact our EU
representative GPSR@taylorandfrancis.com
Taylor & Francis Verlag GmbH, Kaufingerstraße 24, 80331 München, Germany

www.ingramcontent.com/pod-product-compliance
Ingram Content Group UK Ltd.
Pitfield, Milton Keynes, MK11 3LW, UK
UKHW050154110726
473146UK00013B/918